AF706977

DIAGNOSTIC RADIOLOGY PHYSICS: A HANDBOOK FOR TEACHERS AND STUDENTS

The following States are Members of the International Atomic Energy Agency:

AFGHANISTAN
ALBANIA
ALGERIA
ANGOLA
ARGENTINA
ARMENIA
AUSTRALIA
AUSTRIA
AZERBAIJAN
BAHAMAS
BAHRAIN
BANGLADESH
BELARUS
BELGIUM
BELIZE
BENIN
BOLIVIA
BOSNIA AND HERZEGOVINA
BOTSWANA
BRAZIL
BRUNEI DARUSSALAM
BULGARIA
BURKINA FASO
BURUNDI
CAMBODIA
CAMEROON
CANADA
CENTRAL AFRICAN REPUBLIC
CHAD
CHILE
CHINA
COLOMBIA
CONGO
COSTA RICA
CÔTE D'IVOIRE
CROATIA
CUBA
CYPRUS
CZECH REPUBLIC
DEMOCRATIC REPUBLIC OF THE CONGO
DENMARK
DOMINICA
DOMINICAN REPUBLIC
ECUADOR
EGYPT
EL SALVADOR
ERITREA
ESTONIA
ETHIOPIA
FIJI
FINLAND
FRANCE
GABON
GEORGIA
GERMANY
GHANA
GREECE
GUATEMALA
HAITI
HOLY SEE
HONDURAS
HUNGARY
ICELAND
INDIA
INDONESIA
IRAN, ISLAMIC REPUBLIC OF
IRAQ
IRELAND
ISRAEL
ITALY
JAMAICA
JAPAN
JORDAN
KAZAKHSTAN
KENYA
KOREA, REPUBLIC OF
KUWAIT
KYRGYZSTAN
LAO PEOPLE'S DEMOCRATIC REPUBLIC
LATVIA
LEBANON
LESOTHO
LIBERIA
LIBYA
LIECHTENSTEIN
LITHUANIA
LUXEMBOURG
MADAGASCAR
MALAWI
MALAYSIA
MALI
MALTA
MARSHALL ISLANDS
MAURITANIA, ISLAMIC REPUBLIC OF
MAURITIUS
MEXICO
MONACO
MONGOLIA
MONTENEGRO
MOROCCO
MOZAMBIQUE
MYANMAR
NAMIBIA
NEPAL
NETHERLANDS
NEW ZEALAND
NICARAGUA
NIGER
NIGERIA
NORWAY
OMAN
PAKISTAN
PALAU
PANAMA
PAPUA NEW GUINEA
PARAGUAY
PERU
PHILIPPINES
POLAND
PORTUGAL
QATAR
REPUBLIC OF MOLDOVA
ROMANIA
RUSSIAN FEDERATION
RWANDA
SAN MARINO
SAUDI ARABIA
SENEGAL
SERBIA
SEYCHELLES
SIERRA LEONE
SINGAPORE
SLOVAKIA
SLOVENIA
SOUTH AFRICA
SPAIN
SRI LANKA
SUDAN
SWAZILAND
SWEDEN
SWITZERLAND
SYRIAN ARAB REPUBLIC
TAJIKISTAN
THAILAND
THE FORMER YUGOSLAV REPUBLIC OF MACEDONIA
TOGO
TRINIDAD AND TOBAGO
TUNISIA
TURKEY
UGANDA
UKRAINE
UNITED ARAB EMIRATES
UNITED KINGDOM OF GREAT BRITAIN AND NORTHERN IRELAND
UNITED REPUBLIC OF TANZANIA
UNITED STATES OF AMERICA
URUGUAY
UZBEKISTAN
VENEZUELA, BOLIVARIAN REPUBLIC OF
VIET NAM
YEMEN
ZAMBIA
ZIMBABWE

The Agency's Statute was approved on 23 October 1956 by the Conference on the Statute of the IAEA held at United Nations Headquarters, New York; it entered into force on 29 July 1957. The Headquarters of the Agency are situated in Vienna. Its principal objective is "to accelerate and enlarge the contribution of atomic energy to peace, health and prosperity throughout the world".

DIAGNOSTIC RADIOLOGY PHYSICS: A HANDBOOK FOR TEACHERS AND STUDENTS

ENDORSED BY:

AMERICAN ASSOCIATION OF PHYSICISTS IN MEDICINE,
ASIA–OCEANIA FEDERATION OF ORGANIZATIONS FOR MEDICAL PHYSICS,
EUROPEAN FEDERATION OF ORGANISATIONS FOR MEDICAL PHYSICS

INTERNATIONAL ATOMIC ENERGY AGENCY
VIENNA, 2014

COPYRIGHT NOTICE

All IAEA scientific and technical publications are protected by the terms of the Universal Copyright Convention as adopted in 1952 (Berne) and as revised in 1972 (Paris). The copyright has since been extended by the World Intellectual Property Organization (Geneva) to include electronic and virtual intellectual property. Permission to use whole or parts of texts contained in IAEA publications in printed or electronic form must be obtained and is usually subject to royalty agreements. Proposals for non-commercial reproductions and translations are welcomed and considered on a case-by-case basis. Enquiries should be addressed to the IAEA Publishing Section at:

Marketing and Sales Unit, Publishing Section
International Atomic Energy Agency
Vienna International Centre
PO Box 100
1400 Vienna, Austria
fax: +43 1 2600 29302
tel.: +43 1 2600 22417
email: sales.publications@iaea.org
http://www.iaea.org/books

© IAEA, 2014

Printed by the IAEA in Austria
September 2014
STI/PUB/1564

IAEA Library Cataloguing in Publication Data

Diagnostic radiology physics : a handbook for teachers and students. — Vienna : International Atomic Energy Agency, 2014.
p. ; 24 cm.
STI/PUB/1564
ISBN 978–92–131010–1
Includes bibliographical references.

1. Radiology, Medical — Handbooks, manuals, etc. 2. Medical physics — Handbooks, manuals, etc. 3. Radiation dosimetry. 4. Diagnostic imaging. I. International Atomic Energy Agency.

IAEAL 14–00898

FOREWORD

One of the important activities of the IAEA is the education of professionals responsible for the application of radiation. This is no less true in radiation medicine than in other fields, where the physics professional not only needs to understand the physical principles involved, but must also have a sound knowledge of their application to medical practice. Consequently, the IAEA has a long history of supporting education in these areas through the use of guidance documents and, importantly, more directly through cooperation programmes, including the support of Member States in developing their own university infrastructure for postgraduate education programmes in medical physics, the development of clinical training guides and, more recently, web based educational resources.

In 2005, the IAEA published Radiation Oncology Physics: A Handbook for Teachers and Students, as a result of a process of determining a harmonized syllabus for university education of medical physicists in radiation oncology. Following the success of this publication, it was apparent that a similar need existed in the other two specialities of medical physics, namely diagnostic radiology and nuclear medicine. This need has been recognized as a result of the growing importance of medical imaging in all areas of radiation medicine, including radiation oncology, and also because of the growing awareness of the increasing use of complex diagnostic equipment and techniques, such as computed tomography, mammography and interventional radiology. In parallel with this, the past decade has seen the digitization of image based medical technology, with its inherent need for quality processes.

This handbook is intended to provide the basis for the education of medical physicists initiating their university studies in the field of diagnostic radiology. This has been achieved with the contributions of 41 authors and reviewers from 12 different countries. The 24 chapters include a broad coverage of topics relevant to diagnostic radiology physics, including radiation physics, dosimetry and instrumentation, image quality and image perception, imaging modality specific topics, recent advances in digital techniques, and radiation biology and protection. The handbook is not designed to replace the large number of textbooks available on many aspects of diagnostic radiology physics, which will still be necessary to deepen knowledge in the specific topics reviewed here. It is expected that this handbook will successfully fill a gap in the teaching material for medical radiation physics in imaging, providing, in a single volume, the largest possible coverage available today. Its wide dissemination by the IAEA will contribute to the harmonization of education in diagnostic radiology physics and will be the source reference for much of the IAEA clinical training programmes in the field. It will be of value to those preparing for their certification as medical physicists, radiologists and diagnostic radiographers.

Endorsement of this handbook has been granted by following international professional bodies: the American Association of Physicists in Medicine (AAPM), the Asia–Oceania Federation of Organizations for Medical Physics (AFOMP) and the European Federation of Organisations for Medical Physics (EFOMP).

The following international experts are gratefully acknowledged for making major contributions to the development of an earlier version of the syllabus: R. Nowotny (Austria) and M. Sandborg (Sweden). The following individuals made major contributions to this handbook as technical editors: S. Christofides (Cyprus), D.R. Dance (United Kingdom), A.D.A. Maidment (United States of America) and K.-H. Ng (Malaysia). The IAEA scientific officers responsible for the project were (in chronological order) F. Pernička, I.D. McLean and H. Delis.

EDITORIAL NOTE

Although great care has been taken to maintain the accuracy of information contained in this publication, neither the IAEA nor its Member States assume any responsibility for consequences which may arise from its use.

The use of particular designations of countries or territories does not imply any judgement by the publisher, the IAEA, as to the legal status of such countries or territories, of their authorities and institutions or of the delimitation of their boundaries.

The mention of names of specific companies or products (whether or not indicated as registered) does not imply any intention to infringe proprietary rights, nor should it be construed as an endorsement or recommendation on the part of the IAEA.

The IAEA has no responsibility for the persistence or accuracy of URLs for external or third party Internet web sites referred to in this book and does not guarantee that any content on such web sites is, or will remain, accurate or appropriate.

PREFACE

The application of physical principles to reveal internal structures of the body sparked the imagination of the medical profession in the late 19th century and rapidly became the foundation of the practice of diagnostic radiology. The efforts of physical scientists have continued to fuel innovation in medical imaging through a progression of technologies, including the specialization of X ray imaging devices for examination of the breast, blood vessels, moving vessels, teeth and bone density. The use of high frequency sound waves has allowed the instantaneous imaging of soft tissues without the dangers associated with ionizing radiation. The use of mathematical image reconstruction has allowed the visualization of sections of the body, free from the confusion caused by overlying tissue as seen in the computed tomography scanner and the magnetic resonance imager, while the developments in computers have allowed the electronic capture, processing and transfer of medical images.

As was quickly discovered with the application of X rays for medical imaging, the use of radiation on living tissue is not without risk of biological injury. The measurement of radiation, its interaction with matter and its biological effects have led to the studies of radiation dosimetry, radiation biology and epidemiology. These studies are becoming more important in modern radiological imaging as the number, length and complexity of X ray procedures received by the population continues to increase rapidly.

It is in this complex environment that the medical physicist, along with radiologists and radiographers, plays a significant role in the multidisciplinary team needed for medical diagnosis. Medical physicists need to be able to advise on the principles and practice of imaging equipment and assist in purchase processes and quality assurance. They are required to measure the radiation dose received by staff and, most importantly, by the patients undergoing diagnostic examinations. They should be able to advise on the optimal image quality needed for the diagnostic process and to be able to contribute to scientific research. They are also well equipped to assume responsibility for the safe use of radiation at a medical facility.

This book is dedicated to students and teachers involved in programmes that train professionals for work in diagnostic radiology. It teaches the essential physics of diagnostic radiology and its application in modern medicine. As such, it is useful to graduate students in medical physics programmes, residents in diagnostic radiology and advanced students in radiographic technology programmes. The level of understanding of the material covered will, of course, be different for the various student groups; however, the basic language and knowledge for all student groups is the same. The text is also a key reference for medical physics residents undergoing a clinical training programme, as well as those candidates preparing for professional certification examinations.

The text is written to support a set of courses whose content provides the necessary diagnostic and interventional radiological physics knowledge for all of modern diagnostic radiology. While the text is mainly aimed at diagnostic radiology professionals, certain parts may also be of interest to professionals in other branches of medicine that use ionizing radiation for the treatment of disease (radiation therapy and nuclear medicine). The contents are also useful for physicists who are involved in studies of radiation hazards and radiation protection (health physics).

This book represents a collaborative effort by professionals from many different countries, who share a common goal of disseminating their diagnostic radiology physics knowledge and experience to a broad international audience of teachers and students.

S. CHRISTOFIDES
D.R. DANCE
A.D.A. MAIDMENT
I.D. McLEAN
K.-H. NG

CONTENTS

Chapter 1

FUNDAMENTALS OF ATOMIC AND NUCLEAR PHYSICS

K.-H. NG
University of Malaya,
Kuala Lumpur, Malaysia

D.R. DANCE
Royal Surrey County Hospital,
Guildford, United Kingdom

1.1. INTRODUCTION

Knowledge of the structure of the atom, elementary nuclear physics, the nature of electromagnetic radiation and the production of X rays is fundamental to the understanding of the physics of medical imaging and radiation protection. This, the first chapter of the handbook, summarizes those aspects of these areas which, being part of the foundation of modern physics, underpin the remainder of the book.

1.2. CLASSIFICATION OF RADIATION

Radiation may be classified as electromagnetic or particulate, with electromagnetic radiation including visible light, infrared and ultraviolet, X rays and gamma rays (Fig. 1.1), and particulate radiation including electrons, positrons, protons and neutrons.

1.2.1. Electromagnetic radiation

Electromagnetic waves can, like all waves, be characterized by their amplitude, wavelength (λ), frequency (ν) and speed. The amplitude is the intensity of the wave. The wavelength is the distance between identical points on adjacent cycles. The frequency is the number of complete wave oscillations per unit time. The speed of the wave is equal to the product of the frequency and the wavelength, and its magnitude depends upon the nature of the material through which the wave travels and the frequency of the radiation. In a vacuum, however,

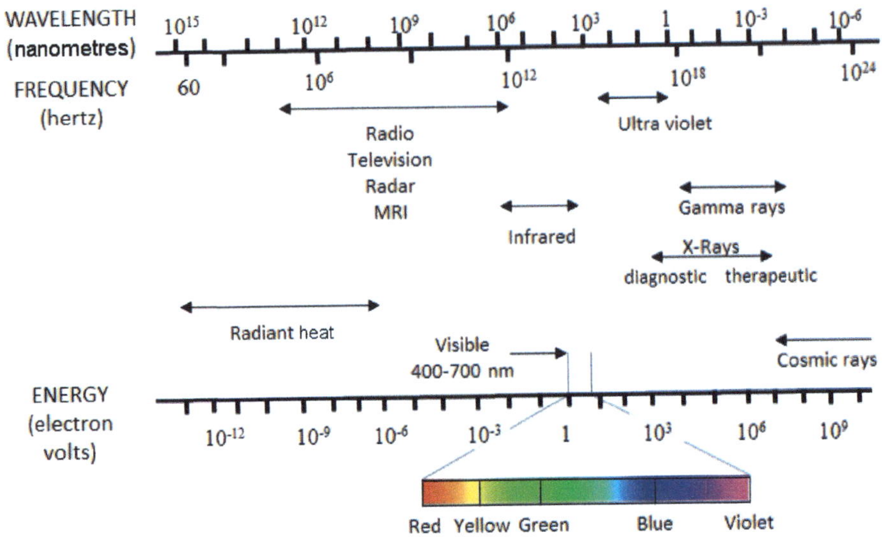

FIG. 1.1. The electromagnetic spectrum. MRI: magnetic resonance imaging.

the speed for all electromagnetic waves is a constant, usually denoted by c, and in which case:

$$c = \lambda \nu \tag{1.1}$$

For X rays, wavelength is usually expressed in nanometres (nm) (1 nm = 10^{-9} m) and frequency is expressed in Hertz (Hz) (1 Hz = 1 cycle/s = 1 s^{-1}).

When interactions with matter are considered, electromagnetic radiation is generally treated as series of individual particles, known as photons. The energy of each photon is given by:

$$E = h\nu \tag{1.2}$$

where the constant h is known as Planck's constant. In diagnostic radiology, the photon energy is usually expressed in units of keV, where 1 electronvolt (eV) is the energy received by an electron when it is accelerated across of a potential difference of 1 V.

1.2.2. Particulate radiation

In diagnostic radiology, the only particulate radiation that needs to be considered is the electron. This has a rest mass of 9.109×10^{-31} kg and a rest energy of 511 keV.

1.2.3. Ionizing and non-ionizing radiations

Radiation is classified as ionizing or non-ionizing, depending on its ability to ionize matter:

- Non-ionizing radiation cannot ionize matter.
- Ionizing radiation can ionize matter, either directly or indirectly:
  - *Directly ionizing radiation*: Fast charged particles that deposit their energy in matter directly, through many small Coulomb (electrostatic) interactions with orbital electrons along the particle track.
  - *Indirectly ionizing radiation*: X or gamma ray photons or neutrons that first transfer their energy to fast charged particles released in one or a few interactions in the matter through which they pass. The resulting fast charged particles then deposit their energy directly in the matter.

The minimum energy required to ionize an atom, i.e. to remove an electron, is known as the ionization potential. For elements, its magnitude ranges from a few electronvolts for alkali metals to 24.5 eV for helium. For water, it is 12.6 eV. Electromagnetic radiation of frequency higher than the near-ultraviolet region of the electromagnetic spectrum is ionizing, whereas electromagnetic radiation with energy below the far-ultraviolet region (e.g. visible light, infrared and radiofrequency) is non-ionizing.

1.3. ATOMIC AND NUCLEAR STRUCTURE

1.3.1. Basic definitions

The atom is composed of a central nucleus surrounded by a cloud of negatively charged electrons. Most of the mass of the atom is concentrated in the atomic nucleus, which consists of Z protons and (A minus Z) neutrons, where Z is known as the atomic number and A the atomic mass number of the nucleus. The proton and neutron have nearly identical rest masses; the proton has a positive charge identical in magnitude to the negative electron charge, and the neutron has no charge. In a non-ionized atom, the number of electrons and number of protons are equal. The radius of an atom is about 0.1 nm, whereas the radius of the nucleus is much smaller, about 10^{-5} nm.

Protons and neutrons are commonly referred to as nucleons; they have identical strong attractive interactions, and are bound in the nucleus with the strong force. In contrast to electrostatic and gravitational forces that are inversely proportional to the square of the distance between two particles, the strong force between two nucleons is a very short range force, active only at distances of the order of a few femtometres. At these short distances, the strong force is the predominant force, exceeding other forces by several orders of magnitude.

Some basic definitions and descriptions are as follows:

- Atomic number Z: number of protons and number of electrons in an atom.
- Atomic mass number A: number of protons Z plus number of neutrons N in an atom ($A = Z + N$).

There is no basic relation between A and Z, but the empirical relationship

$$Z = \frac{A}{1.98 + 0.0155A^{2/3}} \tag{1.3}$$

furnishes a good approximation for stable nuclei.

- A nucleus X with atomic mass number A and atomic number Z is denoted ${}^{A}_{Z}\mathrm{X}$; for example, an isotope of cobalt with 60 nucleons is abbreviated ${}^{60}_{27}\mathrm{Co}$, the ${}^{226}\mathrm{Ra}$ nucleus as ${}^{226}_{88}\mathrm{Ra}$.
- An element may be composed of atoms that all have the same number of protons, but have different numbers of neutrons, i.e. have different atomic mass numbers A. Atoms of identical atomic number Z but differing atomic mass numbers A are called isotopes of a given element.
- Unified atomic mass unit μ: A unit used for specifying the masses of atoms. It is equal to 1/12 of the mass of the ${}^{12}\mathrm{C}$ atom or 931.5 MeV/c^2.
- Atomic weight A_r: A dimensionless physical quantity, the ratio of the average mass of the atoms of an element to the unified atomic mass unit. The average is a weighted mean over all the isotopes of the particular element, taking account of their relative abundance.
- Atomic mass M: Expressed in unified atomic mass units. The atomic mass M is for a particular isotope and is smaller than the sum of the individual masses of constituent particles because of the intrinsic energy associated with binding the particles (nucleons) within the nucleus.
- Atomic g-atom (gram-atom): Number of grams that correspond to N_A atoms of an element, where N_A is Avogadro's constant (6.022×10^{23} atoms/g-atom). The above definition of atomic weight means that A_r g of each element contains exactly N_A atoms. It follows that:

— Number of atoms, N_am, per unit mass of an element:

$$N_\mathrm{am} = \frac{N_\mathrm{A}}{A_\mathrm{r}} \tag{1.4}$$

— Number of electrons, ZN_{am}, per unit mass of an element:

$$ZN_{am} = \frac{Z}{A_r} N_A \tag{1.5}$$

— Number of electrons, ZN_{aV}, per unit volume of an element:

$$ZN_{aV} = \rho Z N_{am} = \rho Z \frac{N_A}{A_r} \tag{1.6}$$

where ρ is the density of the element.

- Note that $Z/A_r \approx 0.5$ for all elements, with the exception of hydrogen, for which $Z/A_r = 1$. Actually, (Z/A_r) slowly decreases from 0.5 for low Z elements to 0.4 for high Z elements.
- If we assume that the mass of a molecule is equal to the sum of the masses of the atoms that make up the molecule, then for any molecular compound there are N_A molecules per g-mole of the compound, where the g-mole (gram-mole or mole) in grams is defined as the sum of the atomic weights of the atoms making up the molecule.

1.3.2. Atomic structure

The modern quantum mechanical model of the atom is built on the work of many physicists. The idea of a dense central nucleus surrounded by orbiting electrons was first proposed by Rutherford in 1911. His model, however, being based on classical physics, had a number of unsatisfactory features. For example, it could not explain the observed emission spectra of the elements. Bohr elaborated Rutherford's atomic model in 1913, based on classical, non-relativistic mechanics, by adding the concept of angular momentum quantization. His model is based on four postulates:

(i) Electrons revolve about the Rutherford nucleus in well defined, allowed orbits (shells), the central attractive Coulomb force of attraction between the electrons and the positively charged nucleus being balanced by the centripetal force arising from the orbital motion.

(ii) While in orbit, the electron does not lose any energy in spite of being constantly accelerated (this postulate is in contravention of classical physics, which predicts that an accelerated charged particle will lose energy in the form of radiation).

(iii) The angular momentum of the electron in an allowed orbit is quantized and only takes values of $n\hbar$, where n is an integer and $\hbar = h/(2\pi)$, where h is Planck's constant.

(iv) An atom or ion emits radiation when an electron makes a transition from an initial orbit with quantum number n_i to a final orbit with quantum number n_f for $n_i > n_f$.

While the work of Bohr was a major breakthrough, successfully explaining aspects of the behaviour of the hydrogen atom, the singly ionized helium atom and the doubly ionized lithium atom, etc., the story did not stop there. Through the work of Heisenberg, Schrödinger, Dirac, Pauli and others, the theory of quantum mechanics was developed. In this theory, the electrons occupy individual energy states defined by four quantum numbers, as follows:

- The principal quantum number, n, which can take integer values and specifies the main energy shell;
- The azimuthal quantum number, l, which can take integer values between 0 and n–1, and which specifies the total rotational angular momentum for the electron;
- The magnetic quantum number, m, which can take integer values between $-l$ and $+l$ and which specifies a component of the angular momentum;
- The spin quantum number, s, which takes values –½ or +½ and specifies a component of the spin angular momentum of the electron.

According to the Pauli exclusion principle, no two electrons can occupy the same state and it follows that the number of electron states that can share the same principal quantum number, n, is equal to $2n^2$.

The energy levels associated with these four quantum numbers can be understood using energy level diagrams such as those shown in Fig. 1.2 for hydrogen and tungsten. In these diagrams, each value of the principal quantum number above $n = 1$ gives rise to a band (or shell) of states of similar energies (shown as a single energy for simplicity). The energy levels associated with the various electron orbits (not drawn to scale) increase with Z and decrease with quantum number n and the average distance from the nucleus. The outer electronic shell (valence shell) determines the chemical properties of the element. The energy bands associated with $n = 1, 2, 3$, etc., are known as the K, L, M, etc., bands. The structure of each band arises from small differences in energy associated with both the l and s quantum numbers.

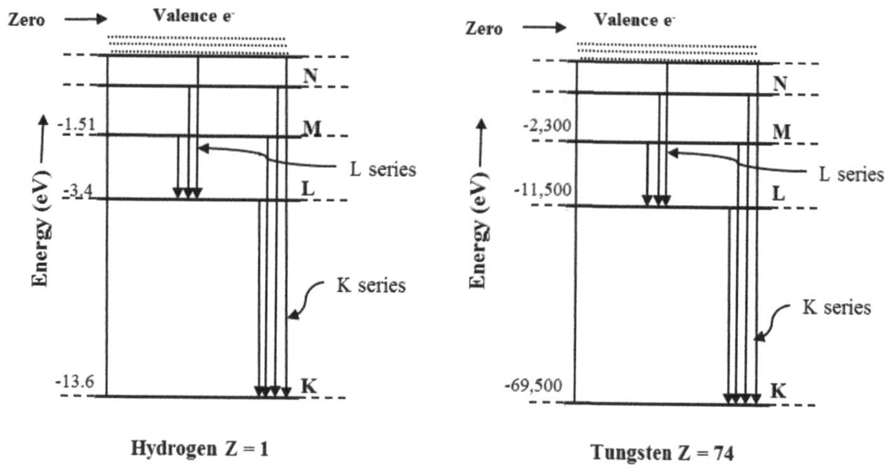

FIG. 1.2. Energy levels for hydrogen and tungsten. Possible transitions between the various energy levels are shown with arrows.

1.4. X RAYS

1.4.1. The production of characteristic X rays and Auger electrons

When charged particles pass through matter, they interact with the atomic electrons and lose energy through the processes of excitation and ionization. Ionization can also be produced as photons pass through matter by interactions such as the photoelectric effect (see Section 2.2.1) and incoherent scattering (see Section 2.2.6). Excitation occurs when there is a transfer of some of the incident particle's energy to electrons in the absorbing material, displacing them to shells further from the nucleus (i.e. to higher energy levels) and leaving a vacancy in the original shell. If the transferred energy exceeds the binding energy of the electron, ionization occurs, resulting in the electron being ejected from the atom. An ion pair, consisting of the ejected electron and the ionized, positively charged atom, is then formed.

While the smallest binding energies (ionization potentials, see Section 1.2.3) for electrons in carbon, nitrogen and oxygen are 11.3, 14.5 and 13.6 eV, respectively, the average energy required to produce an ion pair in dry air (mostly nitrogen and oxygen) is 33.97 eV. The energy difference (approximately 24 eV) is the result of the excitation process.

Whenever a vacancy is created in an inner electronic shell, whether by excitation or ionization, it is filled by an electron from a more distant (outer) shell. This results in a vacancy in this second outer shell, which is then filled by an electron (if available) from an even more distant outer shell, and the whole process repeats, producing a cascade of transitions. The energy released in each transition is carried away by the emission of electromagnetic radiation or by an electron ejected from another outer shell, known as an Auger electron. Depending on the atomic number of the material, and the electronic shells involved, the electromagnetic radiation may be in the visible, ultraviolet or X ray portions of the spectrum. The energy of this radiation is characteristic of the particular atom, since it is equal to the difference in the electron binding energies of the initial and final states for the particular transition, which depends on atomic number. X rays thus emitted are known as characteristic or fluorescent X rays. A naming convention is used in accord with the shell in which the vacancy occurred. X rays emitted in association with an electron transition to the K shell are known as K characteristic X rays, and X rays resulting from an electron transition to the L shell are known as L characteristic X rays, and so forth. Subscripts are used to denote the shell from which the vacancy is filled. The subscript α is used to denote radiation emitted for a transition between neighbouring shells and subscript β to denote radiation emitted for a transition between non-neighbouring shells. Hence, a K_α X ray is emitted for a transition between L and K shells and a K_β X ray for a transition between M or N and K shells (Fig. 1.3). Further subscripts are used as necessary to indicate which subshells are involved in the transition. The lines $K_{\alpha1}$, $K_{\alpha2}$, $K_{\beta1}$ and $K_{\beta2}$ are visible in the X ray spectrum shown in Fig. 5.2 from a tungsten target X ray tube. For X ray spectra from a molybdenum target, however, the energies of the subshells are closer together and the splitting of the K_α and K_β lines is often not resolved (see the molybdenum spectrum shown in Fig. 9.7).

As noted above, the energy carried away is the difference in binding energies between the initial and final states. For example, for tungsten, the energies of the K_α and K_β X rays are given by:

$$E(\mathrm{K}_{\alpha1}) = E_{\mathrm{LIII}} - E_{\mathrm{K}} = -10.2 - (-69.5) = 59.3 \text{ keV} \tag{1.7}$$

$$E(\mathrm{K}_{\alpha2}) = E_{\mathrm{LI}} - E_{\mathrm{K}} = -11.5 - (-69.5) = 58.0 \text{ keV} \tag{1.8}$$

$$E(\mathrm{K}_{\beta1}) = E_{\mathrm{MIII}} - E_{\mathrm{K}} = -2.3 - (-69.5) = 67.2 \text{ keV} \tag{1.9}$$

$$E(\mathrm{K}_{\beta2}) = E_{\mathrm{NIII}} - E_{\mathrm{K}} = -0.4 - (-69.5) = 69.1 \text{ keV} \tag{1.10}$$

When an Auger electron carries away the energy difference between the initial and final states, a further vacancy is created in an outer shell. For example, if the initial transition is from an M to a K shell, and the Auger electron is also emitted from the M shell, there will be two resultant vacancies in the M shell. The kinetic energy of the Auger electron is thus determined by the difference between the binding energy of the shell with the initial vacancy and the sum of the binding energies associated with the two vacancies that are created. For example, for the transition just described for a tungsten target, the energy of the Auger electron is given by:

$$E(\text{Auger}) = E_{\text{K}} - E_{\text{M}} - E_{\text{M}} = -[(-69.5) - (-2.3) - (-2.3)] = 64.9 \text{ keV} \qquad (1.11)$$

For a molybdenum target, the equivalent energy balance for the emission of an Auger electron is shown in Fig. 1.3.

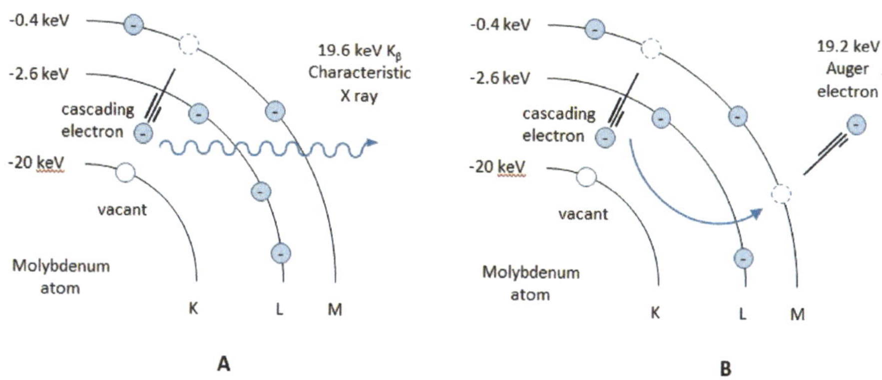

FIG. 1.3. Transition of an electron in the M shell of molybdenum to fill a vacancy in the K shell followed by the emission of (a) a K_β characteristic X ray and (b) an Auger electron.

When considering energy deposition in matter following the creation and subsequent filling of a vacancy, it is important to know whether a fluorescent X ray or an Auger electron is emitted. The probability of emission of a fluorescent X ray is known as the fluorescent yield, ω. Since either a fluorescent X ray or an Auger electron must be emitted, the probability of emitting an Auger electron is $1 - \omega$. Auger electron emission is more important for materials of low atomic number and for transitions amongst outer shells. The K fluorescence yield is close to zero for materials of low atomic number, but increases with increasing atomic number and, for example, is 0.007, 0.17, 0.60 and 0.93 for oxygen, calcium, selenium and gadolinium, respectively.

1.4.2. Radiation from an accelerated charge, bremsstrahlung

Most of the interactions that fast electrons have as they pass through matter are with the atomic electrons. They can, however, also have inelastic interactions with atomic nuclei. In such interactions, the electron path will be deflected and energy transferred to a photon, which is emitted. Because the photon is emitted in association with a slowing down of the electron, it is known as bremsstrahlung, which means 'brake radiation' in German (see Sections 2.4 and 5.2). The energy of the emitted photon can take any value from zero up to the energy of the initial electron, so that the passage of a beam of electrons though matter is accompanied by the emission of a spectrum of photons covering this energy range. Bremsstrahlung photons are the major component of the X ray spectrum emitted by X ray tubes (see Chapter 5).

The probability of bremsstrahlung emission is proportional to the value of Z^2 and is thus higher for higher atomic number materials such as tungsten ($Z = 74$). However, even for this material, the efficiency of bremsstrahlung production is less than 1% for 100 keV electrons. The angle of emission of the bremsstrahlung photons depends upon the electron energy. For electron energies much greater than the rest mass of the electron, the angular distribution is peaked in the forward direction, but as the electron energy decreases, the position of the peak moves so that it is at an angle to the forward direction. When the electron energy is low, the radiation is mainly emitted between 60° and 90° to the forward direction.

BIBLIOGRAPHY

ATTIX, F.H., Introduction to Radiological Physics and Radiation Dosimetry, John Wiley & Sons, New York (1986).

BUSHBERG, J.T., SEIBERT, J.A., LEIDHOLDT, E.M.J., BOONE, J.M., The Essential Physics of Medical Imaging, 2nd edn, Lippincott Williams & Wilkins (2002).

INTERNATIONAL ATOMIC ENERGY AGENCY, Radiation Oncology Physics: A Handbook for Teachers and Students, IAEA, Vienna (2005).

JOHNS, H.E., CUNNINGHAM, J.R., The Physics of Radiology, 4th edn, Charles C. Thomas, Springfield, IL (1983).

Chapter 2

INTERACTIONS OF RADIATION WITH MATTER

J.R. CUNNINGHAM
Camrose, Canada

D.R. DANCE
Royal Surrey County Hospital,
Guildford, United Kingdom

2.1. INTRODUCTION

This chapter deals with the physics of events that occur when photons and electrons interact with matter. These are the radiations that are important for diagnostic radiology, and only those interactions that result in their attenuation, absorption and scattering are dealt with. Other interactions, such as those with nuclei, are not considered here because they only occur for radiation that is higher in energy than that used for diagnostic radiology.

X rays of energy of a few tens of kiloelectronvolts or so have a wavelength of a few nanometres. Since this is also in the general range of atomic dimensions, one would expect interactions to take place between electromagnetic radiation and atoms — and this is indeed the case. Electron dimensions (the 'classical radius of the electron' is 2.8 pm) correspond to the higher end of the diagnostic X ray energy range and one would expect this to be the general region where interactions take place between electromagnetic radiation and the electrons that are constituents of atoms. This is also the case.

The energy range used for diagnostic radiology is generally on the boundary between classical and quantum physics and, following the 'complementarity principle', the numerical details of the interactions will be treated by classical reasoning where appropriate and by quantum mechanical considerations where this gives superior results.

The behaviour of photons and electrons as they traverse matter is very different. Photons in general have zero, one or a few interactions and are exponentially attenuated. Direct computation of the combined effects of several interactions is difficult, and Monte Carlo techniques are usually used to study photon transport through bulk media. Photon interactions are expressed in terms of cross-sections for individual interactions and attenuation coefficients for passage through bulk media. Electrons experience large numbers of interactions and in general gradually lose energy until they are stopped. This is expressed in terms of electron range and material stopping powers.

2.2. INTERACTIONS OF PHOTONS WITH MATTER

The interactions of radiations such as photons and electrons are stochastic and obey the laws of chance. For photon radiation, the concept of cross-section, with its relation to probability, follows directly. This can be explained rather simply by considering a single photon to be incident on a slab of material of area A that contains one target of cross-sectional area σ. The probability of the photon interacting with the target will be the ratio of the two areas: σ/A.

Next, let us say that there are Φ photons and that they are randomly directed at area A, and further, that area A contains n targets, each with area σ. It is easy to see that the expected number of interactions $\Delta\Phi$ between photons and targets will be:

$$\Delta\Phi = \Phi n(\sigma/A) \tag{2.1}$$

Another way of stating this is that the probability of a projectile making a hit is $n(\sigma/A)$, which is just the fraction of the area that is blocked off by the targets.

Now suppose that we change the geometrical description a little and we let the targets be atoms. Their cross-section would be an atomic cross-section. This would not be an actual area of the atom but would be an effective area — effective for an interaction between the photon and the atom that is being considered. Cross-sections are frequently represented by the symbol σ and conventionally expressed in a unit of area called the barn[1].

There are four fundamental X ray interactions that we need to consider; each can be associated with a specific cross-section. It is useful to use different symbols to represent them: τ is used to signify the cross-section for a photon to interact with an atom by the photoelectric effect, σ_{coh} is used to represent the cross-section for interaction by coherent scattering, σ_{incoh} for incoherent scattering and κ for pair and triplet production. The first three of these interactions are important in the diagnostic energy range up to 150 keV, whereas pair and triplet production are only important at much higher energies and are only treated here for completeness.

---

[1] 1 barn = 10^{-28} m^2. This unit is commonly used in nuclear and radiation physics. It is not SI but is, somewhat reluctantly, accepted by that body.

2.2.1. Photoelectric effect

In the photoelectric effect, the incident photon interacts with an atom, which is left in an excited state. The excess energy is released by the ejection of one of the electrons bound to the nucleus. This electron, called a photoelectron, leaves the atom with kinetic energy:

$$T = h\nu - E_s \qquad (2.2)$$

where

E_s is the binding energy of the electron shell from which the electron came;
h is Planck's constant;

and ν is the photon frequency.

The energy transferred to the recoiling atom is very small and can be neglected. The photoelectric effect can only take place if the photon energy, $h\nu$, exceeds the binding energy of the electron in that shell. The most probable electron shell to lose an electron is the one that satisfies this constraint and also has the highest binding energy.

Although this seems like a rather simple process, calculation of the probability of the interaction is very complicated and quantum mechanics is required. This is because it involves the wave function of the whole atomic electron cloud and these functions are available only for relatively simple atoms.

In the diagnostic energy range up to 150 keV, the photoelectric effect cross-section per atom, τ, is given approximately by:

$$\tau(h\nu, Z) = k\frac{Z^n}{(h\nu)^m} \qquad (2.3)$$

where

k is a constant;
Z is the atomic number;
n is an exponent in the range 3.6–5.3, being largest for low atomic numbers;

and m is an exponent in the range 2.5–3.5, again being largest for low atomic numbers.

A typical dependence of τ in the diagnostic photon energy range is thus:

$$\tau \sim \frac{Z^4}{(h\nu)^3} \tag{2.4}$$

This expression indicates a very strong dependence on atomic number as well as a strong inverse dependence on photon energy.

Figure 2.1 shows atomic cross-section data for the photoelectric process for photons irradiating tungsten, molybdenum and copper. They are plotted against photon energy on a log–log scale and cover the energy range from 1 keV to 300 keV. The sharp discontinuities correspond to the positions of the absorption edges for the different materials, which increase in energy with increasing atomic number and shell binding energy. For example, for tungsten, the discontinuity seen at 69.5 keV represents the effect of the K shell. At an energy just less than this, the cross-section is 6.4×10^2 barn/atom, while just above this energy, the coefficient is 3.3×10^3 barn/atom. This represents a sudden increase in cross-section of about a factor of five when the photon energy increases above that of the K shell (K edge). Thus, the major contribution to the cross-section above the K edge comes from interactions with the two K shell electrons. The discontinuities in the tungsten cross-section at energies just greater than 10 keV represent the effect of the L shell, which is more complicated because it comprises three subshells. The effect of the M shell shows up at about 2.5 keV with an even more complex structure. For copper and molybdenum, the K absorption edges are at 8.98 and 20.00 keV, respectively.

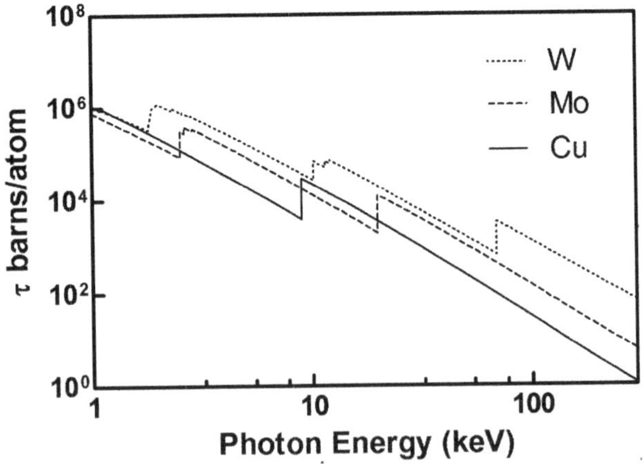

FIG. 2.1. Atomic photoelectric cross-sections for copper (Cu), molybdenum (Mo) and tungsten (W).

The incident photon disappears in the photoelectric interaction. After the interaction, a vacancy is left in the atomic shell structure and this is filled by an electron from a higher shell, with the resulting energy difference being carried off either by a characteristic X ray (also known as a fluorescent X ray) or by another electron from a higher shell, known as an Auger electron. After the initial vacancy is filled, the new vacancy or vacancies will themselves be filled and this process will continue with a cascade of events that may finally leave the atom in a highly ionized state.

2.2.2. Thomson scattering

J.J. Thomson gave the first treatment of the scattering of photons by electrons in the very early years of the 20th century. It was an early attempt to investigate the way the waves described by Maxwell's equations would be expected to interact with the newly discovered electron. His derivation is of historical interest, in that it is based on classical physics and results in a description of photon scattering that is only meaningful at the low energy limit of this interaction.

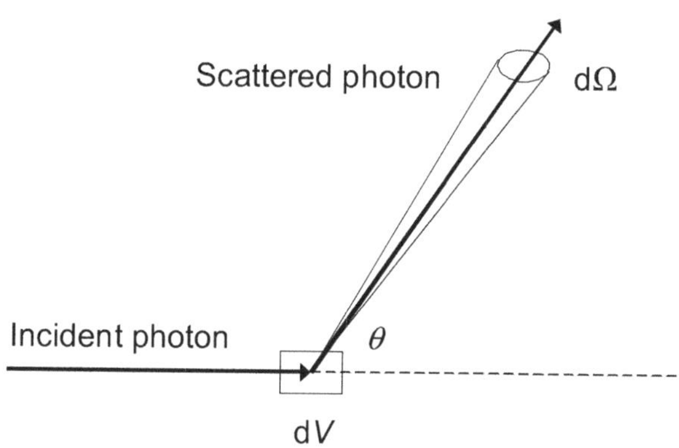

FIG. 2.2. Scattering angles and solid angle. A photon incident on a small volume element dV is scattered though angle θ into the solid angle element dΩ.

We give Thomson's results here as a first step towards the treatment of the coherent and incoherent cross-sections for scattering from atoms. First we need to introduce the concept of the differential cross-section. Whereas the total cross-section is related to the probability that the photon will interact, the differential cross-section, $d\sigma/d\Omega$, is related to the probability that the photon will interact and be scattered into solid angle $d\Omega$ (Fig. 2.2). This probability is proportional to:

$$\frac{d\sigma}{d\Omega}d\Omega \tag{2.5}$$

and the total cross-section is obtained by integrating over all solid angles:

$$\sigma = \int \frac{d\sigma}{d\Omega}d\Omega \tag{2.6}$$

In diagnostic radiology, the shape of the differential cross-section has an important influence on the amount of scattered radiation recorded by the image receptor.

For scattering of a photon by a single free electron, Thomson showed that the differential cross-section, at scattering angle θ, is given by the rather simple expression:

$$\frac{d\sigma_{Th}}{d\Omega} = \frac{r_0^2}{2}\left(1+\cos^2\theta\right) \tag{2.7}$$

In this expression, r_0 is the 'classical radius of the electron' and is given by:

$$r_0 = \frac{ke^2}{m_0 c^2} = 2.81794\times 10^{-15}\ \text{m} \tag{2.8}$$

where

k is a proportionality constant from Coulomb's law;
e is the charge on the electron;
m_0 is the rest mass of the electron;

and c is the speed of light.

It can be seen that Eq. (2.7) predicts that the same amount of energy will be scattered forward as backward, and also that the energy scattered at right angles will be half this amount. Except at low energies, this result does not agree with observation or with the predictions made by quantum mechanics.

Equation (2.7) describes the probability of scattering radiation through a unit solid angle centred on scattering angle θ. An alternative way of expressing the differential scattering cross-section involves describing the probability of scattering into a solid angle described by an annular ring of angular width $d\theta$

centred on angle θ. For unpolarized incident photons, there is no dependence of the scattering probability on the azimuthal scattering angle, and we can then use the relationship between the two solid angles:

$$d\Omega = 2\pi \sin\theta \ d\theta \tag{2.9}$$

The total cross-section for Thomson scattering is thus obtained by using Eq. (2.9) to convert $d\Omega$ to $d\theta$ and by integrating the differential scattering cross-section (Eq. (2.7)) over all scattering angles from 0 to π. The integration is very simple and it gives:

$$\sigma_{Th} = \frac{r_0^2}{2}\int_0^{\pi} 2\pi\left(1+\cos^2\theta\right)\sin\theta d\theta = \frac{8\pi r_0^2}{3} = 66.52\times10^{-30} \ m^2 \tag{2.10}$$

which is constant, predicting that the classical scattering probability is independent of electromagnetic radiation energy. This, of course, is not correct, but it will be seen in Section 2.2.4 that the integrand in Eq. (2.10) is the first term of the much more accurate result obtained by using quantum mechanics (for the Compton effect). In other words, the result obtained by classical physics is the result given by quantum mechanics when the photon energy approaches zero.

2.2.3. Coherent (Rayleigh) scattering

In deriving the expression for Thomson scattering, it was assumed that the electron was free, alone and at rest. In reality, the photon is scattered collectively by the atomic electrons, which are not free, and their proximity to one another is not very different from the wavelength of the radiation. In coherent scattering, essentially no energy is lost by the photon as it transfers momentum to the atom[2] and is scattered through angle θ. The scattering by the different electrons is in phase and the resultant angular distribution is determined by an interference pattern that is characteristic of the atom. The differential cross-section is then given by:

$$\frac{d\sigma_{coh}}{d\Omega} = \frac{d\sigma_{Th}}{d\Omega} F^2(x,Z) \tag{2.11}$$

---

[2] Strictly speaking, the condition of no change in photon energy applies to the inertial frame in which the total momentum of atom plus photon is zero.

where $d\sigma_{Th}/d\Omega$ is the Thomson differential scattering coefficient from Eq. (2.7) and the quantity F is known as the coherent form factor. It may be calculated using quantum mechanical models and is a function of the atomic number of the atom, Z, and the parameter x, which is given by:

$$x = \frac{\sin(\theta/2)}{\lambda} \tag{2.12}$$

where λ is the wavelength of the incident photon. The parameter x is proportional to the transfer of momentum between the initial and scattered photon directions.

For scattering in the forward direction, all the atomic electrons act together, and F is equal to the atomic number and the differential cross-section depends upon Z^2. As the scattering angle increases, F decreases because it becomes increasingly difficult for all the electrons to scatter in phase without any energy transfer. However, for a given value of the scattering angle, the normalized coherent form factor, F/Z, increases with increasing atomic number. Figure 2.3 shows the normalized form factor for three different elements.

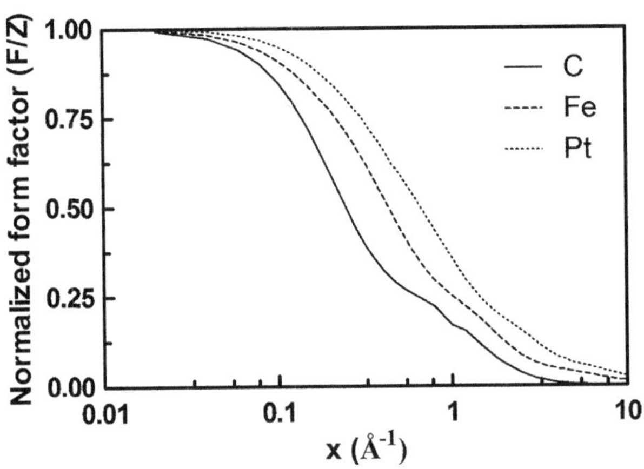

FIG. 2.3. Variation of the normalized form factor F/Z for coherent scattering with the momentum transfer parameter x. Values are shown for carbon (C), iron (Fe) and platinum (Pt) (data from Ref. [2.1]).

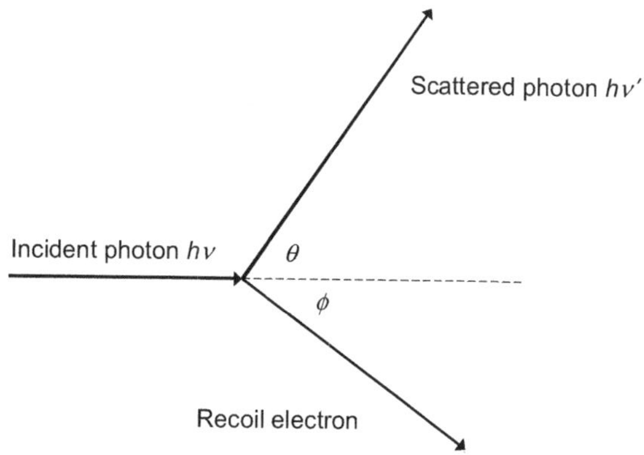

FIG. 2.4. Geometry for Compton scattering.

2.2.4. Compton scattering by free electrons

Compton scattering, like Thomson scattering, is the interaction between electromagnetic radiation and a free electron, but in this case there is an energy transfer to the electron. We consider this case before treating incoherent scattering by an atom. The energy range is such that relativity and quantum mechanics must be used to derive expressions for the cross-section. Both the photon and the electron must be considered as particles. The geometrical arrangement is shown in Fig. 2.4, where the photon is coming in from the left with energy $h\nu$ and momentum $h\nu/c$. It is a billiard-ball-like collision with the electron and is scattered through angle θ, with energy $h\nu'$ and momentum $h\nu'/c$. The electron recoils at angle ϕ with kinetic energy T_e and momentum p_e.

Using conservation of energy and momentum, we can derive several useful relations, such as the ratio of the scattered photon energy, $h\nu'$, to the incident photon energy, $h\nu$:

$$\frac{h\nu'}{h\nu}=\frac{1}{1+\alpha(1-\cos\theta)} \qquad (2.13)$$

where α is the dimensionless ratio $h\nu/m_0c^2$. The relationship between the scattered photon angle and the scattered electron angle is:

$$\cot\phi=(1+\alpha)\tan\left(\frac{\theta}{2}\right) \qquad (2.14)$$

and the scattered electron has kinetic energy given by:

$$T_e = h\nu - h\nu' = \frac{\alpha(1-\cos\theta)h\nu}{1+\alpha(1-\cos\theta)} \tag{2.15}$$

These are the Compton relations. They describe the kinematics of the interaction but say nothing about the probability of interaction, or the cross-section. In the diagnostic energy range, the parameter α is small and, as a consequence, the energy transfer to the recoil electron is also small, being zero in the forward direction and taking its largest value when the photon is backscattered. This is demonstrated in Fig. 2.5, which shows the relationship between the incident and scattered photon energies. For 20 keV, 50 keV and 100 keV incident photons, the maximum energy transfers to the recoil electron are 1.5 keV, 8.2 keV and 28.1 keV, respectively.

The cross-section for the scattering of a photon, with energy $h\nu$ through a given angle θ, was first derived in 1928 by Klein and Nishina using the Dirac theory of the electron.[3] Klein and Nishina obtained the following expression for the differential cross-section for scattering of photons by a single free electron:

$$\frac{d\sigma_{KN}}{d\Omega} = \frac{r_0^2}{2}\left(1+\cos^2\theta\right) f_{KN} \tag{2.16}$$

where

$$f_{KN} = \left\{\frac{1}{1+\alpha(1-\cos\theta)}\right\}^2 \left\{1 + \frac{\alpha^2(1-\cos\theta)^2}{\left[1+\alpha(1-\cos\theta)\right]\left[1+\cos^2\theta\right]}\right\} \tag{2.17}$$

This cross-section reduces to the Thomson cross-section when $\alpha \rightarrow 0$ (that is, $h\nu'/h\nu \rightarrow 1$).

Figure 2.6 shows the differential scattering cross-section plotted as a function of the photon scattering angle plotted in two ways. The lower curve is a graph of the differential coefficient per steradian and the upper curve is a graph of the differential coefficient per unit scattering angle. The differential scattering cross-section $d\sigma/d\theta$ is zero in the forward direction because sin θ is zero (see Eq. (2.9)).

---

[3] The calculation of the interaction probabilities for the Compton effect by Klein and Nishina was one of the first successful tests of quantum mechanics.

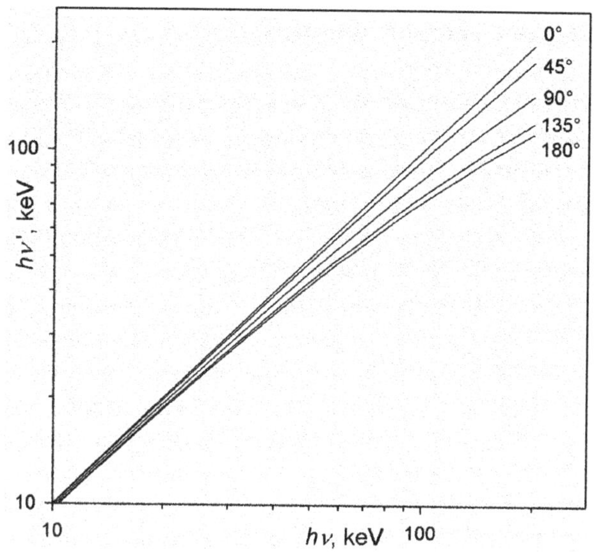

FIG. 2.5. Secondary photon energy hυ' versus primary photon energy hυ for Compton interactions and various scattering angles.

The total Compton cross-section (probability of interaction per electron) for a photon of energy $h\nu$, is obtained by integrating Eq. (2.16) using Eq. (2.9) and the angular range 0 to π for θ. The result is:

$$\sigma_{KN}(h\nu)=2\pi r_0^2\left\{\left(\frac{1+\alpha}{\alpha^2}\right)\left(\frac{2(1+\alpha)}{1+2\alpha}-\frac{\ln(1+2\alpha)}{\alpha}\right)+\frac{\ln(1+2\alpha)}{2\alpha}-\frac{1+3\alpha}{(1+2\alpha)^2}\right\} \tag{2.18}$$

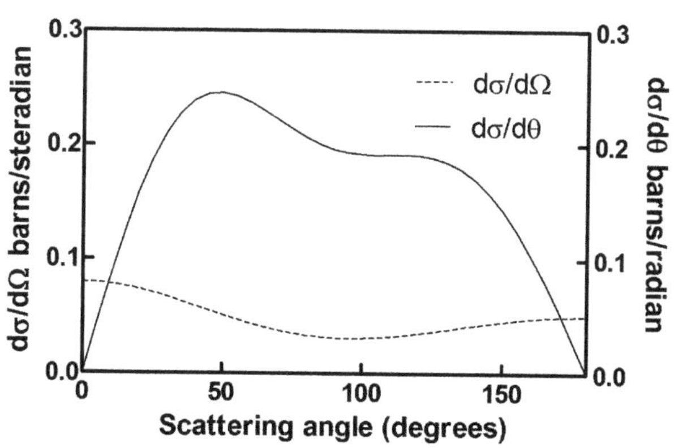

FIG. 2.6. Compton differential cross-sections for scattering of 70 keV photons.

2.2.5. Scattering and energy transfer coefficients

In the incoherent free electron scattering process, the initial photon energy is divided between the scattered photon and the recoiling electron. A differential energy transfer coefficient can be obtained by using the equation:

$$\frac{d\sigma_{tr}}{d\Omega} = \frac{r_0^2}{2}\left(1+\cos^2\theta\right) f_{KN}\left(\frac{\alpha(1-\cos\theta)}{1+\alpha(1-\cos\theta)}\right) \tag{2.19}$$

This can be integrated over all angles to give σ_{tr}, the energy transfer coefficient. The scattering coefficient is then, by definition, the difference between the total cross-section for Compton scattering and the energy transfer coefficient:

$$\sigma_s = \sigma_{KN} - \sigma_{tr} \tag{2.20}$$

2.2.6. Incoherent scattering

For the Compton effect, as with Thomson scattering, it is assumed that the electron is free and at rest. For incoherent scattering by bound atomic electrons, the contributions from individual electrons are added and the differential cross-section takes the form:

$$\frac{d\sigma_{incoh}}{d\Omega} = \frac{r_0^2}{2}\left(1+\cos^2\theta\right) f_{KN} S(x,Z) \tag{2.21}$$

The function S is known as the incoherent scattering function and, as with the coherent form factor, is a universal function of the momentum transfer quantity x and the atomic number. The value for S is zero in the forward direction and increases with increasing momentum transfer, reaching the value of Z, the number of electrons per atom. This increase becomes slower as the atomic number increases. This is illustrated in Fig. 2.7, which shows the normalized incoherent scatter function (S/Z) for three elements.

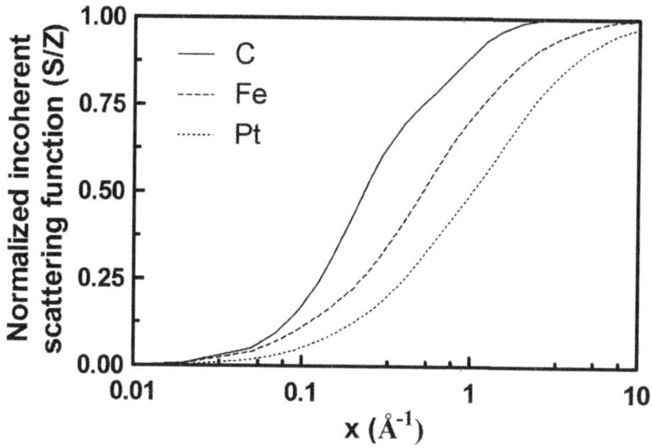

FIG. 2.7. Variation of the normalized incoherent scattering function S/Z with the momentum transfer parameter x. Data are given for carbon (C), iron (Fe) and platinum (Pt) (data from Ref. [2.2]).

The total cross-section for incoherent scattering is obtained by numerical integration of Eq. (2.21). In many situations, it is very nearly equal to the single electron cross-section multiplied by the number of electrons in the atom:

$$\sigma_{\text{incoh}} \approx Z\sigma_{\text{KN}} \tag{2.22}$$

2.2.7. Pair and triplet production

When a high energy photon passes near to an atomic nucleus, the photon may interact with the nuclear coulomb field by a process called pair production. The photon is converted into an electron–positron pair, each with its own kinetic energy. The energy balance is:

$$h\nu = T_+ + T_- + 2m_0c^2 \tag{2.23}$$

on condition that the photon energy exceeds an energy threshold for the interaction of $2m_0c^2$ (1022 keV). Pair production cannot take place for photons with energies less than this. As pair production occurs in the field of the nucleus, the cross-section for this interaction varies approximately as Z^2, where Z is the nuclear charge.

The process can also take place in the field of an electron. It is then called triplet production because the target electron is itself ejected with considerable energy. Two electrons and one positron are thus set in motion. The energy threshold for triplet production is $4m_0c^2$. Thresholds for pair and triplet production are much higher than the photon energies relevant to diagnostic radiology.

2.3. PHOTON ATTENUATION COEFFICIENTS

The discussion above has been concerned with the interaction of photons with individual atoms, but it is also necessary to consider the macroscopic behaviour of photons traversing matter. For this purpose, linear and mass attenuation coefficients are used, which are simply related to the total cross-section. As mentioned in the introduction, photons may undergo more than one interaction as they pass through bulk material. For example, an initial scatter interaction might be followed by a second scattering process, which in turn might be followed by a third scatter, photoelectric absorption or no further interactions, with the photon leaving the bulk material. Linear and mass attenuation coefficients give information about the passage of primary photons through the material. The radiation field at depth in the medium will also include scattered photons, which also contribute to the dose within the medium. The exit beam from the bulk material will also comprise both primary and scattered photons. As noted earlier, such effects are best estimated using Monte Carlo techniques.

2.3.1. Linear attenuation coefficient

Consider a thin uniform slab of material of thickness dx, which is irradiated with a beam of photons incident normally on the slab. Individual photons may pass through the slab without interacting, or they may be absorbed or they may be scattered. From the discussion in Section 2.2, it follows that the probability that an individual photon will interact in this thin section is given by:

$$N_{\mathrm{a}} \sigma \, \mathrm{d}x \tag{2.24}$$

where N_{a} is the number of interaction centres (atoms) per unit volume and σ is the total interaction cross-section per atom.

The quantity $N_{\mathrm{a}}\sigma$ is known as the linear attenuation coefficient and is usually denoted by μ. For scattering by atoms, N_a may be calculated from the Avogadro constant, N_{A}, the atomic weight, A_{r}, and the density, ρ, so that:

$$\mu = N_{\mathrm{a}} \sigma = \frac{1000 N_{\mathrm{A}} \rho}{A_{\mathrm{r}}} \sigma \tag{2.25}$$

This expression is in SI units, so that the dimensions of μ are m^{-1}.

2.3.2. Exponential attenuation

Now consider a thick slab of material and let $\Phi(x)$ represent the fluence[4] of photons that have not interacted in the slab after passage through thickness x. The expected change, dΦ, in this fluence after passage through a further thickness dx is given by:

$$\mathrm{d}\Phi = -\Phi\mu\ \mathrm{d}x \tag{2.26}$$

where the negative sign is used to signify that Φ is decreasing. Equation (2.26) is a restatement of Eq. (2.24). Integration of Eq. (2.26) gives:

$$\Phi = \Phi_0\,\mathrm{e}^{-\mu x} \tag{2.27}$$

where Φ_0 is the initial value of the fluence. This is the equation describing the exponential attenuation of a photon beam. It is known as Beer's law. It should be noted that it describes the number of photons that have not interacted, also known as primary photons. At diagnostic energies, other photons may be present at depth, resulting from photon scattering interactions or the emission of fluorescent photons following a photoelectric interaction.

2.3.3. Mass attenuation coefficient

The linear attenuation coefficient μ is dependent on density, which in turn is dependent on the physical state of the material. As a consequence, μ is not a suitable quantity for data compilations, and the related quantity (μ/ρ), which is independent of density, is used instead. This quantity is known as the mass attenuation coefficient and its dimensions are square metres per kilogram (m^2/kg).

It should be noted that in most data compilations the mass attenuation coefficients are given in units of square centimetres per gram (cm^2/g) because historically they have been expressed in this way and this provides numbers that are convenient to manipulate.

2.3.4. Mass energy transfer coefficients and mass energy absorption coefficients

For dosimetric purposes, it is necessary to know the energy transferred to secondary electrons as a result of the initial interaction. The linear (μ_{tr}) and

---

[4] See Section 3.2.1 for the definition of fluence.

mass energy transfer coefficients (μ_{tr}/ρ) allow calculation of this energy. Both quantities are defined using:

$$\mu_{tr} = \mu \frac{<T>}{h\nu} \tag{2.28}$$

where $<T>$ is the expectation value of the energy converted to secondary electrons.

For photons of energy $h\nu$ traversing a distance dx in a material, the energy $d(h\nu)_{tr}$ transferred by interactions to kinetic energy of electrons is then given by:

$$d(h\nu)_{tr} = \Phi h\nu \mu_{tr}\, dx \tag{2.29}$$

We can use this expression to calculate the kerma in the given material by simply dividing by the mass per unit area traversed ($dm = \rho\, dx$), to obtain:

$$K = \frac{d(h\nu)_{tr}}{dm} = \Phi h\nu \left(\frac{\mu_{tr}}{\rho}\right) \tag{2.30}$$

Some of the energy transferred to the secondary charged particles is lost to radiative processes in the material, mainly bremsstrahlung. To take account of this, we use the mass energy absorption coefficient (μ_{en}/ρ), given by:

$$\left(\frac{\mu_{en}}{\rho}\right) = \left(\frac{\mu_{tr}}{\rho}\right)(1-g) \tag{2.31}$$

where g is the energy fraction lost to radiative processes. For the energies used in diagnostic radiology, g may be taken as zero.

2.3.5. Contribution of individual interactions to the total mass attenuation coefficient

We now consider the four distinctly different mechanisms by which photons may interact with matter. They all compete, in that in any beam of photons they may all occur, each according to its individual probability. The total mass attenuation coefficient is therefore the sum of all the individual mass attenuation coefficients and, using Eq. (2.25), we obtain:

$$\left(\frac{\mu}{\rho}\right)=\left(\frac{\tau}{\rho}\right)+\left(\frac{\mu_{\text{coh}}}{\rho}\right)+\left(\frac{\mu_{\text{inc}}}{\rho}\right)+\left(\frac{\kappa}{\rho}\right)=\left(\tau+\sigma_{\text{coh}}+\sigma_{\text{inc}}+\kappa\right)\frac{N_{\text{A}}}{A_{\text{r}}}1000 \qquad (2.32)$$

The size of each attenuation coefficient will depend on the photon energy and the atomic number of the material. Figure 2.8 shows the mass attenuation coefficients for water for photon energies from 1 keV to 300 keV.

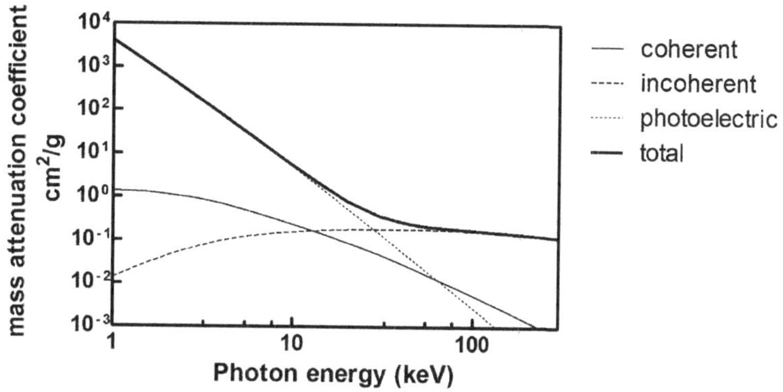

FIG. 2.8. Mass attenuation coefficients of water for each of the interactions discussed. The highest energy shown is below the thresholds for pair and triplet production (data from Ref. [2.3]).

The photoelectric interaction makes the dominant contribution to the total interaction cross-section at the lowest energies. The steep decrease at the lower photon energies is characteristic of the photoelectric effect and ends when incoherent (Compton) scattering becomes dominant, and remains so for the rest of the diagnostic energy range. The crossover position for these two interactions depends on the atomic number, but for water is about 30 keV.

Figure 2.9 shows a comparison of interaction coefficients for various materials of importance in diagnostic radiology. In the energy range up to 100 keV and for the high atomic number materials, the discontinuities arising from the differences in photoelectric absorption at K, L and M edges are evident. The differences in absorption that this creates are important for the design of filters used to shape X ray spectra (particularly in mammography and for imaging using iodinated contrast agents). The position of the K edge(s) for materials used in image receptors can have an important influence on absorption efficiency.

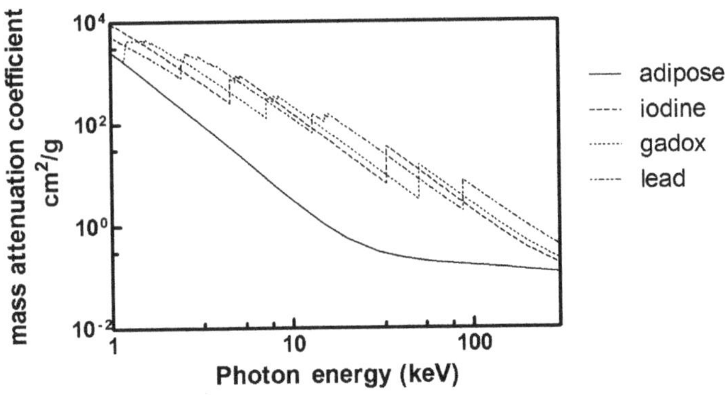

FIG. 2.9. Total mass interaction coefficients for materials relevant to diagnostic radiology (adipose tissue, iodine, gadolinium oxysulphide (gadox) and lead (tissue composition data from Ref. [2.4]; interaction coefficients from Ref. [2.3]).

2.3.6. Coefficients for compounds and mixtures

Mass attenuation coefficients and mass energy transfer coefficients for compounds and intimate mixtures can be obtained by a weighted summation of the coefficients of the constituents:

$$\left(\frac{\mu}{\rho}\right)=\sum_i\left(\frac{\mu}{\rho}\right)_i w_i;\quad \left(\frac{\mu_{\text{tr}}}{\rho}\right)=\sum_i\left(\frac{\mu_{\text{tr}}}{\rho}\right)_i w_i \tag{2.33}$$

where w_i are the normalized weight fractions[5] of the elements i (or mixture components i) present in the absorber. The mass energy absorption coefficient for an element accounts for the radiative losses of the secondary electrons. Since a secondary electron originating from an atom A will also suffer radiative losses from the other constituents, a simple weighted addition can only approximate the mass energy absorption coefficient for a mixture, as long as radiative losses are small, which is then given by:

$$\left(\frac{\mu_{\text{en}}}{\rho}\right)=\sum_i\left(\frac{\mu_{\text{en}}}{\rho}\right)_i w_i \tag{2.34}$$

---

[5] In the case of compounds, the normalized weight fractions are derived from consideration of the chemical composition of the compound and the individual elemental atomic weights (A_r) (see Section 1.3 and also Ref. [2.5]).

2.4. INTERACTIONS OF ELECTRONS WITH MATTER

There are two main mechanisms of energy loss by electrons: ionizational or collisional losses and radiative losses or bremsstrahlung. The principal process for energy loss when electrons pass through matter involves collisions with other electrons. As they have the same mass, the energy losses may be quite large and the changes in direction can also be quite large. Since electrons cannot be distinguished, it is assumed that the electron that leaves the collision with the most energy is the original incident electron. This means that the maximum energy exchange would be half the original energy. In addition, because of the small mass of the electron, it may also interact with the electric field of nuclei and be decelerated so rapidly that some of its energy may be radiated away. This is referred to as bremsstrahlung and is the main process responsible for the production of X rays when an electron beam strikes a target.

Energy lost by charged particles in passing through matter is generally described using a quantity called stopping power, S. This is defined as $S = \mathrm{d}T/\mathrm{d}x$, where $\mathrm{d}T$ is the loss in kinetic energy of the particle as it travels a distance $\mathrm{d}x$. It is more common to express the distance in terms of mass per unit area of the material, giving the mass stopping power S/ρ:

$$\frac{S}{\rho}=\frac{1}{\rho}\frac{\mathrm{d}T}{\mathrm{d}x} \tag{2.35}$$

where ρ is the mass density of the material.

2.4.1. Ionizational (collisional) interactions and ionizational stopping power

This process involves collisions between electrons travelling through matter and the electrons that are part of the material. The result is that the electrons may be dislodged from their atoms, leaving them 'ionized'. It is for this reason that 'ionizing radiation' gets its name. It is rather difficult to measure the rate of energy lost by these interactions but it is relatively easy to calculate it. As electrons are so small, relativistic effects are important even at quite low kinetic energies. Quantum mechanics must also be used and the problem was first solved by Bethe in the early part of the 20th century. The following expression is the Bethe–Bloch formula, extended by Sternheimer, and gives the ionizational mass stopping power:

$$\frac{S_{\mathrm{ion}}}{\rho}=2\pi r_0^2 N_{\mathrm{e}}\frac{\mu_0}{\beta^2}\left[\ln\frac{T^2(T+2\mu_0)}{2\mu_0 I^2}+\frac{T^2/8-(2T+\mu_0)\mu_0\ln 2}{(T+\mu_0)^2}+1-\beta^2-\delta\right],\ \mathrm{MeV\cdot cm^2\cdot g^{-1}} \tag{2.36}$$

Some of this notation has been used before: r_0 is the 'classical radius of the electron', $N_e = N_A(Z/A_r)$, with N_A being the Avogadro constant, Z the atomic number and A_r the atomic weight of the material. The quantity $\mu_0 = m_0c^2$ is the rest mass of the electron multiplied by the speed of light squared, T is the kinetic energy and β is the ratio of the speed of the electron to that of light. The density correction term, δ, was added later by Sternheimer. Its effect is to reduce energy losses, but only at high energies. At 100 MeV, it can be as great as 20%. I is a semi-empirical quantity called the 'mean excitation energy', which is a property of the material and increases as the atomic number of the material increases. Values of I for a large number of materials are given by the National Institute of Standards and Technology.

In the low energy region, below 100 keV or so, the term in front of the square brackets is the most important. The factor of $1/\beta^2$ makes the stopping power nearly inversely proportional to the kinetic energy. For energies above 100 keV, β is essentially 1 and the term in front becomes nearly constant. The terms inside the square bracket increase slowly with energy and the stopping power passes through a minimum in the region of 1 MeV.

Dependence on atomic number is not strong. The factor in front of the square brackets contains the number of electrons per unit mass and since this is given by $N_A(Z/A_r)$, and recalling that Z/A_r is 0.5 or slightly less for all materials except hydrogen, the mass stopping power decreases only slightly as atomic number increases. The mean excitation energy, I, increases as atomic number increases, which also serves to make S_{ion} smaller for high atomic number materials.

2.4.2. Radiative interactions and radiative stopping power

When an electron passes close to a nucleus, it will experience a substantial Coulomb force and will be decelerated. The decelerated charge will radiate energy in the form of electromagnetic radiation. The quantum mechanical solution is complicated and approximations must be made, and the most appropriate form for the result depends on the energy range.

The approximate radiative mass stopping power for diagnostic energies is:

$$\frac{S_{rad}}{\rho} = \sigma_0 \frac{N_A}{A_r} Z^2 (T + \mu_0) \bar{B}, \text{MeV} \cdot \text{cm}^2 \cdot \text{g}^{-1} \tag{2.37}$$

where $\sigma_0 = (1/137)\left(e^2/\mu_0\right)^2 = 0.580$ barns/nucleus. The function $B = B(h\nu/T)$ is a slowly varying function of T and Z with an average for non-relativistic energies, $T << m_0c^2$, of $\bar{B} = 16/3$. Other approximations are valid for relativistic electron energies.

Although this expression looks much simpler than Eq. (2.36), the actual physical problem being solved is much more complicated. The energy loss due to this process is quite strongly dependent on the atomic number, as can be seen by the Z^2 term.

2.4.3. Total stopping power

The total stopping power is the sum of the ionizational and radiative stopping powers, as given by Eqs (2.36, 2.37). That is:

$$S_{\text{tot}} = S_{\text{ion}} + S_{\text{rad}} \tag{2.38}$$

Figure 2.10 shows the ionizational, radiative and total stopping powers for water and for tungsten.

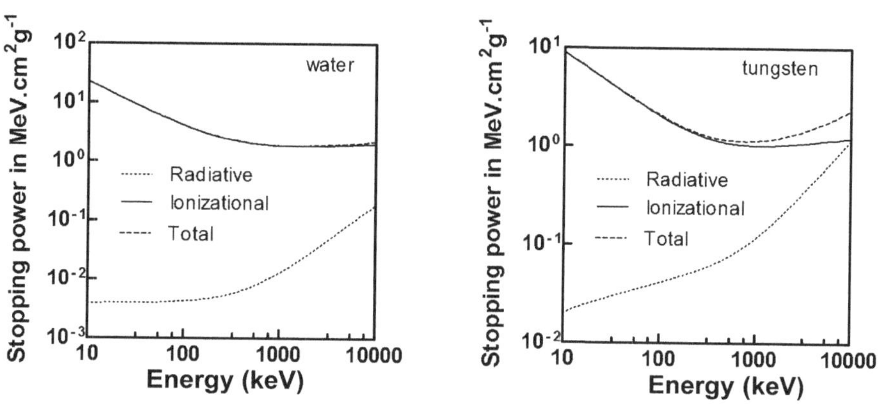

FIG. 2.10. Ionizational, radiative and total stopping powers for water and for tungsten for electron energies from 10 keV to 10 MeV.

From Fig. 2.10 (water), it can be seen that the radiative stopping power for low atomic number materials such as water (or tissue) is almost negligible and the total stopping power can be considered as being due to the ionizational stopping power. Figure 2.10 (tungsten) shows that this is no longer the case for high atomic number materials, such as tungsten. For example, at 100 keV, the radiative stopping power is only slightly more than an order of magnitude less than the ionizational stopping power, and at 10 MeV they are approaching equality.

The properties of the bremsstrahlung spectrum produced at an X ray target are discussed in detail in Chapter 5.

2.4.4. Stopping power in compounds and mixtures

The mass stopping power can be approximated for compounds or intimate mixtures by a weighted addition of the mass stopping power of the elemental constituents, assuming independent contribution to stopping power:

$$\left(\frac{S}{\rho}\right)=\sum_i\left(\frac{S}{\rho}\right)_i w_i \tag{2.39}$$

where w_i are the normalized weight fractions of the elements, i, present in the material. The influence of chemical binding is neglected here.

2.4.5. Linear energy transfer

The mass collision stopping power characterizes the energy loss of the electron resulting from all collisions. The secondary electrons resulting from hard collisions (δ rays) carry part of the energy some distance away from the track and may escape the volume of interest. If phenomena on a microscopic scale are considered (e.g. in radiobiology), the use of the collision stopping power can result in an overestimation of dose. In such situations, the restricted stopping power may be used, which relates to the energy lost in the absorber by secondary particles not exceeding an energy limit, Δ, thus limiting the volume of interest to the range of electrons with energy Δ. A reasonable value for Δ is 10 keV. The restricted stopping power may be expressed as either the mass restricted stopping power $(S/\rho)_{\mathrm{Res}}$ or, for a description of the energy passed to the medium per unit track length, it is common to use the linear energy transfer, L_Δ, given in units of keV/μm:

$$L_\Delta=\frac{\rho}{10}\left(\frac{1}{\rho}\frac{\mathrm{d}T}{\mathrm{d}x}\right)_\Delta \tag{2.40}$$

where the density is given in g/cm^3 and the mass restricted stopping power in MeV·cm^2·g^{-1}.

2.5. DATA SOURCES

Data for photon attenuation coefficients and stopping powers can be found in many textbooks but are conveniently obtained either from web based data sources provided in Refs [2.3, 2.5], or the tabulations in Ref. [2.6]. The composition of body tissues and phantom materials are given in Refs [2.4, 2.7].

REFERENCES

[2.1] HUBBELL, J.H., ØVERBØ, I. Relativistic atomic form factors and photon coherent scattering cross sections, J. Phys. Chem. Ref. Data **8** (1979) 69–106.

[2.2] HUBBELL, J.H., et al., Atomic form factors, incoherent scattering functions, and photon scattering cross sections, J. Phys. Chem. Ref. Data **4** 3 (1975) 471–539.

[2.3] BERGER, M.J., et al., XCOM: Photon Cross Sections Database (version 1.3), National Institute of Standards and Technology, Gaithersburg, MD (2005).

[2.4] INTERNATIONAL COMMISSION ON RADIATION UNITS AND MEASUREMENTS, Tissue Substitutes in Radiation Dosimetry and Measurement, Rep. 44, ICRU, Bethesda, MD (1989).

[2.5] BERGER, M.J., COURSEY, J.S., ZUCKER, M.A., CHANG, J., ESTAR, PSTAR, and ASTAR: Computer Programs for Calculating Stopping-Power and Range Tables for Electrons, Protons, and Helium Ions (version 1.2.3), National Institute of Standards and Technology, Gaithersburg, MD (2005).

[2.6] INTERNATIONAL COMMISSION ON RADIATION UNITS AND MEASUREMENTS, Photon, Electron, Proton and Neutron Interaction Data for Body Tissues, Rep. 46, ICRU, Bethesda, MD (1992).

[2.7] INTERNATIONAL COMMISSION ON RADIOLOGICAL PROTECTION, Basic Anatomical and Physiological Data for Use in Radiological Protection: Reference Values, Publication 85, Pergamon Press, Oxford (2003).

BIBLIOGRAPHY

HUBBELL, J.H., Review of Photon Interaction Cross Section Data in the Medical and Biological Context, Phys. Med. Biol. **44** (1999) R1–22.

INTERNATIONAL COMMISSION ON RADIATION UNITS AND MEASUREMENTS, Stopping Powers for Electrons and Positrons, Rep. 37, ICRU, Bethesda, MD (1984).

JOHNS, H.E., CUNNINGHAM, J.R., The Physics of Radiology, 4th edn, Charles C. Thomas, Springfield, IL (1983).

NATIONAL INSTITUTE OF STANDARDS AND TECHNOLOGY, http://www.nist.gov/index.html

SELTZER, S.M., BERGER, M.J., Evaluation of the collision stopping power of elements and compounds for electrons and positrons, Int. J. Appl. Radiat. Isot. **33** (1982) 1189–1218.

Chapter 3

FUNDAMENTALS OF DOSIMETRY

E.M. YOSHIMURA
Universidade de São Paulo,
São Paulo, Brazil

3.1. INTRODUCTION

Determination of the energy imparted to matter by radiation is the subject of dosimetry. The energy deposited as radiation interacts with atoms of the material, as seen in the previous chapter. The imparted energy is responsible for the effects that radiation causes in matter, for instance, a rise in temperature, or chemical or physical changes in the material properties. Several of the changes produced in matter by radiation are proportional to the absorbed dose, giving rise to the possibility of using the material as the sensitive part of a dosimeter. Also, the biological effects of radiation depend on the absorbed dose. A set of quantities related to the radiation field is also defined within the scope of dosimetry. It will be shown in this chapter that, under special conditions, there are simple relations between dosimetric and field description quantities. Thus, the framework of dosimetry is the set of physical and operational quantities that are studied in this chapter.

3.2. QUANTITIES AND UNITS USED FOR DESCRIBING THE INTERACTION OF IONIZING RADIATION WITH MATTER

Historically, measurement of the ionization produced by radiation was the first choice used to quantify the passage of radiation through matter. Indeed, the quantity exposure, or, more precisely, exposure dose, as defined by the International Commission on Radiation Units and Measurements (ICRU) in 1957, is related to the ability of a photon beam to ionize the air. In recent years, the use of this quantity has been replaced by kerma, a more general quantity that is recommended for dosimeter calibration purposes. Nevertheless, absorbed dose is the quantity that better indicates the effects of radiation on materials or on human beings, and, accordingly, all the protection related quantities are based on it. The use of dosimetric quantities is important in many aspects of the application of radiation. In diagnostic radiology, radiation protection of staff and patients is the most important application of the dosimetric quantities. This section introduces and discusses the main dosimetric quantities, relating them, whenever possible, to the quantities that describe the radiation field.

3.2.1. Radiation fields: Fluence and energy fluence

A radiation field at a point P can be quantified by the physical non-stochastic quantity fluence, which is usually expressed in units of m^{-2} or cm^{-2}, and is given by the relation:

$$\Phi = \frac{dN}{da} \tag{3.1}$$

where dN is the differential of the expectation value of the number of particles (photons or massive particles) striking an infinitesimal sphere with a great circle area, da, surrounding point P, as shown in Fig. 3.1. It is worthwhile noting that the particles included in Φ may have any direction, but correspond to one type of radiation, so that photons and electrons are counted separately, contributing to the photon fluence and the electron fluence, respectively.

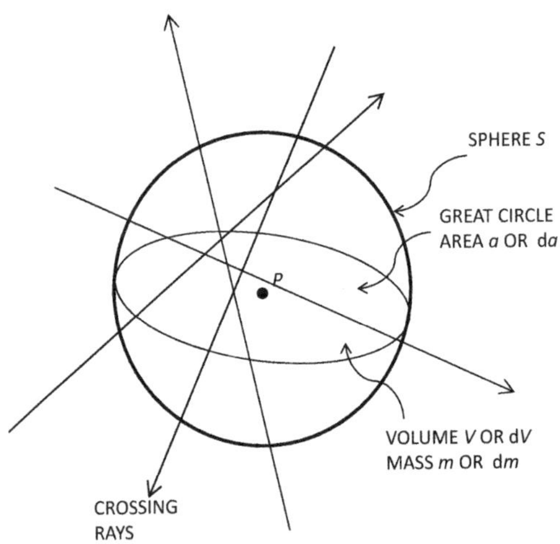

FIG. 3.1. Characterizing the radiation field at a point, P, in terms of the radiation traversing the spherical surface, S.

The concept of energy fluence follows easily, by summing the radiant energy of each particle that strikes the infinitesimal sphere:

$$\Psi = \frac{dR}{da} \tag{3.2}$$

In Eq. (3.2), dR is the differential of the radiant energy R — kinetic energy of massive particles, energy of photons — that impinges on the infinitesimal sphere. The SI unit of energy fluence is joules per square metre (J/m^2).

If the radiation field is composed of particles each with the same energy E, the energy fluence is related to the fluence (or particle fluence) through the simple expression: $\Psi = E\Phi$.

3.2.2. Energy transferred, net energy transferred, energy imparted

3.2.2.1. Energy transferred, net energy transferred

When an uncharged particle — for instance an X ray photon — interacts with matter, part of its energy is transferred in various interaction events. In a volume, V, of material, the energy transferred (ε_{tr}) is given by the sum of all the initial kinetic energies of charged ionizing particles liberated by the uncharged particles in the volume V. For the case where photons in the diagnostic energy range are the uncharged interacting particles, ε_{tr} corresponds to the sum of the kinetic energies of electrons at the moment they are set free in an incoherent scattering or photoelectric interaction in the volume V. For photon energies above the pair production threshold of 1.022 MeV (see Section 2.2.7), kinetic energy may also be transferred to positrons.

As the liberated charged particles interact with matter, part of their initial kinetic energy can be irradiated as photons. There are two main processes responsible for the emission of photons:

(i) The emission of bremsstrahlung radiation by electrons and positrons interacting with nuclei.
(ii) The in-flight annihilation of positrons; the remaining kinetic energy of the positron (T_{ann}) at the moment of the annihilation plus the rest mass energies of the annihilated particles (1.02 MeV) being converted to photon energy.

When the energies of the bremsstrahlung photons ($h\nu_{brem}$) and the annihilation photons (T_{ann}) are subtracted from ε_{tr}, another quantity is defined — the net energy transferred — ε_{tr}^{net}:

$$\varepsilon_{tr}^{net} = \varepsilon_{tr} - \sum h\nu_{brem} - \sum T_{ann} \tag{3.3}$$

For both quantities (energy transferred and net energy transferred), the volume V is the volume where the initial uncharged particles interact. It does not matter if the range of the charged particles is restricted to V or not, their initial kinetic energies are all included in ε_{tr}, and all the bremsstrahlung emissions and excess of energy of the annihilation photons are excluded from ε_{tr}^{net}. For photons in the diagnostic energy range, incident on low Z materials, $\varepsilon_{tr}^{net} = \varepsilon_{tr}$, as there is no pair production, and bremsstrahlung emission by the released electrons is unlikely.

3.2.2.2. *Energy imparted*

A very important concept regarding the deposition of energy from ionizing radiation to matter is the energy imparted. This quantity is defined for any radiation (charged or uncharged) and is related to the part of the radiant energy that can produce effects within an irradiated volume. If V is the irradiated volume, R_{in} is the radiant energy that enters the volume and R_{out} is the energy that leaves the volume, the energy imparted is defined as:

$$\varepsilon = R_{in} - R_{out} + E_{m \to R} - E_{R \to m} \tag{3.4}$$

The quantities $E_{m \to R}$ and $E_{R \to m}$ in Eq. (3.4) are the changes in energy when the rest mass of a particle is converted to radiant energy ($m \to R$) or the energy of a photon is converted to the mass of particles ($R \to m$) inside the volume V. For the energies encountered in diagnostic radiology, both of these terms are negligible and Eq. (3.4) can be rewritten as:

$$\varepsilon = R_{in} - R_{out} \tag{3.5}$$

3.2.3. Kerma and collision kerma

The physical, non-stochastic quantity kerma (K) is related to the energy transferred from uncharged particles to matter. Kerma is the acronym for **k**inetic **e**nergy **r**eleased per unit **ma**ss. It is defined as:

$$K = \frac{d\varepsilon_{tr}}{dm} \tag{3.6}$$

where the quantity $d\varepsilon_{tr}$ is the expectation value of the energy transferred from indirectly ionizing radiation to charged particles in the elemental volume dV of mass dm. The SI unit of kerma is joules per kilogram (J/kg), which is given the special name gray (Gy).

Some important remarks about kerma are:

(i) Kerma may be defined in any material, so it is important that the material is declared when a value of kerma is presented.
(ii) Kerma is defined for indirectly ionizing radiation — uncharged particles such as photons and neutrons — and is related to the first step of transfer of energy from these particles to matter, in which uncharged particles transmit kinetic energy to secondary charged particles.
(iii) The kinetic energy transferred to the secondary particles is not necessarily spent in the volume (dV) where they were liberated. The kerma definition is constrained to the energy the secondary particles receive at the moment of liberation.

3.2.3.1. Components of kerma

The energy transferred from indirectly ionizing radiation to charged particles may be spent in two ways: (i) collisions resulting in ionizations and (ii) conversion to photons. Accordingly, the kerma can be divided into two parts:

$$K = K_{\text{col}} + K_{\text{rad}} \tag{3.7}$$

The collision kerma (K_{col}) is related to that part of the kinetic energy of the secondary charged particles that is spent in collisions, resulting in ionization and excitation of atoms in matter. In terms of the quantities defined before, collision kerma is obtained from the expectation value of the net energy transferred (Eq. (3.3)):

$$K_{\text{col}} = \frac{d\varepsilon_{\text{tr}}^{\text{net}}}{dm} \tag{3.8}$$

The radiative kerma (K_{rad}) is related to that portion of the initial kinetic energy of the secondary charged particles that is converted into photon energy. It is simpler to define radiative kerma as the difference: $K_{\text{rad}} = K - K_{\text{col}}$. The division of kerma in those two components is more didactic than conceptual. It helps the understanding of the relationship between kerma and absorbed dose, which will be treated in the next section.

3.2.4. Kerma for photons

3.2.4.1. Kerma and fluence

Some important relationships for kerma may be obtained for the simple case of a monoenergetic photon beam irradiating matter. The first of these has already been derived in Section 2.3.4. At a point, P, in space where there is a fluence, Φ, of photons of energy $h\nu$, kerma may be calculated as the product of the energy fluence and the mass energy transfer coefficient (μ_{tr}/ρ) of the material:

$$K = \Phi\, h\nu \left(\frac{\mu_{tr}}{\rho}\right) = \left(\frac{\mu_{tr}}{\rho}\right)\Psi \tag{3.9}$$

Similarly, the collision kerma for photons is related to the fluence through the use of the mass energy absorption coefficient (μ_{en}/ρ):

$$K_{col} = \Phi\, h\nu \left(\frac{\mu_{en}}{\rho}\right) = \Psi \left(\frac{\mu_{en}}{\rho}\right) \tag{3.10}$$

Also in Section 2.3.4, a relationship was derived between the energy absorption and energy transfer coefficients (Eq. 2.31). Using this equation, the relationship between collision and total kerma is as follows:

$$K_{col} = K(1-g) \tag{3.11}$$

where g is the average fraction of the energy transferred to the charged particles that is lost to photons when the charged particles are slowed down in the same medium as that in which they were released.

If the photon beam has a spectrum of energies, Eqs (3.9, 3.10) may be generalized through a summation or integration over the range of energies of the discrete or continuous spectrum, respectively.

For photons in the diagnostic energy range (≤150 keV) interacting in low Z material, the differences between the energy absorption and energy transfer coefficients are negligible, as the fraction of the electron energy converted to bremsstrahlung X rays is very small. For these conditions, the radiative kerma is negligible and the collision kerma is numerically equal to the kerma.

3.2.4.2. Kerma and exposure

In the special situation of X ray or gamma ray photons interacting with air, another quantity, exposure, is also defined and is related to collision kerma through a simple expression. According to the ICRU, exposure (X) is defined as the ratio:

$$X = \frac{\mathrm{d}Q}{\mathrm{d}m} \tag{3.12}$$

The quantity dQ in Eq. (3.12) is the absolute value of the total charge of the ions of one sign produced in air when all the electrons and positrons liberated by photons in air of mass dm are stopped in air. The unit of exposure in SI is coulomb per kilogram (C/kg), even though an old non-SI unit (roentgen — R) is still in use. The conversion from R to SI units is 1 R = 2.580 × 10^{-4} C/kg.

From the above definition, the energy spent to produce the charge dQ included in the numerator of Eq. (3.12) corresponds to the expectation value of the net energy transferred to charged particles in air ($\mathrm{d}\varepsilon_{\mathrm{tr}}^{\mathrm{net}}$). The relationship between these two quantities can be expressed in terms of the measurable quantity $\bar{W}_{\mathrm{air}}$, the mean energy spent in dry air to form an ion pair. $\bar{W}_{\mathrm{air}}$ is given by the ratio:

$$\bar{W}_{\mathrm{air}} = \frac{\sum \text{kinetic energies of electrons spent in ionization and excitation}}{\sum \text{ion pairs produced by the secondary electrons in air}} \qquad (3.13)$$

It is important to emphasize that the portion of kinetic energy that is converted into photon energy and the ionizations produced by the bremsstrahlung photons are *not* included in the numerator or the denominator of Eq. (3.13).

The value of $\bar{W}_{\mathrm{air}}$ accepted nowadays was determined by Boutillon and Perroche-Roux in 1987 through the analysis of a set of published experimental values for this quantity:

$$\bar{W}_{\mathrm{air}} = 33.97 \text{ eV/ion pair} = 33.97 \text{ J/C} \qquad (3.14)$$

The relationship between air collision kerma and exposure X may now be obtained:

$$\left(K_{\mathrm{col}}\right)_{\mathrm{air}} = \bar{W}_{\mathrm{air}}\ X = 33.97X \quad \text{(SI)}$$

$$\text{or} \qquad (3.15)$$

$$\left(K_{\mathrm{col}}\right)_{\mathrm{air}} = 0.876 \times 10^{-2} X \quad (X \text{ in R}, K \text{ in Gy})$$

3.2.5. Absorbed dose

Absorbed dose (D), a physical non-stochastic quantity, is defined simply as the ratio:

$$D = \frac{\mathrm{d}\varepsilon}{\mathrm{d}m} \qquad (3.16)$$

where $d\varepsilon$ is the expectation value of the energy imparted by any ionizing radiation to the matter of mass, dm. Absorbed dose is expressed in the same units as kerma, i.e. joules per kilogram (J/kg) or gray (Gy).

Owing to the penetrating character of ionizing radiation, when a large volume is irradiated, energy can be imparted to the matter in a specific volume by radiation that comes from other regions, sometimes very far from the volume of interest. As the absorbed dose includes all the contributions that impart energy in the volume of interest, it is hardly possible to correlate absorbed dose and the fluence of the incident radiation. In fact, knowledge of the radiation fluence in the volume of interest, including scattered radiation, is a necessary condition for calculation of the absorbed dose. The special situation that makes this correlation possible will be dealt with in Section 3.3.

3.2.6. Kerma and absorbed dose

Kerma and absorbed dose are expressed with the same units, and both are related to the quantification of the interaction of radiation with the matter. Apart from the main fact that kerma is used to quantify a radiation field and absorbed dose is used to quantify the effects of radiation, there are some important points in their definitions that should be emphasized. One of the differences is the role of the volume of interest in these quantities; for kerma, it is the place where energy is transferred from uncharged to charged particles; for absorbed dose, the volume of interest is where the kinetic energy of charged particles is spent. For instance, for kerma, only the energy transfer due to interactions of uncharged particles within the volume is included; for absorbed dose, all the energy deposited in the volume is included. Thus, charged particles entering the volume of interest contribute to absorbed dose, but not to kerma. Also, charged particles liberated by a photon in the volume of interest may leave it, carrying away part of their kinetic energy. This energy is included in kerma, but it does not contribute to the absorbed dose.

The largest differences between absorbed dose and kerma appear at interfaces between different materials, as there are differences in ionization density and in scattering properties of the materials. The changes in kerma at the boundaries are stepwise (scaled by the values of the mass energy transfer coefficient), but the changes in absorbed dose are gradual, extending to a region with dimensions comparable to the secondary particle ranges. Figure 3.2 shows graphs of the ratio of the mass energy transfer coefficients of some biological tissues, showing that, for photons with energies in the diagnostic radiology range, the changes in kerma may be very pronounced at the boundary of different tissues, even with the assumption that the photon fluence is constant at the interface. Kerma at the boundaries changes according to these factors.

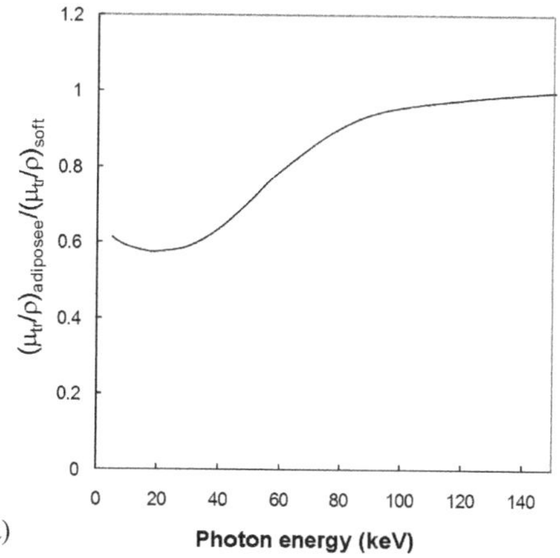

(a)

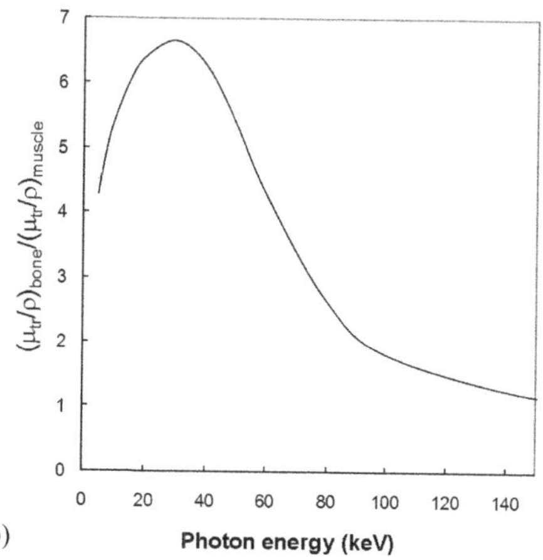

(b)

FIG. 3.2. Ratio of mass energy transfer coefficients for some tissue pairs: (a) adipose to soft tissue; (b) cortical bone to skeletal muscle.

Nonetheless, as Table 3.1 shows, the ranges of electrons set in motion by photons used in diagnostic radiology are small in biological tissues (water has been chosen in this table to simulate all the soft tissues), being less than 1 mm for most of the energies. This indicates that the changes in absorbed dose at the interface between two tissues in the body are limited to small regions. For comparison, Table 3.1 also shows the range of electrons of kinetic energy 1.0 MeV that are released by photons used for external radiotherapy. In this case, the changes in absorbed dose extend to a much greater distance — a few centimetres. This is further discussed in Section 3.3.

TABLE 3.1. RANGE OF ELECTRONS IN WATER AND IN BONE

Electron energy (keV)	Range in water[a]	Range in compact bone[a]
10	2.52 μm	1.49 μm
20	8.57 μm	5.05 μm
50	43.2 μm	25.3 μm
80	97.7 μm	57.1 μm
100	0.143 mm	0.084 mm
150	0.282 mm	0.164 mm
1000	0.437 cm	0.255 cm

[a] Values of continuous slowing down approximation range obtained with the ESTAR program [3.1].

3.2.7. Diagnostic dosimeters

The experimental determination of kerma or absorbed dose and related dosimetric quantities in diagnostic radiology is described in Chapter 22 for patient dosimetry and Chapter 24 for radiation protection purposes. The measurements necessary include: determination of X ray tube output; patient dosimetry through the determination of incident or entrance air kerma, air kerma–area product (KAP) or internal organ doses; and control of doses to staff through area and individual monitoring.

According to the procedure, various dosimeters can be used. Dosimeters are devices used to determine absorbed dose or kerma, or their time rates, based on the evaluation of a detector physical property, which is dose dependent. Besides the detector, which is the sensitive part of the instrument, a dosimeter includes other components that convert the detector signal to the final result of the measurement — the absorbed dose or kerma value. The desired characteristics and the details of functioning of diagnostic dosimeters are treated in Chapter 21.

3.3. CHARGED PARTICLE EQUILIBRIUM IN DOSIMETRY

When a beam of uncharged ionizing particles irradiates a homogeneous material, the ionizing radiation field is transformed to a mixture of the incident beam (attenuated by the material), scattered radiation produced by the interaction of the incident beam in the material, bremsstrahlung radiation and charged particles. Accurate description of the components of the radiation field in a volume where absorbed dose or kerma is to be determined cannot be achieved with analytical methods. Both quantities can be determined using numerical methods (e.g. Monte Carlo simulation), or by experimental means, provided that certain assumptions are fulfilled. The charged particle equilibrium (CPE) in the volume makes experimental determination possible. This section deals with this concept for external photon irradiation.

3.3.1. CPE

An idealized irradiation geometry is shown in Fig. 3.3(a). The radiation field, composed of photons with energy $h\nu$, and normally incident on the block of material under consideration, comes from the left, from vacuum.

As the interactions of photons occur inside the material, electrons are liberated. Assume, for simplicity, that all these charged particles have the same direction in the material, the same energy and a straight track. These tracks are shown in the upper portion of Fig. 3.3(b), where a small volume, dV, inside the

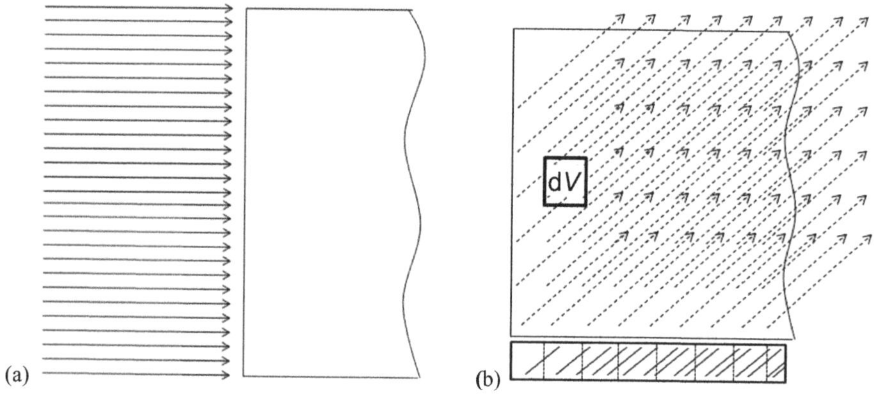

FIG. 3.3. (a) Geometry of a material irradiated from the left with a monoenergetic beam of photons; (b) tracks of the charged particles liberated in the material. The bottom section of the figure shows the path lengths of the charged particles.

material is also shown. Consider what happens as the position of the volume dV moves in a direction parallel to the incoming beam. The number of electron tracks that cross dV is small near the surface of the material, but increases as the volume moves to a greater depth, because more electrons are liberated by photon interactions. This is depicted in the diagram at the bottom of Fig. 3.3(b), where each cell represents the volume dV at a different depth inside the material. As the electron paths have finite lengths in the material (given by their ranges), the number of tracks reaches a maximum at a particular position of dV, and eventually begins to decrease, as the beam is attenuated for greater depths. The total path length of charged particles in each volume can be considered as representing the number of ionizations that occur in the volume. In this idealized situation, the expectation value of the ionization produced in volume dV will vary with depth, as shown in Fig. 3.4(a) for the simplified case where the photon fluence does not vary with depth.

The state of constant ionization shown in Fig. 3.4(a) is termed CPE because in this situation the charged particles that are liberated in the volume dV and leave the volume are balanced, in number and energy, by particles that were liberated elsewhere, and that enter volume dV.

Figure 3.4(b) shows the more realistic situation where the attenuation of the photon beam cannot be neglected and the photon fluence decreases with depth. As a consequence, the expectation value of the total ionization in volume dV increases initially but then decreases slowly with increasing depth in the medium. This state, at depths beyond the maximum of ionization, is called transient charged particle equilibrium. The depth at which the CPE is attained, or the maximum ionization is reached, is of the order of the charged particle range in the material.

Both graphs in Fig. 3.4 assume that the ionization at the surface of the material is zero. In fact, in a more realistic situation, even at the surface, the contribution of scattered radiation has to be taken into account, giving rise to non-zero ionization at the surface. Also, if the photon energies are restricted to the radiology energy range, the maximum ionization is found to be very near to the surface of the irradiated material. There are two main reasons for this: (i) the secondary electrons have a small range in biological tissues (examples are given in Table 3.1) and in solid materials are almost isotropically emitted and (ii) the incoherent scattering of low energy photons produces few photons that are scattered close to the forward direction (see Eqs (2.7, 2.21) and Fig. 2.6).

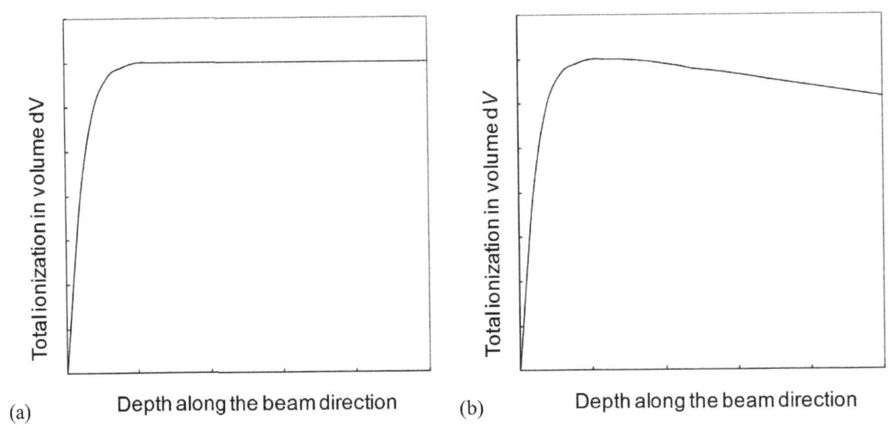

FIG. 3.4. Total ionization inside a volume dV as a function of the depth of the volume in the material, with the following assumptions: (a) the photon fluence is constant; (b) the photon beam is attenuated as it traverses the material.

3.3.2. Relationships between absorbed dose, collision kerma and exposure under CPE

For the photon beam shown in Fig. 3.3(a), the kerma and collision kerma at the entrance of the material are readily obtained by using Eqs (3.8, 3.9). As the material is homogeneous, the collision kerma values within the material vary in proportion to the incident fluence. When the number of interactions is so small that the fluence may be considered constant inside the medium, the variation of K_{col} with depth will be in accordance with Fig. 3.5(a). Usually, however, it is considered that the fluence decreases exponentially with depth in the material, with similar behaviour for K_{col}, as shown in Fig. 3.5(b).

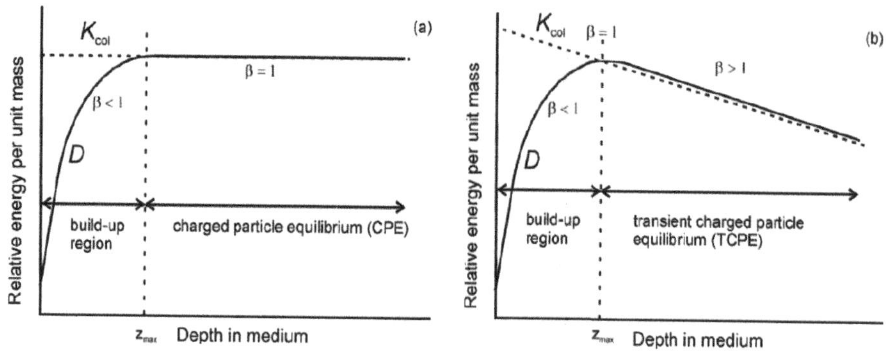

FIG. 3.5. Collision kerma and absorbed dose as a function of depth in a medium, irradiated by a high energy photon beam (see Ref. [3.2]).

As far as absorbed dose is concerned, the calculations are not so simple. First of all, because the absorbed dose, D, depends on the deposition of energy by charged particles, it is smaller at the surface of the material than inside it. As the photon interactions occur, electrons are liberated, contributing to the growth of the energy imparted, according to the growth of the ionization in the material, as shown in Fig. 3.4. Thus, there is a buildup region for the dose, at small depths in the medium, as shown in Fig. 3.5. The buildup region has dimensions (z_{max}) similar to the range of the charged particles in the medium. For high energy photons, this region can extend to 1 or 2 cm, and this effect is responsible for the known effect of skin sparing in external radiotherapy. For diagnostic photon beams, the energies are lower, and the electron ranges are too small to produce this effect: the maximum dose is reached within the skin.

When the absorbed dose reaches the maximum value, assuming that all the radiative photons produced in the medium escape from the volume, there is a coincidence of its value and the collision kerma, as true CPE is achieved. Beyond the buildup region, assuming that the changes in photon fluence are small, as in Fig. 3.5(a), and the volume of interest has small dimensions compared with the electron range, giving rise to true CPE, the relation between absorbed dose and collision kerma remains:

$$D^{\mathrm{CPE}} = K_{\mathrm{col}} = \Phi\, h\nu \left(\frac{\mu_{\mathrm{en}}}{\rho}\right) \tag{3.17}$$

For conditions such as those shown in Fig. 3.5(b), in transient charged particle equilibrium, where the attenuation of the photon beam is not negligible, there is no numerical coincidence between the values of the collision kerma and absorbed dose. Beyond the maximum, the absorbed dose is larger than the collision kerma, as the energy imparted is a result of charges liberated by photon fluences slightly greater than the fluence in the volume of interest. Because there is practically a constant ratio between these quantities, it is usual to write:

$$D = \beta\, K_{\mathrm{col}} \tag{3.18}$$

Usually, the approximation $\beta \approx 1$ can be used for diagnostic radiology and low Z materials.

3.3.3. Conditions that enable CPE or cause its failure

The simplification of the overall situation shown above should not be interpreted as the only condition that can give rise to a true equilibrium of charged particles in the medium. The necessary and sufficient conditions that guarantee the CPE are:

(a) The medium is homogeneous in both atomic composition and mass density.
(b) The photon field is homogeneous in the volume considered.

The first condition avoids changes in the charged particle distribution in the material owing to changes in scattering and absorption properties. The homogeneity in density is not as important as the constancy in composition, according to the Fano theorem (see Section 3.4.2). The second condition requires that the dimensions of the volume of interest are not very large, compared with the mean free path ($1/\mu$) of the photons.

Some examples of practical situations where there is a failure in the conditions, meaning that the CPE cannot be accomplished, are:

(a) Large beam divergence, as with irradiations close to the radiation source;
(b) Proximity of boundaries of the material and any other medium.

3.4. CAVITY THEORY

Determination of absorbed dose in an extended medium generally requires the use of a detector inside the medium. As the sensitive volume of the dosimeter is, in general, not made of the same material as the medium, its presence is a discontinuity. It is called a ‘cavity’ because the early basis of dosimetry was developed for gaseous detectors. Nowadays, the concept has been extended to liquid and solid detectors, but the traditional use of the letters g and w is retained to symbolize the cavity and the medium, respectively, as they come from the gas and the wall that constitute an ionization chamber. The main interests of the cavity theory are to study the modifications of charge and radiation distribution produced in the medium by the cavity, and to establish relations between the dose in the sensitive volume of the dosimeter and the dose in the medium. A brief description of the foundations and the main results of the cavity theory for external irradiation with photons and applications to diagnostic radiology are given next.

3.4.1. Bragg–Gray cavity theory

A cavity can be of small, large or intermediate size compared with the range of the charged particles in the cavity. The Bragg–Gray theory deals with small cavities. W.H. Bragg began the development of the theory, in 1910, but it was L.H. Gray, during his PhD work, co-supervised by Bragg, who formalized the theory. The two main assumptions of this theory are:

(i) The cavity dimensions are so small compared with the range of charged particles within it that the fluence of charged particles inside the cavity is not perturbed by the presence of the cavity.

(ii) There are no interactions of uncharged particles in the cavity, so that the absorbed dose deposited in the cavity is due to the charged particles that cross the cavity.

The first assumption is equivalent to the requirement that the fluence of charged particles in the cavity is equal to that in the medium, w, that surrounds it, and the second assumption means that no charged particle starts or finishes its range in the cavity.

Under these conditions, the ratio of absorbed dose in the medium, w, surrounding the cavity, D_w, to that in the cavity, g (D_g), is given by the ratio of the integrals of the collision stopping powers of electrons in both media, weighted by the same electron fluence — the fluence energy distribution of the electrons in the medium $(\mathrm{d}\Phi/\mathrm{d}T)_w$:

$$\frac{D_w}{D_g} = \frac{\int_{T_{\min}}^{T_{\max}} \left(\frac{\mathrm{d}\Phi}{\mathrm{d}T}\right)_w \left(\frac{\mathrm{d}T}{\rho \mathrm{d}x}\right)_{c,w} \mathrm{d}T}{\int_{T_{\min}}^{T_{\max}} \left(\frac{\mathrm{d}\Phi}{\mathrm{d}T}\right)_w \left(\frac{\mathrm{d}T}{\rho \mathrm{d}x}\right)_{c,g} \mathrm{d}T} = \bar{\bar{S}}_g^w \tag{3.19}$$

The numerator and denominator in Eq. (3.19) are the generalizations of Eq. (3.20), which relates absorbed dose, D, to collision stopping power, $(\mathrm{d}T/\rho\mathrm{d}x)_{c,T}$, for irradiation with a fluence, Φ, of monoenergetic electrons:

$$D = \left(\frac{\mathrm{d}T}{\rho \mathrm{d}x}\right)_{c,T} \Phi \tag{3.20}$$

For the medium and the cavity materials, Eq. (3.19) assumes that there is a distribution of energies of electrons inside the cavity, which is equal to that outside. The symbol $\bar{\bar{S}}_g^w$ has the double bar to indicate that this ratio of average stopping powers considers both the average over the photon generated electron spectrum and the changes in this spectrum that are due to the continuous loss of kinetic energy in the materials.

3.4.2. The Fano theorem

The conditions required by the Bragg–Gray theory are better accomplished if the composition of the cavity is similar to that of the medium, i.e. if both cavity and medium have similar atomic numbers. This was observed in experiments with cavities filled with different gas compositions, and in 1954, U. Fano proved the following theorem:

> *"in a medium of given composition exposed to a uniform field of primary radiation the field of secondary radiation is also uniform and independent of the density of the medium, as well as of the density variations from point to point."*

The Fano theorem is important because it relaxes the requirements on the size of the cavity, which are very hard to meet, for instance, when the photon beam is of low energy. It should be noted, however, that the theorem is valid only for infinite media and in conditions where the stopping power is independent of density.

3.4.3. Other cavity sizes

Consider now the material surrounding the cavity, w, which is large enough to guarantee CPE in almost its entire volume, excluding a very small portion near the boundaries, but which does not disturb the photon fluence in the medium where it is immersed. The region of the medium m that surrounds w is also under

CPE conditions. Under these conditions, the dose to material w and the dose to the medium m are related by the expression:

$$\frac{D_m}{D_w} = \frac{\left(\frac{\mu_{en}}{\rho}\right)_m}{\left(\frac{\mu_{en}}{\rho}\right)_w} \tag{3.21}$$

Equation (3.21) is a simple ratio of mass energy absorption coefficients, and three conditions are implicit:

(i) There is CPE in both media.
(ii) The photon beam is monoenergetic.
(iii) The photon fluence is the same for both media.

If the elemental compositions of w and m are not similar, the backscattering of photons at the boundary can change the photon fluence significantly, regardless of the dimensions of w.

For a spectrum of photons irradiating both materials, Eq. (3.21) may be integrated over the photon energy spectrum, giving:

$$\frac{D_m}{D_w} = \frac{\int_0^{h\nu_{max}} \left(\frac{d\Phi}{dh\nu}\right)_m \left(\frac{\mu_{en}}{\rho}\right)_m h\nu d(h\nu)}{\int_0^{h\nu_{max}} \left(\frac{d\Phi}{dh\nu}\right)_m \left(\frac{\mu_{en}}{\rho}\right)_w h\nu d(h\nu)} \equiv \left(\frac{\overline{\mu}_{en}}{\rho}\right)_w^m \tag{3.22}$$

In this equation, $(\overline{\mu}_{en}/\rho)_w^m$ is an average ratio of mass absorption energy coefficients, which takes into account the photon spectrum that irradiates equally both material, w, considered a large cavity, and m.

Cavities with intermediate sizes are usually treated by Burlin cavity theory, assuming that the cavity and the medium are in CPE and that the elemental compositions of both are similar. The theory will not be developed here, but Burlin's expression is given in Eq. (3.23), where the parameter d assumes values between 0 and 1, according to the cavity dimensions: $d \to 1$ for small cavities and $d \to 0$ for large ones.

$$\frac{D_g}{D_w} = d\overline{\overline{S}}_w^g + (1-d)\left(\frac{\overline{\mu}_{en}}{\rho}\right)_w^g \tag{3.23}$$

3.5. PRACTICAL DOSIMETRY WITH ION CHAMBERS

Ionization chambers are frequently used in diagnostic radiology. They are usually built with a wall that functions as a large cavity; besides containing the gas, it has a thickness that is enough to guarantee CPE in the wall. If the elemental composition of this wall w is similar to the composition of the medium m, where the dose is to be measured, and there is also CPE in the medium, it is possible to relate the dose in the medium to the dose in the wall using Eq. (3.21) or Eq. (3.22). On the other hand, the gas inside the ion chamber is irradiated mainly by the charged particles released in the wall that cross the gas volume according to the Bragg–Gray conditions. In this way, the dose to the material where the chamber is inserted can be obtained through Eq. (3.24), which combines Eqs (3.19, 3.22).

$$D_m = D_g \bar{\bar{S}}_g^w \left(\frac{\bar{\mu}_{\text{en}}}{\rho}\right)_w^m \tag{3.24}$$

If, instead of the dose to the gas, the charge produced in the gas (Q) and the mass of the gas (m_g) are known, Eq. (3.23) reduces to:

$$D_m = \frac{Q}{m_g} \bar{W}_g \bar{\bar{S}}_g^w \left(\frac{\bar{\mu}_{\text{en}}}{\rho}\right)_w^m \tag{3.25}$$

where $\bar{W}_g$ is the mean energy spent in the gas to form an ion pair, as already described for air in Eq. (3.13).

A particularly useful (and common) situation occurs when the wall of the chamber is made of a material with the same atomic composition as the cavity, so that the dose to the cavity and the dose to the wall are considered equal. Under these circumstances, Eqs (3.24, 3.25) are simplified, reducing to:

$$D_m = D_g \left(\frac{\mu_{\text{en}}}{\rho}\right)_g^m \quad \text{for chambers with gas equivalent wall} \tag{3.26}$$

$$D_m = \frac{Q}{m_g} \bar{W}_g \left(\frac{\bar{\mu}_{\text{en}}}{\rho}\right)_g^m \quad \text{for chambers with gas equivalent wall} \tag{3.27}$$

To use Eqs (3.24–3.27) in practice is not trivial, as the spectra of photons and electrons are not known in general, and the charge is not completely collected. Nevertheless, this is done for standard chambers used for calibration of the instruments used in diagnostic radiology, applying correction factors for incomplete charge collection and mismatch of atomic compositions. A standard chamber is compared with the instrument to be calibrated, irradiating both with well characterized photon beams, with qualities comparable to the clinical beams.

REFERENCES

[3.1] NATIONAL INSTITUTE OF STANDARDS AND TECHNOLOGY, ESTAR: Stopping Power and Range Tables for Electrons, http://physics.nist.gov/PhysRefData/Star/Text/ESTAR.html (accessed on 28 August 2012).

[3.2] INTERNATIONAL ATOMIC ENERGY AGENCY, A Syllabus for the Education and Training of RTTs, Training Course Series No. 25, IAEA, Vienna (2005).

BIBLIOGRAPHY

ATTIX, F.H., Introduction to Radiological Physics and Radiation Dosimetry, John Wiley & Sons, New York (1986).

GREENING, J.R., Fundamentals of Radiation Dosimetry, Adam Hilger Ltd, Bristol (1981).

INTERNATIONAL ATOMIC ENERGY AGENCY, Dosimetry in Diagnostic Radiology: An International Code of Practice, Technical Reports Series No. 457, IAEA, Vienna (2007).

INTERNATIONAL ATOMIC ENERGY AGENCY, Radiation Oncology Physics: A Handbook for Teachers and Students, IAEA Vienna (2005).

INTERNATIONAL COMMISSION ON RADIATION UNITS AND MEASUREMENTS, Patient Dosimetry for X Rays used in Medical Imaging, ICRU Rep. 74, ICRU, Bethesda, MD (2005).

JOHNS, H.E., CUNNINGHAM, J.R., The Physics of Radiology, Charles C. Thomas, Springfield, IL (1985).

Chapter 4

MEASURES OF IMAGE QUALITY

A.D.A. MAIDMENT
University of Pennsylvania,
Philadelphia, United States of America

4.1. INTRODUCTION

A medical image is a pictorial representation of a measurement of an object or function of the body. This information can be acquired in one to three spatial dimensions. It can be static or dynamic, meaning that it can also be measured as a function of time. Certain fundamental properties can be associated with all of these data. Firstly, no image can exactly represent the object or function; at best, one has a measurement with an associated error equal to the difference between the true object and the measured image. Secondly, no two images will be identical, even if acquired with the same imaging system of the same anatomic region; this variability is generally referred to as noise.

There are many different ways to acquire medical image data; the various mechanisms of acquisition are described in detail in the subsequent chapters. However, regardless of the method of image formation, one must be able to judge the fidelity of the image in an attempt to answer the question "How accurately does the image portray the body or the bodily function?" This judgement falls under the rubric of 'image quality'. In this chapter, methods of quantifying image quality are described.

Knowledge of image quality allows one to compare various imaging system designs for a given modality and to compare the information contained in images acquired by different imaging modalities. The impact of image quality on an imaging task, such as detection of a lesion in a particular organ, can also be determined. Various imaging tasks require differing levels of image quality; an image may be of sufficient quality for one task, but inadequate for another task.

The metrics introduced here are much used in the following chapters in this handbook, as the design, performance and quality control of different imaging systems are discussed. First, however, one needs to learn the meaning of 'image quality'.

4.2. IMAGE THEORY FUNDAMENTALS

4.2.1. Linear systems theory

In all imaging systems, the output, g, is a function of the input, f. The function, H, is usually called the transfer function or system response function. For a continuous two dimensional (2-D) imaging system, this relationship can be written as:

$$g(x, y) = H\{f(x, y)\} \quad (4.1)$$

In Eq. (4.1) and most equations that follow, the functions have been expressed with two dependent variables to represent a 2-D image; this convention has been chosen to ensure consistency throughout the chapter. However, the imaging problem can be treated in any number of dimensions. The simple concept represented by Eq. (4.1), and shown in Fig. 4.1, implies that we can predict the output of an imaging system if we know the input and the characteristics of the system. That is, g is the image of the scene, f.

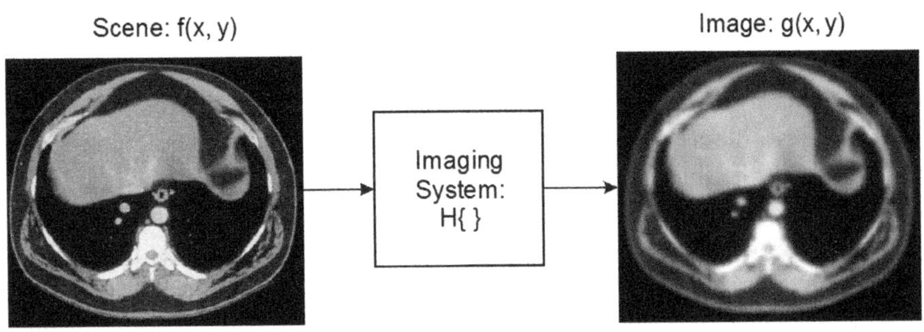

FIG. 4.1. The image, g(x, y), portrays a cross-section of the thorax, f(x, y), blurred by the transfer function, H, of the imaging system.

Unfortunately, this general approach to image analysis is very difficult to use; it is necessary to compute the transfer function at each location in the image for each unique object or scene. This analysis is greatly simplified when two fundamental assumptions can be made: linearity and shift invariance (abbreviated jointly as LSI).

4.2.1.1. Linearity

A linear system is one in which the output of the system can be expressed as a weighted sum of the input constituents. Thus, if a system presented with input f_1 results in output $g_1(x, y) = H\{f_1(x, y)\}$ and input f_2 results in output $g_2(x, y) = H\{f_2(x, y)\}$, then:

$$H\{af_1(x, y) + bf_2(x, y)\} = H\{af_1(x, y)\} + H\{bf_2(x, y)\} = ag_1(x, y) + bg_2(x, y) \quad (4.2)$$

In general, most imaging systems are either approximately linear or can be linearized, or can be treated as being linear over a small range of signals. The assumption of linearity lets us formulate the transfer function as an integral of the form:

$$g(x, y) = \int\int f(x', y')H(x, y, x', y')\,\mathrm{d}x'\,\mathrm{d}y' \quad (4.3)$$

However, most modern imaging systems are digital. As a result, images consist of measurements made at specific locations in a regular grid. With digital systems, these measurements are represented as an array of discrete values. In the discrete case, Eq. (4.3) can be reformulated as multiplication of a matrix $\boldsymbol{H}$, where the input scene and output image are given as vectors (for one dimensional (1-D) images) or matrices (for higher dimension images):

$$\boldsymbol{g} = \boldsymbol{H}\boldsymbol{f} \quad (4.4)$$

In this formulation, each element in $\boldsymbol{g}$ is called a 'pixel' or 'pixel element', while each element in $\boldsymbol{f}$ is called a 'del' or 'detector element'. A pixel represents the smallest region that can uniquely encode a single value in the image. By similar reasoning, the term 'voxel' or 'volume element' is used in three dimensional (3-D) imaging. In both Eqs (4.3, 4.4), $\boldsymbol{g}$ is expressed as a weighted sum, $\boldsymbol{H}$, of the source signals, $\boldsymbol{f}$. It is important to note that H or $\boldsymbol{H}$ is still quite complicated. If $\boldsymbol{g}$ and $\boldsymbol{f}$ have $m \times n$ elements, then $\boldsymbol{H}$ has $(mn)^2$ elements; that is, there is a unique transfer function for each pixel in the image because the value of each pixel arises from a different weighted sum of the dels.

4.2.1.2. Shift invariance

A system is shift invariant if the system response function, H, does not change as a function of position in the image. By further adding the stipulation of shift invariance, it is possible to formulate the transfer function without reference to a specific point of origin. This allows us to write the integration in Eq. (4.3) as a convolution:

$$g(x, y) = \int\int f(x', y')h(x - x', y - y')\,\mathrm{d}x'\,\mathrm{d}y' \tag{4.5}$$

where h is now a function of two variables, while H was a function of four variables in the case of a 2-D imaging system.

In the discrete formulation of a shift invariant system, the matrix $\boldsymbol{H}$ in Eq. (4.4) now has a unique property; it is Toeplitz. As a practical measure, we often use a circulant approximation of the Toeplitz matrix. This approximation is valid, provided the point spread function is small compared with the size of the detector. The discrete Fourier transform (FT) of the circulant approximation of $\boldsymbol{H}$ is a diagonal matrix. This property has particular appeal in analysing LSI systems, as we have gone from a formulation in which $\boldsymbol{H}$ has as many as $(mn)^2$ non-zero elements to one that has exactly mn distinct elements.

As a result, it is possible to construct a new matrix, $\boldsymbol{h}$, from $\boldsymbol{H}$ such that Eq. (4.4) can now be rewritten as:

$$\boldsymbol{g} = \boldsymbol{h} * \boldsymbol{f} \tag{4.6}$$

where * is the circulant convolution operator and, in the case of 2-D detectors, the images $\boldsymbol{f}$ and $\boldsymbol{g}$, and the response function $\boldsymbol{h}$ are each matrices with $m \times n$ distinct elements, which are cyclically extended in each direction. The assumptions of linearity and shift invariance are key to making most imaging problems tractable, as there is now a common transfer function, $\boldsymbol{h}$, that applies to each pixel in the image.

Recalling that for FT pairs, the convolution in one domain corresponds to multiplication in the other domain, we can now rewrite Eq. (4.6) as:

$$\tilde{\boldsymbol{g}} = \tilde{\boldsymbol{h}}\tilde{\boldsymbol{f}} \tag{4.7}$$

where the tilde (~) denotes the discrete FT. This implies that an object with a given spatial frequency referenced at the plane of the detector will result in an image with exactly the same spatial frequency, although the phase and amplitude may change.

Now, with few exceptions, most systems are not truly shift invariant. For example, consider a simple system in which a pixel in the image is equal to the average of the matching del in the scene and the eight immediate neighbouring dels. The transfer function will be identical for all the pixels in the interior of the image. However, pixels on the four edges and four corners of the image will have different transfer functions because they do not have a full complement of neighbouring pixels upon which to calculate this average. That said, most systems can be treated as shift invariant (with regard to this boundary problem), provided the blurring (or correlation) between pixels is small compared with the size of the image. A second strategy to ensure shift invariance is to consider the transfer function locally, rather than globally. This strategy allows one to ignore differences in the detector physics across the full field of the detector, such as the oblique incidence of X rays.

4.2.2. Stochastic properties

In all real imaging systems, it is necessary to consider both the degradation of the image from blurring, given by the transfer characteristics, and the degradation of the image from the presence of noise. Noise can arise from a number of sources, including the generation of the signal carriers, the propagation and transformation of these carriers through the imaging process, and the addition of extraneous noise from various sources, such as the imaging electronics. Thus, it is necessary to modify the image transfer equation (Eq. (4.4)) to include a term for the noise, $\boldsymbol{n}$. Noise is generated from a random process; as a result, the noise recorded in each image will be unique. Any given image, $\dot{\boldsymbol{g}}$, will include a single realization of the noise, $\dot{\boldsymbol{n}}$, so that:

$$\dot{\boldsymbol{g}} = \boldsymbol{Hf} + \dot{\boldsymbol{n}} \tag{4.8}$$

Actually, some noise (e.g. X ray quantum noise) will be generated in the process of forming the scene, $\boldsymbol{f}$, and hence will be acted upon by the transfer function, $\boldsymbol{H}$, while other noise (e.g. electronic readout noise) will not have been acted upon by the transfer function. In Eq. (4.8) we ignore this distinction. Also, all quanta do not necessarily experience the same transfer function; variability in the transfer of individual quanta leads to the well known Swank and Lubberts' effects.

The introduction of noise in images means that imaging systems have to be evaluated statistically. The exact treatment of the images is dependent upon both the nature of the noise present when the image is recorded and the imaging system. System linearity (or linearizability) will help to make the treatment of images in the presence of noise tractable. In general, however, we also need to assume that the noise is stationary. A stochastic noise process is stationary if the process does not change when shifted either in time or in space. That is, the moments of a stationary process will not change based upon the time when observations begin. An example is X ray quantum noise, where the probability of generating an X ray does not depend upon when the previous or subsequent

X ray quanta are created. Similarly, in a shift invariant imaging system, it does not matter which point on the detector is used to calculate the moments of a stationary process, as each point is nominally the same.

A wide sense stationary process is one in which only the mean and covariance are stationary. Since a Poisson process is fully characterized by the mean, and a Gaussian process is fully characterized by the mean and variance, it is typical to require an imaging process to be wide sense stationary. It is, in fact, common to treat the noise as being Gaussian and having zero mean. In practice, this is sufficient for almost all imaging systems.

It should be noted that digital images consisting of discrete arrays of pixels or voxels are not strictly stationary. Shifts of the origin that are not commensurate with the pixel spacing will potentially result in different images being acquired. However, a system is said to be cyclostationary if the statistical properties are unchanged by shifts in the origin of specific amounts (i.e. multiples of the pixel or voxel pitch). A system is 'wide sense cyclostationary' if the mean and covariance are unchanged by specific shifts in the origin. In general, we can assume most digital imaging systems are wide sense cyclostationary, at least locally.

To measure the signal in a pixel, exclusive of the noise, we may simply average the value in that pixel over many images of the same scene to minimize the influence of the noise on the measurement. In a similar fashion, we can estimate the noise in a pixel by calculating the standard deviation of the value of that pixel over many images of the same scene (see, for example, Fig. 4.10). Calculations that involve a large number of images are clearly time consuming to acquire and process in order to estimate the mean and standard deviation with sufficient accuracy. However, this problem is tremendously simplified if one can additionally assume ergodicity.

An ergodic process is one in which the statistical properties of the ensemble can be obtained by analysing a single realization of the process. For example, X ray quantum noise is frequently referred to as 'white noise', implying that in different realizations all spatial frequencies are equally represented, or equivalently, that the noise from individual quanta is uncorrelated. White noise is ergodic. This means, for example, that we can calculate the average fluence of an X ray beam by either averaging over a region or averaging over multiple images. When an appropriate imaging system is used to image an ergodic process (such as a uniform scene imaged with X rays), calculations performed from a number of sample images can be replaced by calculations from one image. For example, the noise in a particular pixel that was originally measured from image samples can now be measured from a region of a single image.

4.2.3. Sampling theory

With few exceptions (notably screen film radiography), modern imaging systems are digital. A digital image is only defined as discrete points in space, called sampling points. The process of sampling by a del generally involves the integration of continuous signal values over a finite region of space around the sampling point. The shape of these regions is defined by the sampling aperture; the distance between the sampling points is called the sampling pitch.

In an idealized 2-D detector, the sampling aperture of each del is represented by a square of dimension a'. Such dels are repeated with pitch a to cover the entire detector (Fig. 4.2). It is not strictly necessary for the aperture and pitch to have the same size, nor to be square. For example, active matrix X ray detectors (see Section 7.4.3) can have regions that are not radiation sensitive, such as the data and control lines and del readout electronics. The fill factor of an active matrix detector is typically defined as the ratio $(a'/a)^2$. The fill factor is commonly less than unity. It is also possible for the del aperture to be larger than a^2. For example, in computed radiography, the scanning laser will typically stimulate fluorescence from a circular region having a diameter greater than the sampling pitch. As discussed later, this has benefit in reducing aliasing.

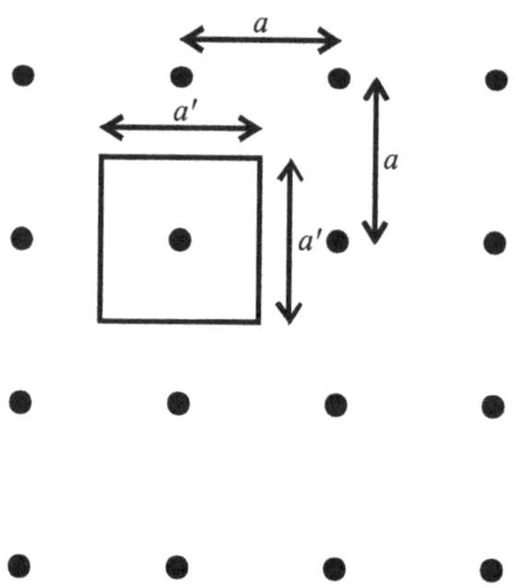

FIG. 4.2. Rectangular array of dels in which a single del with a square aperture of dimensions a' × a' is shown centred upon a series of sampling points with pitch a in orthogonal directions.

The process of sampling a continuous signal, f, by a single del is given by:

$$f(x_i, y_j) = \iint f(x, y) A(x - x_i a, y - y_j a)\, dx\, dy \tag{4.9}$$

where A is the aperture function and (x_i, y_j) are integer indices of the del. In practice, the aperture function is non-zero over a limited area, thus providing finite limits to the integral in Eq. (4.9).

It is clear from Eq. (4.9) that if one were to shift the sampling points by a non-integer amount (i.e. incommensurate with the pixel pitch), the recorded image would vary. For this reason, digital systems are only cyclostationary. In general, these changes are small, especially for objects that are large relative to the sampling pitch. However, for small objects, these changes can be significant.

Sampling a continuous signal $f(x, y)$ on a regular grid with grid spacing a, is equivalent to multiplying f by a comb function, comb_a. The comb function is an infinite sum of Dirac delta functions centred at the sampling points. Multiplication by the comb function in the image domain is equivalent to convolution by the FT of the comb function in the Fourier domain. The FT of the comb function is also a comb function, but with a grid spacing of $1/a$. This convolution has the form:

$$(\tilde{f} * \text{comb}_{1/a})(u, v) = \sum_{j=-\infty}^{\infty} \sum_{k=-\infty}^{\infty} \tilde{f}\left(u - \frac{j}{a}, v - \frac{k}{a}\right) \tag{4.10}$$

This implies that the FT of f is replicated at each point on a grid with a spacing of $1/a$, and an infinite sum of all the replicates is taken.

The frequency $1/a$ is called the sampling rate. The Nyquist–Shannon sampling theorem provides guidance in determining the value of a needed for a specific imaging task. Ideally, the Fourier spectrum of f should not have components above the frequency $1/2a$. This frequency is called the Nyquist frequency (see Fig. 4.3(a)). When this condition is not met, the Fourier spectra will contain components with spatial frequencies that exceed the Nyquist frequency and, as a result of Eq. (4.10), the infinite sum of spectra will overlap, as shown in Fig. 4.3(b). This overlap between the superimposed spectra will result in aliasing. Aliasing degrades the sampled image because it incorrectly portrays high frequency information present in the scene as lower frequency information in the image (Fig. 4.3(b), black curve). To avoid aliasing, the Nyquist frequency must be greater than or equal to the maximum frequency in the image prior to sampling. In many system designs, however, it is impossible to avoid aliasing.

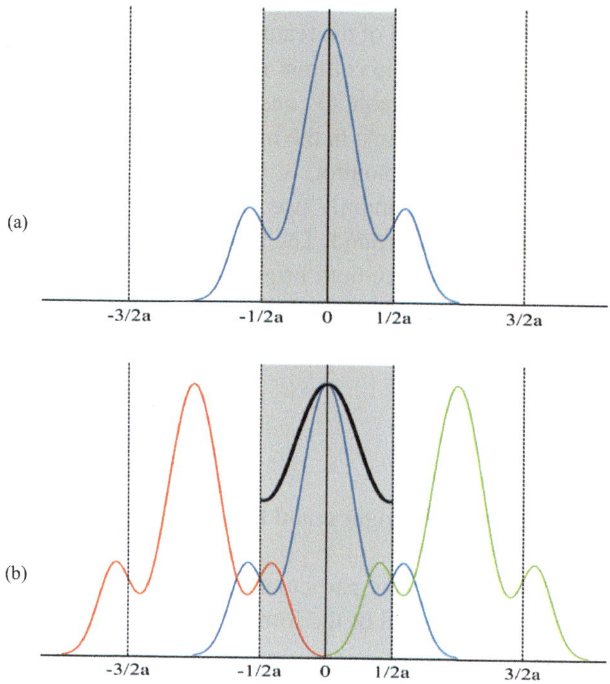

FIG. 4.3. The discrete space FT is uniquely defined on a region of the Fourier domain from −1/2a to 1/2a (shaded grey). The input signal (a) is replicated in the Fourier domain; (b) only three replicates are shown for clarity. Frequencies outside the shaded region are aliased and summed, so that the discrete space FT is given by the black line in (b).

4.3. CONTRAST

4.3.1. Definition

Contrast is defined as the ratio of the signal difference to the average signal. The rationale behind this is that a small difference is negligible if the average signal is large, while the same small difference is readily visible if the average signal is small. In general, in medical imaging, we will want to achieve a high contrast to visualize disease features well.

There are two common definitions of contrast in medical imaging. The Weber contrast, or the local contrast, is defined as:

$$C = \frac{f_{\mathrm{f}} - f_{\mathrm{b}}}{f_{\mathrm{b}}} \tag{4.11}$$

where f_f and f_b represent the signal of the feature and the background, respectively. Note that here we are defining the contrast in terms of the scene f, as shown in Section 4.3.2. It is equally acceptable to consider the contrast of the image, g, or the contrast measured at other points in the image chain, such as the contrast of a feature displayed on a computer monitor.

The Weber contrast is commonly used in cases where small features are present on a large uniform background. The modulation or Michelson contrast is commonly used for patterns where both bright and dark features take up similar fractions of the image. The modulation contrast is defined as:

$$C_M = \frac{f_{max} - f_{min}}{f_{max} + f_{min}} \tag{4.12}$$

where f_{max} and f_{min} represent the highest and lowest signals.

The modulation contrast has particular interest in the Fourier analysis of medical images. Consider a signal of the form:

$$f(x, y) = A + B \sin(2\pi ux) \tag{4.13}$$

Substituting Eq. (4.13) into Eq. (4.12) gives:

$$C_M = \frac{A + B - (A - B)}{A + B + A - B} = \frac{B}{A} \tag{4.14}$$

Thus, we see that the numerator expresses the amplitude or difference in the signal $B = (f_{max} - f_{min})/2$, while the denominator expresses the average signal $A = (f_{max} + f_{min})/2$.

Care should be taken as to which definition of contrast is used. The correct choice is situation dependent. In general, the local contrast is used when a small object is presented on a uniform background, such as in simple observer experiments (e.g. two alternative forced choice experiments, see Section 18.4.2). The modulation contrast has relevance in the Fourier analysis of imaging systems.

4.3.2. Contrast types

In medical imaging, the subject contrast is defined as the contrast (whether local or modulation) of the object in the scene being imaged. For example, in

X ray imaging, the subject contrast depends upon the X ray spectrum and the attenuation of the object and background. In radionuclide imaging, the subject contrast depends upon radiopharmaceutical uptake by the lesion and the background, the pharmacokinetics and the attenuation of the gamma rays by the patient. Similarly, one can define the subject contrast for computed tomography (CT), magnetic resonance imaging (MRI) and ultrasound.

The image contrast depends upon the subject contrast and the characteristics of the imaging detector. For example, in radiographic imaging, the image contrast is affected by the X ray spectrum incident upon the X ray converter (e.g. the phosphor or semiconductor material of the X ray detector), the converter composition and thickness, and the greyscale characteristics of the converter, whether analogue (such as film) or digital.

The display contrast is the contrast of the image as displayed for final viewing by an observer. The display contrast is dependent upon the image contrast and the greyscale characteristics of the display device and any image processing that occurs prior to or during display.

4.3.3. Greyscale characteristics

In the absence of blurring, the ratio of the image contrast to the subject contrast is defined as the transfer function of the imaging system. As discussed in Section 7.2.3, the greyscale characteristic of film is non-linear (the characteristic or Hurter and Driffield (H&D) curve is shown in Section 7.3.4.3). Thus, to stay within the framework of LSI systems analysis, it is necessary to linearize the response of the film. This is typically done using a small signals model in which the low contrast variations in the scene recorded in the X ray beam, $\Delta I/I_0$, produce linear changes in the film density, ΔD, such that:

$$\Delta D = \frac{\gamma \lg(\mathrm{e}) \Delta I}{I_0} \tag{4.15}$$

where γ is called the film gamma, and typically has a value of between 2.5 and 4.5 (Fig. 4.4). The greyscale characteristic, Γ, can now be calculated as:

$$\Gamma = \frac{\Delta D}{\Delta I} = \frac{\gamma \lg(\mathrm{e})}{I_0} \tag{4.16}$$

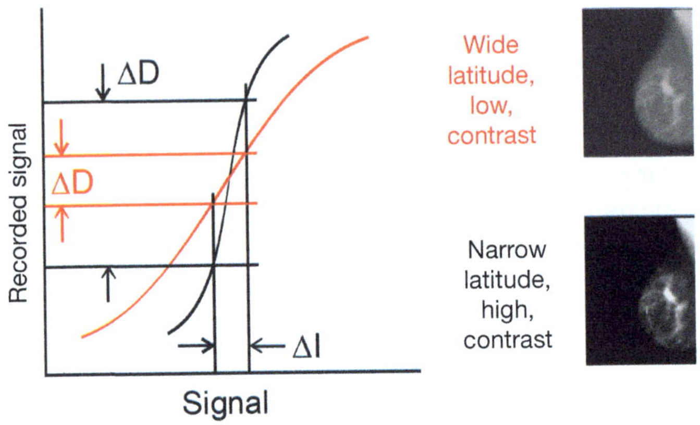

FIG. 4.4. Two greyscale response functions are shown; one with low contrast and wide latitude, the other with high contrast and narrow latitude.

In a similar fashion, the greyscale characteristic of a digital system with a digital display can be defined. In general, digital displays have a non-linear response with a gamma of between 1.7 and 2.3.

It should be noted that Γ does not consider the spatial distribution of the signals. In this sense, we can treat Γ as the response of a detector which records the incident X ray quanta, but does not record their location. Equivalently, we can consider it as the direct current (DC) (static) response of the imaging system. Given that the FT of a constant is equal to a delta function at zero spatial frequency, we can also consider this response to be the zero spatial frequency response of the imaging system.

4.4. UNSHARPNESS

In the preceding discussion of contrast, we considered large objects in the absence of blurring. However, in general, we cannot ignore either assumption. When viewed from the spatial domain, blurring reduces the contrast of small objects. The effect of blurring is to spread the signal laterally, so that a focused point is now a diffuse point. One fundamental property of blurring is that the more the signal is spread out, the lower the intensity of the point image, and thus the lower the contrast.

The effect of blurring on local contrast is illustrated in Fig. 4.5. An image of a point is shown blurred by convolution with Gaussian kernels of diameters 16 pixels, 32 pixels and 64 pixels. As can be seen, the intensity of the signal decreases as the amount of blurring increases, because the signal of the object is being spread over an increasingly large area. This also means that the peak signal is only degraded if the size of the object is smaller than the width of the blurring function; the contrast of larger objects will not be affected (Fig. 4.1).

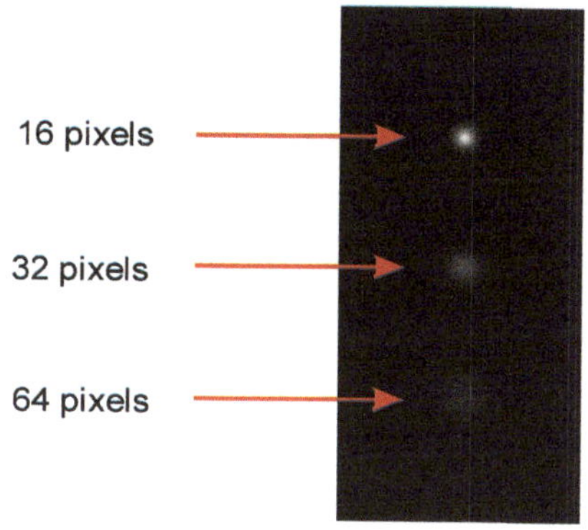

FIG. 4.5. A point object is blurred with increasingly larger kernels. The larger the kernel, the greater the blurring and the lower the contrast for small objects.

4.4.1. Quantifying unsharpness

Consider the operation of an impulse function on an imaging system. If an imaging system is characterized by an LSI response function $h(x - x', y - y')$, then this response can be measured by providing a delta function as input to the system. Setting $f(x, y) = \delta(x, y)$ in Eq. (4.5) gives:

$$g(x, y) = \int\int \delta(x', y')h(x - x', y - y')\,\mathrm{d}x'\,\mathrm{d}y' = h(x, y) \tag{4.17}$$

We refer to the system transfer function as the point spread function (PSF) when specified in the spatial domain. In fact, the blurring of a point object, shown in Fig. 4.5, is a pictorial display of the PSF. It is common to consider the PSF as either being separable:

$$h(x, y) = h(x)h(y) \tag{4.18}$$

or circular symmetric:

$$h(r) = h(x, y) \tag{4.19}$$

where

$$r = \sqrt{x^2 + y^2} \tag{4.20}$$

depending upon the imaging system.

While it is possible to calculate the blurring of any object in the spatial domain via convolution with the LSI system transfer function, h, the problem is generally better approached in the Fourier domain (Eq. (4.7)). To this end, it is informative to consider the effect of blurring on modulation contrast. Consider a sinusoidal modulation given by:

$$f(x, y) = A + B \sin(2\pi(ux + vy)) \tag{4.21}$$

The recorded signal will be degraded by the system transfer function $\tilde{h}(u,v)$, such that:

$$g(x, y) = A\tilde{h}(0,0) + B|\tilde{h}(u, v)|\sin(2\pi(ux + vy)) \tag{4.22}$$

Here, any phase shift of the image relative to the scene is ignored for simplicity. We see, therefore, that the modulation contrast of object f is:

$$C_f = \frac{B}{A} \tag{4.23}$$

and the modulation contrast of the image, g, is:

$$C_g = \frac{B|\tilde{h}(u, v)|}{A\tilde{h}(0,0)} \tag{4.24}$$

We can now define a new function, T, called the modulation transfer function (MTF), which is defined as the absolute value ratio of C_g/C_f at a given spatial frequency (u, v):

$$T(u, v) = \frac{|\tilde{h}(u, v)|}{\tilde{h}(0,0)} \tag{4.25}$$

The MTF quantifies the degradation of the contrast of a system as a function of spatial frequency. By definition, the modulation at zero spatial frequency, $T(0, 0) = 1$. In the majority of imaging systems, and in the absence of image processing, the MTF is bounded by $0 \leq T \leq 1$. In addition, it should also be noted that, based on the same derivation, the greyscale characteristic $\Gamma = \tilde{h}(0,0)$.

The measurement of the 2-D PSF for projection or cross-sectional imaging systems, or 3-D PSF for volumetric imaging systems (and hence the corresponding 2-D or 3-D MTF) requires that the imaging system be presented with an impulse function. In practice, this can be accomplished by imaging a pinhole in radiography, a wire in cross-section in axial CT, or a single scatterer in ultrasound. The knowledge of the MTF in 2-D or 3-D is useful in calculations in signal detection theory (see Section 4.6.3.1).

It is more common, however, to measure the MTF in a single dimension. In the case of radiography, a practical method of measuring the 1-D MTF is to image a slit formed by two metal bars spaced close together. Such a slit can be used to measure the line spread function (LSF). Among other benefits, imaging a slit will provide better resilience to quantum noise, and multiple slit camera images can be superimposed (bootstrapped) to improve definition of the tails of the LSF.

The LSF is, in fact, an integral representation of the 2-D PSF. For example, consider a slit aligned vertically in an image (which here we assume corresponds to the y axis). Then the LSF, $h(x)$, is given by:

$$h(x) = \int h(x, y)\, \mathrm{d}y \tag{4.26}$$

The integral in Eq. (4.26) can be simplified if we assume that the PSF is separable, as in video based imaging systems. For example, substituting Eq. (4.18) into Eq. (4.26) gives:

$$\int h(x)h(y)\, \mathrm{d}y = h(x) \tag{4.27}$$

This is based on the fact that the area under the PSF, $h(y)$, is unity. It should be clear from this that the LSF and the 1-D MTF are FT pairs. If we assume a rotationally symmetrical PSF (Eqs (4.19, 4.20)), as might be found in a phosphor based detector, the PSF is related to the LSF by the Abel transform:

$$h(x) = 2\int_x^{\infty} \frac{h(r)r}{\sqrt{x^2 - r^2}}\, \mathrm{d}r \tag{4.28}$$

and

$$h(r) = \frac{-1}{\pi}\frac{\mathrm{d}}{\mathrm{d}r}\left(\int_r^{\infty} \frac{h(x)r\, \mathrm{d}x}{x\sqrt{x^2 - r^2}}\right) \tag{4.29}$$

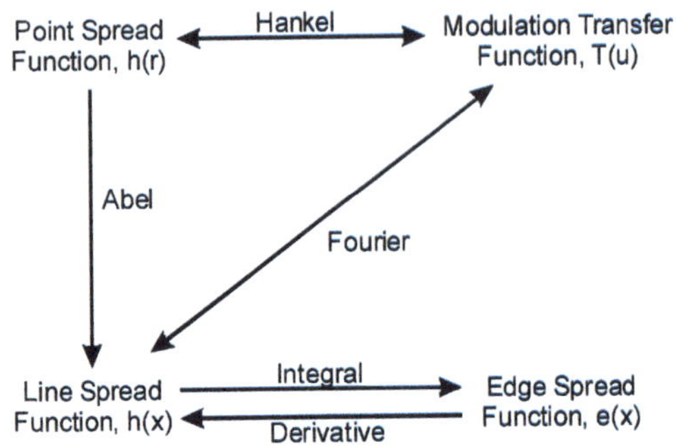

FIG. 4.6. The 1-D forms of the system response function are shown, together with the functional relationship. Included are the rotationally symmetrical PSF, the LSF and the ESF shown for illustration as a function of x, and the MTF.

Note that while the forward Abel transform is tractable, the inverse transform is not. However, the inverse transform can be calculated by first applying the FT and then the Hankel transform (Fig. 4.6).

A further simplification is to image an edge, rather than a line. The edge spread function (ESF) is simply an integral representation of the LSF, so that:

$$e(x) = \int_{-\infty}^{x} h(x)\mathrm{d}x \tag{4.30}$$

and

$$h(x) = \frac{\mathrm{d}}{\mathrm{d}x} e(x) \tag{4.31}$$

Nowadays, the ESF is the preferred method of measuring the system response function of radiographic systems. There are two clear benefits. Firstly, an edge is easy to produce for almost any imaging system, although issues such as the position of the edge need to be carefully considered. Secondly, the ESF is amenable to measuring the presampled MTF of digital systems (see Section 4.4.2.2).

4.4.2. Measuring unsharpness

4.4.2.1. Limiting spatial resolution

The spatial resolution is a metric to quantify the ability of an imaging system to display two unique objects closely separated in space. The limiting spatial resolution is typically defined as the maximum spatial frequency for which modulation is preserved without aliasing.

The limiting resolution can be measured by imaging line patterns or star patterns in radiography and arrays of cylinders imaged in cross-section in cross-sectional imaging systems such as CT and ultrasound. All of these methods use high contrast, sharp edged objects. As such, the limiting spatial resolution is typically measured in line pairs per unit length. This suggests that the basis functions in such patterns are rect[1] functions. By contrast, the MTF is specified in terms of sinusoids and thus is specified in terms of spatial frequencies in cycles per unit length.

There is no strict relationship between a particular MTF value and the limiting spatial resolution of an imaging system. The Coltman transform can be used to relate the square wave response measured with a bar or star pattern and the sinusoidal response measured by the MTF. Ultimately, however, the ability to detect an object (and hence resolve it from its neighbour) is related to the signal to noise ratio (SNR) of the object (see Section 4.6). As a rule of thumb, the limit of resolution for most imaging systems for high contrast objects (e.g. a bar pattern) occurs at the spatial frequency where the MTF ≈ 0.05 (5%).

4.4.2.2. MTF

In practice, it is difficult to measure the MTF of an analogue system (such as film) without first digitizing the analogue image. As such, it is important that the digitization process satisfies the Nyquist–Shannon sampling theorem to avoid aliasing. This is possible in some instances, such as digitizing a film, where the digitizer optics can be designed to eliminate aliasing. In this instance, however, the MTF that is measured is not the MTF of the film but rather is given by:

$$T_{\mathrm{m}} = T_{\mathrm{a}} T_{\mathrm{d}} \tag{4.32}$$

where T_{m} is the measured MTF; T_{a} is the MTF of the analogue system; and T_{d} is the MTF of the digitizer. With Eq. (4.32), it is possible to recover T_{a}, provided $T_{\mathrm{d}} > 0$ over the range of frequencies of interest.

In many systems, however, it is not possible to avoid aliasing. For example, in a digital radiography detector that consists of an amorphous selenium (a-Se) photoconductor coupled to a thin film transistor array, the selenium has very high limiting spatial resolution — much higher than can be supported by the pixel pitch of the detector (see Chapter 7 for details). The resolution pattern

[1] The rect function is covered in more detail in Sections 12.4.3, 13.2.2 and 17.2.1.

shown in Fig. 4.7 is made with such a system. In such instances, there are some important facts to understand. First, aliasing will occur with such a system, as can be seen in Fig. 4.7(b). It is unavoidable. This means that predicting the exact image recorded by a system requires knowledge of the location of the objects in the scene relative to the detector matrix with subpixel precision, as well as knowledge of the blurring of the system prior to sampling. The latter can be determined by measuring what is known as the presampling MTF. The presampling MTF is measured using a high sampling frequency, so that no aliasing is present in the measurement. It is important to realize that, in spite of its name, the presampling MTF does include the blurring effects of the sampling aperture.

The presampling MTF measurement starts with imaging a well defined edge placed at a small angle (1.5–3°) to the pixel matrix/array. From this digital image, the exact angle of the edge is detected and the distance of individual pixels to the edge is computed to construct a supersampled (SS) ESF. Differentiation of the SS-ESF generates an LSF, whose FT produces the MTF. This is the preferred method for measuring the MTF nowadays.

(a) (b)

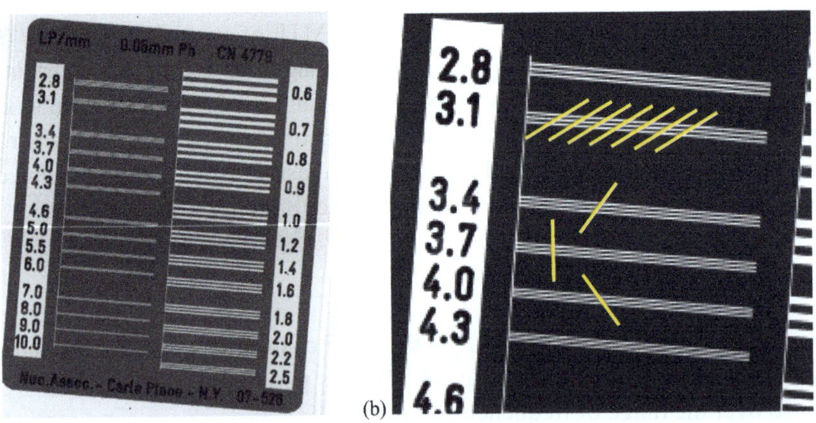

FIG. 4.7. (a) A digital radiograph of a bar pattern is shown. Each group in the pattern (e.g. 0.6 lp/mm) contains three equally spaced elements. In (b), a magnified region of the pattern is shown. Here, we can deduce that the limiting resolution is 3.4 lp/mm. Higher frequencies are aliased as shown by the reversal of the bands (highlighted in yellow) that arise from the digital sampling process.

4.4.3. Resolution of a cascaded imaging system

In the previous section, we dealt with the special situation in which an analogue image, such as film, is digitized by a device such as a scanning photometer. In this situation, the measured MTF is the product of the film MTF and the MTF of the scanning system. This principle can be extended to more generic imaging systems that are composed of a series of individual components (Fig. 4.8). A classic example is to compare the blurring of the focal spot and imaging geometry with that of the detector. Another classic example is of a video fluoroscopic detector containing an X ray image intensifier. In this instance, the MTF of the image is determined by the MTFs of the X ray image intensifier, the video camera and the lenses coupling the intensified image to the camera. This is true because the image passes sequentially through each of the components, and each successive component 'sees' an increasingly blurred image. The one caveat to this concept is that aliasing must be addressed very carefully once sampling has occurred. The principle of cascaded systems analysis is frequently used, as it allows determination of the impact of each component on spatial resolution, and provides a useful tool for analysing how a system design can be improved.

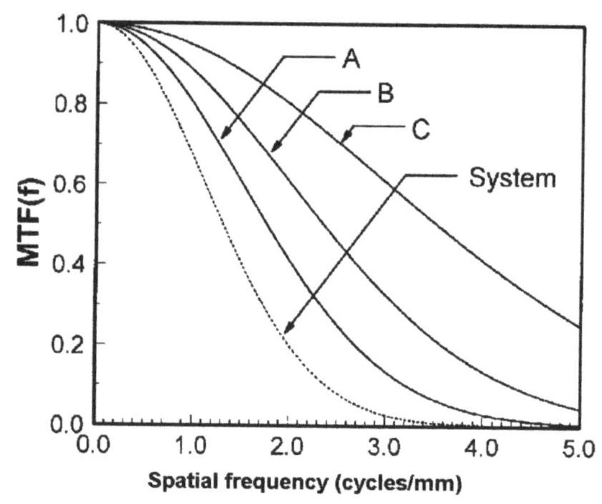

FIG. 4.8. Example of how a system MTF is the product of its components. The overall or system MTF is the product of the MTFs of the three components A, B and C.

4.5. NOISE

The Greek philosopher Heraclitus (c. 535 B.C.) is claimed to have said that "you cannot step twice into the same river". It can similarly be asserted that one can never acquire the same image twice. There lies the fundamental nature of image noise. Noise arises as random variations in the recorded signal (e.g. the number of X ray quanta detected) from pixel to pixel. Noise is not related to anatomy; rather, it arises from the random generation of the image signal. Note, however, that noise is related, for example, to the number of X ray quanta; thus, highly attenuating structures (e.g. bones) will appear noisier than less attenuating structures.

In a well designed X ray imaging system, X ray quantum noise will be the limiting factor in the detection of objects. As illustrated in Fig. 4.9, the ability to discern the disc is degraded as the magnitude of the noise is increased. The optimal radiation dose is just sufficient to visualize the anatomy or disease of interest, thus minimizing the potential for harm. In a seminal work, A. Rose showed that the ability to detect an object is related to the ratio of the signal to noise. We shall return to this important result in Section 4.6.3; however, first we must learn the fundamentals of image noise.

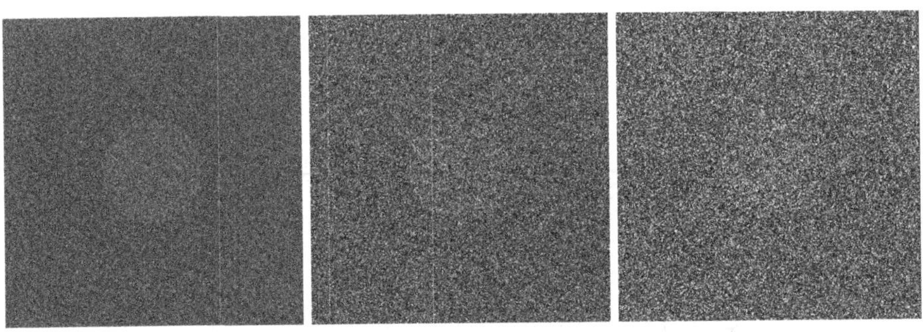

FIG. 4.9. The ability to detect an object is dependent upon both the contrast of the object and the noise in the image.

4.5.1. Poisson nature of photons

The process of generating X ray quanta is random. The intrinsic fluctuation in the number of X ray quanta is called X ray quantum noise. X ray quantum noise is Poisson distributed. In particular, the probability of observing n photons given α, the mean number of photons, is:

$$P(n,\alpha)=\frac{\alpha^n e^{-\alpha}}{n!} \tag{4.33}$$

where α can be any positive number and n must be an integer.

A fundamental principle of the Poisson distribution is that the variance, σ^2, is equal to the mean value, α. When dealing with large mean numbers, most distributions become approximately Gaussian. This applies to the Poisson distribution when a large number of X ray quanta (e.g. more than 50 per del) are detected.

The mean variance equality for X ray quantum noise limited systems is useful experimentally. For example, it is useful to test whether the images recorded by a system are limited by the X ray quantum noise. Such systems are said to be X ray quantum noise limited, and the X ray absorber is called the 'primary quantum sink' (see also Section 7.2.2), to imply that the primary determinant of the image noise is the number of X ray quanta recorded. In the mean variance experiment, one measures the mean and standard deviation parametrically as a function of dose. When plotted log–log, the slope of this curve should be ½. When performed for digital X ray detectors, including CT systems, this helps to determine the range of air kerma or detector dose over which the system is X ray quantum noise limited.

4.5.2. Measures of variance and correlation/covariance

Image noise is said to be uncorrelated if the value in each pixel is independent of the values in other (neighbouring) pixels. If this is true and the system is stationary and ergodic, then it is trivial to achieve a complete characterization of the system noise; one simply needs to calculate the variance (or standard deviation) of the image on a per pixel basis.

Uncorrelated noise is called 'white noise' because all spatial frequencies are represented in equal amounts. All X ray noise in images starts as white noise, since the production of X ray quanta is uncorrelated both in time and in space. Thus, the probability of creating an X ray at any point in time and in any particular direction does not depend on the previous quanta that were generated, nor any subsequent quanta.

Unfortunately, it is rare to find an imaging system in which the resultant images are uncorrelated in space. This arises from the fact that each X ray will create multiple secondary carriers that are necessarily correlated, and these carriers diffuse from a single point of creation. Thus, the signal recorded from a single X ray is often spread among several pixels. As a result, the pixel variance is reduced and neighbouring pixel values are correlated.

Noise can also be correlated by spatial non-uniformity in the imaging system, i.e. non-stationarity. In most real imaging systems, the condition of stationarity is only partially met. One is often placed in a situation where it must be decided if the stationarity condition is sufficiently met as to allow treating the system as shift invariant. An example of this is shown in Fig. 4.10, in which an imaging system consisting of an array of charge coupled device (CCD) detectors is shown. The image on the right is a variance image, obtained by estimating the variance in each pixel using multiple images (i.e. multiple realizations from the ensemble). The image shows that there are strong spatial variations in the variance owing to differences in the coupling efficiency of the fibre optics and the sensitivity differences of the CCDs.

FIG. 4.10. An early digital X ray detector prototype is shown, which consisted of a phosphor screen coupled to an array of fibre optic tapers and CCD cameras. The image on the right is a measurement of the per pixel variance on a small region indicated by yellow on the detector face. The pattern seen is caused by the junctions of the fibre optics and optical vignetting and other optical aberrations in the fibre optics, which result in strong spatial dependencies of both the signal and noise properties of the detector.

Noise can be characterized by the autocorrelation at each point in the image, calculated as the ensemble average:

$$R(x, y, x+\Delta x, y+\Delta y) = \langle \dot{g}(x, y)\dot{g}(x+\Delta x, y+\Delta y)\rangle \tag{4.34}$$

Here, we use the notation $\dot{g}$ to denote that g is a random variable. Correlations about the mean:

$$\Delta\dot{g}(x, y) = \dot{g}(x, y) - \langle \dot{g}(x, y)\rangle \tag{4.35}$$

are given by the autocovariance function:

$$K(x, y, x+\Delta x, y+\Delta y) = \langle \Delta\dot{g}(x, y)\Delta\dot{g}(x+\Delta x, y+\Delta y)\rangle \tag{4.36}$$

On the basis of the assumption of stationarity, $\langle \dot{g}(x,y)\rangle = g$ is a constant, independent of position. If the random process is wide sense stationary, then both the autocorrelation and the autocovariance are independent of position (x, y) and only dependent upon displacement:

$$R(\Delta x, \Delta y) = R(x, y, x+\Delta x, y+\Delta y) \tag{4.37}$$

$$\boldsymbol{K}(\Delta x, \Delta y) = K(x, y, x+\Delta x, y+\Delta y) \tag{4.38}$$

If the random process is ergodic, then the ensemble average can be replaced by a spatial average. Considering a digital image of a stationary ergodic process, such as incident X ray quanta, the autocovariance forms a matrix:

$$\boldsymbol{K}(\Delta x, \Delta y) = \frac{1}{2X}\frac{1}{2Y}\sum_{-X}^{+X}\sum_{-Y}^{+Y}\Delta\dot{g}(x, y)\Delta\dot{g}(x+\Delta x, y+\Delta y) \tag{4.39}$$

where the region over which the calculation is applied is $2X \times 2Y$ pixels. The value of the autocovariance at the origin is equal to the variance:

$$\boldsymbol{K}(0, 0) = \langle \Delta\dot{g}(x, y)\Delta\dot{g}(x, y)\rangle = \sigma_A^2 \tag{4.40}$$

where the subscript A denotes that the calculation is performed over an aperture of area A, typically the pixel aperture.

4.5.3. Noise power spectra

The correlation of noise can be determined either in the spatial domain using autocorrelation (as we have seen in the previous section) or in the spatial frequency domain using the noise power spectrum (NPS), also known as the Wiener spectrum, after N. Wiener, who pioneered its use. There are a number of requirements that must be met for the NPS of an imaging system to be tractable. These include: linearity, shift invariance, ergodicity and wide sense stationarity. In the case of digital devices, the latter requirement is replaced by wide sense cyclostationarity. If the above criteria are met, then the NPS completely describes the noise properties of an imaging system. In point of fact, it is impossible to meet all of these criteria exactly; for example, all practical detectors have finite size and thus are not strictly stationary. However, in spite of these limitations, it is generally possible to calculate the *local* NPS.

By definition, the NPS is the ensemble average of the square of the FT of the spatial density fluctuations:

$$\boldsymbol{W}(u, v) = \left\langle \lim_{X,Y\to\infty} \frac{1}{2X}\frac{1}{2Y}\left| \int_{-X}^{+X}\int_{-Y}^{+Y} \Delta\dot{g}(x, y)\mathrm{e}^{-2\pi\mathrm{i}(ux+vy)}\mathrm{d}x\,\mathrm{d}y \right|^2 \right\rangle \tag{4.41}$$

The NPS and the autocovariance function form an FT pair. This can be seen by taking the FT of Eq. (4.36) and applying the convolution theorem.

The NPS of a discrete random process, such as when measured with a digital X ray detector, is:

$$\boldsymbol{W}(u,v)=\left\langle \lim_{N_x,N_y\to\infty}\frac{a_x}{N_x}\frac{a_y}{N_y}\left|\sum_{x,y}\Delta\dot{\boldsymbol{g}}(x,y)\mathrm{e}^{-2\pi\mathrm{i}(ux+vy)}\right|^2\right\rangle \tag{4.42}$$

N_x and N_y are the number of dels in x and y, and a_x and a_y are the pitch of each del in x and y. Equation (4.42) requires that we perform the summation over all space. In practice, this is impossible as we are dealing with detectors of limited extent. By restricting the calculation to a finite region, it is possible to determine the Fourier content of the fluctuations in that specific region. We call this simple calculation a sample spectrum. It represents one possible instantiation of the noise seen by the imaging system, and we denote this by $\dot{\boldsymbol{W}}$:

$$\dot{\boldsymbol{W}}(u,v)=\frac{a_x}{N_x}\frac{a_y}{N_y}\left|\sum_{m,n}\Delta\dot{\boldsymbol{g}}(x_m,y_n)\mathrm{e}^{-2\pi\mathrm{i}(ux_m,vy_n)}\right|^2 \tag{4.43}$$

An estimate of the true NPS is created by averaging the sample spectra from M realizations of the noise:

$$\ddot{\boldsymbol{W}}(u,v)=\frac{1}{M}\sum_{i=1}^{M}\dot{\boldsymbol{W}}_i(u,v) \tag{4.44}$$

Ideally, the average should be done by calculating sample spectra from multiple images over the same region of the detector. However, by assuming stationarity and ergodicity, we can take averages over multiple regions of the detector, significantly reducing the number of images that we need to acquire.

Now, the estimate of the NPS, $\ddot{\boldsymbol{W}}$, has an accuracy that is determined by the number of samples used to make the estimate. Assuming Gaussian statistics, at frequency (u, v), the error in the estimate $\ddot{\boldsymbol{W}}(x, y)$ will have a standard error given by:

$$\sqrt{\frac{c}{M}}\ddot{\boldsymbol{W}}(u,v) \tag{4.45}$$

where $c = 2$ for $u = 0$ or $v = 0$, and $c = 1$ otherwise. The values of c arise from the circulant nature of the FT.

Typically, 64 × 64 pixel regions are sufficiently large to calculate the NPS. Approximately 1000 such regions are needed for good 2-D spectral estimates.

Remembering that the autocorrelation function and the NPS are FT pairs, it follows from Parseval's theorem that:

$$\boldsymbol{K}(0,0) = \frac{1}{x_0 y_0 N_x N_y} \sum_{u,v} \ddot{\boldsymbol{W}}(u,v) \tag{4.46}$$

This provides a useful and rapid method of verifying a NPS calculation. There are many uses of the NPS. It is most commonly used in characterizing imaging device performance. In particular, the NPS is exceptionally valuable in investigating sources of detector noise. For example, poor grounding often causes line frequency noise (typically 50 or 60 Hz) or its harmonics to be present in the image. NPS facilitates the identification of this noise; in such applications, it is common to calculate normalized noise power spectra, since the absolute noise power is less important than the relative noise power. As we shall see in Section 4.6, absolute calculations of the NPS are an integral part of detective quantum efficiency (DQE) and noise equivalent quanta (NEQ) measurements, and the NPS is required to calculate the SNR in the application of signal detection theory.

Unlike the MTF, there is no way to measure the 'presampling NPS'. As a result, high frequency quantum noise (frequencies higher than those supported by the sampling grid) will be aliased to lower frequencies, in the same way that high frequency signals are aliased to lower frequencies. Radiation detectors with high spatial resolution, such as a-Se photoconductors, will naturally alias high frequency noise. Radiation detectors based on phosphors naturally blur both the signal and the noise prior to sampling, and thus can be designed so that both signal and noise aliasing are not present. There is no consensus as to whether noise aliasing is beneficial or detrimental. Ultimately, the role of noise aliasing is determined by the imaging task, as we shall see in Section 4.6.3.1.

As with the MTF, it is sometimes preferable to display 1-D sections through the 2-D (or 3-D) noise power spectrum or autocovariance. There are two presentations that are used, the central section:

$$\boldsymbol{W}_{\mathrm{C}}(u) = \boldsymbol{W}(u,0) \tag{4.47}$$

and the integral form:

$$W_{\mathrm{I}}(u) = \sum\nolimits_{v} W(u, v) \tag{4.48}$$

Similarly, if the noise is rotationally symmetrical, it can be averaged in annular regions and presented radially. The choice of presentation depends upon the intended use. It is most common to present the central section. Regardless, the various 1-D presentations are easily related by the central slice theorem, as shown in Fig. 4.11.

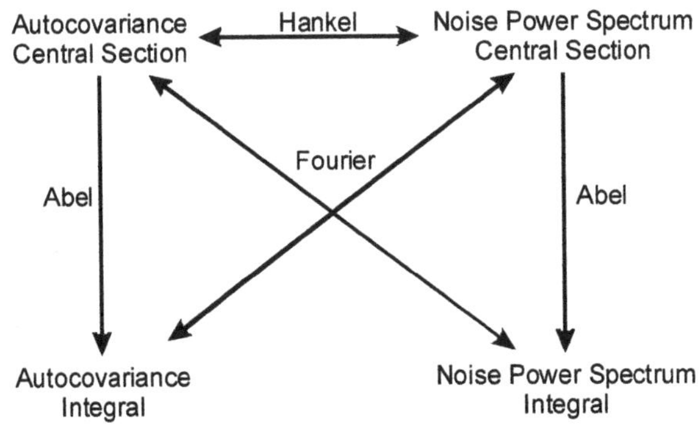

FIG. 4.11. Both 1-D integral and central sections of the NPS and autocovariance can be presented. The various presentations are related by integral (or discrete) transformations. Here, the relationships for rotationally symmetrical 1-D noise power spectra and autocovariance functions are shown.

4.5.4. Noise power spectra of a cascaded imaging system

The propagation or cascade of noise is substantially more complicated than the composition of the MTF. A proper analysis of noise must account for the correlation of the various noise sources. These can be numerous, including the primary X ray quantum noise, the noise arising from the transduction of the primary quanta into secondary quanta (such as light photons in a phosphor or carriers in a semiconductor), and various additive noise sources such as electronic noise from the readout circuitry of digital detectors. While the general theory of noise propagation is beyond the scope of this work, the two simple examples that follow may be illustrative.

4.5.4.1. Image subtraction

It is common to add or subtract or otherwise manipulate medical images. A classic example is digital subtraction angiography (see Section 8.4.2), where a projection image with contrast agent is subtracted from a precontrast mask image to produce an image that shows the difference in attenuation between the two images that arises from the contrast agent (actually, the logarithms are subtracted). In the absence of patient motion, the resultant image depicts the contrast enhanced vascularity.

The effect of the subtraction is to increase the image noise. This arises because, for a given pixel in the image, the pixel values in the mask and the contrast enhanced images are uncorrelated. As a result, the subtraction incorporates the noise of both images. Noise adds in quadrature; thus, the noise in the subtracted image is $\sqrt{2}$ greater than the noise in the source images. To ameliorate the noise increase in the subtraction image, it is typical to acquire the mask image at much higher dose, thereby reducing the contribution of the mask noise to the subtraction.

4.5.4.2. Primary and secondary quantum noise

Consider the simple imaging system shown in Fig. 4.12. In this system, X ray quanta are incident on a phosphor screen (stage 1). A fraction of those quanta are absorbed by the screen to produce light (stage 2). A substantial number of light quanta (perhaps 300–3000) are produced per X ray quantum (stage 3). A small fraction of the light quanta is collected by the lens (stage 4), and a fraction of the collected light quanta produces carriers (electrons and holes) in the optical image receptor (e.g. a CCD camera) (stage 5).

The process of producing an electronic image from the source distribution of X rays will necessarily introduce noise. In fact, each stage will alter the noise of the resultant image. In this simple model, there are two primary sources of noise: (i) the X ray (or primary) quantum noise and (ii) secondary quantum noise. By secondary quantum noise, we refer to noise arising from the production of light in the phosphor, the transmission of light through the optical system and the transduction of light into signal carriers in the optical image receptor. Both the light quanta and signal carriers are secondary quanta.

Each stage involves a random process. The generation of X ray quanta is governed by a Poisson process. In general, we can treat the generation of light quanta from individual X ray quanta as being Gaussian. Stages 3–5 involve the selection of a fraction of the secondary quanta and are thus governed by binomial processes. The cascade of these processes can be calculated mathematically. However, a simple approach to estimating the dominant noise source in a medical image is to determine the number of quanta at each stage of the imaging cascade; the stage with the minimum number of quanta will typically be the dominant noise source.

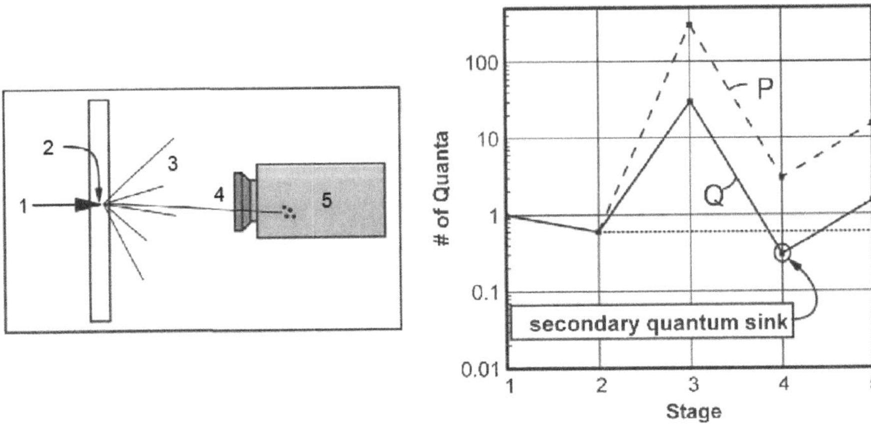

FIG. 4.12. The concept of quantum accounting is illustrated. A simple X ray detector is shown. At each stage of the imaging system, the number of quanta per incident X ray is calculated, to determine the dominant noise source.

4.6. ANALYSIS OF SIGNAL AND NOISE

4.6.1. Quantum signal to noise ratio

There is a fundamental difference between the high contrast and low contrast resolution of an imaging system. In general, the high contrast resolution is limited by the intrinsic blurring of the imaging system. At some point, the system is unable to resolve two objects that are separated by a short distance, instead portraying them as a single object. However, at low contrast, objects (even very large objects) may not be discernible because the signal of the object is substantially lower than the noise in the region containing the object.

Generally, the SNR is defined as the inverse of the coefficient of variation:

$$\mathrm{SNR} = \langle g \rangle / \sigma_g \tag{4.49}$$

where $\langle g \rangle$ is the mean value and σ_g is the standard deviation of a variable.

This definition of the SNR requires that a single pixel (or region) be measured repeatedly over various images (of the ensemble), provided that each measurement is independent (i.e. there is no correlation with time). In an ergodic system, the ensemble average can be replaced by an average over a region. This definition is of value for photonic (or quantum) noise, because in a uniform X ray field, X ray quanta are not spatially correlated. However, most imaging systems do blur the image to some degree, and hence introduce correlation in the noise. As a result, it is generally inappropriate to calculate pixel noise by analysing pixel values in a region for absolute noise calculations. It is also important to note that many imaging systems have some lag or ghosting in which some residual signal is present over time. Lag will similarly introduce correlation over time and hence make absolute noise calculations more difficult.

The definition of SNR in Eq. (4.49) is only useful when the image data are always positive, such as photon counts or luminance. In systems where positivity is not guaranteed, such as an ultrasound system, the SNR is defined as the power ratio, and is typically expressed in decibels (dB):

$$\mathrm{SNR}_{\mathrm{dB}} = 10\,\mathrm{lg}\frac{P_s}{P_n} = 10\,\mathrm{lg}\left(\frac{A_s}{A_n}\right)^2 = 20\,\mathrm{lg}\frac{A_s}{A_n} \qquad (4.50)$$

where P is the average power and A is the root mean square amplitude of the signal, s, or noise, n.

4.6.2. Detective quantum efficiency

From the work of Rose, it is clear that the image quality of X ray imaging systems is determined by the number of quanta used to produce an image. This leads to the definition of the detective quantum efficiency, DQE, a measure of the fraction of the quantum SNR of the incident quanta that is recorded in the image by an imaging system. Thus, the DQE is a measure of the fidelity of an imaging system. It is common to define the DQE as:

$$\mathrm{DQE} = \frac{\mathrm{SNR}_{\mathrm{out}}^2}{\mathrm{SNR}_{\mathrm{in}}^2} \qquad (4.51)$$

where the SNR^2 of the image is denoted by the subscript *out* and the SNR^2 of the incident X ray quanta is given by:

$$\mathrm{SNR}_{\mathrm{in}}^2 = \phi \qquad (4.52)$$

where ϕ is the average number of X ray quanta incident on the detector.

R. Shaw introduced the concept of the DQE to medical imaging and also introduced the term noise equivalent quanta, NEQ. The NEQ is the effective number of quanta needed to achieve a specific SNR in an ideal detector. As such, Eq. (4.51), together with Eq. (4.52), can be rewritten as:

$$\mathrm{DQE} = \frac{\mathrm{NEQ}}{\phi} \tag{4.53}$$

so that:

$$\mathrm{NEQ} = \mathrm{SNR}_{\mathrm{out}}^{2} \tag{4.54}$$

In some sense, the NEQ denotes the net worth of the image data in terms of X ray quanta, and the DQE defines the efficiency with which an imaging system converts X ray quanta into image data.

The definition of DQE given in Eq. (4.51) or Eq. (4.53) is the zero spatial frequency value of the DQE (see also Section 7.2.2). Zero spatial frequency refers to a detector that counts X ray quanta but does not produce a pixelated image (i.e. we only care about the efficiency of counting the X ray quanta). Thus, ϕ is a simple count of the X ray quanta incident on the detector. By this definition, an imaging system that perfectly absorbs each X ray and that does not introduce any other noise will perfectly preserve the SNR of the X ray quanta, and hence $\mathrm{NEQ} = \phi$ and $\mathrm{DQE} = 1$.

If we consider an X ray detector that is perfect in every way except that the quantum detection efficiency, η, is <1.0, we observe that while the incident number of quanta is again ϕ, only $\eta\phi$ quanta are absorbed. As a result, the $\mathrm{NEQ} = \eta\phi$ and $\mathrm{DQE} = \eta$. Thus, in this special instance, the DQE is equal to the quantum detection efficiency, η, the efficiency with which X ray quanta are absorbed in the detector.

The DQE can be more generally expressed in terms of spatial frequencies:

$$\mathrm{DQE}(u,v) = \frac{\mathrm{SNR}_{\mathrm{out}}^{2}(u,v)}{\mathrm{SNR}_{\mathrm{in}}^{2}(u,v)} \tag{4.55}$$

where we rely upon the property that X ray quantum noise is white, leading to the $\mathrm{SNR}_{\mathrm{in}}$ being a constant (Eq. (4.52)):

$$\mathrm{SNR}_{\mathrm{in}}^{2}(u,v) = \Phi \tag{4.56}$$

where Φ is the photon fluence and has units of inverse area:

$$\mathrm{DQE}(u,v) = \frac{\mathrm{NEQ}(u,v)}{\Phi} \tag{4.57}$$

The DQE(u, v) tells us how well the imaging system preserves the SNR_{in} at a specific spatial frequency, (u, v). In a similar fashion, NEQ(u, v) denotes the effective number of quanta that the image is worth at that frequency.

The NEQ and DQE can be calculated from measurable quantities. Specifically:

$$\mathrm{DQE}(u,v)=\frac{\Phi\Gamma^2 T^2(u,v)}{W(u,v)} \tag{4.58}$$

and

$$\mathrm{NEQ}(u,v)=\frac{\Phi^2\Gamma^2 T^2(u,v)}{W(u,v)} \tag{4.59}$$

From Eqs (4.58, 4.59), it is clear that in an ideal system, the NPS is proportional to the MTF squared, $W(u, v) \propto T^2(u, v)$.

Standard measurement conditions for the NEQ and DQE have been specified by the International Electrotechnical Commission. Most typically, an RQA 5 spectrum is used for radiography and RQA-M for mammography. Tabulations of fluence as a function of air kerma are used in conjunction with measurements of kerma to calculate Φ.

4.6.3. Signal to noise ratio

As defined above, the quantum SNR is related to the relative variation of pixel values in a uniform region. However, it is often necessary to compare the amplitude of a specific signal to the background noise. An alternative definition of the SNR is the difference in the means of two regions to the noise in those regions:

$$\mathrm{SNR}=\frac{|<x_a>-<x_b>|}{\sigma} \tag{4.60}$$

where $\bar{x}_a$ and $\bar{x}_b$ are the mean values in the region of an object (a) and background (b) and σ is the standard deviation of the background (Fig. 4.13).

The choice of the background region is important; the standard deviation should be calculated using the region that yields a meaningful result. For example, if image processing (such as thresholding) is used to force the background to a uniform value, then SNR as defined by Eq. (4.60) will be indefinite. Note that the

SNR, as defined in Eq. (4.60), goes by a number of names, including the signal difference to noise ratio and the contrast to noise ratio.

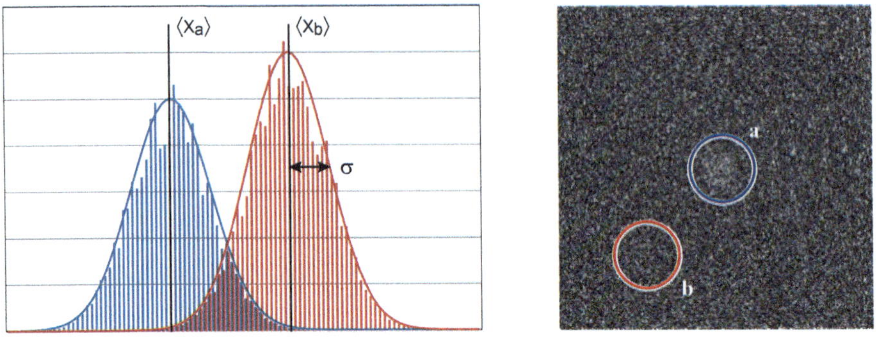

FIG. 4.13. A uniform disk (a) is shown on a uniform background (b) in the presence of X ray quantum noise. The SNR of the object is calculated as the difference in the mean signals divided by the noise characterized by the standard deviation (σ) of the background.

The value of the SNR was first explained by Rose, who was interested in quantifying the quality of television images. Rose showed that an object is distinguishable from the background if the SNR ≥ 5. This can be related to a simple *t*-test in which an error rate of less than 1 in 10^6 occurs when the difference in the means is equal to five standard deviations. Nowadays, this is known in imaging research as the Rose criterion. It should be noted that the requirement of SNR ≥ 5 is actually quite strict. Depending upon the image task, it is possible to operate successfully at lower SNR values.

The assumption of the Rose model is that the limiting factor in the detection of an object is the radiation dose (and hence number of X ray quanta) used to produce the image. In an ideal imaging system, this is true. In fact, the design of all imaging systems is driven by the goal of being X ray (primary) quantum noise limited, as discussed in Section 4.5.4.2. However, Wagner has proposed a taxonomy of noise limitations that is worth noting.

In the Wagner taxonomy, there are four potential limitations in terms of the detection of objects, namely, (i) quantum noise limited, (ii) artefact limited, (iii) anatomy limited and (iv) observer limited. X ray quantum noise limited performance is the preferred mode of operation, because the ability to detect or discriminate an object is determined solely by the radiation dose. Ideally, this is how all detectors should operate. Artefact limitation is the case in which the imaging system introduces artefacts that limit detection. Classic examples include CT and MRI, where acquisition artefacts can predominate over the signal of interest. Anatomy limited detection occurs when the normal anatomy (e.g. ribs in chest radiography or the breast parenchyma in mammography) mask the detection of objects, thereby reducing observer performance. Finally, there are situations in which the observer is the limiting factor in performance. For example, a lesion may be readily visible, but the observer is distracted by an obvious benign or normal finding. Thus, detection was possible but did not occur. In this chapter, we deal exclusively with quantum noise limited performance, which can be calculated using 'ideal observer' methods. In Chapter 18, the modelling of real observers is discussed.

4.6.3.1. Task specific

The MTF, NPS, NEQ and DQE are frequency dependent characterizations of the detector. However, these allow us to calculate the image of a scene. In particular, we can now use the SNR to quantify the ability of the detector to be used in signal known exactly and background known exactly tasks, assuming an ideal observer working with Gaussian statistics. In this scenario, the observer is challenged with the task of deciding between two hypotheses based upon a given set of data. Under the first hypothesis, the expected input signal is present, $\boldsymbol{f}_{\mathrm{I}}$, and the image, $\boldsymbol{g}_{\mathrm{I}}$, is described by an appropriate Gaussian probability distribution. Under an alternative hypothesis, the expected input signal is absent, $\boldsymbol{f}_{\mathrm{II}}$.

The SNR of this task is given by:

$$\mathrm{SNR}_{\mathrm{I}}^2 = \left\langle \Gamma^2 \iint \frac{|\Delta \boldsymbol{f}(u,v)|^2\, \boldsymbol{T}(u,v)^2}{\boldsymbol{W}(u,v)} \mathrm{d}u\, \mathrm{d}v \right\rangle \tag{4.61}$$

where $\Delta \boldsymbol{f}(u,v) = \boldsymbol{f}_{\mathrm{I}}(u,v) - \boldsymbol{f}_{\mathrm{II}}(u,v)$ is the difference between the signal absent and present. For a digital detector, $\boldsymbol{T}$ is the presampled MTF, and thus we must account for aliasing by summing over all aliases of the signal:

$$\mathrm{SNR}_{\mathrm{I}}^2 = \left\langle \Gamma^2 \iint \frac{\sum_{j,k} |\Delta \boldsymbol{f}(u+u_j, v+v_k)|^2\, \boldsymbol{T}(u+u_j, v+v_k)^2}{\boldsymbol{W}(u,v)} \mathrm{d}u\, \mathrm{d}v \right\rangle \tag{4.62}$$

where the indices j and k are used to index the aliases (in 2-D). In this way, we can calculate the SNR of the ideal observer for the detection of any object in a signal known exactly/background known exactly task. In Chapter 18, methods are described that extend this model to include characteristics of real observers.

4.6.4. SNR^2/dose

The ultimate goal of radiation safety in medical imaging is to obtain a figure of merit based on the maximum benefit to the patient for the smallest detriment. We can now calculate the SNR for the detection of a known object (e.g. a tumour) on a known background. This calculation is based upon parameters of a specific detector, so that detectors can be compared or optimized. This calculation can act as a useful surrogate of the benefit, since a disease once detected can be treated. It is necessary, therefore, to relate this benefit to some metric of risk.

A useful metric is the ratio SNR^2/E, where E is the effective dose (see Section 22.3.4). This metric is formulated with the SNR^2, based on the fact that in quantum noise limited imaging, $\mathrm{SNR} \propto \sqrt{\Phi}$; thus, the ratio is invariant with dose. Other descriptors of patient dose may also be useful; for example, optimization in terms of skin dose. Using this formation, it is possible, for example, to determine the optimal radiographic technique (tube voltage, filtration, etc.) for a specific task.

BIBLIOGRAPHY

BARRETT, H.H., MYERS, K.J., Foundations of Image Science, John Wiley & Sons, Hoboken, NJ (2004).

BEUTEL, J., KUNDEL, H.L., VAN METTER, R. (Eds), Handbook of Medical Imaging: Vol. 1, Physics and Psychophysics, SPIE Press, Bellingham, Washington, DC (2000).

INTERNATIONAL COMMISSION ON RADIATION UNITS AND MEASUREMENTS, Medical Imaging – The Assessment of Image Quality, ICRU Rep. 54, ICRU, Bethesda, MD (1996).

INTERNATIONAL ELECTROTECHNICAL COMMISSION, Medical Electrical Equipment — Characteristics of Digital X-ray Imaging Devices: Part 1: Determination of the Detective Quantum Efficiency, Rep. IEC 62220-1, IEC, Geneva (2003).

PRINCE, J.L., LINKS, J.M., Medical Imaging: Signals and Systems, Prentice Hall, Upper Saddle River, NJ (2006).

Chapter 5

X RAY PRODUCTION

R. NOWOTNY
Medical University of Vienna,
Vienna, Austria

5.1. INTRODUCTION

The differential absorption of X rays in tissues and organs, owing to their atomic composition, is the basis for the various imaging methods used in diagnostic radiology. The principles in the production of X rays have remained the same since their discovery. However, much refinement has gone into the design of X ray tubes to achieve the performance required for today's radiological examinations. In this chapter, an outline of the principles of X ray production and a characterization of the radiation output of X ray tubes will be given. The basic processes producing X rays are dealt with in Section 1.4.

5.2. FUNDAMENTALS OF X RAY PRODUCTION

The production of X rays involves the bombardment of a thick target with energetic electrons. These electrons undergo a complex sequence of collisions and scattering processes during the slowing down process, which results in the production of bremsstrahlung and characteristic radiation. A simplification of this process, based on classical theory, is described next.

5.2.1. Bremsstrahlung

Energetic electrons are mostly slowed down in matter by collisions and excitation interactions. If an electron comes close to an atomic nucleus, the attractive Coulomb forces cause a change of the electron's trajectory. An accelerated electron, or an electron changing its direction, emits electromagnetic radiation, given the name bremsstrahlung (braking radiation), and this energy of the emitted photon is subtracted from the kinetic energy of the electron. The energy of the bremsstrahlung photon depends on the attractive Coulomb forces and hence on the distance of the electron from the nucleus.

Using classical theory to consider the electron bombardment of a thin target yields a constant energy fluence from zero up to the initial electron kinetic energy (Fig. 5.1(a)). A thick target can be thought of as a sandwich of many thin target layers, each producing a rectangular distribution of energy fluence. As the electron is slowed down in each layer, the maximum energy in the distribution becomes less, until the electron comes to rest. The superposition of all these rectangular distributions forms a triangular energy fluence distribution for a thick target, the 'ideal' spectrum (Fig. 5.1(b)). Indeed, this model is a simplification, as quantum mechanical theory shows that the distribution for a thin layer is not rectangular and a stepwise reduction of the electron energy from layer to layer does not conform to the slowing down characteristics of electrons.

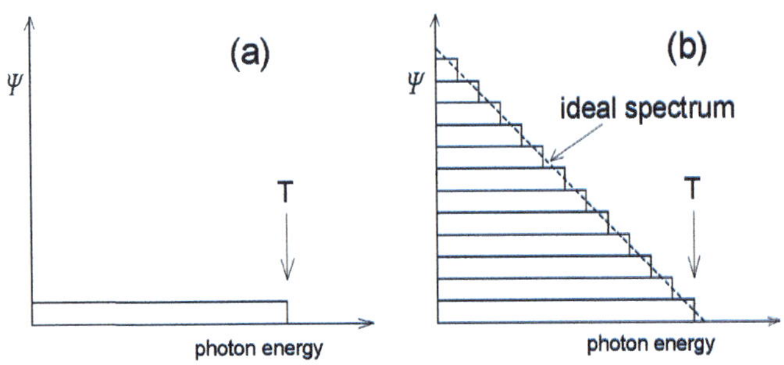

FIG. 5.1. (a) Rectangular distribution of the X ray energy fluence Ψ for a thin target bombarded with electrons of kinetic energy T. (b) For a stack of thin targets, each target layer contributes a rectangular fluence distribution, assuming a uniform electron energy decrease from layer to layer. The superposition forms a triangular ideal spectrum.

The triangular spectrum does not include any attenuation effects. Following the concept of the model, an increase in electron energy increases the number of thin layers each radiating X rays. The triangular area grows proportionally to the square of the electron energy. Considering that the total energy fluence is proportional to the triangular area, and as the X ray tube voltage U_A defines the kinetic energy of the electrons bombarding the anode, the radiation output of an X ray tube is proportional to U_A^2. This relationship only holds if spectral changes due to attenuation and emission of characteristic radiation are ignored. However, this is a reasonable rule of thumb.

5.2.2. Characteristic radiation

A fast electron colliding with an electron of an atomic shell could knock out the electron, provided its kinetic energy exceeds the binding energy of the electron in that shell. The binding energy is highest in the most inner K shell and decreasing for the outer shells (L, M, etc.). The scattered primary electron carries away the difference of kinetic energy and binding energy. The vacancy in the shell is then filled with an electron from an outer shell, accompanied by the emission of an X ray photon with an energy equivalent to the difference in binding energies of the shells involved. For each element, binding energies, and the monoenergetic radiation resulting from such interactions, are unique and characteristic for that element.

K radiation denotes characteristic radiation for electron transitions to the K shell, and likewise, L radiation for transitions to the L shell. The origin of the electron filling the vacancy is indicated by suffixes (α, β, γ, etc.), where α stands for a transition from the adjacent outer shell, β from the next outer shell, etc. K_α radiation results from L to K shell transitions; K_β radiation from M to K shell transitions, etc. Energies are further split owing to the energy levels in a shell, indicated with a numerical suffix. Further, each vacancy in an outer shell following from such a transition gives rise to the emission of corresponding characteristic radiation causing a cascade of photons.

Table 5.1 gives the binding energies and the K radiation energies for the common anode materials used in diagnostic radiology.

Instead of characteristic radiation, the energy available could be transferred to an electron that is ejected from the shell (Auger electron). The probability of Auger electron production decreases with atomic number.

TABLE 5.1. BINDING ENERGIES AND K RADIATION ENERGIES OF COMMON ANODE MATERIALS

Element[a]	Binding energy (keV)		Energies of characteristic X rays (keV)			
	L shell	K shell	$K_{\alpha 1}$	$K_{\alpha 2}$	$K_{\beta 1}$	$K_{\beta 2}$
Mo	2.87/2.63/2.52	20.00	17.48	17.37	19.61	19.97
Rh	3.41/3.15/3.00	23.22	20.22	20.07	22.72	23.17
W	12.10/11.54/10.21	69.53	59.32	57.98	67.24	69.07

[a] Mo: molybdenum; Rh: rhodium; W: tungsten.

5.2.3. X ray spectrum

The electrons are slowed down and stopped in the target, within a range of a few tens of micrometres, depending on the tube voltage. As a result, X rays are not generated at the surface but within the target, resulting in an attenuation of the X ray beam. This self-filtration appears most prominent at the low energy end of the spectrum (Fig. 5.2). Additionally, characteristic radiation shows up if the kinetic electron energy exceeds the binding energies. L radiation is totally absorbed by a typical filtration of 2.5 mm Al (aluminium equivalent, see Section 5.6.2). The K edge in the photon attenuation of tungsten can be noticed as a drop of the continuum at the binding energy of 69.5 keV. For tungsten targets, the fraction of K radiation contributing to the total energy fluence is less than 10% for a 150 kV tube voltage.

As shown in Section 2.4.4, the radiative mass stopping power of electrons is proportional to Z^2, where Z is the atomic number of the absorber. Integration of the radiative mass stopping power along the electron path gives the total X ray energy fluence, Ψ, as $\Psi \sim ZIU^2$, where I denotes electron current and U the tube voltage. If a high bremsstrahlung yield is required, metals with high Z are preferable. Tungsten (Z = 74) is commonly chosen, as it also withstands high temperatures (2757°C at 1.3×10^{-2} Pa vapour pressure). The efficiency for the conversion of electrical power to bremsstrahlung radiation is proportional to UZ. At 100 kV, the efficiency is as low as ~0.8%. This is the cause of most of the technical problems in the design of X ray tubes, as practically all electrical power applied in the acceleration of electrons is converted to heat.

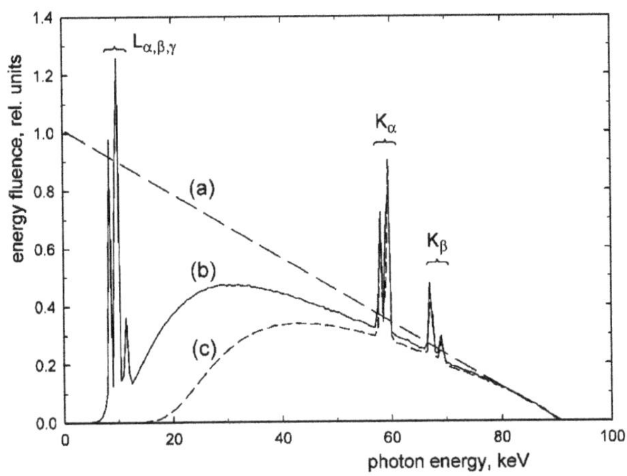

FIG. 5.2. (a) Ideal bremsstrahlung spectrum for a tungsten anode (tube voltage 90 kV), (b) an actual spectrum at the beam exit port, including characteristic X rays (anode angle 20°, inherent filtration 1 mm Be) and (c) the spectrum filtered with an equivalent of 2.5 mm Al.

The ideal spectrum appears triangular, with the energy fluence taken as the quantity describing the spectral intensity. The photon fluence is a more practical quantity for calculations using spectral data and is, therefore, used in the following sections. More refined models for the generation of X ray spectra have been developed using Monte Carlo methods. For practical purposes, a semi-empirical approach gives satisfactory results, which are useful in simulations [5.1].

5.3. X RAY TUBES

5.3.1. Components of the X ray tube

The production of both bremsstrahlung and characteristic radiation requires energetic electrons hitting a target. Accordingly, the principal components of an X ray tube are an electron source from a heated tungsten filament, with a focusing cup serving as the tube cathode, an anode or target, and a tube envelope to maintain an interior vacuum (Fig. 5.3). The filament is heated by a current that controls the thermionic emission of electrons, which, in turn, determines the electronic current flowing from the cathode to the anode (tube or anode current). The accelerating potential difference applied between the cathode and anode controls both X ray energy and yield. Thus, two main circuits operate within the X ray tube — the filament circuit and the tube voltage circuit.

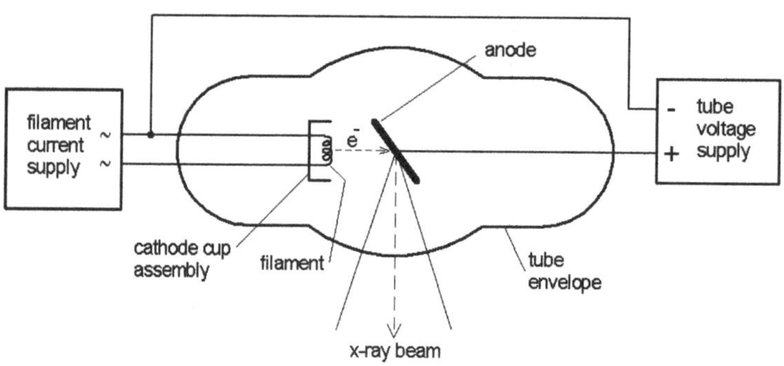

FIG. 5.3. Principal components of an X ray tube.

Typical anode currents, depending on the examination mode, are <10 mA in fluoroscopy and 100 mA to >1000 mA in single exposures. The typical range of tube voltages is 40–150 kV for general diagnostic radiology and 25–40 kV in mammography.

5.3.2. Cathode

The arrangement of the filament, the focusing cup, the anode surface and the tube voltage generates an electric field accelerating the electrons towards the focal spot at the anode (Fig. 5.4). X ray tubes with two focal spots usually employ two separate filament/cup assemblies (cathode blocks).

The degree of focusing depends on the potential difference or bias voltage between the filament and focusing electrode. The focal spot will be largest if both are at the same potential. With an increasing negative bias voltage at the focusing cup, the focus size will decrease and finally the electron current will be pinched off. This effect is sometimes used to control the focus size electronically, or for a fast switching of the anode current (grid controlled tubes) when short radiation pulses are required, as in pulsed fluoroscopy. Some bias can simply be achieved by connecting the filament and cup with a high resistance grid leak resistor. Some electrons emitted from the filament surface will hit and charge the cup. The cup is discharged via the grid leak resistor, maintaining the cup at a negative potential difference.

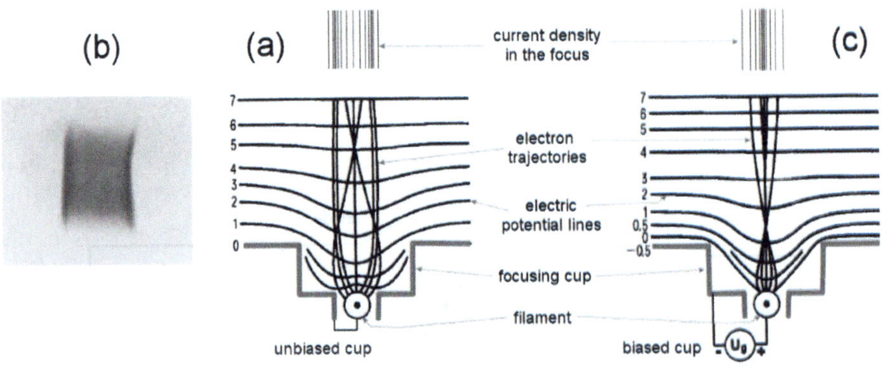

FIG. 5.4. (a) The effect of an unbiased focusing cup on the electric field and electron trajectories. The typical bimodal distribution of the current density can also be seen in a pinhole image of the focus (b). Biasing the focusing cup (c) leads to a compression of the trajectories, giving a smaller focus. Numbers indicate potential difference near the cup in kilovolts.

The spiral wound filament is typically made from tungsten wire of 0.2–0.3 mm diameter and operates at around 2700 K. To minimize the evaporation of tungsten from the hot surface, the filament temperature is kept at a lower level, except during exposure, when it is raised to operational levels. Thermionic emission of electrons increases with temperature (Richardson's law) and produces a cloud of electrons (space charge) enclosing the filament. This space charge shields the filament from the anode voltage. Increasing the filament temperature increases the space charge, and at some point the electric field cannot remove all electrons produced, but the anode current remains steady (space charge limited current) (Fig. 5.5). An attempt to increase the anode current by increasing the filament current might eventually end in filament failure. Generator control usually prevents this situation.

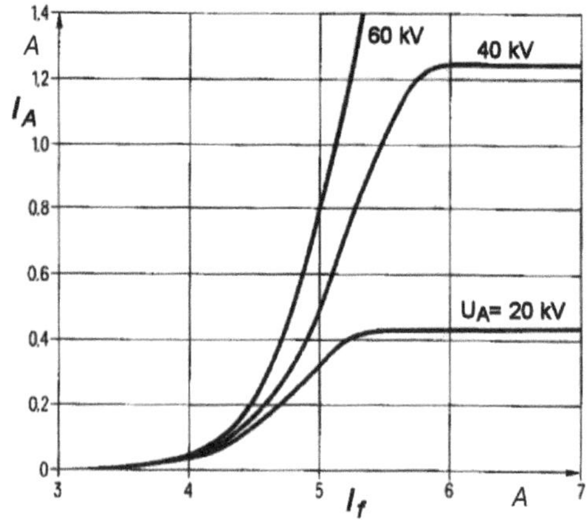

FIG. 5.5. Anode current I_A versus filament current I_f for various anode voltages U_A, showing space charge limited anode currents for the lower tube voltages.

For high anode voltages, all electrons boiled off the filament are accelerated to the anode, giving an anode current that is fairly independent of tube voltage (saturation current) (Fig. 5.6).

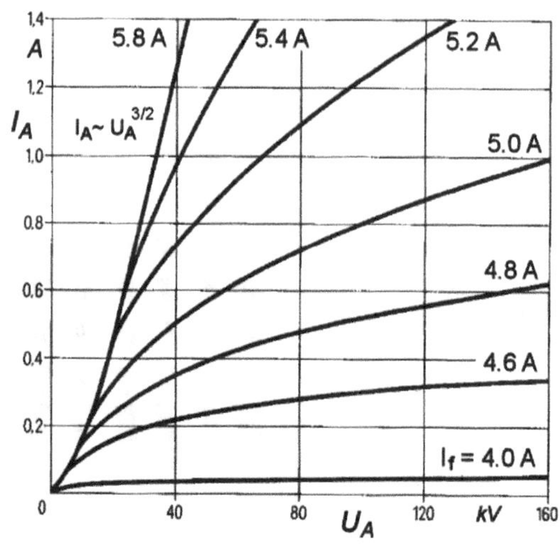

FIG. 5.6. Tube current I_A versus tube voltage U_A depending on filament current I_f. Note: current saturation occurs for the lower filament currents.

5.3.3. Anode

5.3.3.1. Choice of material

For common radiographic applications, a high bremsstrahlung yield is mandatory, requiring materials with high atomic numbers (Z). Additionally, because of the low effectiveness of X ray production, it is also essential that the thermal properties are such that the maximum useful temperature determined by melting point, vapour pressure, heat conduction, specific heat and density is also considered. Tungsten ($Z = 74$) is the optimum choice here.

For mammography, other anode materials such as molybdenum ($Z = 42$) and rhodium ($Z = 45$) are frequently used. For such anodes, X ray spectra show less contribution by bremsstrahlung but rather dominant characteristic X rays of the anode materials. This allows a more satisfactory optimization of image quality and patient dose. In digital mammography, these advantages are less significant and some manufacturers prefer tungsten anodes.

5.3.3.2. Line focus principle (anode angle)

For measurement purposes, the focal spot size is defined along the central beam projection. For high anode currents, the area of the anode hit by the electrons should be as large as possible, to keep the power density within acceptable limits. To balance the need for substantial heat dissipation with that of a small focal spot size, the line focus principle is used (Fig. 5.7(a)).

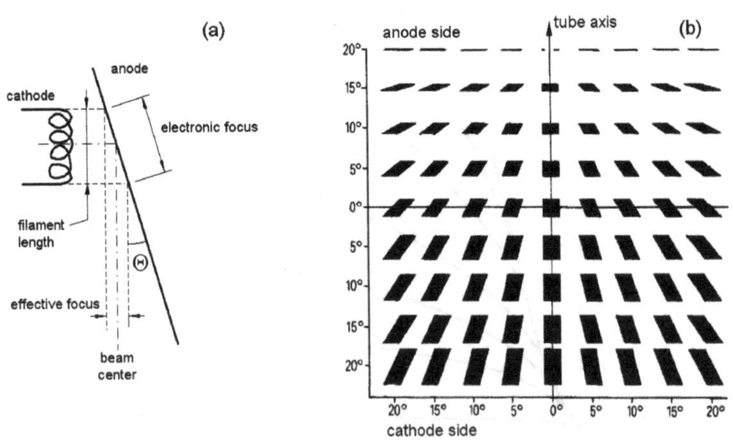

FIG. 5.7. (a) Line focus principle: the length of the filament appears shortened in the beam direction; (b) graphic representation of the focal spot shape at different locations in the radiation field (anode angle 20°).

The anode is inclined to the tube axis, typically with the central ray of the X ray field perpendicular to the tube axis. The electrons hit the anode in the electronic focus, largely determined by the length of the cathode filament. The electronic focus appears shortened in beam direction by sin θ as the effective focus (Fig. 5.7(a)). Anode angles in diagnostic tubes range from 6° to 22°, depending on their task, with 10–16° used for general purpose tubes. The radial dimension of the focus size is given by the filament coil diameter and the action of the focusing cup. The size of the focal spot of an X ray tube is given for the central beam in the X ray field running perpendicular to the electron beam or the tube axis. The actual focal spot size depends on the position within the field of view, increasing from the anode side of the tube to the cathode (Fig. 5.7(b)).

The reduction of anode angles to achieve smaller effective focus sizes is limited by the size of the field of view required as the X ray beam is cut off by the anode. A further limit is given by the heel effect (Fig. 5.8). Here, the X rays produced at depth within the anode suffer some absorption losses according to the distance passed in the anode material. For X rays emerging near the anode side of the X ray field, the losses are higher, resulting in an inhomogeneous X ray intensity across the beam. In projection radiography, this effect can be verified by measuring the air kerma across the beam, although it is often barely noticeable in radiographs. In mammography, the heel effect is used to create a decrease in the incident air kerma from the chest wall to the nipple, matching the decrease in organ thickness.

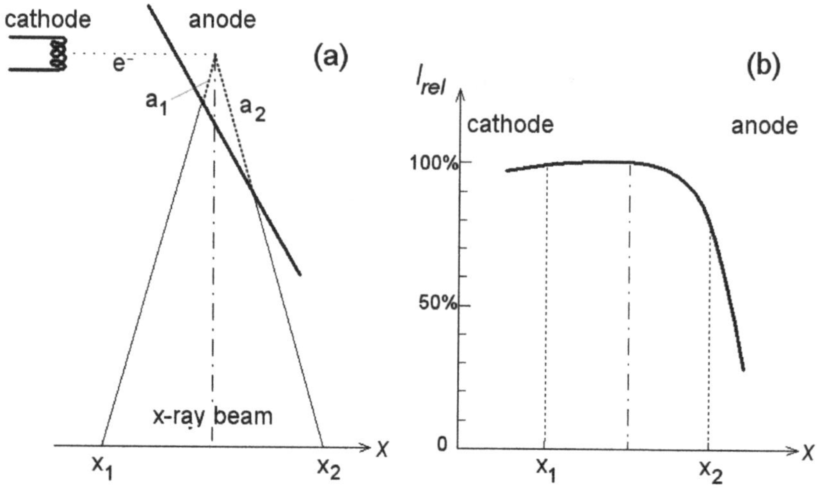

FIG. 5.8. (a) Absorption of X rays at the cathode side of the X ray field is less (a_1) than at the anode side (a_2). (b) The steep drop in intensity I_{rel} at the anode side reflects the increased absorption (heel effect).

In addition to X rays produced in the primary focus, some off focus radiation results from electrons scattered from the anode, which are then accelerated back and hit the anode outside of the focal area. Extrafocal radiation can contribute up to ~10% of the primary X ray intensity. Since the effective focal spot size for off focus radiation is substantially larger than for the primary focus, it has little impact on image quality such as background fog and blurring. Sometimes, parts of the body are imaged outside the collimated beam by off focus radiation (e.g. the ears in frontal projections of the skull). Off focus radiation also increases patient dose. The best position for a diaphragm to reduce off focus radiation is close to the focus.

5.3.3.3. Stationary and rotating anodes

For X ray examinations that require only a low anode current or infrequent low power exposures (e.g. dental units, portable X ray units and portable fluoroscopy systems), an X ray tube with a stationary anode is applicable (Fig. 5.9). Here, a small tungsten block serving as the target is brazed to a copper block to dissipate the heat efficiently to the surrounding cooling medium. As the focal spot is stationary, the maximum loading is determined by the anode temperature and temperature gradients.

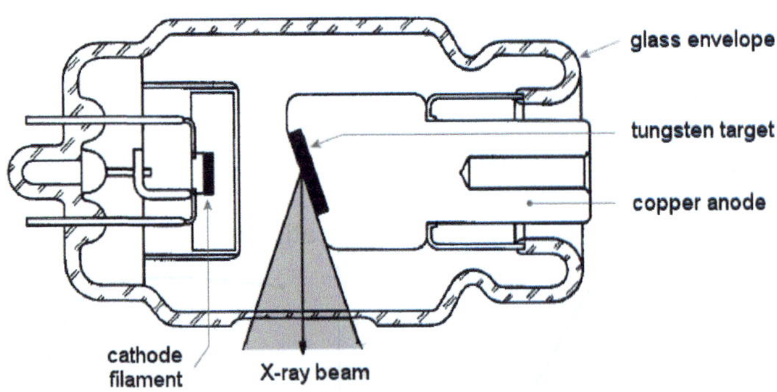

FIG. 5.9. Dental X ray tube with a stationary anode.

Most X ray examinations need photon fluences that cannot be obtained with stationary anodes, as bombarding the same spot with higher anode currents leads to melting and destruction of the anode. In a tube with a rotating anode, a tungsten disc rotates during an exposure, thus effectively increasing the area bombarded by the electrons to the circumference of a focal track. The energy is dissipated to a much larger volume as it is spread over the anode disc (Fig. 5.10). The anode disc is fixed to a rotor and a spindle with a short stem. The spindle is supported by two ball bearings. In newer developments, floating bearings with liquid metal have been introduced.

The rotating anode is attached to the rotor of an asynchronous induction motor. The rotor is mounted within the tube housing on bearings (typically ball bearings). The squirrel cage rotor is made up of bars of solid copper that span the length of the rotor. At both ends of the rotor, the copper bars are connected through rings. The driving magnetic fields are produced by stator windings outside the tube envelope. The rotational speed of the anode is determined by the frequency of the power supply and the number of active windings in the stator. The speed can be varied between high (9000–10 000 rev./min) and low (3000–3600 rev./min) values using all three or one phase only. In examinations requiring rather low anode currents, as in fluoroscopic applications, the tube is run at low speed. Rotor bearings are critical components of a rotating anode tube and, along with the whole assembly, cycling over a large temperature range results in high thermal stresses.

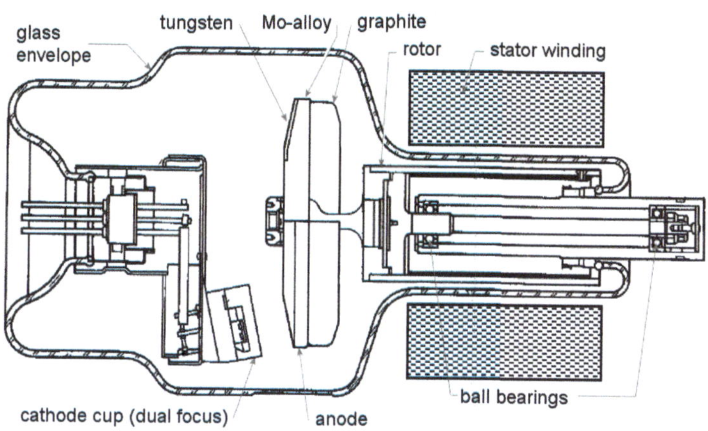

FIG. 5.10. X ray tube with a rotating compound anode and a glass envelope. Mo: molybdenum.

5.3.3.4. Thermal properties

The main limiting factor in the use of X ray tubes is the thermal loading capacity of the anode. Figure 5.11 shows a typical diagram of maximum permissible load versus time. Within the first 100 ms, the maximum load is determined by mechanical stress in the anode material developing from temperature gradients near the surface of the focal spot (A). As a consequence, cracks can develop, leading to an increase in anode surface roughness. This effect can be reduced by the use of a more ductile alloy as the focal track (e.g. tungsten/rhenium alloys) or by an increase in the size of the focal spot or the rotational speed of the anode.

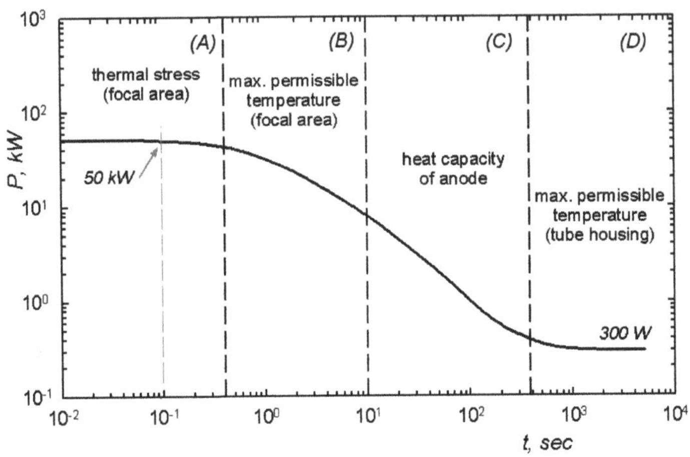

FIG. 5.11. Maximum permissible tube load versus time for a single exposure, constant current, 100 kV tube voltage and a 50 kW tube. The nominal power is determined for an exposure time of 0.1 s.

The energy released in the focal spot raises the temperature to a maximum permissible level (2757°C for tungsten) for exposures up to a few seconds, thus limiting the maximum load ((B) in Fig. 5.11). In computed tomography (CT) and fluoroscopic procedures, longer exposure times are needed (10 s to >200 s). Here, the dissipation of heat across the entire anode disc becomes important. The important physical properties are then the heat conduction and heat capacity of the anode disc ((C) in Fig. 5.11). The heat capacity is the energy stored in the anode disc with the anode at its maximum permissible temperature. It depends on the specific heat and mass of the anode materials. Molybdenum is superior to tungsten in this respect. Increasing the mass of the anode (diameter, thickness) has its limitations, as balancing the rotating anode becomes difficult for the wide range of temperatures occurring. Since graphite has a higher specific heat than molybdenum or tungsten at higher temperatures, the heat capacity can be increased by attaching graphite heat sinks to the back of the anode disc. Graphite enhances the dissipation of heat by black body thermal radiation.

The maximum permissible load for long or continuous exposures is determined by the effectiveness of heat removal from the anode ((D) in Fig. 5.11). Most of the heat is removed by thermal radiation and absorbed in the tube envelope and the surrounding insulating oil. The maximum permissible temperature and the heat capacity of the tube housing are then the limiting factors of applicable power. Heat conduction in traditional tubes from the anode disc via the stem, spindle, bearings and bearing support is not very efficient. In some tube designs, the ball bearings have been replaced by special bush bearings (spiral groove bearings), with a liquid gallium alloy for lubrication. The thermal resistance of such bearings is much lower than that of ball bearings, which enhances the heat flow and increases the continuous power rating of the tube. In the latest generation of X ray tubes (rotational envelope tube), the removal of heat from the anode is increased considerably by directly exposing the back of the anode disc to the cooling oil, enabling long exposures with high anode currents, as required in CT scans.

5.3.3.5. Tube envelope

The tube envelope maintains the required vacuum in the X ray tube. A failing vacuum, resulting from leakage or degassing of the materials, causes increased ionization of the gas molecules, which slows down the electrons. Further, a current of positive ions flowing back could impair or destroy the cathode filament. The envelope is commonly made of glass but high performance tubes increasingly have glass–metal or ceramic–metal envelopes.

The X ray beam exits the tube through a window in the envelope. To reduce absorption, the thickness of the glass is reduced in this area. If low energy X rays are used, as in mammography, the exit port is a beryllium window, which has less absorption than glass because of its low atomic number.

5.3.3.6. Tube housing

The X ray tube (often referred to as the insert) is installed in a tube housing that provides the structural support required (Fig. 5.12). The space between the housing and the envelope is filled with transformer oil, serving as electrical insulation and for heat removal from the envelope surface, which is heated by the infrared radiation from the anode. The change of oil volume with varying temperature is taken care of by the expansion bellows. The oil carries the heat away to the housing by convection, sometimes enhanced by forced cooling with a ventilator or heat exchangers.

The housing also provides radiation shielding to prevent any radiation except the primary beam from leaving the housing. The inside of the housing is lined with lead sheets to minimize leakage radiation. The maximum acceptable exposure due to leakage radiation is limited by regulation. Further, tube housings provide mechanical protection against the impact of envelope failure.

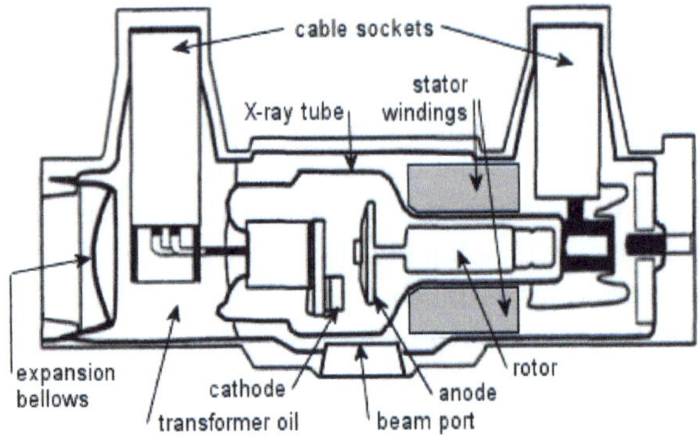

FIG. 5.12. Typical housing assembly for a general purpose X ray tube.

5.4. ENERGIZING AND CONTROLLING THE X RAY TUBE

The X ray generator provides all the electrical power sources and signals required for the operation of the X ray tube, and controls the operational conditions of X ray production and the operating sequence of exposure during an examination. The essential components are a filament heating circuit to determine anode current, a high voltage supply, a motor drive circuit for the stator windings required for a rotating anode tube, an exposure control providing the image receptor dose required, and an operational control (Fig. 5.13). The operational control is often accomplished by a microprocessor system but electromechanical devices are still in use. Modern generators provide control of the anode temperature by monitoring the power applied to the tube and calculating the cooling times required according to the tube rating charts

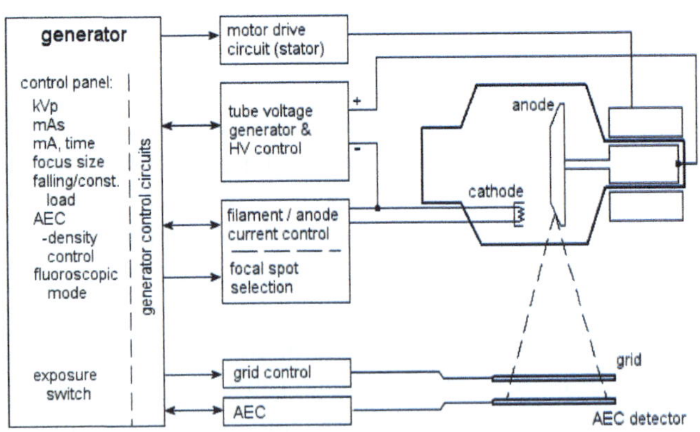

FIG. 5.13. Schematic diagram of a basic X ray generator. AEC: automatic exposure control; kVp: peak voltage.

5.4.1. Filament circuit

An isolated transformer supplies the filament heating current. The generator is programmed to set the heating current according to the tube characteristics. Heating currents range up to 10 A with voltages of <15 V AC (alternating current). To minimize thermal stress and increase durability, the filament is permanently preheated to a temperature at which thermionic emission is negligible.

The thermal inertia of the filament limits the speed of change in the tube current (e.g. falling load). The thermal time constants range from 50 to 200 ms. For a frequency of heating currents of 100 or 120 Hz, some tube current ripple is a result of the temperature variations induced. For high frequency generators, the thermal inertia of the filament suppresses fluctuations of thermionic emission.

5.4.2. Generating the tube voltage

Irrespective of the waveform, the tube voltage is defined as the peak voltage, kVp, of the voltage train. The voltage ripple, R, is given as the relative difference of the minimum voltage, kV_{min}, from the peak voltage, $R = (kVp - kV_{min})/kVp$. In X ray units, the tube voltage is supplied symmetrically to the tube, i.e. a net potential difference of 150 kV is achieved by feeding –75 kV to the cathode and +75 kV to the anode. This is accomplished electrically by grounding the centre tap of the secondary coil of the high voltage transformer. The requirements for electrical isolation are then less stringent. In mammography with tube voltages <40 kV, and with some high performance tubes, one electrode is kept at ground potential.

Except for grid controlled tubes, the length of an exposure is determined by the provision of high voltage to the tube by switching in the primary circuit. Electromechanical relays used to be employed in single-phase and three-phase generators, but electronic switching components, such as thyristors, are now used. In single-phase generators, timing is only possible in multiples of pulses, giving inaccurate timing for short exposures. Three-phase generators use a prepulse of low current to avoid magnetic saturation of the transformer core. When the high voltage is turned off, the charge stored in the cable capacitance and the circuit is discharged via the X ray tube. The end of the voltage waveform, therefore, shows some tailing, an effect that impairs the production of short pulses.

5.4.2.1. Single-phase generators

Single-phase generators use a single-phase mains supply and a step up transformer with a fixed winding ratio. The high voltage is set by a variation of the primary voltage with a switched autotransformer. Half-wave rectification of the transformed voltage gives a 1-pulse waveform, where a pulse is a half wave per period of mains frequency (50 or 60 Hz). Some low power X ray units use the tube as a self-rectifying diode with current only flowing from the cathode to the anode, but reverse current flow, as a result of a hot anode, is a limiting factor. Nowadays, solid state diodes are used as rectifiers. A full wave rectification yields two half waves per period (2-pulse waveform). Voltage ripple of 1- and 2-pulse waveforms is 100%.

5.4.2.2. *Three-phase generators*

With a three-phase mains supply, three AC voltages, each with a phase shift of 120°, are available. Full wave rectification then gives six half waves per period (6-pulse waveform), with a nominal ripple of 13.4%. Owing to imbalances in transformer windings and voltages, the ripple might, in practice, approach 25%. Adding another secondary winding to the transformer gives two secondary voltages. Combining the full wave rectified secondary voltages using delta and wye connections yields a total of six phases with a phase shift of 60° each. Full wave rectification then gives a total of 12 pulses per period, with a nominal ripple of 3.4% (in practice, less than 10% is achieved). Three-phase generators are more efficient and allow for much higher tube output than single-phase generators.

5.4.2.3. *High frequency generators*

This type of generator includes a stabilized power supply in the front end of the device. First, the mains supply is rectified and filtered to produce a direct current (DC) supply voltage needed for an inverter circuit. The inverter generates pulses that are transformed, rectified and collected in a capacitor to give the high voltage for the tube. The inverter pulse rate is used to control the tube voltage. The actual voltage on the tube is sensed by the generator and compared with the voltage set on the console. The difference is then used to change the pulse rate of the inverter until the set voltage is achieved. Similarly, a separate inverter system is used for the tube current.

The pulse shape of a single X ray exposure pulse resembles a fundamental frequency of several tens of kilohertz, giving rise to the generator's name. Transformers for such frequencies are much smaller than for 50/60 Hz voltages, reducing the weight substantially. In low power generators, the whole generator could be included in the tube housing, avoiding any high voltage cabling.

The voltage ripple depends on many technical factors, but for low power applications it is typically ~13%, dropping to ~4% at higher currents. The time constants relevant for voltage and current control are typically <250 μs, enabling better timing control of the exposure than with single and three phase generators.

5.4.2.4. *Capacitor discharge generators*

In places with inadequate mains supply, or in remote locations, capacitor discharge generators are helpful. A capacitor is charged to a high voltage just before an exposure. The capacitor is connected to the X ray tube with the start and length of exposure controlled by a grid. High tube currents and short exposure times can be obtained. However, discharging a capacitor implies a falling tube voltage during exposure. Typically, voltage drops of ~1 kV/mAs are usual. As kerma drops with voltage, the appropriate exposure of thick body parts can be problematic.

5.4.2.5. *Constant voltage generators*

Constant voltage generators achieve a DC high voltage with minimal ripple through the use of a closed loop linear voltage controller (e.g. high voltage triodes) in series with the tube. High frame rates and voltage stability are achieved. Constant potential generators use a complex technology with high costs of investment and operation and, as a consequence, have lost popularity.

5.4.2.6. *Comparison of generator technologies*

Figure 5.14 shows a comparison of voltage waveforms together with their associated kerma output. In radiology, it is desirable to keep exposure times as low as achievable. One-pulse waveforms produce radiation in only half of a cycle, and double the exposure time compared with 2-pulse voltages. As the kerma output rises approximately with the square of the tube voltage, there is a substantial amount of time in a half wave of 1- and 2-pulse waveforms, with little or no contribution to kerma output, again effectively increasing the exposure time.

The 1- and 2-pulse waveforms also yield softer X ray spectra, which implies increased radiation dose to the patient. Both the exposure time and the patient dose indicate that the optimum waveform would be a DC voltage with essentially no ripple, but 12-pulse and high frequency generators are near optimum.

Generators transforming mains AC voltages suffer from external voltage instabilities. Devices for compensating for these fluctuations are often integrated into the generator design; high frequency generators that provide tube supplies with higher stability and accuracy are currently the state of the art.

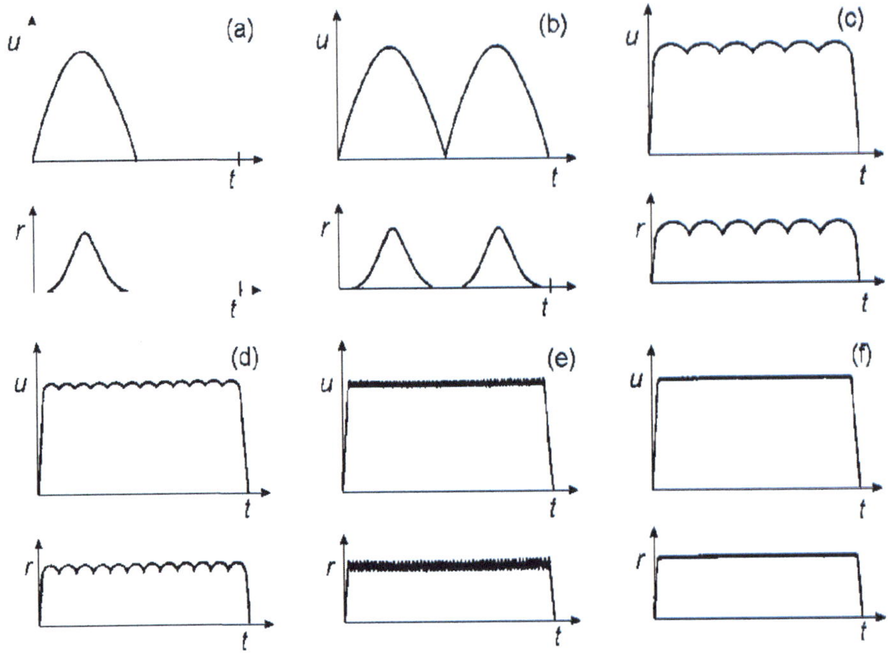

FIG. 5.14. Voltage waveforms (u) and associated tube output (dose rate (r)) versus time (t) for (a) 1-pulse, (b) 2-pulse, (c) 6-pulse, (d) 12-pulse, (e) high frequency and (f) constant voltage generators.

5.4.3. Exposure timing

Exposure of a radiograph can be set manually by choosing tube current and exposure time. Except in examinations with little variability in body dimensions (e.g. extremities), an AEC is mandatory to achieve a consistent image quality or film density. The AEC terminates an exposure when the image receptor has received a preset level of radiation.

The AEC system consists of one to three radiation detectors (ionization chambers or solid state detectors). The signal from these detectors is amplified and integrated, corrected for response in photon energy and dose rate, and finally, compared with the preset dose level. The exposure is terminated when the chosen level is attained. In case the AEC does not terminate the exposure, a backup timer sets a time limit. On installation of a radiographic unit, the dose levels are set, taking into consideration all the components of the imaging chain, i.e. film and screens, imaging plates, film development, preferred tube voltage and filtration, acceptable image noise, etc. This process needs to be carried out for all tube voltages, image receptor and examination types in question. Some products allow for fine manual adjustment to the preset dose level by a density control on the console adapting the density in steps of 10–20%.

Radiographic devices commonly have ionization chambers as AEC detectors positioned immediately in front of the radiographic cassette. The detectors must show no visible radiographic contrast on the image. For low energy X ray units (e.g. mammography, paediatric units), this is difficult to achieve and detectors are therefore positioned behind the image receptor. Solid state detectors are mostly employed in this case.

The position of the detectors is delineated on the table top or wall stand, to assist the operator in patient positioning. As absorption in the patient's body can vary substantially across the beam, the operator can select a detector or a combination of detectors for exposure control, to obtain optimal exposure in the dominant part of the image. As an example, for a chest X ray in posterior–anterior projection, the two lateral detectors positioned under the lung regions are chosen, while in lateral projection, the central detector is selected.

5.4.4. Falling load

To avoid image blurring due to patient motion, short exposure times are mandatory. To produce the shortest possible exposure, the generator starts with the maximum permissible current and, in the course of the exposure, lowers the tube current consistent with tube ratings (falling load). Thus, the tube is operating at the maximum permissible power rating during the entire exposure. In some products, an exposure with falling load can be run at a reduced power setting (e.g. 80% of the maximum power) to lower the stresses. The operator sets the tube voltage, focus size and, if not in AEC mode, the mAs value, but not mA or time.

5.5. X RAY TUBE AND GENERATOR RATINGS

5.5.1. X ray tube

The nominal voltage gives the maximum permissible tube voltage. For most tubes this will be 150 kV for radiography. For fluoroscopy, another nominal voltage might be specified. The nominal focus is a dimensionless figure characterizing the focal size (IEC336). For each nominal focus, a range of tolerated dimensions is given for the width and length of the focus, e.g. a nominal focus of 1.0 allows for a width of 1.0–1.4 mm and a length of 1.4–2.0 mm.

The power rating, P, for a given focus is the maximum permissible tube current for a 0.1 s exposure at a tube voltage of 100 kV. A more practical quantity is the power rating obtained with a thermal preload of the anode (thermal anode reference power) of typically 300 W (see also Fig. 5.11). P depends on the focal spot size and ranges from ~100 kW for 1.5 mm down to ~1 kW for 0.1 mm focus size.

Physical data for the anode include the target angle, anode material and diameter of the disc. The anode drive frequencies determine the rotational speed of the anode. High power loading of the anode requires high rotational speeds. To save the bearings from wear and damage, the speed is reduced for low power settings, as in fluoroscopy. The inherent filtration of the tube is given in equivalents of millimetres Al (see Section 5.6.2).

The heat capacity, Q, of the anode is the heat stored in the anode after arriving at the maximum permissible temperature. Q is equivalent to electrical energy, where $Q = wU_A I_A t$, with tube voltage U_A and current I_A, and the exposure time, t. U_A given as a peak voltage is multiplied with a waveform factor, w, to obtain the effective tube voltage (root mean square voltage). W has values of 0.71 for 1- and 2-pulse generators, 0.96 for 6-pulse generators and 0.99 for 12-pulse generators. Q is then given in joules (J). Since early generators were based on single-phase supplies, w was simply set to 1.0 for 1- and 2-pulse generators and 1.35 for 6-pulse generators, giving the heat capacity in another unit, the heat unit (HU), where 1 J = 1.4 HU. The heat capacity of general purpose tubes starts at ~200 kJ, ranging up to >1000 kJ for high performance tubes.

The maximum anode heat dissipation indicates the maximum rate of heat loss typically available at maximum anode temperature. These data depend on temperature and tube type. Tube data also include cooling and heating characteristics (Fig. 5.15).

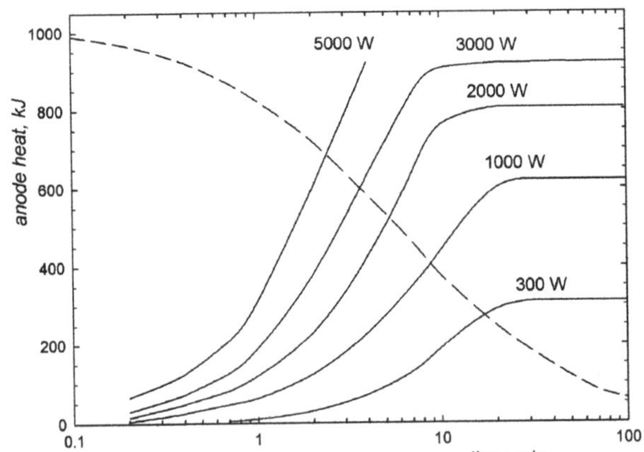

FIG. 5.15. Cooling of the anode (dashed curve) and heat buildup for several constant input heat rates. The rated heat capacity of the anode is 1000 kJ.

5.5.2. Tube housing

The maximum heat capacity for a tube assembly is typically in the range of 1000–2000 kJ. Maximum continuous heat dissipation describes the steady state of heat flowing in and cooling off. Figure 5.16 shows typical heating and cooling characteristics.

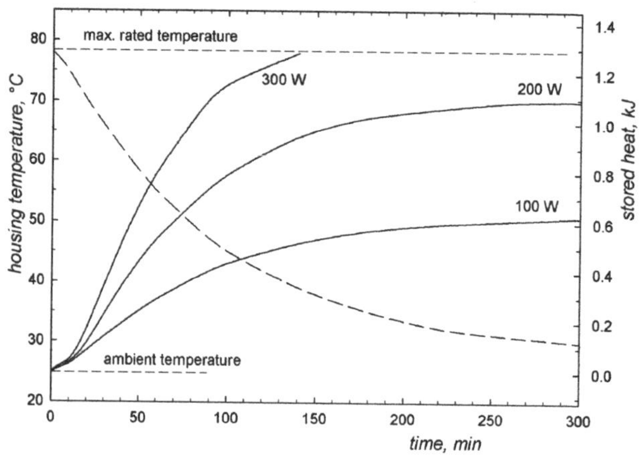

FIG. 5.16. Typical cooling characteristics of a passively cooled tube housing (dashed curve) and heating curves for a constant power input of 100, 200 and 300 W.

The patterns of loading the tube in an examination vary from single radiographic exposures to long high current CT scans, from simple fluoroscopic examinations to long interventional procedures. The tube rating charts contain basic data to estimate the required cooling times. These limits have to be observed, particularly if the control panel gives no indication on actual tube loading or required cooling times. Exposures made by physicists in their measurements can be repeated much more frequently than within the course of a patient examination, and several such high power exposures without observation of the appropriate cooling times can damage the anode and bearings.

5.6. COLLIMATION AND FILTRATION

5.6.1. Collimator and light field

The limitation of the X ray field to the size required for an examination is accomplished with collimators. The benefits of collimating the beam are twofold — reduction in patient dose and improvement of image contrast due to a reduction in scattered radiation. A collimator assembly is typically attached to the tube port, defining the field size with adjustable parallel opposed lead diaphragms or blades (Fig. 5.17). To improve the effectiveness of collimation, another set of blades might be installed at some distance from the first blades in the collimator housing. Visualization of the X ray field is achieved by a mirror reflecting the light from a bulb. The bulb position is adjusted so that the reflected light appears to have the same origin as the focal spot of the tube. The light field then 'mimics' the actual X ray field. The congruency of light and X ray field is subject to quality control. One must be aware that some of the penumbra at the edges of the radiation field is due to extra focal radiation.

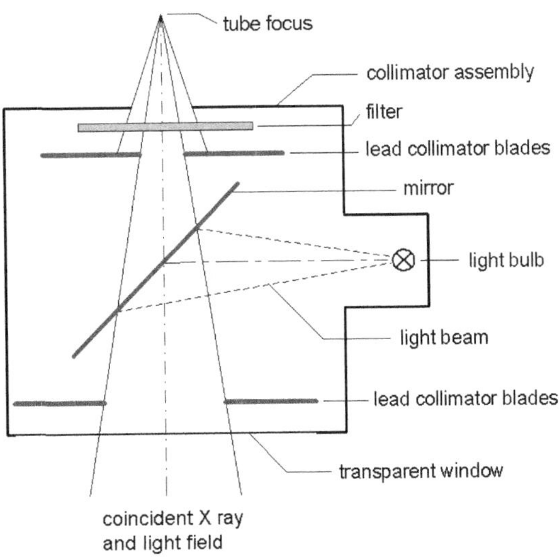

FIG. 5.17. Typical X ray field collimator assembly.

Adjustment of the field size is done manually by the operator, but with a positive beam limitation system, the size of the imaging detector is automatically registered and the field size is adjusted accordingly.

For fluoroscopy, other collimator types are in use, with variable circular and slit diaphragms. In some applications (dental and head examinations), beam restrictors with a fixed field size are typically used.

5.6.2. Inherent filtration

X rays generated in the anode pass various attenuating materials before leaving the tube housing. These materials include the anode, tube envelope exit port (glass or metal), insulating oil and the window of the tube housing. This inherent filtration is measured in aluminium equivalents (unit: mm Al). Aluminium does not perfectly mimic the atomic composition of the attenuating materials present, thus, measurement of the Al equivalent is usually made at 80 kVp (or otherwise the kVp settings should be stated). Typically, the inherent filtration ranges from 0.5 to 1 mm Al. The mirror and the window in the collimator housing also contribute to inherent filtration with an Al equivalent of about 1 mm.

5.6.3. Added filtration

Since filtration effectively reduces the low energy component in the X ray spectrum, a minimum total filtration of at least 2.5 mm Al is required to reduce unnecessary patient dose. Additional filter material is positioned between the tube window and collimation assembly as required. Typical filter materials include aluminium and copper, and in some cases, rare earth filters such as erbium that utilize K edge attenuation effects. Individual filters may be manually selected on some units. In modern fluoroscopy units, filters are inserted automatically, depending on the examination programme chosen.

The effect of added filtration on the X ray output is an increase in the mean photon energy and half value layer (HVL) (see Section 5.7.1) of the beam. As the X rays become more penetrating, less incident dose at the patient entrance is required to obtain the same dose at the image receptor, giving a patient dose reduction. Since image contrast is higher for low energy X rays, the addition of filters reduces image contrast and optimum conditions must be established, depending on the type of examination. Added filtration also increases tube loading, as the tube output is reduced and must be compensated for by an increase in mAs to obtain the image receptor dose required.

In mammography, special provisions concerning filtration are required to obtain the optimum radiation qualities.

5.6.4. Compensation filters

In some examinations, the range of X ray intensities incident upon the image receptor exceeds the capabilities of the detector. Compensation or equalization filters can be used to reduce the high intensities resulting from thinner body parts or regions of low attenuation. Such filters are usually inserted in the collimator assembly or close to the tube port. Examples of compensation filters include wedge filters for lateral projections of the cervical spine, or bowtie filters in CT.

5.7. FACTORS INFLUENCING X RAY SPECTRA AND OUTPUT

5.7.1. Quantities describing X ray output

Total photon fluence is not a satisfactory quantity to describe X ray output; rather, it is the spectral distribution of the photon fluence as a function of photon energy that is useful for research in X ray imaging. Spectral data are rarely available for individual X ray units, although computer programs exist that give useful simulations.

X ray tube output can be expressed in terms of the air kerma and measured free in air (see Chapter 22). A measure of the penetration and the quality of the X ray spectrum is the HVL. The HVL is the thickness of absorber needed to attenuate the X ray beam incident air kerma by a factor of two. In diagnostic radiology, aluminium is commonly chosen as the absorber, giving the HVL (unit: mm Al).

5.7.2. Tube voltage and current

Figure 5.18 shows the effect of tube voltage on spectral distribution. Both maximum and mean photon energy depend on the voltage (kV). The shape of the low energy end of the spectrum is determined by the anode angle and the total filtration. Note the appearance of characteristic radiation in the 100 kV beam and the increase in photon yield with increasing tube voltage. Tube current has no

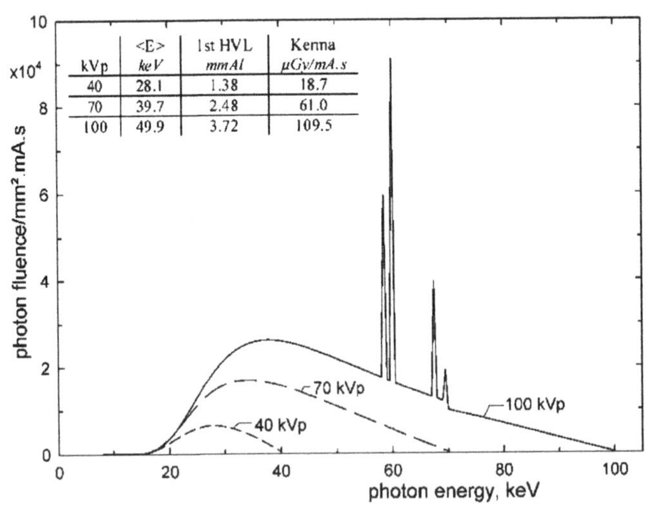

FIG. 5.18. X ray spectra for various tube voltages and a tungsten target (constant voltage, anode angle 12°).

influence on the photon distribution; however, photon intensities are proportional to mAs.

5.7.3. Tube voltage ripple

Figure 5.19 shows spectral variations for a tube voltage of 70 kV for various voltage ripples. A DC voltage gives the hardest spectrum with maximum photon yield. With an increase in ripple, the yield drops and the spectrum softens.

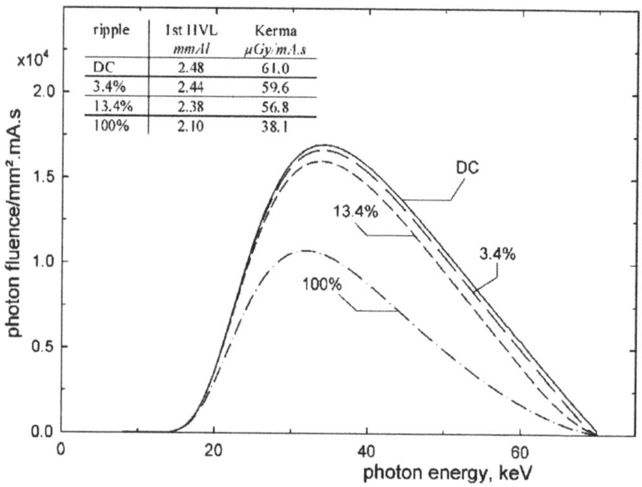

FIG. 5.19. Variation of X ray spectra from a tungsten target with tube voltage ripple at 70 kVp tube voltage. DC: constant potential; 3.4%: 12-pulse or converter generator; 13.4%: 6-pulse generator; 100%: 2-pulse generator.

5.7.4. Anode angle

The anode angle determines the degree of X ray absorption in the anode material. A decrease in anode angle causes an increase in the absorption length within the target. Accordingly, the maximum photon energy remains unchanged, but hardness increases and yield drops with decreasing anode angle (Fig. 5.20).

5.8. FILTRATION

As low energy photons do not contribute to the formation of an image, filters are used to reduce the low energy component. Figure 5.21 illustrates the effect of added filters on an X ray spectrum (90 kV, 3.4% ripple). Again,

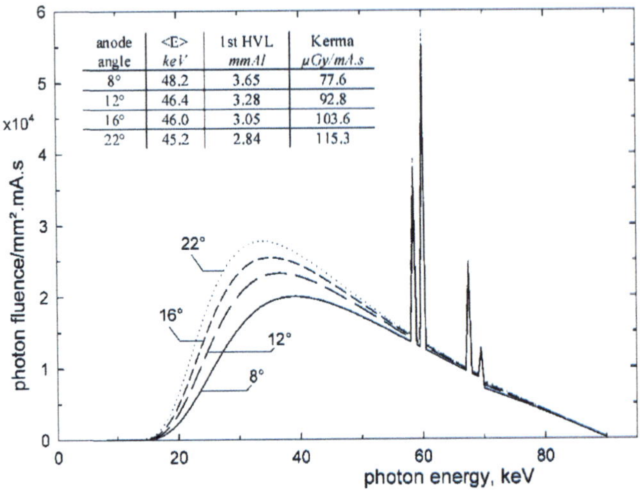

FIG. 5.20. X ray spectra obtained for various anode angles and a tube voltage of 90 kV (DC).

increasing filtration gives spectral hardening and reduction in tube output. X ray contrast declines with spectrum hardness, which should be considered in the selection of optimal exposure parameters.

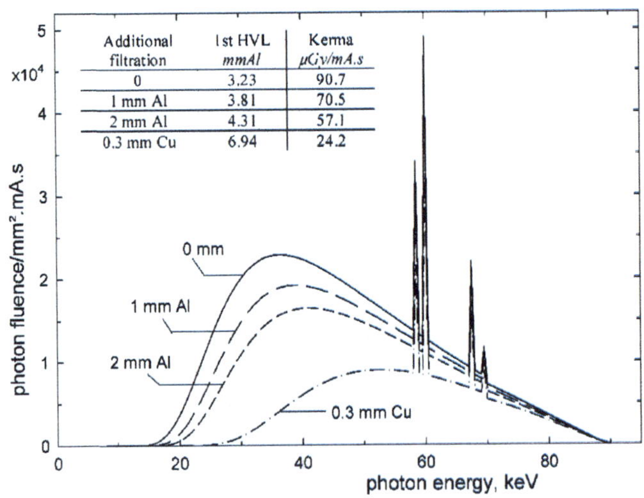

FIG. 5.21. Effect of additional filtration on the X ray spectrum.

Anode roughness increases with total tube workload and increases self-filtration. Hence, tubes tend to show a slight increase in X ray hardness and a decrease in kerma output over operational tube life.

REFERENCE

[5.1] BIRCH, R., MARSHALL, M., ARDRAN, G.M., Catalogue of Spectral Data for Diagnostic X rays, Rep. 30, Health Protection Agency, London (1979).

BIBLIOGRAPHY

BUSHBERG, J.T., SEIBERT, J.A., LEIDHOLDT, E.M.J., BOONE, J.M., The Essential Physics of Medical Imaging, 2nd edn, Lippincott, Williams & Wilkins, Baltimore, MD (2002).

CRANLEY, K., GILMORE, B.J., FOGARTY, G.W.A., DESPONDS: Catalogue of Diagnostic X ray Spectra and Other Data, IPEM Report 78, Institute of Physics and Engineering in Medicine, York (1997).

NOWOTNY, R., HÖFER, A., A computer code for the calculation of diagnostic X ray spectra, Fortschr. Geb. Roentgenstr. Nuklearmed. **142** 6 (1985) 685–689.

Chapter 6

PROJECTION RADIOGRAPHY

J.L. POLETTI
UNITEC Institute of Technology,
Auckland, New Zealand

6.1. INTRODUCTION

In its simplest form, X ray imaging is the collection of attenuation shadows that are projected from an ideal X ray point source on to an image receptor. This simple form is true for all X ray imaging modalities, including complex ones that involve source and receptor movement, such as computed tomography (CT). This simplified view, however, is made vastly more complex by the non-ideal point source, by the consequences of projecting a 3-D object on to a 2-D detector and by the presence of scattered radiation, generated within the patient, which will degrade any image that is captured.

6.2. X RAY IMAGE FORMATION

6.2.1. Components of an imaging system

The principal components of a system for X ray projection radiography are illustrated in Fig. 6.1. Of these components, the grid and the automatic exposure control (AEC) are optional, depending on the imaging task. Further components such as shaped filtration, compression devices or restraining devices may be added for special cases. The X ray tube and collimation device are described in Chapter 5 and the image receptor systems are described in Chapter 7.

When considering such systems, the concept of an ideal imaging task is often useful, as illustrated in Fig. 6.1. These concepts are covered in detail in Chapter 4 and are discussed only briefly here. When considering the ideal imaging task — the detection of a detail against a uniform background — the ideal X ray spectrum is monochromatic when the three constraints of patient dose, image quality and X ray tube loading are considered. Any particular projection may consist of more than one such task, each with a different ideal monochromatic energy. The choice of X ray spectrum for each task is, therefore, always a compromise, so that the actual bremsstrahlung and characteristic radiation spectrum provide the best approximation to the ideal monochromatic spectrum for the particular projection.

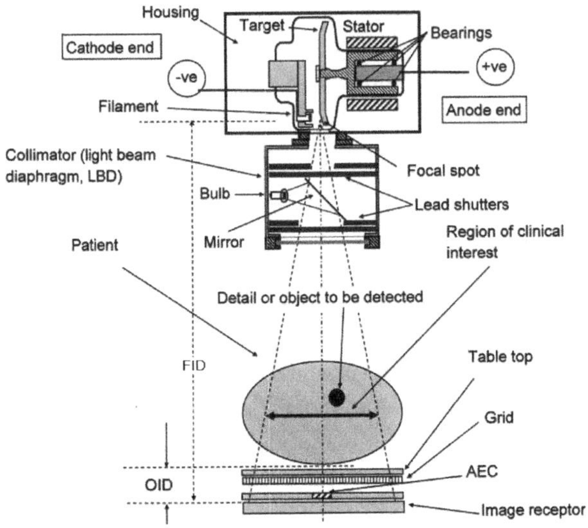

FIG. 6.1. Components of a projection radiography system, including an ideal imaging task, for the detection of a detail against a background. FID: focus to image distance; OID: object to image distance.

Considering an ideal imaging task, as illustrated in Fig. 6.1, contrast may be defined simply as $C = \Delta B/B$, where B is the image brightness (or shade of grey) in a background region and ΔB is the difference in brightness for a small detail. For small values of ΔB, linearity of brightness with X ray intensity (I) is assumed, so the contrast is $\Delta I/I$. This is generally valid for a particular monochromatic spectrum. For a real polychromatic spectrum, a monochromatic spectrum with the average energy of the actual spectrum may be used as an approximation, or the result can be integrated over all spectral energies. Since the X ray intensity is related to thickness by the attenuation law, it follows that the primary contrast for a detail of thickness x_d and linear attenuation coefficient μ_d embedded in a material of linear attenuation coefficient μ_b is given by:

$$C_P = 1 - e^{-(\mu_d - \mu_b)x_d} \tag{6.1}$$

To find the average contrast for a particular detail, Eq. (6.1) should be integrated over the detail. For thin spherical details (e.g. microcalcifications in mammography, solitary pulmonary nodules in chest X rays), this is straightforward and results in a contrast that is 2/3 the contrast obtained for a ray passing through the centre of the detail.

With regard to the relationship between the linear attenuation coefficient and the mass attenuation coefficient (see Section 2.3.3), contrast will exist for details that differ in mass attenuation coefficient, or in density, or both. The contrast will depend on the thickness of the detail, but not the thickness of surrounding tissue. Since the values of μ reduce as photon energy increases, the contrast is seen to be inversely related to the kV setting. Thus, kV may be considered to be the contrast control, where contrast is strictly the detail contrast. For screen film imaging, the difference in optical density (OD) due to the detail is proportional to the subject contrast multiplied by the gamma of the screen film system (see Section 7.3.4). For digital image receptors, the relationship is more complex, since the contrast of the displayed image is independently adjustable.

If the energy absorbed in a small region of the image receptor due to primary rays is E_p, and that due to secondary rays is E_s, then scatter may be quantified by the scatter fraction:

$$\mathrm{SF} = E_s/(E_p + E_s) \tag{6.2}$$

or by the scatter to primary ratio:

$$\mathrm{SPR} = E_s/E_p \tag{6.3}$$

The relationship between the two is:

$$\mathrm{SF} = ((\mathrm{SPR}^{-1}) + 1)^{-1} \tag{6.4}$$

In the presence of scattered radiation, Eq. (6.1) becomes:

$$C_\mathrm{P} = 1 - \mathrm{e}^{-(\mu_\mathrm{d} - \mu_\mathrm{b})x_\mathrm{d}} \frac{1}{1+\mathrm{SPR}} \tag{6.5}$$

Clearly, minimization of scatter is important, leading to the use of antiscatter techniques (see Sections 6.3.4 and 6.3.5). Accurate collimation to the region of clinical interest also minimizes scatter, as well as reducing the dose to the patient.

6.2.2. Geometry of projection radiography

From Fig. 6.1, it is clear that the primary effect of projection radiography is to record an image of a 3-D object (the patient) in 2-D, resulting in superposition of the anatomy along each ray. This leads to a number of effects that need to be considered in the design of equipment, the production of the images and their interpretation. In particular, for each projection there will be a region of clinical interest, somewhere between the entrance and exit surface of the region to be imaged. Considerable training and experience is required for the radiographer to choose correctly the geometrical variables to image this region, based on superficial visible or palpable landmarks. These variables include the FID, OID, projection direction (lateral, craniocaudal, etc.) or angulation, centring point and beam collimation area. In some cases, the correct projection of joint spaces also needs to be considered.

6.2.3. Effects of projection geometry

6.2.3.1. Superposition

As noted in Section 6.2.2, radiographs are a 2-D representation of a 3-D object. This superposition leads to a significant loss of image contrast, which provided one of the prime motivations for the development of CT scanners. Superposition also leads to loss of all depth information, and ambiguity in the relative sizes of objects at different depths. Furthermore, it directly overlays objects in such a way that it can become difficult or impossible to distinguish one from the other, or even to identify some of the objects.

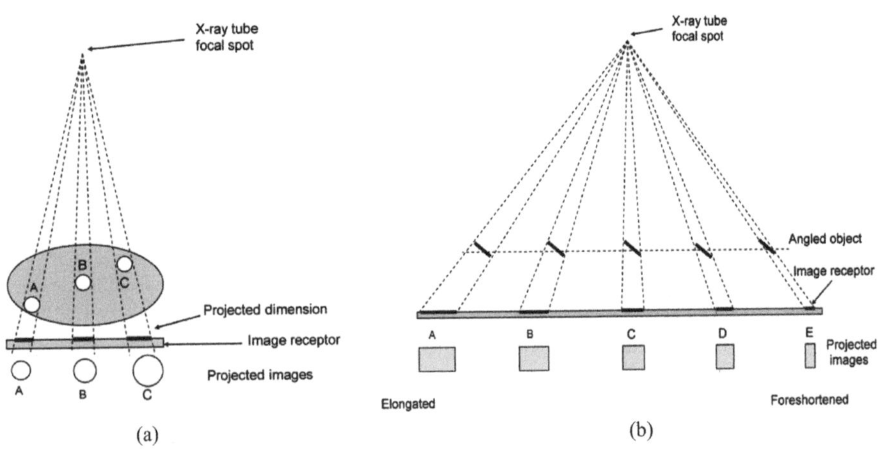

FIG. 6.2. (a) Effect of depth of objects on their projected size; (b) effect of angulation on the projected length of an angled object.

6.2.3.2. *Geometrical distortion*

Geometrical distortion can be considerable and confusing in projection radiographs. The first effect is that all objects are magnified in the image. The further from the image receptor the object is placed, the greater the OID and the greater the magnification. The image size of objects, therefore, depends on their actual size and on the OID and projection direction, leading to ambiguity. This effect is illustrated in Fig. 6.2(a). The three spheres A, B and C are the same size, but are projected at different sizes owing to their OIDs. Furthermore, projection leads to shape distortion. In Fig. 6.2(b), a tilted object is shown projected at a range of angles, illustrating the increasing degree of foreshortening as the angle increases.

6.2.3.3. *Inverse square law*

For an isotropic point source, the X ray beam intensity is inversely proportional to the square of the distance from the source. An X ray tube with its attached collimator is a good approximation to a point source for distances greater than about 50 cm from the focal spot, and obeys the inverse square law (ISL) almost exactly at distances greater than this. Only at low kV settings, such as those typical of mammography, does air attenuation affect the inverse square relationship. This is illustrated in Fig. 6.3, where the air kerma per unit mAs is shown over the FID range of 50–250 cm. Figure 6.3 also presents the calculated curve assuming the ISL.

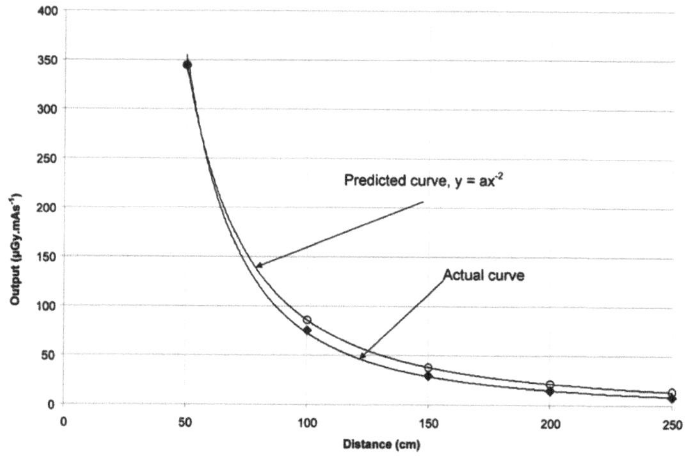

FIG. 6.3. Deviation from the ISL due to air attenuation for a tungsten target X ray beam with 0.5 mm Al added filtration at a voltage setting of 30 kV and no compression paddle.

The ISL results in the need for an increase in the mAs as the FID is increased in order to maintain the same air kerma at the image plane. The increase required is given by:

$$\frac{\text{mAs}_2}{\text{mAs}_1} = \left(\frac{d_{\text{FID}_2}}{d_{\text{FID}_1}} \right)^2 \tag{6.6}$$

Furthermore, the air kerma at the patient entrance surface is greater than that at the image receptor (neglecting attenuation), by the ratio:

$$\left(\frac{d_{\text{FID}}}{d_{\text{FSD}}} \right)^2 \tag{6.7}$$

In these expressions, d_{FID} is the FID and d_{FSD} is the focus to skin distance (FSD). It is easy to show that as the FID is increased, the incident air kerma (K_{i}) may be decreased, keeping the same kerma at the image plane; the formula for this is:

$$K_{\text{i}_2} = K_{\text{i}_1} \left(\frac{d_{\text{FID}_2}}{d_{\text{FID}_1}} \frac{d_{\text{FSD}_1}}{d_{\text{FSD}_2}} \right)^2 \tag{6.8}$$

This relationship can be used to prevent excessive skin doses; generally, an FID of 100 cm or greater is sufficient. It does not, however, result in a similar reduction in the overall dose to the patient because an increase in the entrance surface X ray beam size is required as the FID is increased in order to prevent cut-off of the region of clinical interest. The effective dose is approximately proportional to the dose–area product. The dose reduces at longer FID according to Eq. (6.8), but the area increases. Therefore, there is little or no change in effective dose [6.1].

6.2.3.4. *Geometrical unsharpness*

Ideal image sharpness would be produced by a point source, the spatial resolution in such a case being limited by the image receptor factors such as phosphor layer thickness, lateral spread of light in scintillators, and the image matrix. However, owing to the restriction on the permissible temperature of the focal spot and the focal track of the anode, typical focal spot sizes of 0.6–2.0 mm are required. Most X ray tubes also have a fine focal spot for high resolution images of small body parts; typically, the fine focal spots are 0.3–1.0 mm, but must be operated at lower mAs to protect the X ray tube from heating effects.

The spatial resolution depends on the focal spot size and the image receptor, and both need to be considered. For the demagnified image, the width of the penumbra, or more correctly the edge gradient, caused by a focal spot of size X_F, is given by the geometric unsharpness (U_g) divided by the magnification, where U_g is given by:

$$U_g = X_F \frac{d_{OID}}{d_{FID}} \tag{6.9}$$

where d_{OID} is the OID. Since the magnification, m, of the object at the image receptor is given by:

$$m = \frac{d_{FID}}{d_{FID} - d_{OID}} \tag{6.10}$$

then Eq. (6.9) is equivalent to:

$$U_g = X_F\ (m-1)/m \tag{6.11}$$

If the FID were to be changed, then to maintain the same focal spot resolution, the new focal spot size may be determined using Eq. (6.9) for the old and new cases and equating. This gives:

$$X_{F_{new}} = X_{F_{old}} \frac{d_{FID_{new}}}{d_{FID_{old}}} \tag{6.12}$$

However, the change in FID will change the magnification, which will affect the overall image sharpness because of the effect of the image receptor blur. The overall unsharpness is given by Dance et al. [6.2] as:

$$U = F\left[\frac{1}{m^2} + \frac{\left(1-\frac{1}{m^2}\right)FS^2}{F^2}\right]^{1/2} \qquad U = U_r\sqrt{\left(\frac{1}{m^2} + \left(1-\frac{1}{m}\right)^2\left(\frac{X_F}{U_r}\right)^2\right)} \tag{6.13}$$

In this expression, U_r is the intrinsic image receptor unsharpness (that for $m = 1$) and it is assumed that the geometric and receptor unsharpness can be added in quadrature. The overall unsharpness U is scaled to a magnification of 1.

Optimization of projection radiographs involves choosing an appropriate focal spot size. This requires a compromise between the exposure time and the resolution. For example, a very small focal spot will provide good spatial resolution, but only permit a low tube current, therefore requiring a long exposure time, leading to increased risk of motion blur. While it may be considered that quantum noise limits the detectability of fine details, there is some evidence that smaller focal spots than are currently employed may lead to improved spatial resolution. This is because the system detective quantum efficiency (see Chapter 4) is affected by the focal spot modulation transfer function (MTF).

The focal spot MTF may be measured using a pinhole to determine the point spread function, or a slit to determine the line spread function, and calculating the normalized modulus of the Fourier transform of the spread function. Figure 6.4 shows a pinhole image of a focal spot and a 2-D representation of the MTF.

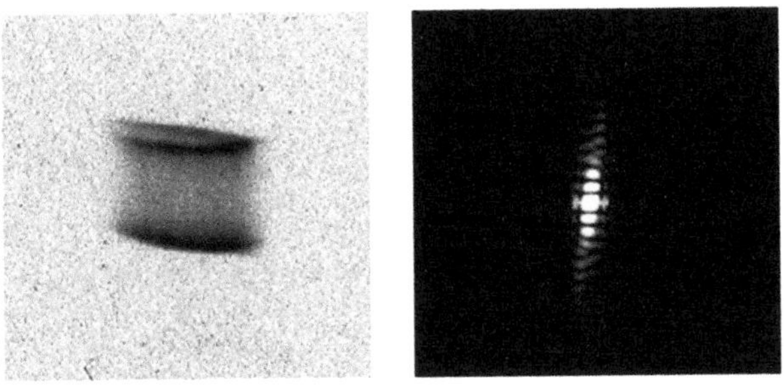

FIG. 6.4. Typical distribution of the X ray intensity of a 2.0 mm focal spot (left) and the corresponding 2-D MTF (right).

Note that the MTF of a focal spot is given by convention for a magnification factor of 2.0. To correct the MTF for the true magnification, the frequency axis must be scaled as follows (where the symbols have the obvious meanings):

$$f_{\mathrm{new}} = f_{\mathrm{old}} \left(\frac{m_{\mathrm{old}} - 1}{m_{\mathrm{old}}} \times \frac{m_{\mathrm{new}}}{m_{\mathrm{new}} - 1} \right) \tag{6.14}$$

6.2.4. Magnification imaging

Magnification is a relatively uncommon technique. Probably the most important example of its use is in magnification mammography (see Section 9.3.5). However, there are instances when significant magnification is unavoidably present in standard radiographic projections. These include the lateral hip and lateral cervical spine projections. Magnification is achieved by increasing the OID, which generally requires an increase in the FID as well. The actual magnification achieved varies with depth in the patient. For example, if the patient thickness is 20 cm, the FID 140 cm and the FSD 80 cm, then the magnification varies between 1.4 at the exit side of the patient and 1.75 at the entrance side. Magnification requires employment of a larger image receptor. For large body regions this may not be possible. The use of magnification has consequences for dose, spatial resolution and signal to noise ratio (SNR).

6.2.4.1. Dose

A number of effects occur when increasing the OID. There is a substantial reduction in the scatter fraction at the image receptor because the scattered rays are generally directed away from the receptor. To maintain the dose to the image receptor, an increase in the mAs, and hence the patient dose, would be required, mainly because of the loss of scatter but also because of the increase in FID owing to the ISL. Owing to the reduction in scatter fraction, magnification may usually be performed without the use of an antiscatter grid. This leads to a reduction in mAs in proportion to the Bucky factor, which is the ratio mAs with a scatter reduction method divided by mAs without a scatter reduction method (see Sections 6.3.4 and 6.3.5). This factor is typically between about three and six.

6.2.4.2. Unsharpness

Any increase in the OID leads to a reduction in image sharpness due to the geometric blur of the focal spot, as given by Eqs (6.9, 6.11). Consequently, use of magnification techniques requires a significant reduction in focal spot size, compared with contact methods. Any improvement in the overall sharpness of the complete system is generally because of the increase in size of the image compared with the unsharpness of the image receptor, owing to effects such as light spread for screen film systems and the pixel size for digital systems. Magnification can, therefore, improve spatial resolution, compared with the result of a simple zoom of a digital image, which enlarges the pixels as well as the image.

6.2.5. Contrast agents

The subject contrast for many X ray examinations is low, owing to the similarity in the atomic number and density of soft tissues and blood. The contrast of organs and of the circulatory system may be substantially increased with the use of higher atomic number contrast agents. These generally employ barium compounds for study of the gastrointestinal tract, and iodine compounds for soft tissues and the circulatory system. These two elements have considerable photoelectric attenuation because their K edges are in the diagnostic X ray energy range — iodine at 33 keV and barium at 37 keV. The maximum contrast that can be achieved will occur for photon energies just above the K edge of these elements (see Section 2.2.1). This, in turn, requires the choice of a kV setting that produces exit spectra with the majority of photon energies within the appropriate region of the spectrum. Optimum kV settings are between 60 and 70 kV for iodine contrast and up to 80 kV for barium. Examinations of the gastrointestinal tract sometimes employ air as well as barium.

It should be noted that there is a small but unavoidable incidence of adverse reactions to contrast media, which are generally minor but occasionally serious or even fatal.

Recently, targeted contrast agents have been developed based on gold nanoparticles that have superior contrast enhancement to traditional iodine based agents, with minimal toxicity and negligible negative reactions.

6.2.6. Dual energy imaging

An alternative and sometimes an additional method of improving image contrast is with the use of two quasi-simultaneous images of the body using different X ray spectra and processing them into separate images, one reflecting the photoelectric process and one the Compton effect. By combining these images using specific weightings, differences between bone tissue and soft tissue or between air and soft tissue can be displayed (details can be found in Section 10.4 on dual energy X ray absorptiometry). In order to make these separations, X ray images acquired at different tube voltages and/or filtrations are required.

6.2.7. Technique selection

With screen film systems, technique selection is relatively straightforward. The choice of kV setting is based largely on the required contrast, and the mAs is then chosen to produce a suitable OD for the region of clinical interest, generally about 1.0 net OD. With digital systems, the direct link between technique setting and image appearance has been lost, making correct technique selection much more difficult.

6.2.7.1. *Effect of tube voltage on contrast, noise and dose*

To determine whether a detail will be detectable in the image, noise must be considered. The primary source of noise is generally the random arrival of photons at the image receptor, which may be considered as a Poisson process. From Rose's expression [6.3], the number of detected photons required per unit area, to image a detail of size d and contrast C with an SNR of k, is:

$$N = k^2/C^2d^2 \tag{6.15}$$

The value of k required to be certain that an observed detail is real and not due to chance fluctuations in the number of photons is often taken to be five. Thus, as C is increased, the number of photons required at the image receptor is reduced, so that a reduction in kV will produce an image of satisfactory quality at a lower image receptor dose, provided that the contrast range does not exceed the dynamic range of the image receptor. However, this reduction in kV will require an increase in the mAs, leading to an increase in patient dose. The dose to the image receptor depends approximately on kV, and is linear with mAs. The patient dose (K_i) is proportional to mAs and approximately to kV^2. The overall effect on patient dose, therefore, is approximately proportional to kV^{-3}.

For example, consider a setting of 60 kV at 40 mAs. Using the 15% rule (see Section 6.2.7.3), this could be changed to 69 kV at 20 mAs. The patient dose will then be reduced to $(69/60)^{-3}$ = 66%. However, the increase in kV will result in a reduction in the contrast to noise ratio (CNR), which may be acceptable, in which case a worthwhile reduction in dose will have been achieved. If the image receptor dose is considered to be a variable, then there is a wide range of kV and mAs combinations that will produce a diagnostically acceptable image, but at a wide range of patient dose levels. In order to manage digital imaging systems, suitable levels of image receptor dose have been determined by all manufacturers of such systems, expressed in a variety of proprietary exposure indices to represent the dose to the image receptor. Generally, there will be a selection of indices suitable for imaging the extremities, trunk and chest. For CT, these correspond approximately to screen film system speeds[1] of 100, 200 and 400, respectively. Direct and indirect digital systems are somewhat faster,

---

[1] Speed is defined as the inverse of the exposure required to produce a film OD of 1.0 above base + fog. A speed in the range of 100–1000 is typical for most radiographic procedures.

allowing higher nominal speeds. For all digital systems, the choice of suitable kV and mAs combinations requires that for each projection, the kV and mAs produce the correct value of the exposure index and that the maximum value of the kV is chosen that will allow diagnostically acceptable CNR. This is readily demonstrated in practice. If a suitable phantom is radiographed at a low kV and with suitable mAs, and a range of further images is obtained at increased kV settings at the same mAs, the images will appear very similar. This is because the reduction in contrast with increasing kV is matched by the increased number of photons detected, resulting in a similar CNR for each image. Each increase in kV will cause an increase in patient dose by kV^2, so such a procedure is clearly clinically unacceptable. If, instead, each kV increase is accompanied by a reduction in mAs to maintain the image receptor dose and exposure index, then the image quality will become steadily worse as kV is increased, until a point is reached at which the image quality is no longer acceptable. A corollary to this is that images exposed with a kV below the optimum level and an mAs above the optimum level look better, leading to the phenomenon of 'exposure creep'.

6.2.7.2. Matching technique to study

Given the procedure established in the previous section (6.2.7.1), the choice of suitable kV setting for any body region involves two steps. The first is to choose a suitable image receptor dose and speed to produce acceptable levels of image noise. For example, a wrist X ray may require a nominal speed of 100, whereas a posterior–anterior chest may permit a speed of 400. Regions of low subject contrast such as the abdomen then require a relatively low kV setting, whereas regions of high contrast such as the chest require a high kV setting. Guideline kV settings are widely available, such as those given in the quality criteria documents of the European Union. The kV setting should then be increased gradually, with appropriate mAs reduction to maintain the CNR or film OD, until the loss of image quality is just tolerable.

For screen film imaging, this also requires matching the dynamic range of the image receptor system to the range of the input signal. This is illustrated in Figs 6.5 and 6.6 for the two extreme cases of chest radiography, which has high subject contrast with a wide latitude image receptor system, and mammography, which features low subject contrast with a narrow latitude image receptor system.

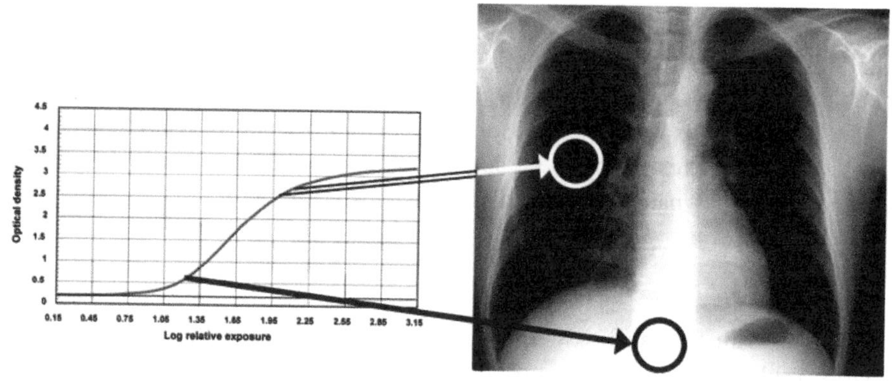

FIG. 6.5. Matching of the kV setting and mAs to the dynamic range of the image receptor for a study of a high contrast region of the body (a chest X ray).

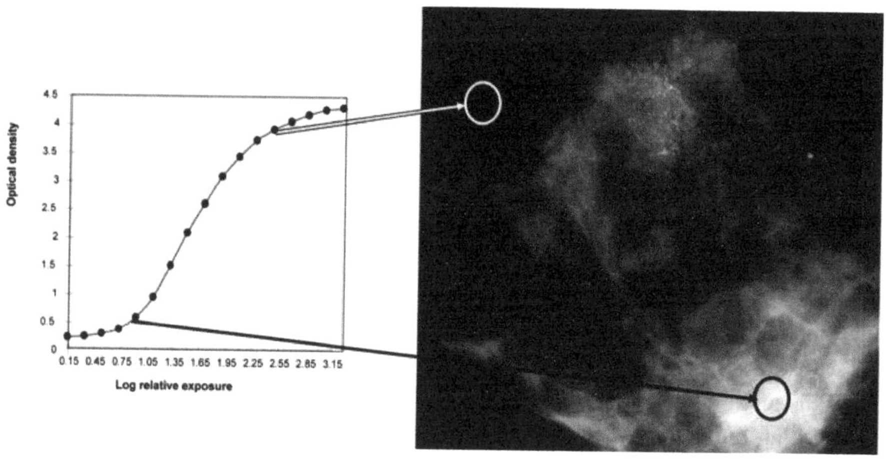

FIG. 6.6. Matching of the kV setting and mAs to the dynamic range of the image receptor for a study of a low contrast region of the body (a mammogram).

6.2.7.3. *Relationship between kV and mAs*

Given that the image receptor dose is proportional to mAs and to kV, some simple exposure rules may be derived. Firstly, it is observed that an increase in kV of 15% results in an increase in image receptor dose by a factor of two — hence the so-called '15% rule', that an increase in kV of 15% is equivalent to a doubling of the mAs and a reduction by 15% is equivalent to halving the mAs. Furthermore, an increase in kV of 5% results in an increase in image receptor dose of 30%, leading to the '5% rule' that a 5% increase in kV is equivalent to a 30% increase in mAs and a reduction of 5% in kV is equivalent to a reduction in mAs by 30%. Finally, since a 15% increase in kV is about 10 kV between 60 and 80 kV, another commonly used rule is that a 10 kV increase is equivalent to doubling the mAs, and a 10 kV reduction is equivalent to halving the mAs. None of these rules are exact, but their use is satisfactory because of the tolerance for small exposure errors owing to the latitude of screen film systems, and because of the very wide dynamic range of digital systems.

6.2.7.4. AEC

Even with the most skilled of practitioners, manual setting of technique factors results in inconsistent exposures, so that ODs vary in screen film imaging and image noise levels vary with digital systems. In addition, a number of rejects and repeats are unavoidable, because of exposure errors. AEC systems are intended to increase exposure consistency and reduce reject and repeat rates. The principle is to measure the X ray flux at the image receptor and to terminate the exposure when sufficient energy has been absorbed (see Section 5.4.3).

However, the advantages of AEC systems may be achieved only if the systems are correctly calibrated and properly used. Calibration is required for a number of reasons, including energy dependence and beam hardening. Energy dependence is due to the varying sensitivity of the AEC detectors and the image receptor system at different kV settings. Correction factors are included in the control system to allow for kV setting. The beam hardening caused by the patient is more difficult. The system is not able to measure the amount of beam hardening and hence cannot correct for beam hardening errors. Therefore, AEC systems include controls for manual correction by the radiographer. These generally include compensation settings such as –3, –2, –1, 0, +1, +2 and +3; each setting increasing the mAs delivered by a constant factor such as $\sqrt{2}$. There is also a patient size button, with settings for a thin patient, an average patient and a large patient.

6.3. SCATTERED RADIATION IN PROJECTION RADIOGRAPHY

It is often stated that photon scattering is of no benefit for projection radiography, leading only to fogging of the image. However, this is incorrect, as the appropriate contrast for every projection is chosen by setting a suitable kV to provide the correct proportion of photoelectric and scatter interactions. At low kV settings, the contrast is high, owing to the predominance of the photoelectric effect, while at high kV it is low, owing to the predominance of scattering interactions (see Chapter 2). These effects are illustrated in Fig. 6.7, which shows the contribution of photoelectric and scattering interactions for the primary contrast of a 1 mm sphere of calcium embedded in water, over a range of energies from 20 keV to 100 keV, calculated using Eq. (6.1).

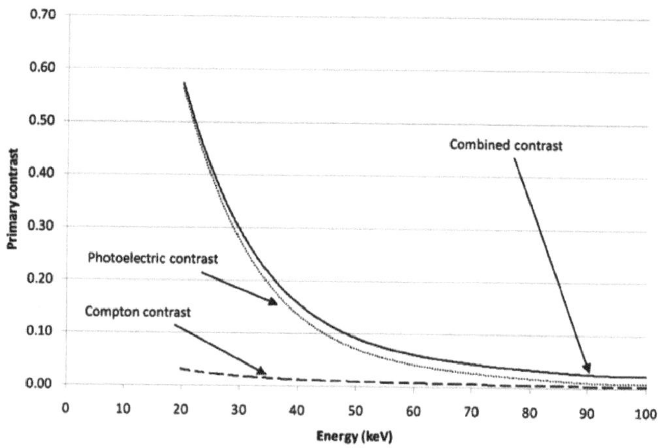

FIG. 6.7. Calculated primary contrast for a 1 mm sphere of calcium embedded in water, from 20 keV to 100 keV.

6.3.1. Origins of scattered radiation

Both the incoherent and coherent interactions lead to scattered radiation impinging on the image receptor. The former is by far the more significant. For example, for a 10 cm patient thickness and a 20 cm × 20 cm X ray field, only 19% of the scatter fraction is due to coherent scatter at 50 kV and 9% at 120 kV.

6.3.2. Magnitude of scatter

The magnitude of the scatter depends on many variables. The dependences on radiographic procedure, X ray beam size, patient thickness and position in three dimensions are described in the following sections. The magnitude is also widely claimed to depend on the kV setting, but as shown in Section 6.3.2.5, this is only the case for very inefficient image receptors.

6.3.2.1. Dependence upon radiographic procedure

The radiographic procedure itself has a strong influence on the proportion of scatter, depending on whether the subject is a region consisting largely of bone, or soft tissue, or some intermediate combination. This is because the scatter interactions are predominant for soft tissue, but the photoelectric interaction is predominant for bone over much of the diagnostic energy range (except mammography).

6.3.2.2. Dependence on field size

As the field size increases from a narrow beam with almost no scatter to very large, the scatter fraction increases until a saturation level is reached, beyond which little or no further increase in scatter fraction occurs. Figure 6.8(a) shows this effect for phantom thicknesses of 5, 10, 20 and 30 cm.

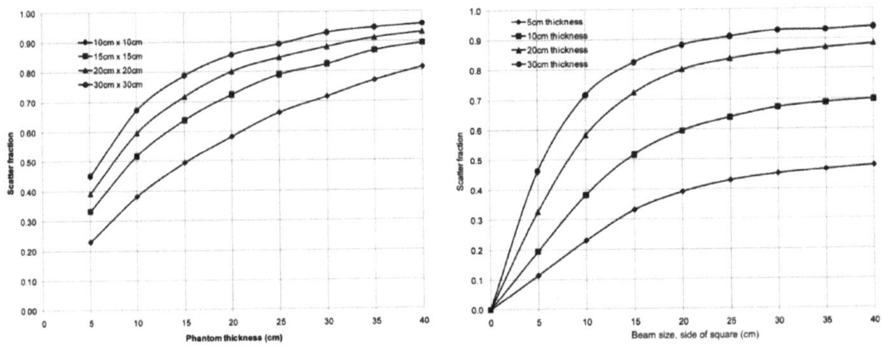

(a) (b)

FIG. 6.8. (a) Dependence of scatter fraction on beam area for four phantom thicknesses at 80 kV. (b) Dependence of scatter fraction on phantom thickness for four X ray field sizes at 80 kV. Both parts of the figure show the scatter fraction at the centre of image.

6.3.2.3. Dependence on thickness

Figure 6.8(b) shows the effect of tissue thickness on scatter fraction for tissue thicknesses ranging from 5 cm to 30 cm and four X ray beam field sizes. The scatter fraction increases rapidly with patient thickness, but tends to saturate for very large patients. These data also demonstrate the necessity for scatter reduction methods, especially for large patients.

6.3.2.4. Dependence on position (in 3-D)

While the scatter fraction (at the image receptor) has been considered so far, it is also important to quantify the scattered radiation in all directions from the patient, as this affects the dose to personnel. Knowledge of scatter levels is required in order to determine appropriate radiation shielding levels. It is also useful to consider the proportion of backscatter at the patient's entrance surface, as this may contribute significantly to the skin dose and complicates measurements of patient dose when free in air measurements are used. Scatter may be categorized as forward scatter, affecting mainly the image receptor, and side scatter and back scatter, affecting the dose to personnel.

If a small volume of soft tissue is considered, and we make the simplifying assumption that the scatter angular distribution is given by the Klein–Nishina formula (Eq. (2.16)), the proportions of forward scatter and back scatter can be seen to be similar for photons in the diagnostic energy range, with the proportion of forward scatter increasing as the photon energy is increased. At 90°, the differential cross-section $d\sigma/d\Omega$ is approximately half that at 0° and at 180°. However, for the large volumes of tissue typical in projection radiography, there is considerable attenuation, so that back scatter can be significant. For larger body regions, the proportion of photons that undergo more than one scatter is significant, leading to a more isotropic scatter distribution than predicted by the Klein–Nishina formula.

It is found experimentally that, to a close approximation, the magnitude of scattered radiation follows the ISL at distances greater than about 500 cm from the patient, and that the magnitude is directly proportional to the X ray field size. A rule of thumb for scatter levels is that the scatter dose at 1 m from the patient is 1/1000 of the dose in the primary beam at the patient entrance surface.

With respect to position in the image, the proportion of scattered radiation is found to be greatest in the centre, reduced at the edges and reduced further still at the corners of the field. This is illustrated in Table 6.1, which shows the scatter fraction (SF) and scatter to primary ratio (SPR) at intervals of 2.5 cm along the axis and diagonal of a 30 cm × 30 cm X ray field.

6.3.2.5. Dependence on energy

In soft tissue, as the energy is increased, the photoelectric cross-section reduces approximately as $1/E^3$, whereas the scattering cross-section reduces as $1/E$. Hence, the probability of scattering events increases relative to photoelectric events as the energy increases, but the overall probability of scattering events decreases. Conversely, the energy of the scattered photons increases as the energy increases, so that they are less likely to be attenuated in the body and are more likely to escape. The overall effect is that side scatter, back scatter and forward scatter all increase as energy is increased. However, the primary beam attenuation also decreases as energy is increased. Consequently, the scatter fraction reaching the detector is found to show little dependence on energy. In the case of efficient image receptors, almost all of the primary beam and scatter are absorbed, so the scatter fraction for the receptor is similar to that reaching the receptor. However, inefficient image receptors are more sensitive to scattered photons than to

TABLE 6.1. DEPENDENCE OF SF AND SPR ON POSITION IN X RAY FIELD
(30 cm × 30 cm PMMA phantom, 20 cm thick, at 80 kV, 3 mm Al equivalent, total filtration, 100 cm FID and 5 cm OID, 30 cm × 30 cm X ray field at image receptor)

Distance from centre (cm)	On axis		Diagonal	
	SF	SPR	SF	SPR
0	0.856	5.93	0.856	5.93
2.5	0.854	5.84	0.854	5.87
5.0	0.845	5.47	0.848	5.56
7.5	0.831	4.92	0.834	5.04
10.0	0.806	4.17	0.824	4.68
12.5	0.781	3.57	0.794	3.85
15.0	—	—	0.750	3.00
17.5	—	—	0.697	2.30
20.0	—	—	0.671	2.04

primary photons. This is because the primary beam is incident approximately perpendicular to the receptor, so the path length is similar to the phosphor thickness. However, the scattered photons are incident obliquely, resulting in greater path length on average and a greater probability of absorption. They may also be of lower energy, which will generally also increase the probability of absorption. This effect becomes more significant as energy is increased, so the detected scatter fraction also increases with energy. Figure 6.9 shows scatter fractions for a Kodak Lanex Regular and a calcium tungstate ($CaWO_4$) par speed screen over a range of kV settings from 50 kV to 120 kV. The scatter fraction for the Kodak Lanex Regular, an efficient rare earth screen, is independent of kV setting, whereas the scatter fraction increases with energy for the (obsolete) par speed $CaWO_4$ screen.

6.3.3. Effect of scatter

Scattered radiation, as shown in the previous sections, comprises the majority of the radiation detected by image receptor systems for examination of large body regions. This section considers the effect of scatter on contrast and noise, and methods of scatter reduction.

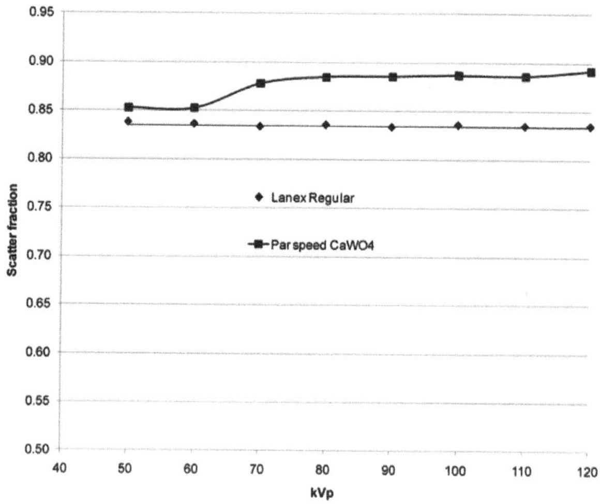

FIG. 6.9. Dependence of scatter fraction on kV setting for two systems for 20 cm phantom thickness and 30 cm × 30 cm field size for the Kodak Lanex Regular, an efficient rare earth system, and par speed calcium tungstate, an inefficient system.

6.3.3.1. Contrast reduction

The effect of scatter on contrast is quantified by the contrast degradation factor (CDF), which, from Eq. (6.5), is given by:

$$\text{CDF} = \frac{1}{1+\text{SPR}} \tag{6.16}$$

6.3.3.2. Noise

The quantum noise in the image arises from both primary and scattered photons and both of these image contributions, therefore, affect the SNR. For example, Ullman et al. [6.4] have calculated the SPR and SNR per pixel (SNR_p) for digital chest imaging with a grid, and show that the SPR varies from about 2 in the region of the mediastinum, to about 0.6 in the lung fields, leading to a value of SNR_p behind the mediastinum of about 20, and about 60 in the lung fields.

6.3.4. Methods of scatter reduction — antiscatter grids

The scatter component of the image may be considered to consist of the primary image convolved with a 'scatter spread function', which gives a highly blurred version of the image. The resulting image may be considered to be the sum of these two images. Efforts are being made to employ this idea for computerized scatter removal, rather than using grids or other methods. This approach is complicated because the scatter spread function is not shift invariant (see Section 6.3.2.4).

In the absence of computerized methods, the use of antiscatter grids is routine for the vast majority of radiographic projections, apart from those of the extremities. Grids vary greatly in terms of the degree of scatter rejection, and in the increase in dose to the patient that their use requires. All are designed to allow a large proportion of the primary photons to reach the image receptor, while removing a good proportion of the scattered photons from the radiation field.

6.3.4.1. Grid construction

Apart from the special cellular grids used in some mammography systems, a grid generally consists of an array of thin lead strips aligned to allow passage of the primary beam, as shown in Fig. 6.10. The lead strips are separated by an interspace material and have protective covers on the top and bottom (not shown in the diagram). The number of lead strip lines per centimetre is known as the strip frequency, and the ratio of the height of the lead strips to the width of the interspace material is known as the grid ratio, r, which is given by:

$$r = \frac{h}{d} \tag{6.17}$$

where the distances h and d are shown in Fig. 6.10.

From consideration of this figure, it will be seen that all scattered photons whose angle of incidence is less than $\tan^{-1} r$ will hit a lead strip, so that scatter rejection will increase with increasing grid ratio. On the other hand, the transmission of primary photons through the grid will decrease because of the increased thickness of the interspace. If the lead strips are too thin, they can be penetrated by the scattered X ray photons, and if they are too thick, they will stop too many primary photons. Thus, the design of the grid is a compromise between the requirements of good scatter rejection and high primary photon transmission.

Usually, the lead strips are tilted to match the divergence of the primary beam, at a chosen distance from the focus called the focal length of the grid, and the grid is then referred to as a focused grid. Such grids must be used at the correct focal distance within a permissible tolerance. If a grid is used at the wrong distance, the tilt of the lead strips will not match the angle of divergence of the primary beam. The primary beam will then be attenuated progressively more towards the edge of the image. This is termed grid cut-off. Some types of grid have parallel strips and are known as parallel grids. These always have some degree of grid cut-off and should not be used at short distances. The degree of cut-off will be affected by the field size used and the grid ratio.

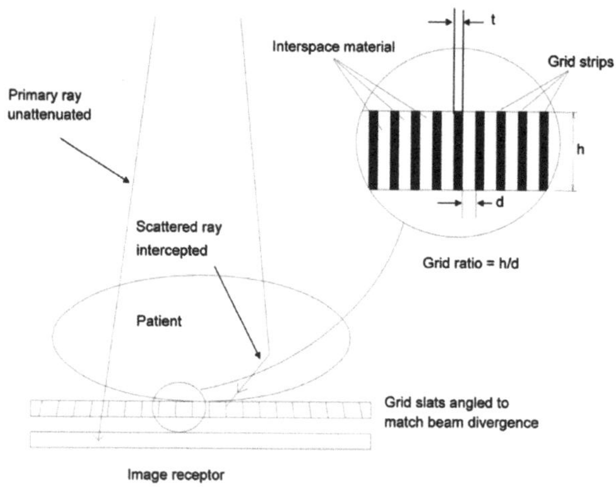

FIG. 6.10. Construction and principle of operation of a focused antiscatter grid (not to scale).

The grid interspace material is plastic, carbon fibre or other low atomic number material. Older grids used aluminium. The material of the grid covers should also be of low atomic number. Since the interactions of the X rays with the lead strips will mainly be photoelectric, for photons of energy above the K edge of 88 keV, the emission of K fluorescent X rays must be considered in grid design. Table 6.2 gives the geometrical construction details for a range of typical grids.

TABLE 6.2. TYPICAL CONSTRUCTION DATA FOR THREE ANTISCATTER GRIDS

Grid ratio	Strip frequency (cm^{-1})	h (mm)	d (mm)	t (mm)
10:1	33	2.5	0.25	0.050
10:1	40	2.0	0.20	0.050
12:1	57	1.56	0.13	0.045

The appearance of the image of the grid lines in radiographic images is generally unacceptable for diagnosis. Furthermore, for digital systems, the image of the grid lines may interfere with the pixel matrix, generating Moiré patterns in the image. Consequently, a mechanism is usually provided to oscillate the grid in a direction perpendicular to the grid lines, to blur them out during the exposure.

6.3.4.2. *Measures of grid performance*

Grid performance is specified in terms of parameters that relate to the associated dose increase that is necessary and the improvement in contrast that is achieved. A good grid will eliminate 80–90% of the scatter, while transmitting at least 75% of the useful beam. The following quantities are used:

- The primary transmission, T_p, is a narrow beam measurement of the ratio of X ray intensity with and without the grid present.
- The secondary transmission, T_s, is a broad beam measurement of the scattered radiation intensity with and without the grid.
- The total transmission, T_t, is a measurement of the total intensity of X rays with and without the grid.
- The grid factor or Bucky factor is the dose increase factor associated with the use of the grid:

$$\text{Bucky factor} = \frac{\text{exposure (mAs) required with grid}}{\text{exposure (mAs) required without grid}} = \frac{1}{T_t} \quad (6.18)$$

- The selectivity is a measure of the effectiveness of the grid, given by:

$$\Sigma = \frac{T_p}{T_s} \quad (6.19)$$

- The contrast improvement factor, CIF, is given by:

$$\text{CIF} = \frac{\text{contrast with grid}}{\text{contrast without grid}} = \frac{T_p}{T_t} \quad (6.20)$$

It should be noted that, as well as removing scatter, the grid will harden the X ray beam and calculations of the CIF should, in principle, allow for this effect. Usually, however, the correction is no more than a few per cent and, to a good approximation, the CIF can be calculated as:

$$\mathrm{CIF} = \frac{1 - \mathrm{SF_g}}{1 - \mathrm{SF_{ng}}} \tag{6.21}$$

where, SF_g and SF_{ng} are the scatter fractions with and without the grid.

Table 6.3 gives the Bucky factor and CIF for a range of grids for the commonly used LucAl patient equivalent phantoms.

6.3.4.3. Grid selection

From the data in Table 6.3, it is clear that for each radiographic projection there is an optimum grid ratio that will provide adequate scatter reduction, with an acceptable increase in dose to the patient. For example, chest and abdomen projections on adults would require ratios of 10:1 or 12:1. Practically, however, the grid is permanently fitted to the cassette holding device in radiographic tables and wall mounted devices, and these generally have a ratio of at least 10:1. Consequently, grid use is generally far from optimized.

6.3.4.4. Grid artefacts and alignment

There are several possible misalignments that will lead to artefacts in projection images. Additionally, a damaged grid will generate artefacts and must be replaced. Figure 6.11 illustrates the possible misalignments of the grid. In practice, it is possible for a number of these to be present at once. Note that the moving grid is laterally decentred during operation, although the degree of offset is small on average.

6.3.5. Other methods of scatter reduction

While the use of a grid is effective at reducing scattered radiation, image quality may be further improved by careful collimation and by patient compression. Alternatives to antiscatter grids include the use of air gaps and slit scanning systems.

TABLE 6.3. BUCKY FACTOR AND CIF AT THE CENTRE OF THE IMAGE RECEPTOR FOR A SELECTION OF RADIOGRAPHIC GRIDS, USING A 25.4 cm × 25.4 cm X RAY FIELD WITH THE LucAl CHEST AND ABDOMEN PATIENT EQUIVALENT PHANTOMS, 120 kV FOR THE CHEST DATA AND 70 kV FOR THE ABDOMEN DATA

Grid ratio	SF	SPR	Bucky factor[a]	CIF
LucAl chest phantom				
No grid	0.390	0.640	—	1.0
6:1	0.188	0.231	1.53	1.33
8:1	0.150	0.176	1.62	1.39
10:1	0.123	0.141	1.69	1.44
12:1	0.103	0.115	1.75	1.47
16:1	0.076	0.082	1.85	1.51
LucAl abdomen phantom, soft tissue region				
No grid	0.712	2.472	—	1.0
6:1	0.371	0.588	2.79	2.18
8:1	0.302	0.433	3.20	2.42
10:1	0.256	0.345	3.53	2.58
12:1	0.221	0.283	3.81	2.71
16:1	0.174	0.211	4.27	2.87
LucAl abdomen phantom, spine region				
No grid	0.837	5.155	—	1.0
6:1	0.472	0.894	—	3.25
8:1	0.388	0.636	—	3.76
10:1	0.327	0.486	—	4.14
12:1	0.276	0.382	—	4.45
16:1	0.203	0.254	—	4.91

[a] The low Bucky factor is due to the scatter level in chest radiography being low, as the lungs act as a large air gap, and the thickness of other tissues is relatively low. The Bucky factor for the abdomen is constant for each grid.

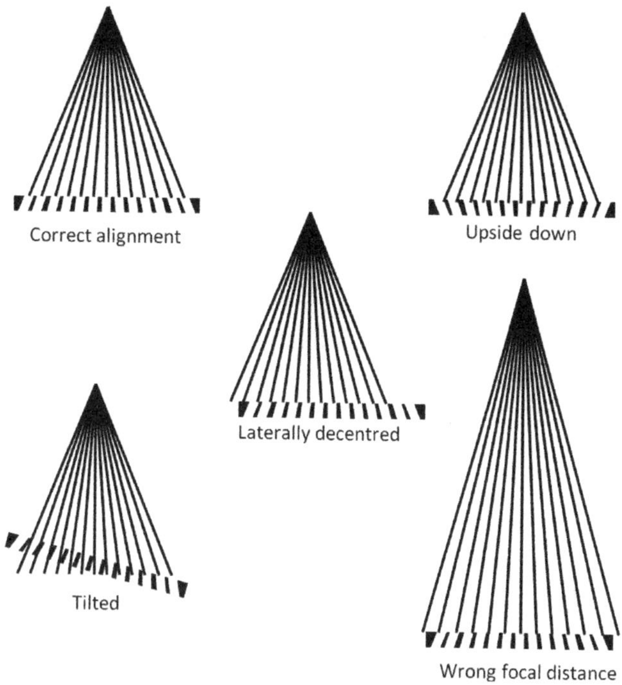

FIG. 6.11. Grid misalignments leading to cut-off.

6.3.5.1. Collimation

Section 6.3.2.2 describes the effect of X ray field size on scatter fraction. Since smaller field sizes reduce the scatter fraction, it is good practice to collimate the X ray beam to as small an area as possible for each projection, thereby improving image quality and reducing patient dose. This requires radiographers with a good knowledge of anatomy and good judgement of the wide variety of patients to ensure that the region of clinical interest is included in each image, without needlessly irradiating tissues that are not of interest. Care is required not to collimate too tightly, which would increase the possibility of the region of clinical interest being cut off, resulting in a repeated exposure and increased dose.

6.3.5.2. Compression

The data in Section 6.3.2.3 clearly demonstrate the increase in scatter fraction with patient thickness. Therefore, if the patient thickness can be reduced during exposure, by applying a compression band for example, then the amount of scatter will be reduced. This has further benefits, because a shorter exposure time can be used, reducing movement blur and patient dose. Alternatively, the kV setting may be reduced, improving the contrast or SNR in the image. For these reasons, among others, compression is routinely used in mammography (see Section 9.3.2).

6.3.5.3. *Air gap*

As described in Section 6.2.4, the use of an increased OID, or air gap, results in magnification and a reduction in scatter fraction. This is because the divergent scattered rays will be increasingly less likely to strike the image receptor as the OID is increased, and are therefore much less likely than the primary rays to strike the image receptor.

The effect of the OID on the scatter fraction is illustrated in Fig. 6.12.

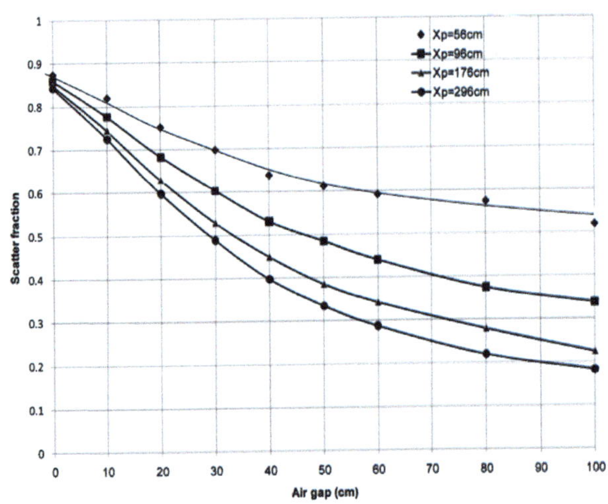

FIG. 6.12. Effect of air gaps (OID) on scatter fraction for a range of FIDs. X_p is the source–patient exit distance. PMMA phantom of 20 cm thickness, X ray field size of 20 cm × 20 cm, 90 kV.

6.3.5.4. *Scanning slit*

As explained in Section 6.3.2.2, the scatter fraction depends on the X ray field size. Slot scanning systems take advantage of this to minimize the scatter fraction by using a very small area X ray field with a slit aperture, which must be scanned across the patient to produce the image. Such systems generally feature a pre-patient collimator to define the fan shaped beam, and a post-patient collimator to intercept any scattered radiation produced.

For digital systems, the area detector may be replaced by silicon or similar strip detectors, in which case the post-patient collimation is not required. These systems may feature a single fan beam scanned across the region of interest, or a number of fan beams. The latter allows for faster scanning and shorter exposure times. The basic principle is illustrated in Fig. 6.13. These systems are capable of good scatter rejection without the necessity for a grid, but require smooth and precise movement of the collimator systems and stable X ray generator performance. There is increased probability of artefacts compared with conventional methods, and the longer exposure time required increases the risk of movement blur in the images and reduces the life of the X ray tube.

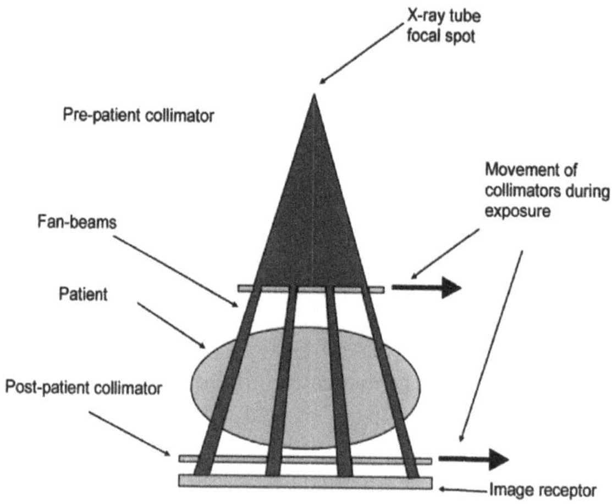

FIG. 6.13. Principle of operation of a multislit scanning system for scatter reduction. The image receptor and post-patient collimator may be replaced by strip detectors in digital systems.

REFERENCES

[6.1] POLETTI, J.L., McLEAN, I.D., The effect of source to image-receptor distance on effective dose for some common X ray projections, Br. J. Radiol. **78** (2005) 810–815.

[6.2] DANCE, D.R., EVANS, S.H., SKINNER, C.L., BRADLEY, A.G., "X-ray transmission computed tomography", The Physics of Medical Imaging (WEBB, S., Ed.), Adam Hilger Ltd, Bristol (1988).

[6.3] ROSE, A., Vision: Human and Electronic, Plenum Press, New York (1974).
[6.4] ULLMAN, G., SANDBORG, M., DANCE, D.R., HUNT, R., ALM CARLSSON, G., Distributions of scatter-to-primary and signal-to-noise ratios per pixel in digital chest imaging, Radiat. Prot. Dosim. **114** (2005) 355–358.

BIBLIOGRAPHY

BUSHBERG, J.T., SEIBERT, J.A., LEIDHOLDT, E.M.J., BOONE, J.M., The Essential Physics of Medical Imaging, 2nd edn, Lippincott, Williams & Wilkins, Baltimore, MD (2002).

DOI, K., et al., Physical and clinical evaluation of new high-strip-density radiographic grids, Radiology **147** (1983) 575–582.

EUROPEAN COMMISSION, European Guidelines on Quality Criteria for Diagnostic Radiographic Images, Rep. EUR 16260 EN, Office for Official Publications of the European Communities, Luxembourg (1996), ftp://ftp.cordis.lu/pub/fp5-euratom/docs/eur16260.pdf (accessed on 23 August 2012).

FETTERLY, K.A., SCHUELER, B.A., Experimental evaluation of fiber-interspaced antiscatter grids for large patient imaging with digital X ray systems, Phys. Med. Biol. **52** (2007) 4863–4880.

INTERNATIONAL ELECTROTECHNICAL COMMISSION, Diagnostic X-ray Imaging Equipment: Characteristics of General Purpose and Mammographic Anti-scatter Grids, IEC-60627, IEC, Geneva (2001).

MILNE, E.N.C., "The role and performance of minute focal spots in roentgenology with special reference to magnification", CRC Crit. Rev. Radiol. Sci. **2** (1971) 269–310.

Chapter 7

RECEPTORS FOR PROJECTION RADIOGRAPHY

J.A. ROWLANDS
Thunder Bay Regional Research Institute,
Thunder Bay, Canada

U. NEITZEL
Philips Healthcare,
Hamburg, Germany

7.1. INTRODUCTION

X ray images are formed as shadows of the interior of the body. Since it is not yet practical to focus X rays, an X ray receptor has to be larger than the body part to be imaged. Thus, the first challenge in making an X ray receptor is the need to image a large area. A second challenge is to make a system that has an image quality as good as that allowed by the physics, i.e. permits the detection of objects whose size and contrast are limited only by the quantum statistics (see Section 4.6). This means absorbing most of the X ray quanta and using these in an efficient, i.e. a quantum noise limited, manner, while providing adequate spatial resolution simultaneously.

The capture of an X ray image may conceptually be divided into three stages. The first is the interaction of the X ray with a suitable detection medium to generate a measurable response. The second is the temporary storage of this response with a recording device. The third is the measurement of this stored response. As an example, the stages for a screen film system are: (i) the interaction of an X ray in a phosphor material followed by generation of visible light photons, (ii) the subsequent creation of a latent image in the photographic film by these photons, and, finally, (iii) the development of a fixed photographic image. A fourth stage required for reusable systems (i.e. those not requiring consumables such as film) is (iv) the erasure of all previous images within the detection system in order to prepare for a fresh image.

The four steps for a digital direct conversion flat panel imaging system are: (i) the absorption of an X ray followed by the release of multiple secondary electrons in a photoconductor, (ii) drifting of the electrons and holes to individual electrodes where they are stored, until (iii) the readout phase, when the charges are transferred to amplifiers where they are digitized line by line. This approach is interesting, as the readout simultaneously, and without further effort, performs the essential step (iv) erasure. Breaking up the stages in this manner is helpful to the understanding of the physics of image acquisition, which itself is key to the optimization of the receptor design and the understanding of fundamental limitations on image quality. It is also key to developing an understanding of the complementary strengths of the various approaches used in the past and currently and which are likely to be used in the future.

7.2. GENERAL PROPERTIES OF RECEPTORS

Before describing in more detail the properties of the different types of image receptor used for projection radiography, it is necessary to consider the various physical properties and quantities that are used to specify their performance. Some of these have already been discussed in Chapter 4.

7.2.1. Receptor sensitivity

The initial image acquisition operation is identical in all X ray receptors. In order to produce a signal, the X ray quanta must interact with the receptor material. The probability of interaction or quantum detection efficiency, A_Q, for an X ray of energy E is given by:

$$A_Q = 1 - \exp(-\mu(E,Z)T) \tag{7.1}$$

where

μ is the linear attenuation coefficient of the receptor material;
Z is the material's atomic number;

and T is its thickness.

As virtually all X ray sources for radiography emit X rays over a spectrum of energies (see Chapter 5), the quantum detection efficiency must be specified either as a function of energy or as an effective value over the spectrum of X rays incident on the receptor. Figure 7.1 shows A_Q plotted as a function of energy for various T values of selected receptor materials. A_Q will, in general, be highest at low values of E, decreasing with increasing E. At diagnostic X ray energies, the main interaction process is the photoelectric effect because of the relatively high Z of most receptor materials. If the material has a K atomic absorption edge, E_K,

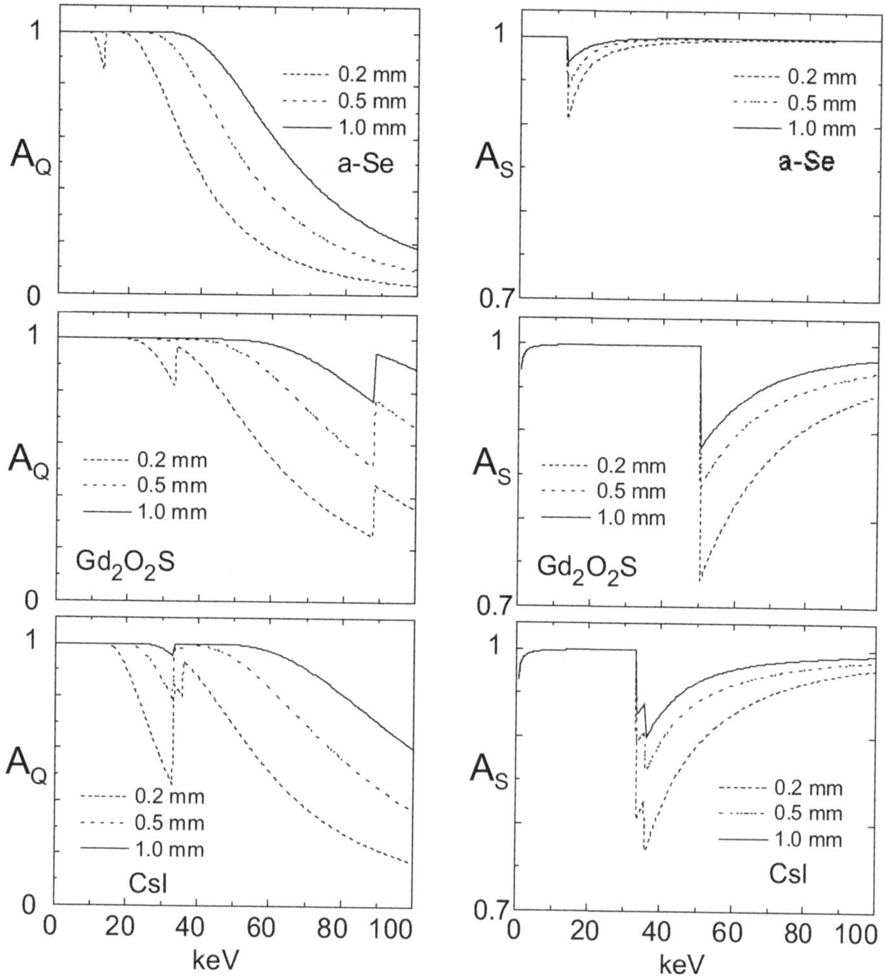

FIG. 7.1. Quantum detection efficiency, A_Q, and Swank factor, A_S, for representative examples of an X ray photoconductor a-Se, a screen phosphor, Gd_2O_2S, and the scintillator, CsI. The curves are for the primary interaction using the photoelectric coefficient only; other components adding to the Swank factor are discussed in the text. The thicknesses are for an assumed 100% packing fraction of the material, which is realistic for a-Se and should be increased by ~2 for a powder screen such as Gd_2O_2S, and by ~1.1–1.2 for an evaporated structured CsI layer.

in the energy region of interest, then A_Q increases dramatically at E_K, causing a local minimum in A_Q for $E < E_K$.

The photoelectric interaction of an X ray quantum with the receptor generates a high speed photoelectron. During the subsequent loss of kinetic energy of the electron in the receptor, excitation and ionization occur, producing the secondary signal (optical quanta or electronic charge). The sensitivity of any imaging system therefore depends both on A_Q and the primary conversion efficiency (the efficiency of converting the energy of the interacting X ray to a more easily measurable form such as optical quanta or electrical charge). Conversion efficiency can be re-expressed as the conversion factor, i.e. in terms of the number of secondary particles (light photons in a phosphor or electron hole pairs (EHPs) in a photoconductor) released per X ray. For a surprising number of materials and systems, this is ~1000 quanta or EHPs per 50 keV X ray. The conversion factor is closely related to the intrinsic band structure of the solid from which the receptor is made, as shown in Fig. 7.2. In all receptor materials, the valence band is almost fully populated with electrons and the conduction band is practically empty. The forbidden energy gap, E_g, governs the energy scale necessary to release a mobile EHP, i.e. to promote an electron from the valence band to the conduction band. Although E_g is the minimum permitted by the principle of conservation of energy, this can be accomplished only for photons of energy E_g. For charged particles releasing energy (e.g. through the slowing down of high energy electrons created by an initial X ray interaction), the requirements of conserving both energy and crystal momentum, as well as the presence of competing energy loss processes, necessitate ~$3E_g$ to release an EHP. Thus, in Fig. 7.2(a) for a photoconductor (generally a material with E_g ~ 2eV — the minimum energy at which the thermally excited dark current is negligible), the maximum number of EHPs is 50 000/(2 eV × 3) ~ 8000 EHP. This is possible for good photoconductors, but for the only practical photoconductor used in commercial systems at this time (a-Se), there are other losses (primarily to geminate recombination, i.e. the created EHPs recombine before separation by the applied electric field), which limit it to ~1000–3000 EHPs, depending on the applied electric field.

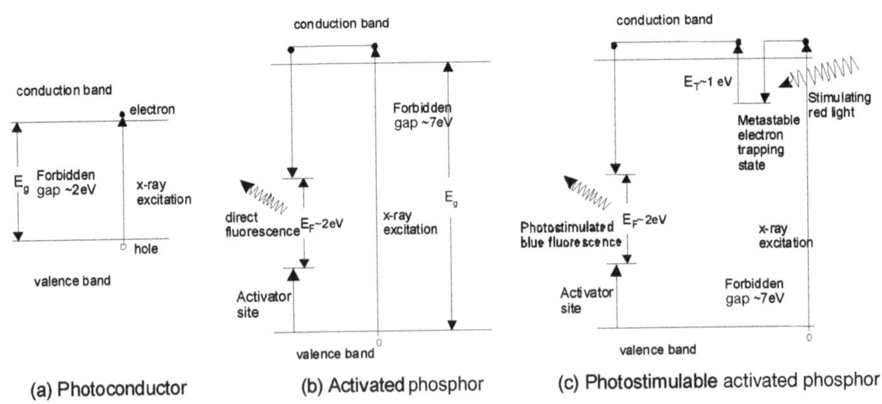

FIG. 7.2. Band structure of photoconductors, activated phosphors and photostimulable phosphors.

In phosphors (scintillators are another name for phosphors, usually used when in a crystalline or dense mass), the band gap is usually much higher (~8 eV), so the intrinsic conversion factor is typically lower — only ~2000 EHPs are released (50 000/8 eV × 3), which, however, in an activated phosphor results in emission of only slightly fewer (1800) light photons. Additional optical losses due to light absorption in the phosphor layer (sometimes with an intentionally included dye) and/or non-reflective backing, dead space between the photoreceptors (fill factor) and non-ideal quantum detection efficiency of the photoreceptors, further reduce the light per X ray. This typically results in ~1000 EHPs collected in the photoreceptor for our prototypical 50 keV X ray.

7.2.2. Receptor X ray noise

All images generated by quanta are statistical in nature, i.e. although the image pattern can be predicted from the attenuation properties of the patient, it will fluctuate randomly about the mean predicted value. The fluctuation of the X ray intensity follows Poisson statistics, so that the variance, σ^2, about the mean number of X ray quanta, N_0, falling on a receptor element of a given area, is equal to N_0. Interaction with the receptor can be represented as a binomial process, with probability of success, A_Q, and the distribution of interacting quanta is still Poisson, with variance:

$$\sigma^2 = N_0 A_Q \tag{7.2}$$

If the detection stage is followed by a process that provides a mean gain, g, then the distribution will not be Poisson even if g is Poisson distributed. It is also possible that other independent sources of noise will contribute at different stages of the imaging system. Their effect on the variance will be additive. A complete linear analysis of signal and noise propagation in a receptor system must also take into account the dependence on spatial frequency of both the signal and the noise.

It is important that the number of secondary quanta or electrons at each stage of image production is considerably greater than N_0, to avoid having the receptor noise dominated by a secondary quantum sink. Consideration of the propagation of noise is greatly facilitated by consideration of a quantum accounting diagram, examples of which are shown in Fig. 7.3. The concept is that the noise from each stage of the imaging system is related to the number of secondary quanta or electrons at each stage; so ideally there should, for all stages, be many more (if possible exceeding 1000) such secondary quanta or particles representing each primary quantum (i.e. X ray). The point at which this number is lowest is the secondary quantum sink. It will be seen that the examples given, all of which are

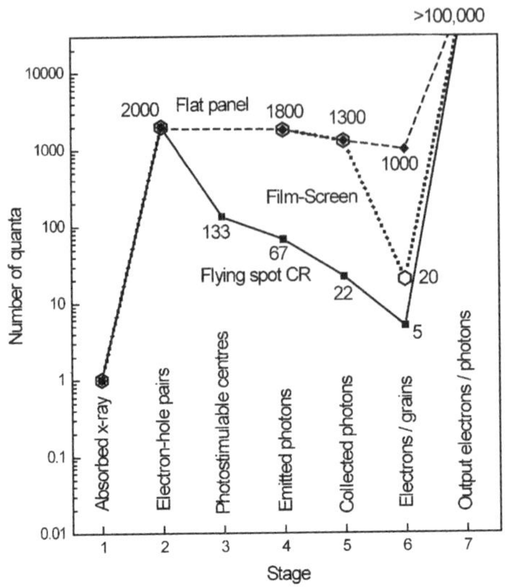

FIG. 7.3. Quantum accounting diagrams for screen film, CR (computed radiography) with a flying spot scanner and flat panel digital radiography. The vertical axis represents the average number of quanta or individual particles (electrons or film grains) representing the initial absorbed X ray (assumed to be of 50 keV) at each stage in the imaging system. The critical point for each modality is where the minimum number of quanta or EHPs represents a single X ray. For flat panel systems this is ~1000, while for screen film it is 20 and only 5 for CR. This is the weakest link.

used commercially, have at least five secondaries per primary, but it is very easy to find systems in which this is not the case, such as non-intensified fluoroscopy or some optically (lens) coupled radiographic X ray systems.

The noise in X ray images is related to the number of X rays per pixel in the image and hence to the X ray exposure to the receptor. However, the relative noise can be increased by lack of absorption of the X rays, as well as by fluctuations in the response of the receptor to those X rays that are absorbed. There are also unavoidable fluctuations in the signal produced in the detection medium, even when X rays of identical energy interact and produce a response. These are caused by the statistical nature of the competing mechanisms that occur as the X ray deposits energy in the medium. Together, they give rise to a category of noise known as gain fluctuation or Swank noise. The gain fluctuation noise can be determined experimentally using the pulse height spectrum (PHS). From this, the Swank factor, A_S, is obtained as a combination of the zeroth, first and second moments (M_N) of the PHS, using the formula:

$$A_S = \frac{M_1^2}{M_0 M_2} \quad (7.3)$$

The ideal PHS, obtained when all absorbed X rays give rise to equal amounts of signal, results in a delta function and a Swank factor of unity. However, in practice, there are a number of effects that may broaden this spectrum, resulting in a Swank factor of less than unity. Figure 7.1 shows A_S values for various receptor materials and thicknesses chosen to encompass a range that is technologically possible and therefore may be encountered in practice. These are calculated values and are for the photoelectric effect only, which effectively means that only K escape is accounted for. K escape is the emission of a K fluorescent X ray following a photoelectric interaction, which then escapes from the receptor without depositing further energy. As the energy of the K fluorescent X ray is below the K edge, it has a smaller interaction probability than the original incident X ray photon (see Section 2.2.1). The range of values of A_S varies from 0.7 to 1.0. Further losses due to optical effects will be seen in screens, and other losses in photoconductors due to trapping of charge. Swank demonstrated that, for many situations, the combination of factors can be performed simply by multiplying the component factors. For example, the K escape Swank factor can be multiplied by the optical Swank factor to obtain the overall Swank factor. Theoretically, for an exponential PHS, which can occur for screens with very high optical absorption, the optical value can be as poor as 0.5, resulting in a range of 0.5–1.0 for the optical effects. Overall, therefore, the possible range of receptor Swank factors for screens is 0.35–1.0.

The noise due to both quantum absorption and gain fluctuations can be combined to create the zero spatial frequency detective quantum efficiency (DQE)(0), which is given by:

$$\mathrm{DQE}(0) = A_Q A_S \tag{7.4}$$

Owing to its importance, it is worth reiterating that the DQE(0) is the effective quantum detection efficiency obtained when the noise in the measured image is compared with what it would be if it were an ideal (perfect) absorber and there was no gain fluctuation noise.

7.2.3. Greyscale response and dynamic range

The greyscale response used for an imaging system has to do with the physics of X ray detection, the imaging task to be performed and, the most difficult part, the response of the human eye–brain system to optical images. In practice, many of the decisions made by system designers are empirical rather than entirely from theoretical analysis. However, some rules of thumb can be helpful.

Regarding human vision, there is a spatial frequency range in which the human eye is most acute. This is an intermediate frequency range, neither too low nor too high. Regarding intensity, it is, as for all human senses, essentially logarithmic in its response, i.e. it is fairly good at seeing fractional differences, provided these are directly juxtaposed. Otherwise, the human eye is quite poor at quantitative evaluations of intensity. In order to separate the subjective eye–brain response from the more quantitative issues, it is usual to leave out (in the case of inherently non-linear systems such as film), or to correct for, the optical display part of the system that has a non-linear response (e.g. cathode ray tube viewing monitors or liquid crystal flat panel displays). Only then can most systems be, for practical purposes, modelled as being linear.

The greyscale response is usually expressed as the characteristic curve, a plot of the response of the system to a stimulus; for example, in fluoroscopy this would be the optical intensity at the video monitor plotted as a function of the irradiation of the sensor at the corresponding point in the image.

The range of intensities that can be represented by an imaging system is called the dynamic range and depends on the pixel size that is used, in a manner that depends on the modulation transfer function (MTF). However, for any pixel size, the dynamic range for an X ray imaging task can be broken down into two components. The first describes the ratio between the X ray attenuation of the most radiolucent and the most radiopaque paths through the patient appearing on the same image. The second is the required precision of the X ray signal measured in the part of the image representing the most radiopaque anatomy. If, for example, there is a factor of 10 in attenuation across the image field and it is desired to have 10% precision in measuring the signal in the most attenuating region, then the dynamic range requirement for the receptor would be 100.

The dynamic range that can be achieved by a practical linear imaging system can be defined in terms of the response at the output referred back to the input in terms of the X ray exposure:

$$\text{Dynamic range} = \frac{X_{\text{max}}}{X_{\text{noise}}} \tag{7.5}$$

where X_{max} is the X ray exposure providing the maximum signal that the receptor can respond to before saturation (i.e. that point where the receptor output ceases to respond sensibly to further input), and X_{noise} is the root mean square receptor noise in the dark (i.e. no X ray exposure).

X rays are attenuated exponentially; thus, an extra tenth-value layer thickness of tissue will attenuate the beam by 10, while a lack of the same tenth-value thickness will increase the X ray exposure by 10. Thus, when a mean exposure value, X_{mean}, for the system is established by irradiating a uniform phantom, we are interested in multiplicative factors above and below this mean value; i.e. X_{mean} is a geometric rather than arithmetic mean. For a dynamic range of 100, the correct range is that given by the geometric mean of $10X_{mean}$ and $0.1X_{mean}$, *not* that given by the arithmetic mean which would be ~$2X_{mean}$ and $0.02X_{mean}$.

7.2.4. Receptor blur

Spatial resolution in radiography is determined both by the receptor characteristics and by factors unrelated to the receptor. The latter includes unsharpness (blurring) arising from geometrical factors such as: (i) penumbra (partial X ray shadow) due to the effective size of the X ray source and the magnification between the anatomical structure of interest and the plane of the image receptor (see Section 6.2.3.4) and (ii) motion blurring due to relative motion of the patient with respect to the receptor and X ray focal spot. In the overall design of an imaging system, it is important that these other physical sources of unsharpness be considered when the aperture size and sampling interval are chosen. If, for example, the MTF is limited by unsharpness due to the focal spot, it would be of little value to attempt to improve the system by designing the receptor with much smaller receptor elements.

The major issues relating to receptor blur are summarized in Fig. 7.4. They include the fundamental issues arising within any material, which are (i) geometrical blurring due to oblique incidence of X rays, which is especially marked far from the central ray (i.e. the X ray that strikes the receptor normally), (ii) the range of the primary electron — the primary electron usually gives up its energy in small amounts, ~100–200 eV at a time, but this is sufficient to scatter the electron at any angle, thus the path of the primary is usually a random walk and it does not go so far from its initial point of interaction, with a separation of ~1 μm for 10 keV electrons and ~50–100 μm for 100 keV electrons, depending on the medium in which it is absorbed, and (iii) the reabsorption of K fluorescence X rays some distance from the primary photoelectric interaction, something that is likely because of the general rule that a material is relatively transparent to its own K fluorescence, owing to the minimum in attenuation below the K edge. In addition, there are material dependent effects specific to direct conversion and indirect conversion receptors, shown in Figs 7.4(d)–(f) and (g)–(l), respectively. Also to be considered in digital systems are the effects of the del aperture and sampling interval.

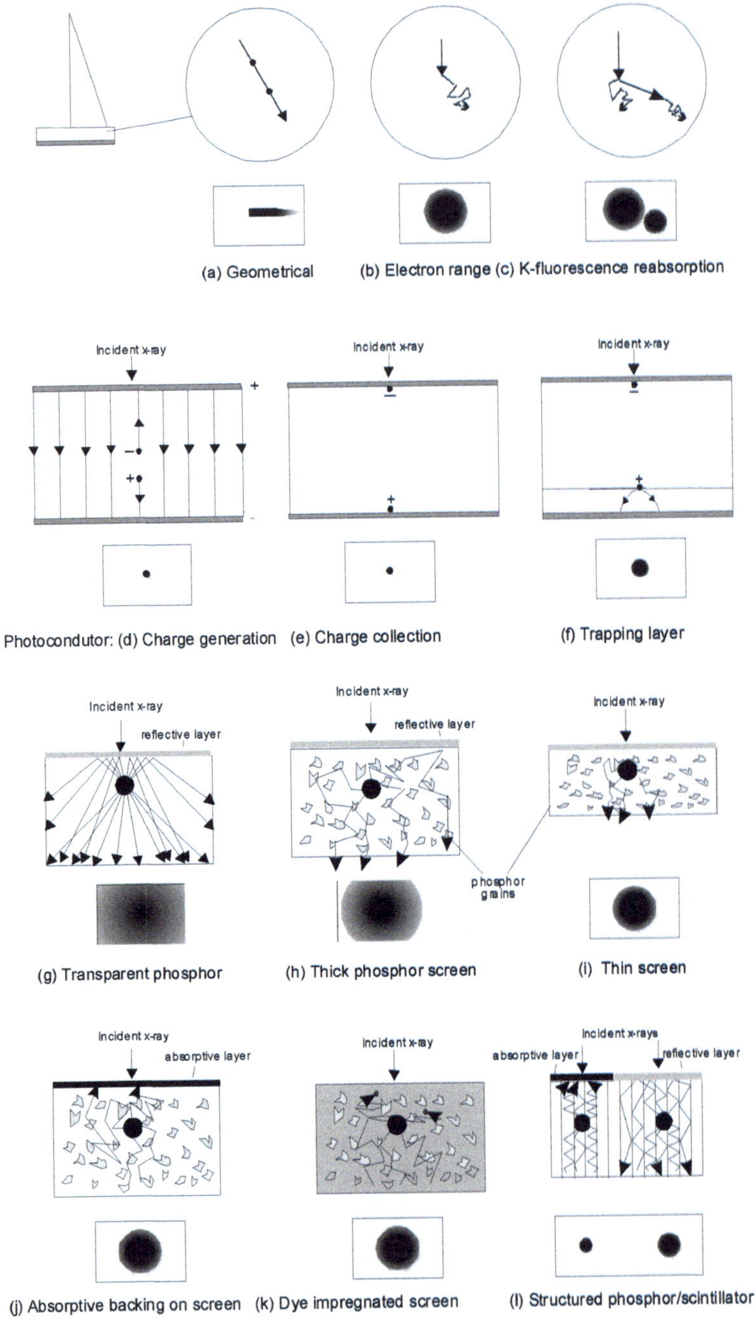

FIG. 7.4. Mechanisms of resolution loss in layer receptors. Images: (a)–(c) general to all receptors; (d)–(f) for direct conversion layers (photoconductors); (g)–(l) indirect conversion types — powder phosphor screens and structured phosphors (usually called scintillators).

7.2.5. Fixed pattern noise

It is important that the radiographic imaging system provides uniformity, i.e. the sensitivity is constant over the entire area of the image. If this is not the case, patterns that might disrupt the interpretation of the image will result. This is fixed pattern noise. In an analogue imaging system, great pains must be taken in the design and manufacture of receptors to ensure that they provide a uniform response. In digital systems, post-processing can be often be used to alleviate manufacturing limitations in uniformity.

In a linear imaging system, the fixed pattern noise, which can be expressed as a pixel to pixel variation in gain and offset (often because of a dark current in the related sensor), can, in principle, be corrected and the effect completely eliminated. The procedure is to correct patient images using dark field (unirradiated) and bright field (uniformly irradiated) images.

7.3. FILM AND SCREEN FILM SYSTEMS

In using a screen film receptor to take an X ray image, a radiographer must load a film into a cassette, carry the cassette to the examination room, insert the cassette into the X ray table, position the patient, make the X ray exposure, carry the cassette back to the processor to develop the film, wait for the film to be developed and, finally, check the processed film for any obvious problems, to ensure that the film is suitable for making a medical diagnosis before returning to the X ray room.

7.3.1. Systems

7.3.1.1. The screen film combination and cassette

The screen film combination consists of phosphor screen(s) and film designed to work together enclosed within a cassette. The cassette can be opened (in a dark environment) to allow the film to be inserted. When the cassette is closed, the film is kept in close contact with the screen, or, more commonly, a pair of screens, facing towards the film, as shown in Fig. 7.5.

Incident X rays first pass through the front of the cassette before reaching the screens. When they interact in the screen, some of the energy deposited is converted to light, which can travel from the interior of the screen to the screen surface, where it enters the optically sensitive part of the film called the emulsion and transfers its information into a latent image in the film. The film is then removed from the cassette and developed, so that the latent image is converted to a permanent image in the form of silver deposited in the emulsion layer of the film.

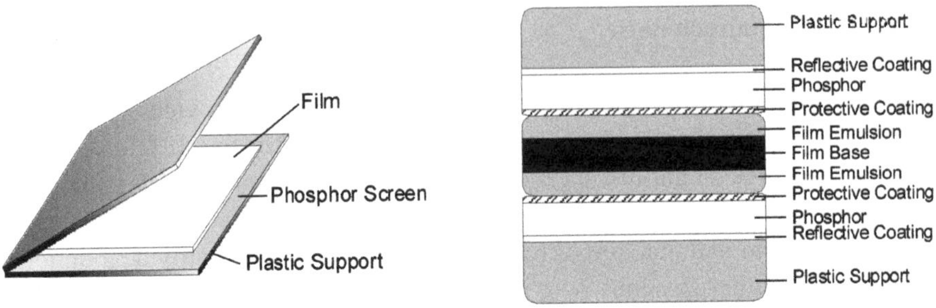

(a) (b)

FIG. 7.5. Screen film receptor: (a) opened cassette showing placement of film and position of screens, and (b) cross-sectional view through a dual screen system used in general purpose radiography with the film sandwiched between two screens.

In Fig. 7.5(a), a film is shown in the open cassette. In most cases, as shown in Fig. 7.5(b), two screens face the film, which has two emulsions — one on either side of the film base with an antihalation layer placed between the two emulsions. During X ray exposure, the antihalation layer is opaque and prevents light crossing over from one emulsion to the other, thus reducing 'cross-talk' and hence blurring. The opaque antihalation layer is removed during the film development, rendering the film transparent for subsequent viewing. For the highest resolution (e.g. mammography), a single screen in the back of the cassette may be used in contact with a single emulsion film (see Section 9.4.1).

Film emulsions can be used as direct receptors for X ray images. The earliest X ray images were taken with film alone. In fact, film was used in this way for mammography up to the 1960s. The sole remaining clinical application for film without screens is in dental radiography using intraoral films. However, the X ray absorption efficiency of such films is relatively poor (~1–5%). Thus, currently, all diagnostic X ray film images are obtained using screen(s) in conjunction with the film.

7.3.2. The screen

7.3.2.1. Screen structure

The screen structure is shown in Fig. 7.5(b). Phosphor grains are combined with a polymer binder and deposited on a substrate or backing. The ratio of binder volume to phosphor volume in the mixture controls the fractional volume of the phosphor layer finally occupied by air pockets or voids. Typically, the binder is nitrocellulose, polyester, acrylic or polyurethane and the plastic support or backing material is also a polymer, e.g. polyethylene terephthalate,

200–400 μm thick. The use of a black or white backing permits adjustment of the reflectivity and absorptivity at the phosphor interface. In most screens, the typical phosphor grain size is 3–8 μm. Usually, a very thin transparent protective layer is subsequently applied to complete the screen structure.

7.3.2.2. Phosphor

Screen film systems employ a phosphor in the initial stage to absorb X rays and produce light. As shown in Fig. 7.2(b), phosphors work by exciting electrons from the valence band to the conduction band, so creating EHPs, which are free to move within the phosphor. Some of these will recombine without giving off any radiant energy. However, in an activated phosphor, most (>90%) EHPs will recombine at an activation centre (created by atomic impurities called activators) and, in the process, emit light. The specific colour of the light emitted is related to the optical transitions in the activator. By changing the activator, the light colour can be changed. Since light photons each carry only a small amount of energy (~2–3 eV), many light photons can be created from the absorption of a single X ray, resulting in significant quantum amplification.

This quantum amplification is the conversion gain of the phosphor. The original screens used until the 1970s were calcium tungstate ($CaWO_4$), which is naturally activated and hence not particularly efficient and emits light in the deep blue and UV radiation. More recently, rare earth phosphors with explicit centres for the emission of light at the activator site have resulted in the most commonly used material, gadolinium oxysulphide (Gd_2O_2S:Tb with Tb in dilute amounts (0.1–1% as an activator)). For Gd_2O_2S, an X ray photon energy of 50 keV is equivalent to that of ~20 000 green light quanta (E = 2.4 eV), although, as a result of losses, typically only 1800 are produced in practice. The green emission from the rare earth phosphors has also required a change from conventional film in two regards. Firstly, ordinary film is sensitive only in the blue and requires additional sensitization to be green sensitive; it is then called orthochromatic. Secondly, green light is far more penetrating than blue light and so requires an antihalation layer to prevent crossover of images between emulsions.

7.3.2.3. Thickness

The choice of the thickness of a radiographic screen has to balance (i) the increase in A_Q with thickness, which favours a thick screen, and (ii) the efficient escape of light and, usually more importantly, blurring due to spreading of light, which favours a thin screen. In order to create a sharp X ray image, a transparent phosphor screen would be ineffective, since light could move large distances within the phosphor and cause excessive blurring, as shown in

Fig. 7.4(g). Instead, X ray screens are made to be highly scattering (turbid). This is accomplished by using high refractive index phosphor grains embedded in a low refractive index binder. Once a light photon exits a grain, it tends to reflect off the neighbouring grain surfaces rather than passing through them. Thus, the lateral spread of the light is confined by diffusion (multiple scattering), which helps to maintain the spatial resolution of the phosphor layer. Items (h) and (i) of Fig. 7.4 illustrate schematically the effect of phosphor thickness on the spatial resolution of a turbid phosphor screen.

Owing to the presence of the binder material, the amount of phosphor present in a screen is usually quoted in terms of the screen loading or areal density, which is the mass of phosphor per unit area of the screen. Typical values for Gd_2O_2S phosphor screens or screen pairs range from 20 to 200 mg/cm^2, depending on the application.

7.3.2.4. Optical design

The possible optical designs for the screen structure are shown in Figs 7.4(g)–(k), which illustrate how the resolution is related to the design. Screen film combinations are not very efficient at absorbing X rays because there is such a severe trade-off between resolution and quantum detection efficiency. Only if it is acceptable to have a relatively blurred image is it possible to use a screen thick enough to be efficient in absorbing X rays. High resolution screen film combinations absorb no more than ~10–20% of the X rays, whereas general purpose screens may absorb ~30–40%. Since the X rays also pass through the film, some darkening will be developed as a result of direct interactions of X rays with the film emulsion. Usually, this is so small that it can be ignored.

The optical design of a phosphor screen critically affects its imaging performance. Factors such as phosphor grain size, size distribution, bulk absorption and surface reflectivity, as well as intentionally entrained tiny bubbles to increase scattering, can have significant effects on image quality. The thickness of the protective overcoat layer can also affect the spatial resolution of the screen. Optical effects are used to change the imaging properties of the screen. For example, as shown in Fig. 7.4(j), an absorptive backing helps reduce blurring by preferentially absorbing the photons with a long path length, but at the cost of reduced overall conversion efficiency. With a reflective backing, most of the light escapes the front of the screen and is available to be recorded. Light absorbing dye can be added to the screen to enhance the resolution, as shown in Fig. 7.4(k). This is similar but not identical in effect to an absorptive backing.

7.3.3. Photographic film and the photographic process

Photographic film is a unique material that is sensitive to a very few quanta of light. While at normal ambient temperature, it can record a latent optical image from an exposure of a fraction of a second, maintain this latent image for months, and eventually be developed without significant loss of information. It is also used as a display and archiving medium.

7.3.3.1. Film structure

The photographic process uses a thin layer (called the emulsion) of silver halide crystals (called grains) suspended in gelatine and supported on a transparent film base, as shown in Fig. 7.5(b). The grains are primarily silver bromide (~95%) with the balance being silver iodide and sometimes trace amounts of silver chloride. The grains are of variable size (micrometre) and shape (cubic or tabular, i.e. flat), depending on the application. The film base thickness is standardized to about 180 μm, to allow smooth transport through automatic film processors; the emulsion is typically 3–5 μm thick and can be on one (single sided) or both (double sided) sides of the base. During the manufacturing process, the grains are sensitized by introducing sensitivity specks on to the grains by reaction with sulphur compounds.

7.3.3.2. The photographic process

The key feature which gives unexposed film its long shelf life is that more than one light photon must impinge on an individual silver halide crystal grain in order to create a stable latent image. A single light photon creates an electron that is trapped for a short time (about a second) in a unique point on the grain called the sensitivity speck. If no other photons are absorbed by this grain, then the electron will escape from the grain. However, if a few more electrons (the exact number depends on various factors) are released in the same grain within this time, the electrons stabilize each other at the sensitivity speck and a latent image (i.e. a quasi-permanent precursor to an optical image) is established. The multielectron process is key to understanding not only the long shelf life and the non-linear response of film to light, but other behaviours, described later, such as reciprocity failure.

7.3.3.3. Development of the latent image

After exposure of the film to light, the latent image is formed at the sensitivity specks on the individual film grains. Film processing converts the latent image to a viewable permanent image. Film processing can be split into three phases — development, fixing and wash. These processes are facilitated by suspension of the grains in a thin layer of water-permeable gelatine supported on a flexible substrate. Chemicals are transported to the crystals without disturbing their positions when the film is dipped into chemical solutions. The development process turns a sensitized transparent crystal of silver halide grain into a speck of metallic silver that absorbs light and is, therefore, opaque. Since these grains are very small (<1 μm), light absorption dominates over reflection, because of multiple scattering, and they appear black in the same way as any finely powdered metal. The gain of the system in terms of the number of silver halide molecules converted into metallic silver per absorbed light photon is an extremely large number ($>10^8$), which is the basis of its uniquely high sensitivity.

7.3.3.4. Fixing of the image

After the latent image has been developed, the unexposed, and therefore undeveloped, transparent silver halide crystals remain within the gelatine layer. Thus, the emulsion is still sensitive to light and, if further exposed, would sensitize the grains, which could self-develop and change the image. During the fixing stage, these undeveloped silver halide crystals are dissolved and removed chemically, thereby fixing the image.

7.3.3.5. Wash — making it archival

Next after fixing is the water wash, where the processing chemicals and any remaining dissolved silver halide are removed, leaving only the insoluble silver grains embedded in pure gelatine. Drying removes the excess water solvent from the gelatine and results in a completely permanent archival material known as photographic film.

7.3.3.6. Automated processor design

While the development process can be performed manually with trays filled with chemical solutions for a few images per day, or deep tanks for more images, these are very labour intensive processes that are difficult to control. Automated processors are therefore used, which are much faster in operation (typically 90 s from introducing the exposed film to receiving the dried image) and are suitable for maintaining consistent image quality by keeping not only the speed point but also the complete characteristic curve consistent (see Section 7.3.4.3).

The basic concept is to use deep processing tanks kept at a constant temperature, and rollers immersed in the tanks to transport the film and ensure that it is subjected to consistent processing conditions. A key additional process is the drying step, which means that the film can be viewed as soon as it emerges from the processor. In practice, the simplest arrangement is for the processor to be built into the wall of the dark room with film removed from the cassette by a darkroom technician and a final image deposited in a tray outside the darkroom. The cassette is brought into the darkroom by an interlocked pass through or light lock, which permits darkroom technicians to keep their eyes dark adapted and reduces the risk of accidental exposure. In more advanced departments, the darkroom can be completely eliminated and the film removed from the cassette and reloaded by an automated system. Artefacts related to the automatic film processors include picking up of unwanted debris, resulting in dust specks on the developed film, and periodic variations in the intensity of the film (roller marks), which arise primarily from the variation in speed in the tanks arising from the difficulty of manufacturing and maintaining the rollers to be exactly round and smooth. In some dry environments, electrostatic discharges can cause characteristic artefacts.

7.3.4. Greyscale characteristics of film images

7.3.4.1. Optical density

A film radiograph is an image of the incident pattern of X radiation in which the distribution of the number of developed grains is related to the distribution of the number of X rays incident on the receptor. When viewed on a viewbox (source of back illumination), the developed grains reduce the light transmission through the film so that the viewed image is a negative. The transmission of light through a film, or more precisely its inverse, the attenuation, is expressed as the optical density (OD), which is defined by:

$$\mathrm{OD} = \lg \frac{I_\mathrm{I}}{I_\mathrm{T}} \tag{7.6}$$

where I_T is the intensity of light transmitted through the film resulting from the intensity of light, I_I, incident on it and lg is the common logarithm (base 10). Thus, for $I_\mathrm{I}/I_\mathrm{T} = 10$, the corresponding OD = 1. The transmissions of two absorbers placed in sequence in an optical path are multiplicative, so that the logarithms are additive. Thus, two films each of OD = 1 would have a total OD = 2.

Although the reason for a negative image is purely historical and determined by the nature of the photographic process, the display of images on digital systems where there is freedom to change is usually very similar to that found with film. This is no accident, as the look of radiographic film has been tailored over the years to be optimum for its purpose of displaying the image most efficiently to the human observer.

7.3.4.2. Densitometry, sensitometry

In a radiology department dependent on screen film, the instability of film processing systems is the single most problematic aspect. Maintaining a match in characteristics between films processed from one day to the next, from morning to afternoon, and from one film processor to another is an essential task called film quality control (see also Chapter 19). This requires daily monitoring of the processor with the exposure of a film from a standard batch using a standardized step wedge with a sensitometer, and measurement of the developed film with a densitometer. Essential to this approach is the consistent testing at the same time each day, measurement of the temperature of the solutions and keeping them chemically fresh. It is also essential to establish predetermined quantitative action points and to take associated actions to correct any deficiencies.

7.3.4.3. Hurter and Driffield curve

The curve that relates the OD of the film to the logarithm of the exposure is known as the characteristic curve or the H&D curve, after Hurter and Driffield, who first introduced its use. Figure 7.6 shows the H&D curve for a screen film receptor. The form of the curve can be explained as follows. For photographic films there is some film density, known as base plus fog, even in the absence of any exposure. The base component is the transmission of the film substrate or base. The fog is the absorption in the emulsions due, primarily, to unwanted self-development of grains unexposed to light. Together, base plus fog in fresh films are ~0.3 OD or less. A small amount of light will have difficulty in creating any kind of developable image, because of the metastability of single electron excitations. Thus, small amounts of light cause little darkening of the film. When enough light is incident, the film starts to develop (the toe), then it responds more rapidly and the approximately straight line part of the curve emerges. The gradient or gamma of the curve (i.e. its slope) actually varies continuously with OD. Once there is sufficient light to develop most of the grains, then saturation begins, giving rise to the shoulder of the curve, i.e. a flattening at high exposure. The characteristic curve can be modified in many ways. The film designer can adjust the slope and/or sensitivity of the curve, allowing adaptation to different applications by changing the grains, size distribution and grain loading of the film and the development process. The latitude of a film is the range of exposures for which a sensible response of OD occurs. In practice, it is measured between the end of the toe and the beginning of the shoulder. It is desirable that the gamma be as large as possible, to show low contrast objects while simultaneously maximizing the latitude. Satisfying these conflicting requirements is the art of producing a satisfactory film design. Typical values of gamma for radiography are in the range of 2–3.

When X rays are used to irradiate film directly, an X ray interacting with a silver grain in the emulsion can deposit sufficient energy to create a primary electron, which will, in turn, deposit its energy in the immediate neighbourhood. This will potentially create enough electrons within each of the 10–100 grains near to the primary interaction to sensitize each to the point that it will be developable. Thus, the optical H&D response is bypassed and the initial response to X ray absorption is linear in exposure. In addition, there is little risk of image fading or reciprocity failure, which accounts for its usefulness as an imaging radiation dosimeter.

7.3.5. Reciprocity

An image receptor that produces the same response for a given exposure, independent of the exposure time, is said to exhibit reciprocity. Film has remarkably good reciprocity in the range of exposures normally used in photographic cameras, i.e. from ~0.001 to 0.1 s. This also covers most exposure times encountered in medical radiography. However, for the very long exposure times of 2–5 s used in conventional film tomography and mammography, reciprocity failure can be important. The reason for reciprocity failure lies in the photographic process — the photographic grain is designed not to be sensitized unless sufficient light falls on it in a short time. Although this saves it from fogging in the dark, the by-product is reciprocity failure at long exposure times. Specifically, this means that a long exposure will need a longer time than extrapolation by reciprocity from shorter exposure would indicate. This can be of the order of a 30–40% increase in time for exposures of 2–5 s.

7.3.6. Screen film imaging characteristics

There is a dependence of speed (i.e. relationship between darkening of film to radiation) and beam quality for screen film systems, which is usually compensated for by the phototimer having ‘knowledge’ of the generator settings. Thus, it is not usually evident on a well calibrated system. For example, for a Gd_2O_2S screen with a K edge of ~50 keV and absorption of ~40% at its ‘sweet spot’ of 80 kV and usual filtration, the speed is maximized. Speed drops somewhat with increased kV, but more significantly at lower kV. Similar results

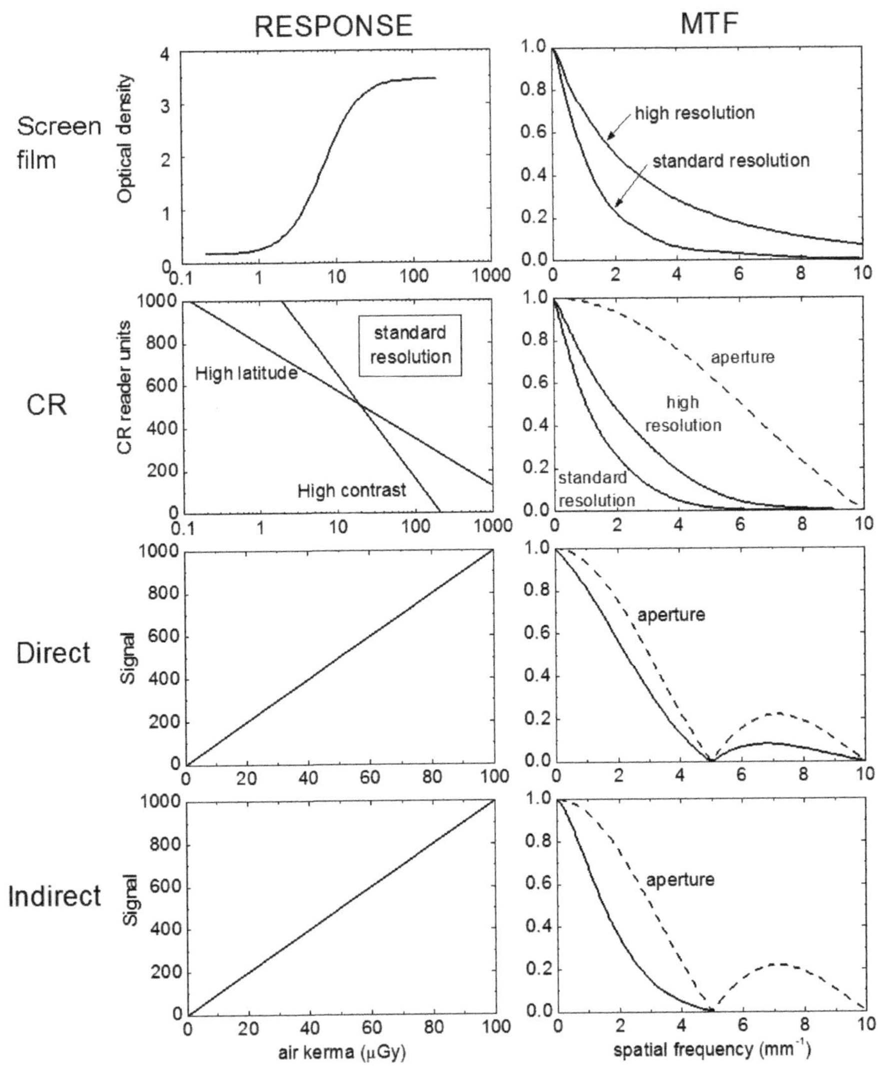
RESPONSE
MTF
Screen film
CR
Direct
Indirect
Optical density
CR reader units
Signal
high resolution
standard resolution
standard resolution
High latitude
High contrast
aperture
high resolution
standard resolution
air kerma (μGy)
spatial frequency (mm⁻¹)

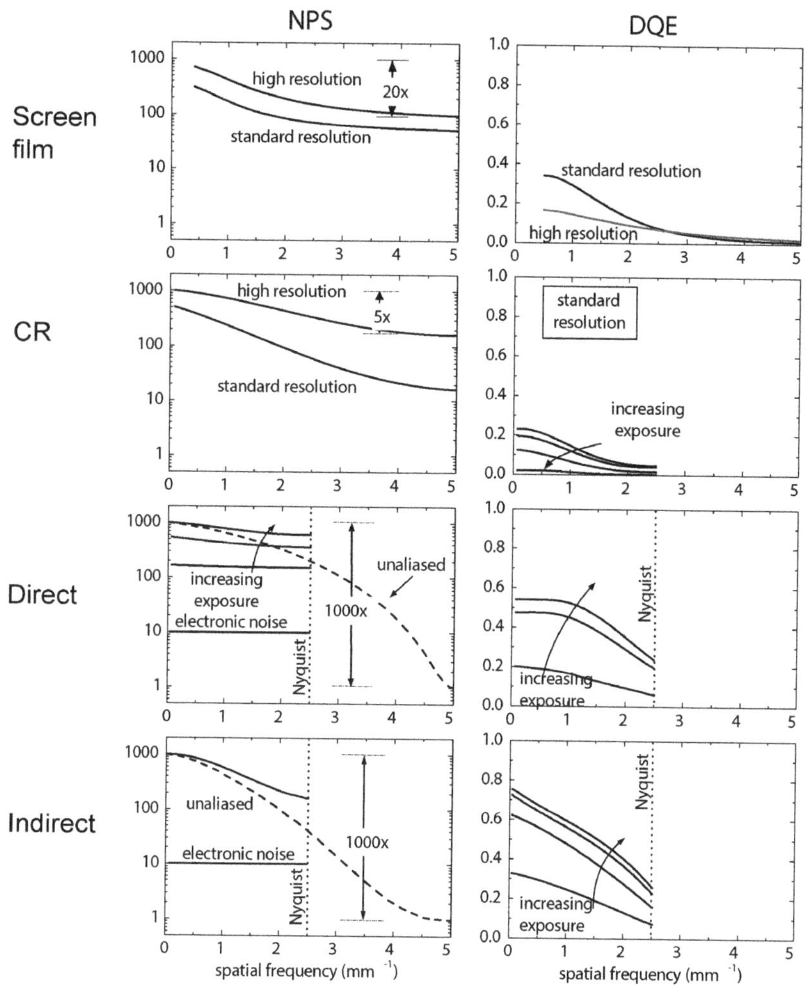

FIG. 7.6. Characteristic curves, MTF, noise power spectrum (NPS) and DQE for screen film, computed radiography (CR), direct conversion flat panel systems and indirect conversion flat panel systems.

are found for most screen film combinations, with the sweet spot depending on the K edge of the primary X ray absorber(s) in the screen.

Key factors in the design of a screen film cassette are, firstly, excellent contact between the screen and the film to prevent blurring or loss of light and, secondly, the front surface of the cassette must be easily penetrated by X rays and not cause scatter. Often there is a lead layer at the back of the cassette to control X ray backscatter from external objects whose highly blurred image could otherwise be superimposed. Film requires 3–10 photons per grain before a developable grain can be created, corresponding to an effective DQE (see Section 4.6.2) for light of a few per cent. Fortunately, the gain of the screen in terms of the light photons released per complete absorption of an X ray photon is of the order of 800 (for mammography ~20 keV) and 2000 (for diagnostic energies ~50 keV). Thus, the effective quantum gain of X rays to developable grains in the film is 800–2000 × 1–2% or 8–16 grains per X ray for mammography and 20–40 grains per X ray for general radiography. This is the information on which the quantum accounting diagram for screen film was established (Fig. 7.3).

Each X ray interacting with the screen creates many light photons, but because of the small size of the film grains (<1 μm) compared with the screen blurring (~100–500 μm), usually only one, or at most a few, light photons from the same X ray will interact with each individual grain. Thus, there will be essentially no correlation between individual X ray events and individual grains, resulting in the same shape of the response curve for the screen film system when irradiated by X rays as for the film when irradiated by light (i.e. the optical H&D curve as measured with a sensitometer colour matched to the emission of the phosphor). This is a key point in validating film sensitometry for radiography.

The variation of the probability of X ray interaction with depth in the phosphor screen is exponential, so that the number of interacting quanta and the amount of light created will be proportionally greater near the X ray entrance surface. The highest resolution screen film systems are therefore generally configured from a single screen placed such that the X rays pass through the film before impinging on the phosphor. This back screen configuration improves the spatial resolution of the final image compared with the alternative front screen configuration. (It can be noted that owing to the thickness (~0.7 mm) of the standard glass substrate currently used for active matrix fabrication, and its consequent significant absorption of X rays, flat panel systems are all configured in the front screen orientation.) However, for all but the highest resolution requirements, dual screens are used, as they can make a better trade-off between high quantum detection efficiency and high resolution. Screen nomenclature always stresses the positive. A high resolution and low A_Q screen is referred to as a high resolution screen; a low resolution screen has a high A_Q.

7.3.6.1. *MTF, NPS and DQE of screen film systems*

The MTF of screen film primarily depends on the need of the application. It can be excellent, especially as measured by the single criterion of limiting resolution (usually defined at the frequency, f, for which the MTF(f) drops to 4%). The variation of MTF with screen thickness/loading is very significant.

Figure 7.6 shows the variation of MTF with spatial frequency f for two double screen systems for different applications (Lanex Fine for high resolution — bone — and Lanex Regular for general radiography) and hence different screen thicknesses and optical design. Digital systems have only recently been able to approach their capabilities. It was controversial for many years as to whether or not high frequency information available only in screen films was necessary in mammography. The answer to this question is obtained by looking at the NPS for screen film, also shown in Fig. 7.6, which demonstrates that the noise at high spatial frequency reaches an asymptote where the noise is white and quantum noise is negligible, owing to the negligible MTF of the screens at these frequencies. This is seen to be a factor about 20 times lower than the noise power extrapolated to zero spatial frequency.

This result could have been predicted from the quantum accounting diagram (Fig. 7.3), which shows that the secondary quantum sink is 20 silver grains per absorbed X ray. The serious consequence for the DQE of screen film, also shown in Fig. 7.6, is that the DQE shows an extremely rapid drop off with f, which results in merging of the standard and high resolution screens above $f = 2.5$ mm^{-1}, to an essentially negligible value, despite the high resolution screen showing a noticeable MTF, even at $f = 10$ mm^{-1}.

7.4. DIGITAL RECEPTORS

7.4.1. Digital imaging systems

In a digital imaging system, at some stage, the incident X ray image must be sampled in both the spatial and intensity dimensions. In the spatial dimension, samples are obtained as averages of the intensity over receptor elements or dels. These are usually square and spaced at equal intervals throughout the plane of the receptor. The pixel is the corresponding elemental region of the image. In the intensity dimension, the signal is digitized into one of a finite number of levels, which are expressed in binary notation as bits. To avoid degradation of image quality, it is essential that the del size and the bit depth, n (when n is given by 2^n), are appropriate for the requirements of the imaging task. The matrix size or the coverage of the array is different, depending on the application and the size of the body part to be imaged and the magnification.

The fractional area of the del that is active has an upper limit of unity but can often be smaller than that, because of a reduced fill factor. The linear dimension of the active portion of each del defines an aperture. The aperture determines the spatial frequency response of the receptor. If the aperture is square with dimension d, then the MTF of the receptor will be proportional to $|\text{sinc}(\pi fd)|$ times the intrinsic MTF of the equivalent analogue receptor, where f is the spatial frequency along the x or y direction, and the MTF will have its first zero at the frequency $f = 1/d$, expressed in the plane of the receptor.

For example, a receptor with d = 200 μm will have an MTF with its first zero at f = 5 cycles/mm. Also of considerable importance is the sampling interval, p, of the receptor, i.e. the pitch in the receptor plane between corresponding points on adjacent dels. The sampling theorem states that only frequencies, f, in the object less than $<1/2p$ (the Nyquist frequency, f_N) can be faithfully imaged. Thus, if the pattern contains higher frequencies, then a phenomenon known as aliasing occurs, wherein the frequency spectrum of the image beyond the f_N is mirrored or folded about the f_N in 'accordion-like' fashion and added to the spectrum of lower frequencies, increasing the apparent spectral content of the image below f_N (see Section 4.2.3).

It is important to realize that both signal and noise can show aliasing effects. In a receptor composed of discrete elements, the smallest possible sampling interval in a single image acquisition is $p = d$. Even in this most favourable case, $f_N = 1/2d$, while the aperture response only falls to zero at $2f_N$. If the del dimension is less than the sampling interval, then the zero is at $>2f_N$, further increasing the aliasing.

In Fig. 7.7, the phenomenon of aliasing is demonstrated with an image of a radiographic bar pattern. Aliasing can be avoided by band limiting the image before sampling, i.e. attenuating the higher frequencies such that there is no appreciable image content beyond f_N. This may be accomplished by other blurring effects intrinsic to the receptor. These mechanisms can have different effects on the noise and signal and may not necessarily reduce noise and signal aliasing in the same manner.

7.4.2. Computed radiography

In computed radiography (CR) (Fig. 7.8), the imaging plate (a screen made using a photostimulable phosphor) is positioned in a lightproof cassette exposed to a patient X ray image and then produces a digital image in a system that extracts the exposed plate from the cassette while protecting it from ambient light, reads it out, erases the image and returns it to the user in the cassette ready to be reused. CR belongs to a class of systems that could be called reusable plate technologies, and directly replaces screen film. There are currently no competing technologies for the reusable plate, despite the fact that the image quality of CR is poorer than for digital radiography (DR) (see Section 7.4.3) and that it requires a longer exposure to produce acceptable images.

FIG. 7.7. Aliasing effect on the image of a bar pattern. Top: original. Centre: the pixel pitch is sufficiently small and all bar patterns are resolved correctly despite some blurring. Bottom: the pixel pitch is too large to resolve the finest bar pattern and aliasing occurs.

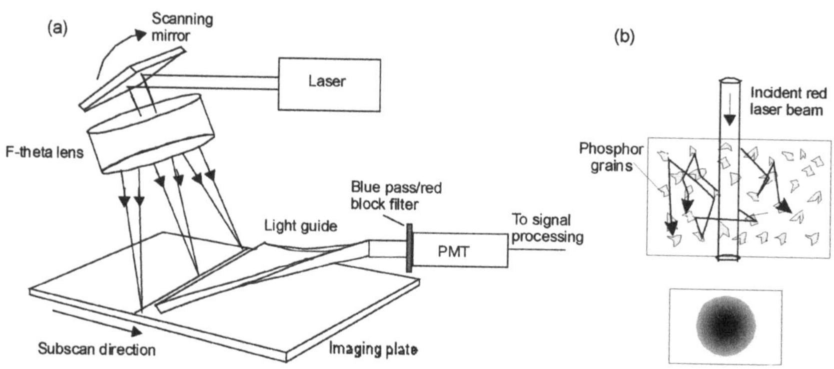

FIG. 7.8. CR system based on the use of reusable photostimulable phosphor plates housed in cassettes. (a) Readout of the plate in a laser scanner with photostimulated light collected in a light guide and detected by a photomultiplier tube. (b) The primary source of light blurring in CR; the light that comes out of the phosphor grains as shown in Fig. 7.4 is irrelevant and does not contribute to resolution loss, but rather the spreading by scattering of the incident exciting light gives rise to the blurring.

7.4.2.1. Method of latent image formation

The photostimulable phosphor screen in the imaging plate is very similar to a conventional X ray screen, except it uses a phosphor that contains traps for excited electrons as shown in the band structure diagram in Fig. 7.2. Most photostimulable phosphors are in the barium fluorohalide family, typically BaFBr:Eu. X ray absorption mechanisms are identical to those of conventional phosphors. The photostimulable phosphors differ in that the useful optical signal is not derived from the light emitted in prompt response to the incident radiation as in conventional screen film systems, but rather from subsequent stimulated light emission when EHPs are released from traps. The initial X ray interaction with the phosphor crystal causes EHPs to be generated. Some of these electrons produce blue/green light in the phosphor in the normal manner but this is not used for imaging. Instead, the phosphor is intentionally designed to contain metastable EHP traps that store a latent image as a spatial distribution of trapped electrons and holes. By stimulating the phosphor by irradiation with red light, these EHPs are released from the traps and are free to move in the valence band (holes) and conduction band (electrons). These mobile EHPs subsequently trigger the emission of shorter wavelength (blue) light. CR screens are exposed in a cassette with only one screen. This reduces the absorption efficiency compared with a dual screen combination and this is an intrinsic disadvantage of CR compared with screen film imaging.

7.4.2.2. Image readout

The readout system shown in Fig. 7.8(a) for photostimulable phosphor plates uses a red laser beam flying spot scanning system to stimulate the screen on a point by point basis, exciting photostimulated light (usually blue) from the screen in proportion to the previous X ray irradiation of that point. In practice, depending on the laser intensity, readout of a photostimulable phosphor plate yields only a fraction of the stored signal. The blue light is collected by a light guide that is a critically important component in the avoidance of a secondary quantum sink, i.e. the light is funnelled to a photomultiplier tube that detects and amplifies the signal. Note that in contrast to the situation with screen film, the scattering of the blue light does not degrade the image resolution, unlike the scattering of the red light on its way to the photocentres does, as shown in Fig. 7.8(b). Finally, the image plate is then flooded with light to erase any residual image and is then ready for reuse. The advantage of photostimulable phosphor systems is that they are digital systems with a very high dynamic range.

Examination of the quantum accounting diagram in Fig. 7.3 for scanned readout of the flying spot laser readout of CR shows that the incident absorbed

X ray, although originally represented by multiple quanta, is at a critical stage in the chain represented by only about five electrons in the photomultiplier. This is not adequate in keeping the additional noise from the system negligible when the relatively poor MTF is also factored in. This can be seen in the NPS for CR in Fig. 7.6.

7.4.2.3. Image properties

The underlying physical phenomenon of the photoluminescence process is expected to be linear and this is demonstrated over the exposure range relevant for medical imaging in Fig. 7.6. The variation of the signal with exposure is essentially linear over the four orders of magnitude relevant for diagnostic radiography. In practice, the photomultiplier signal in most CR readers is processed logarithmically before being output as shown in Fig. 7.6. It appears at first sight that the characteristic curve of CR is radically different from all other imaging systems. This is because images using both screen film and CR are designed for radiography and it is conventional to use negative images (sometimes called white bone). The conventional mode of displaying the H&D curve hides this. Depending on the application and the plate type used, the relationship is different but consistent between settings. Various latitude and speed points can be set to mimic the behaviour of screen film systems.

The MTF is shown in Fig. 7.6 for high resolution and normal resolution imaging plates. It can be seen that they are comparable to those for corresponding screen film systems also seen in Fig. 7.6. This is not surprising as they are based on similar concepts to those for screens, and manufacturers can make any resolution simply by changing the thickness. The real test of equivalence will only come when we compare DQE.

The overall image quality for CR can be evaluated from the DQE shown in Fig. 7.6 for a normal resolution plate. It is, however, interesting to see how closely the screen film systems match the results for CR. This is related to the high level of secondary quantum noise evident in the NPS of CR, worse in fact than for screen film, but not unexpected when the quantum accounting diagrams for the two modalities are compared, as in Fig. 7.3, where it can be seen that both have secondary quantum sinks in the range 5–20 quanta per X ray.

All screens are made in similar ways and it is impossible to make them completely uniform in properties, despite best efforts. As well as quantum noise, therefore, the image from an entirely uniform exposure will contain noise due to the non-uniformity and granularity of the underlying structure. How does this manifest itself in images? Suppose there was a 1% variation in effective thickness of the plate, then all images would reflect that change, giving an effective signal to noise ratio (SNR) of 1%. On the other hand, the SNR due to quantum mottle decreases as the square root of exposure. Thus, we reach the apparently surprising conclusion that structural noise will be most important when the exposure is high. This is shown experimentally for a CR system in Fig. 7.6, where the DQE decreases as the exposure increases. The same effect will be present in screen film systems, although here variations in screen uniformity and film granularity will both be involved. For screen film systems, it is impossible to correct for structure noise, though it could perhaps be accomplished for CR if efforts were made to ensure that the imaging plate was always positioned exactly within the reader. However, this is not done and is probably unnecessary, as the loss of DQE due to structural noise occurs at relatively high exposure levels, generally outside the clinically important regions of the image.

7.4.3. Digital radiography

The key digital technology permitting an advance in medical X ray applications is the flat panel active matrix array, originally developed for laptop computer displays. The unique technology underlying active matrix flat panel displays is large area integrated circuits called active matrix arrays because they include an active switching device — the thin film transistor.

Active matrix arrays are an example of an important class of readout methods that are called self-scanned. A self-scanned readout structure may be defined as one in which the image created in a certain plane is read out in the same plane. The advantage of such structures is that they are thin in the third dimension. Active matrix arrays use hydrogenated amorphous silicon as the semiconductor deposited on a thin (~0.7 mm) glass substrate, which facilitates large area manufacture. The current scale on which the substrate is made exceeds several metres along one side, thus permitting several sensors to be made on a single substrate, and facilitating mass production of monolithic devices. Although long promised, it is not yet feasible to make all the readout components on the glass substrate; gate drivers, readout amplifiers and analogue to digital converters, for example, are still made separately and bonded to the substrate. A complete imaging system on glass remains elusive. Another large class of self-scanned receptors — charge coupled devices (CCDs) and complementary metal oxide semiconductor sensors (CMOS) — although used extensively in optical systems incorporating demagnification, such as those for X ray image intensifiers (XRII), needs to be made from single crystal silicon. Recent advances in CMOS are making possible the economic manufacture of large area (wafer scale) devices, which are replacing some XRII/camera systems in fluoroscopic applications.

Active matrix technology allows the deposition of semiconductors, across large area substrates in a well controlled fashion, such that the physical and electrical properties of the resulting structures can be modified and adapted for many different applications. Coupling traditional X ray detection materials, such as phosphors or photoconductors, with a large area active matrix readout structure forms the basis of flat panel X ray imagers.

In Fig. 7.9, the layout of a group of pixels on an active matrix array is shown. All the switches along a particular row are connected together with a single control line (gate line). This allows the external circuitry to change the state of all the switching elements along the row with a single controlling voltage. Each row of pixels requires a separate switching control line. The signal outputs of the pixels down a particular column are connected to a single data line with its own readout amplifier. This configuration allows the imager to be read out one horizontal line at a time. Unlike the charge transfer readout method used in many modern CCDs, active matrix arrays do not transfer signal from pixel to neighbouring pixel but from the pixel element directly to the readout amplifier via the data line; this is similar to complementary metal oxide semiconductor imaging devices.

A distinction is made between flat panel X ray imaging devices that incorporate a photoconductor to produce electrical charges on detection of an X ray (direct conversion) and those that use a phosphor to produce visible wavelength photons on detection of an X ray (indirect conversion).

In the direct conversion approach shown in Fig. 7.9(a), a photoconductor is directly evaporated on to an active matrix array. The charge released in the bulk of the photoconductor is collected by a large applied field, which brings the electrons and holes to their respective electrodes. Those reaching the upper

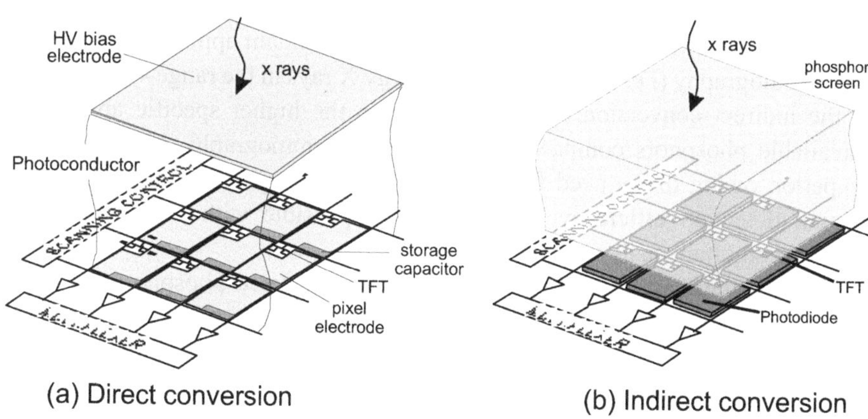

FIG. 7.9. Large area active matrix array concept which is applicable for both (a) direct conversion (using a photoconductor layer) and (b) indirect conversion systems (using a phosphor layer), depending on whether a pixel electrode or photodiode is used on the active matrix array. In both cases, thin film transistors are made using semiconductor hydrogenated amorphous silicon (a-Si:H).

continuously biased electrode are neutralized by the power supply providing the bias, effectively replenishing the loss of field that would otherwise be caused. The charge reaching the readout pixel electrode is stored temporarily on the capacitance of the pixel until readout. The magnitude of the signal charge from the different pixels contains the imaging information inherent in the intensity variations of the incident X ray beam.

In the indirect conversion approach shown in Fig. 7.9(b), a phosphor layer (e.g. a structured scintillator such as CsI:Tl) is placed in intimate contact with an active matrix array. The intensity of the light emitted from a particular location of the phosphor is a measure of the intensity of the X ray beam incident on the surface of the receptor at that point. Each pixel on the active matrix has a photosensitive element that generates an electrical charge whose magnitude is proportional to the light intensity emitted from the phosphor in the region close to the pixel. This charge is stored in the pixel until the active matrix array is read out.

There are, in principle, two advantages of direct compared with indirect conversion: (i) the lower number of conversion stages makes it possible to have a significantly higher conversion efficiency of X ray energy to EHPs on the active matrix (~8000 direct versus 1000–2000 indirect), and (ii) much higher resolution due to the elimination of blurring during the charge (direct) or photon (indirect) collection phase. Proposed photoconductors for this purpose include HgI_2, CdTe and PbO. However, there is only one currently practical photoconductor, a-Se; its conversion efficiency is, in fact, comparable to phosphors, and technical issues limit its thickness, so it is used mostly in mammography where its low Z is ideally matched to the absorption requirements and high resolution is attained (see Section 9.4.2). Thus, at the current time, the dominant approaches to digital general radiography (i.e. using diagnostic energy X rays in the range 40–150 keV) use the indirect conversion approach owing to the higher specific absorption of available phosphors compared with a-Se. In mammography, however, a-Se is superior, owing to the need for a low energy X ray spectrum, increasing the quantum detection efficiency beyond what is possible with a phosphor and because it simultaneously increases resolution.

One of the main issues with the design of powdered phosphor screens is the balance between spatial resolution and X ray detection efficiency. As the phosphor is made thicker to absorb more X rays, the emitted light can spread further from the point of production before exiting the screen. This conflict is significantly eased by the use of a structured phosphor such as CsI. When evaporated under the correct conditions, a layer of CsI will condense in the form of needle-like, closely packed crystallites (see also Section 8.2.1). In this form, the resolution is better than for a powder phosphor screen. However, resolution may be further enhanced by fracturing into thin pillar-like structures by exposure to thermal shock. This has the disadvantage of reducing the effective density of the structured layer to about 80–90% that of a single CsI crystal. This pillar or columnar structure is illustrated in Fig. 7.4(l). The hope was that these columns would act as fibre optic light guides, owing to the difference in refractive index, n, between CsI ($n = 1.78$) and the inert gas or air ($n \sim 1$) that fills the gaps between the pillars.

Taking this model literally, light photons produced by the absorption of an incident X ray will be guided towards either end of the pillar if they are emitted within the range of angles that satisfy conditions for total internal reflection. Theoretical calculations predict that ~83% of the isotropically emitted light will undergo internal reflection within a perfectly uniform pillar. The other ~17% will scatter between pillars and cause a reduction in the spatial resolution. Actual layers of CsI have a somewhat reduced light collimating capability, owing to the unavoidable non-uniformity of the surface of the pillars, unavoidable contacts between adjacent pillars at various points within the layer, and defects in the cracking.

In addition, in practice, the layers form with an initial layer which is not columnar, and the columns develop beyond a certain thickness (~50 μm). However, they maintain significantly higher resolutions for a given thickness of phosphor than powder screens. As a rule of thumb, they seem to have twice the resolution of a powder phosphor screen of the same physical thickness and optical design (e.g. presence or absence of reflective backing). This corresponds to 3–4 times the mass loading for the structured phosphor layer when packing density and atomic number are accounted for.

To increase the light collection capabilities of the layer, a reflective backing can also be added to the X ray entrance surface of the CsI, to redirect the light photons emitted in this direction back towards the exit surface. This significantly increases the light output of the layer but at the cost of a reduced spatial resolution.

7.4.3.1. MTF, NPS and DQE of DR systems

The MTFs of the direct and indirect systems show a distinct qualitative difference. The overall MTF of each system is the product of the MTF of the X ray detection medium and the del aperture function, which, for a uniform response, is the sinc function. The intrinsic spatial resolution of a photoconductor is extremely high, which means that the system MTF for a direct conversion receptor will be, to all intents and purposes, the sinc function, as all other components are close to unity.

In a system that utilizes a phosphor as the X ray detection medium, the spatial response of the phosphor is much poorer than the sinc function. The MTF of the combination will then be practically equivalent to the phosphor MTF. In other words, for the a-Se system, it is the aperture function that defines the presampling MTF, whereas for the phosphor based system (even columnar CsI), the phosphor blurring dominates and defines the overall MTF. Furthermore, for the parameters usually chosen in practical medical systems, the MTF at the f_N is ~60% for the direct system and closer to 10% for the indirect system. This implies that, in general, noise aliasing will be severe for the direct and almost negligible for the indirect detection approaches.

The degree of aliasing permitted by the system designer can be established from an evaluation of the presampled MTF and a calculation of the area above f_N compared with the area below. See Fig. 7.6 for examples of MTFs for well matched systems using 200 μm pixels and a 200 μm sampling pitch for both direct and indirect conversion systems at the same incident energy.

The NPSs of direct and indirect receptors are shown in Fig. 7.6 and demonstrate striking differences. The NPS of the direct receptor, owing to its minimal spatial filtration prior to sampling, combined with aliasing of the frequencies above the f_N, is almost white (i.e. is independent of spatial frequency). In contrast, the indirect receptor shows a marked drop in NPS with increasing frequency, owing to the greater presampling blurring inherent in the phosphor layer demonstrated by comparing the MTFs. The NPS for higher frequencies shows a linear exposure dependence but it does not go to zero at zero exposure — this is the electronic noise that, at the pixel level, is of the order of several thousand electrons root mean square. This noise is to be compared with the signal per X ray, which is of the order of 1000.

Finally, by combining MTF and NPS, the DQE(f) can be obtained (see Fig. 7.6). In the particular direct conversion receptor illustrated, the DQE(0) is somewhat smaller, owing to the relatively poor X ray absorption efficiency of the photoconductor, whereas DQE drops very little with spatial frequency, owing to the very high MTF (which is close to the ideal sinc function). In contrast, the DQE(0) of the indirect conversion system is higher, owing to better X ray absorption efficiency, but the DQE drops more rapidly with increasing spatial frequency, owing to the poorer MTF (which is mitigated to some degree by the reduction of NPS with frequency, owing to the drop in MTF).

7.4.4. Other systems

7.4.4.1. Optically coupled systems

Indirect conversion flat panel systems use optical coupling of an imaging screen to an active matrix array. Earlier systems used smaller receptor arrays in conjunction with a fibre optic or a lens to couple the image to other optical devices, such as a CCD or a video camera (Figs 7.10(a) and (b)). However, there is then a considerable loss of light, depending on the collection angle of the optical system compared with the broad angle over which light is emitted from screens.

These methods, therefore, have significant problems in maintaining good noise properties. This is because the efficiency of collecting the light from the screen by the imaging device is generally rather poor. For example, even in the best optically coupled lens system with a pair of relay lenses of the same focal length placed back to back, the coupling efficiency cannot practically exceed 20% and this is only possible with very high aperture lenses (*f*/0.75).

This is the case for 1:1 imaging. With demagnification M (the ratio of the side of the screen to the corresponding image sensor dimension) greater than unity, the coupling efficiency drops by M^2. Thus, for commonly seen demagnifications of 20, the coupling efficiency is 0.05%. Under these conditions, only ~1 light photon on average represents the interaction of each X ray. This is a serious secondary quantum sink. This may manifest itself in two general ways, the first being the inability to represent fully the position of the X ray, resulting in a decrease in DQE(0) and a much more rapid decrease in DQE with f, and the second being due to the very small gain resulting in amplifier noise becoming dominant at much higher exposure levels than it would otherwise.

In the presence of these kinds of secondary quantum sinks, the transfer of light from the screen to the receptor becomes the limiting stage in the imaging chain and significantly degrades the overall performance of the system. The use of a fibre optic taper can alleviate, but not eliminate, the loss due to demagnification, for the reason that the acceptance angle for light decreases in the same way as for lenses. Thus, a major technical advantage for the flat panel receptor (in an indirect detection configuration) is that it can be placed in direct contact with the emission surface of the screen, as shown in Fig. 7.10(c). Its collection efficiency for the emitted light is consequently much higher (~50% and approaching 100% for special configurations) than with the demagnifying approaches.

It is interesting to compare the situation for an X ray image intensifier (Fig. 7.10(d)). Here, by converting light photons from the phosphor to electrons in the photocathode in direct contact with the phosphor, the ability to bend the path of electrons, which is impossible for light, is critical in maintaining much higher collection efficiency and so avoid a secondary quantum sink. This is a critical reason why X ray image intensifiers were important for so long, despite their other problems (see Chapter 9).

7.4.4.2. Photon counting

Photon counting receptors, and their close relatives photon energy discriminating receptors, interact with incident photons one by one and

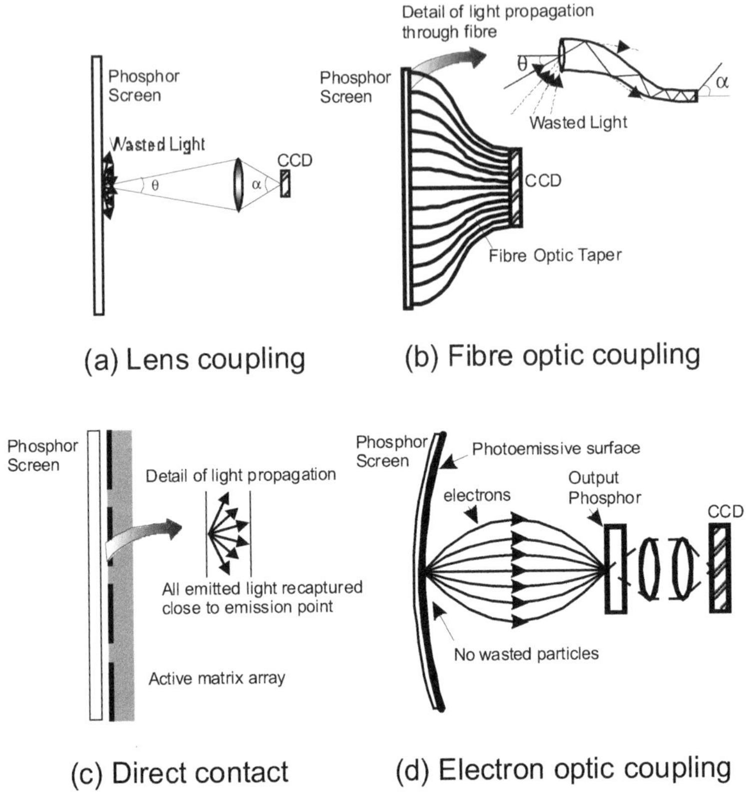

FIG. 7.10. Optical coupling through (a) lenses, (b) demagnifying fibre optic taper, (c) direct contact and (d) electron optical demagnification in an X ray image intensifier.

'report back' to the system that a photon has been detected (a count), or a photon within a specific energy range has been detected (energy discriminated count). The potential advantage in image quality of these systems is significant. There are two reasons; the first is that they can entirely eliminate the effect of amplifier noise. This is because the signal from a single X ray emerges in a very short time and can thus be made to be >5 times the noise. This advantage makes it possible to work at very low exposure rates where other systems would be dominated by amplifier noise. The second reason is that by knowing the size of the signal, the energy of the incident X ray may be estimated, and thus correction for Swank noise performed, and in the context of the complete system, better weighting of the importance of high energy (highly penetrating, low contrast) versus lower energy (less penetrating, higher contrast) X rays can be accomplished. This can perhaps increase the SNR by factors of the order of two. Unfortunately, these improvements pale in comparison to the increase in complexity of the circuitry necessary at each pixel, which may generally be of the order of 1000 to 100 000 fold. This practical factor has, to this point, limited the application to mammography, but improvements in microelectronic circuitry are reaching the point where more general application may be feasible in the near future.

7.4.4.3. Scanning geometries

Answering the question of what X ray geometry should be used is the first fundamental decision facing the designer of an X ray imaging system. Conventionally, producing an X ray image involves exposing the entire area of interest simultaneously and detecting it with an area sensor as in screen film radiography. Other approaches are possible, but the simplest is to obtain a pencil beam of radiation (accomplished by collimating the broad area flux from the X ray tube by means of a lead blocker with a small hole in it) and scanning it over the patient, one point at a time. A single sensor aligned with the pencil beam creates an image of the patient.

Other variations between pencil and area beams can be seen in Fig. 7.11. Slit irradiation is obtained with a fan beam of radiation and an aligned single line receptor that is scanned perpendicularly to the line across the patient. Both pencil beam and slit beam scanning are extremely inefficient in the utilization of X rays. Most of the X rays are removed by the collimator and a full scan imposes an enormous heat load on the tube. It is possible to improve the efficiency of such systems by employing a multiline or slot receptor, where the X ray beam extends across the full image field in one dimension and is several lines wide in the other (see also Section 6.3.5.4). There are two types of slot receptor. Firstly, a slot is moved discontinuously across the width, a single exposure made and the multiple lines read out. This process is repeated until the entire area is covered. Secondly, in time domain integration, the slot beam receptor moves continuously and the image is read out one line at a time.

Why use these complicated scanning methods to produce images when it appears that irradiating a static area receptor is much simpler? Several concepts must be balanced. The first is relative simplicity of construction. In an X ray scanning system designed to image patients, there must be no significant wasted radiation, which means that very accurate precollimation of the X ray beam must be used. This is made more difficult by the requirements of scanning the system. In the early development of digital radiographic systems, it was only technically feasible to create linear receptor arrays and no practical area arrays existed; thus, the mechanical complexities were acceptable since there were no alternatives.

However, history has shown that when area receptors are feasible they are to be preferred. Each method has image quality advantages and disadvantages but the most important consideration is scattered radiation. A reduced area receptor can be much more efficient than area receptors in eliminating scatter. However,

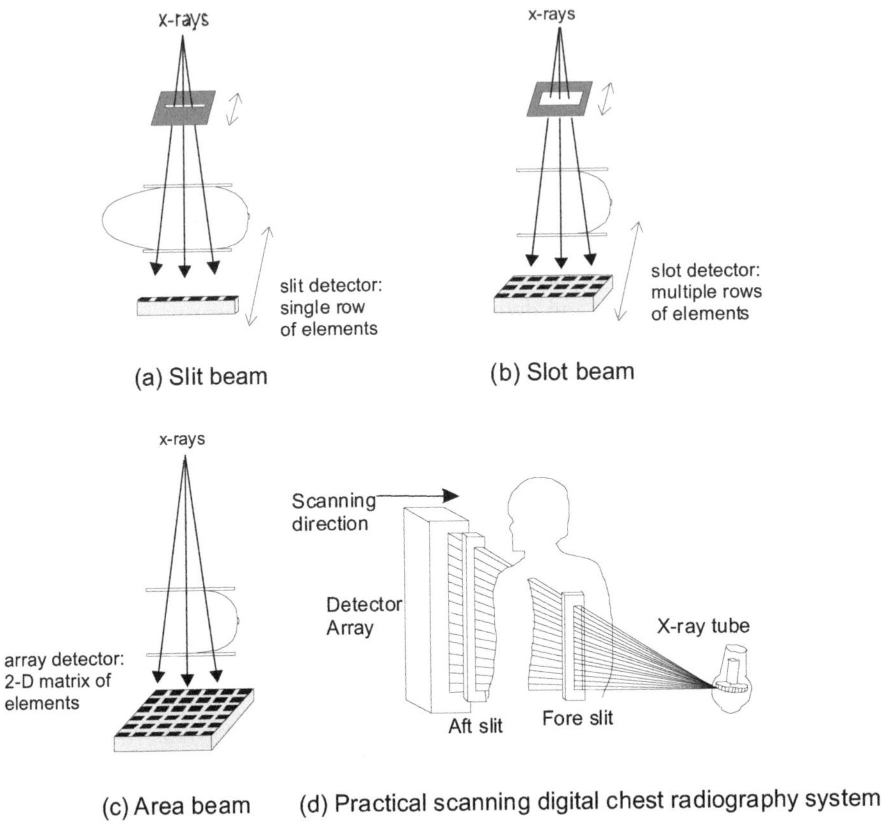

FIG. 7.11. Geometry of radiation possible in radiography in addition to pencil beam (not shown). (a) Slit beam; (b) slot beam; (c) area beam; (d) practical scanning DR system.

the shortest exposure and the least loading of the tube are huge strengths of the area receptor, which make it the preferred option unless control of scatter is paramount. Existing systems that use scanning systems are: (i) highly specialized devices where scatter control to ensure quantitative imaging is essential, for example, dual energy imaging for bone densitometry where the complete and exact elimination of scatter overwhelms other concerns (see Section 10.4), and (ii) devices that permit photon counting approaches where the technical demands of the additional counters and discriminators needed at a per pixel basis are currently prohibitive for an area receptor but are feasible, for example, by using several silicon receptors in an edge-on configuration.

7.4.5. Artefacts of digital images

The raw image information acquired from the current generation of flat panel receptor systems is unsuitable for immediate image display. It must be processed to remove a number of artefacts to obtain a diagnostic quality radiograph.

7.4.5.1. Moiré

A particularly visually disturbing effect, which is to be avoided at all costs, is Moiré fringing arising from spatial interference between the periodic structure of flat panel receptors and a stationary grid. Moving the grid perpendicularly to the grid lines during the exposure, using a Potter–Bucky grid arrangement, should eliminate this problem.

7.4.5.2. Ghosting and lag

Corrections for image carry over or lag (these are effects seen in dark field exposure after prior exposure) or ghosting (these are effects producing change in gain related to prior images and so are seen in flood field images) may sometimes be necessary. These phenomena may be particularly problematic after large exposures to the imager, or when the imager is used in mixed mode (i.e. receptor designed to be capable of both fluoroscopic and radiographic imaging) and the system is moved to fluoroscopy after a large radiographic exposure.

7.4.6. Comparisons of digital and analogue systems

Advantages of digital over analogue systems for radiography include:

Advantages related to image quality and dose:

- Generally a lower dose is needed;
- Higher resolution in some situations;
- Greater dynamic range.

Advantages related to convenience in use:

- Elimination of handling and carrying of cassettes;
- Immediate evaluation of images for image quality and positioning;
- Transmission of digital images;

- Digital archiving, searching picture archiving and communications systems;
- Elimination of unique image (film);
- Image processing to present the image information more optimally to the reader;
- Elimination of distortion and shading (cf. X ray image intensifiers);
- Enabling advanced applications (e.g. digital tomosynthesis, cone beam CT, dual energy imaging and computer aided detection).

The image quality of radiographic detectors has experienced a quantum jump in the last decades as flat panel imagers have become feasible. However, there is still no completely satisfactory system and the cost is very high compared with the systems they have replaced. There is still much to be done to provide quantum limited performance for all radiographic imaging receptors at reasonable cost.

BIBLIOGRAPHY

DAINTY, J.C., SHAW, R., Image Science, Principles, Analysis and Evaluation of Photographic-type Imaging Processes, Academic Press, London (1974).

ROWLANDS, J.A., Topical review: The physics of computed radiography, Phys. Med. Biol. **47** (2002) R123–R166.

ROWLANDS, J.A., YORKSTON, J., "Flat panel detectors for digital radiography", Medical Imaging Vol. 1, Physics and Psychophysics (BEUTEL, J., KUNDEL, H.L., VAN METTER, R.L., Eds), SPIE Press, Bellingham, Washington, DC (2000) 223–228.

Chapter 8

FLUOROSCOPIC IMAGING SYSTEMS

A.K. JONES
University of Texas MD Anderson Cancer Center,
Houston, United States of America

8.1. INTRODUCTION

Fluoroscopy refers to the use of an X ray beam and a suitable image receptor for viewing images of processes or instruments in the body in real time. Fluoroscopic imaging trades the high signal to noise ratio (SNR) of radiography for high temporal resolution, as factors that maintain patient dose at an acceptable level must be used.

8.2. FLUOROSCOPIC EQUIPMENT

Fluoroscopic imaging systems use much of the same technology as radiographic systems, with some modifications and additions. Depending on the intended use, a fluoroscopic system may require a high power generator and a high heat capacity X ray tube. The major difference between radiographic and fluoroscopic equipment is the image receptor. Early fluoroscopic systems used an intensifying screen, similar to that used in radiographic screen film imaging, that was viewed directly by the radiologist. However, direct view systems produced dim images that required the radiologist's eyes to be dark adapted, and frequently resulted in high doses to both patient and radiologist. The development of the X ray image intensifier (XRII) was essential to the success of modern fluoroscopic imaging.

8.2.1. The fluoroscopic imaging chain

Fluoroscopic imaging systems commonly include an antiscatter grid as the first element in the imaging chain. The focused antiscatter grid serves the same purpose in fluoroscopic imaging as it does in radiographic imaging, namely, removing contrast degrading scattered radiation from the X ray beam (see Chapter 6).

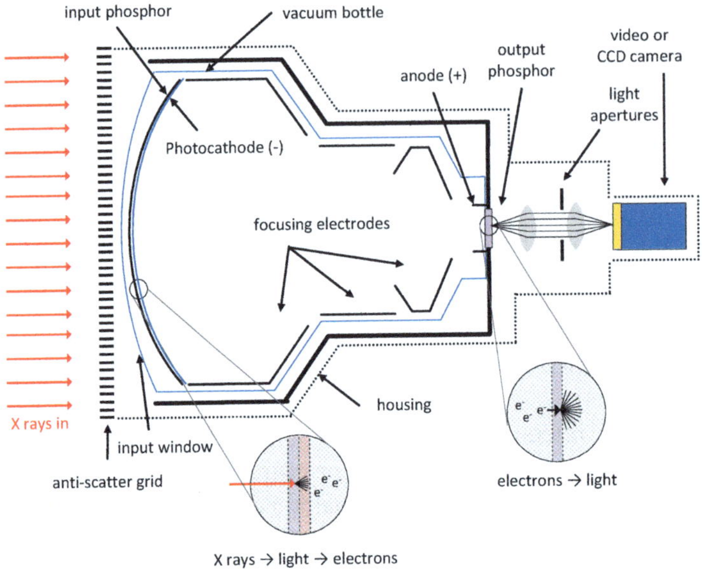

FIG. 8.1. Basic structure of an XRII.

The XRII converts the low intensity X ray photon fluence exiting the patient into a high fluence of visible photons by using multiple conversion layers and a series of electrodes inside a vacuum container (Fig. 8.1). X rays that have passed through the patient, antiscatter grid and the metal support structure of the XRII are incident on an input phosphor. Caesium iodide (CsI:Tl) is the most commonly used input phosphor material used to convert X rays to light. The needle-like structure of crystalline CsI (Fig. 8.2) minimizes the lateral spread of light within the input phosphor (see Section 7.4.3). These light photons strike a very thin bi- or multi-alkali photocathode[1], where electrons are released through the photoelectric effect, repulsed from the photocathode, and accelerated towards the positively charged anode at the output phosphor. The electrons are converted back to light at the output phosphor by a thin powder phosphor, typically ZnCdS:Ag (P-20). The electron beam is accelerated to energies of between 25 and 30 keV, and is focused on to the anode by a series of electrodes located at the sides of the XRII (Fig. 8.1). Typical incident air kerma rates (IAKRs) for fluoroscopic imaging using a 30 cm field of view (FOV) range from 15 to 40 μGy/min (8.8–22 nGy/frame) and may vary, based on the selected frame rate.

[1] The photocathode (about 20 nm thick) is very sensitive to surface contamination that occurs at a slow rate with the loss of vacuum. This decrease is typically between 5% and 10% in the first year and will progressively reduce to a few per cent per year. This loss is independent of the photocathode current and will occur whether the image intensifier is used or not.

FIG. 8.2. Top portion of a ~750 μm thick film of CsI:Tl, demonstrating well separated columnar growth (courtesy V. Nagarkar and co-workers at RMD, Inc.).

Intensification of the X ray image occurs through two mechanisms. The first is electronic or flux gain, which is a result of the kinetic energy gained by electrons as they are accelerated between the photocathode and the output phosphor (anode). Electronic gain has a typical value of 50.

The second source of intensification is minification gain. Minification gain is a result of the reduction of a large X ray image at the input phosphor (e.g. 40 cm) on to the smaller diameter output phosphor, typically 2.5 cm. For example, the minification gain of an XRII using a 40 cm FOV and a 2.5 cm output phosphor would be $40^2/2.5^2$ or 256. Minification gain decreases as electronic magnification increases.

Brightness gain is the product of the electronic gain and minification gain and is a measure of the overall system gain. Brightness gain ranges from 2500 to 7000, depending on the FOV (see Section 7.4.4.1).

The conversion factor is another measure of the gain of an XRII and is most commonly used for the specification of XRII performance. The conversion factor is the ratio of the luminance in units of candela per square metre (cd/m^2), measured at the output phosphor to the X ray IAKR, in units of μGy/s, measured at the input phosphor using standardized beam conditions. The conversion factor ranges from 9 to 27 $cd\cdot m^{-2}\cdot \mu Gy^{-1}\cdot s^{-1}$ in a new XRII. The conversion factor decreases over time and may ultimately fall to a level that requires replacement of the XRII.

The optical image from the output of the XRII is routed to other components for display on a viewing monitor or for recording of the images. In its simplest form, the optical system consists of three elements: (i) a collimating lens to shape the divergent light from the output phosphor into an almost parallel beam, (ii) an aperture to limit the amount of light reaching a video (or TV) camera and (iii) a lens to focus the image on to the video camera (Fig. 8.1). The aperture can either be fixed or variable, the latter usually being under automatic control. As in photography, the aperture is adjusted in *f* stops. A decrease of one *f* stop allows double the amount of light through, while an increase of one *f* stop allows half the amount of light through.

A video camera is commonly used to capture the output image from an image intensifier. The optical image is captured by the camera and converted to an analogue electrical signal that conforms to a recognized video format[2]. Initially, video cameras utilized a scanning electron beam that interrogated a photoconducting target and were collectively known as vidicon cameras[3]. The resistivity of the photoconducting target changes, based on the amount of light striking the target, creating a latent image of the output phosphor on the vidicon camera target. As the electron beam is scanned rapidly across the target, its intensity is modulated by the charge image present on the target. The small current passing through the target is integrated across a large resistance and converted to a voltage that is amplified. The intensity modulation of the electron beam thus converts a 2-D optical image into an analogue voltage waveform. Vidicon cameras for this application are increasingly being replaced by charge coupled device (CCD) cameras.

Video cameras are selected on the basis of several fundamental characteristics, including lag and SNR. Cameras with low SNR contribute to increased noise levels in fluoroscopic images, although temporal integration can reduce this effect. Lag describes the speed of response of the video camera to a changing signal, and cameras with high lag will retain some residual signal from the current frame for several subsequent frames and will also require several frames to build up to full signal. High lag can result in blurred images of moving objects, but noise will be reduced through temporal integration. For this reason, moderate lag can be advantageous when not imaging rapidly moving objects. Maximum SNR is achieved when a video camera is operated near its maximum signal level, and it is important that the aperture is set accordingly.

---

[2] There are three main video formats used in TV (video) systems around the world. National Television System Committee (NTSC) format is used in North America, much of South America and some Asian countries; the phase alternate line (PAL) system is used extensively in Europe, Africa and Asia; and the French system SECAM (sequential colour with memory) is used in French and Russian speaking countries.

[3] The vidicon camera uses a specific target material, Sb_2S_3. A considerable number of other target materials have been used, giving a variety of names to such cameras, including plumbicon (PbO), chalnicon (CdSe), etc. However, all share the same principle of operation and are considered as a part of the 'vidicon family'.

The analogue video waveform from the video camera can be displayed directly on a video monitor. Digital image processing, however, requires that the analogue waveform be digitized with the use of an analogue to digital converter. The degree to which the digital image represents the analogue image depends on the bit depth (see Chapter 16) and sampling rate. In order to perform advanced image processing such as recursive filtering, images must be stored in a video buffer.

The CCD camera is a solid state device composed of many discrete photoconducting cells. Optical light from the output phosphor is converted to electrons in the amorphous silicon photoconducting layer of the CCD (see Section 7.4.3). The electrons are stored in potential wells created by applying a voltage between rows and columns of cells. Stored charge that has accumulated during an exposure is read out using parallel and serial shift registers that move charge from column to column and row to row in a 'bucket brigade' fashion, creating an analogue signal that is amplified and output as a video signal, or digitized directly.

The CCD holds several advantages over the vidicon camera, including the absence of lag[4] and a wider dynamic range. It can also reduce or eliminate image blooming, an image distortion caused when the input signal exceeds the dynamic range of the video camera. In a CCD, this is accomplished by incorporating drains in each cell that direct excess charge to ground, preventing blooming. This modification is at the expense of the fill factor, and reduces the overall quantum detection efficiency (QDE) of the camera.

Recent fluoroscopic systems incorporate flat panel image receptors, which possess several advantages over XRIIs. These advantages include their larger size, less bulky profile, absence of image distortions, and a higher QDE at moderate to high IAKRs to the receptor (Fig. 8.3), owing to the absence of a metal support structure. Flat panel image receptors now allow new applications, including rotational angiography and cone beam CT (CBCT). However, flat panel image receptors suffer from additive noise sources, including read noise (see Section 8.3.2), and therefore perform poorly compared with XRIIs at low IAKR (Fig. 8.3). Typical IAKR values for fluoroscopic imaging with a full FOV flat panel receptor (30 cm × 40 cm) range from 27 to 50 μGy/min (30–55 nGy/pulse) and may vary with the selected frame rate.

After all image processing has been performed, the image must be converted from digital to analogue form, for image display on a viewing monitor. Early television standards[5] determined that at least 525 video scan lines of the image were necessary to display moving images adequately. Alternating current power

[4] Note that the absence of lag also results in images with higher noise levels. For this reason, recursive filtering (see Section 8.6.3) is often applied to add some lag artificially.

[5] NTSC system in this case.

supply frequencies of 60 Hz required that all scan lines be displayed during an integral number of cycles (1/60 s, 1/30 s, etc.). Since all scan lines could not be displayed in one cycle at a 60 Hz frame rate, because of video bandwidth restrictions and unacceptable image flicker at slower scan rates, it was necessary to scan two frames or video fields, each containing one half (262½) of the scan lines, in an interlaced fashion. This combined the advantages of both the 60 Hz and 30 Hz frame rates and interlaced scanning became the television standard. Interlaced scanning provides a refresh rate of 60 Hz, while only requiring the bandwidth of a 30 Hz progressive scanning video.

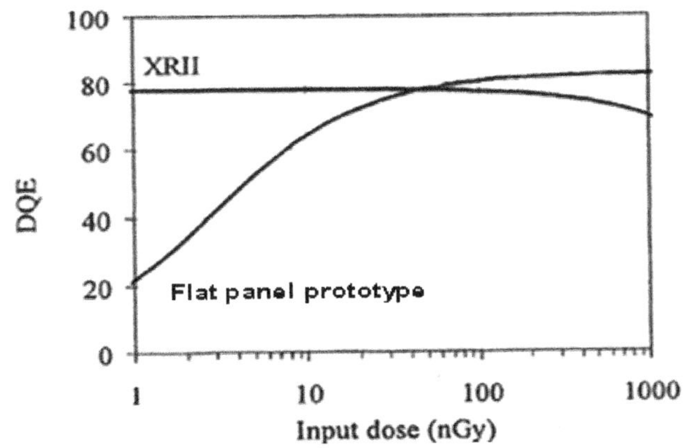

FIG. 8.3. DQE(0) as a function of input dose for prototypical flat panel and XRII image receptors.

The resolution of a scanning video system is limited in the vertical direction by the number of effective lines used to make the image. The effective number of lines is the number of scan lines in the image multiplied by the Kell factor. The Kell factor is an empirically determined factor that describes the vertical image degradation and is device specific, ranging from 0.7 for scanned pixel video pickups (e.g. vidicon camera) and display devices (e.g. cathode ray tube) to as high as 0.9–0.95 for fixed pixel devices such as CCD cameras and liquid crystal display monitors. The causes of vertical image degradation include the finite size of the scanning electron beam, low pass filtering of interlaced scan lines in scanned pixel devices, and the fact that both scanned and fixed pixels may not align exactly with a scanned signal. In the horizontal direction, the resolution of a video system is limited by the bandwidth of the video system. In most systems, the bandwidth is adjusted to give equal resolution in both the vertical and horizontal directions.

8.2.2. Automatic exposure control[6]

Radiographic systems use automatic exposure control (AEC) devices that automatically adjust radiographic technique factors (most often the mAs) to deliver a constant signal intensity at the image receptor in response to differences in patient thickness, X ray tube energy, focus to detector distance and other technical factors. Similarly, in fluoroscopic systems, the AEC controls the IAKR to the XRII, to prevent fluctuation in image brightness and SNR that would make diagnosis or navigation of instruments difficult.

Fluoroscopic AEC may use the signal from a sensor such as a photodiode or a photomultiplier tube or, more commonly, the signal from the video camera or directly from a flat panel image receptor, to determine necessary adjustments of fluoroscopic technique factors such as tube voltage and tube current. The selection of fluoroscopic technique factors follows predetermined curves that are stored in the generator and which usually allows for some choices, including a standard curve, low dose curve and high contrast curve (Fig. 8.4). The complexity of fluoroscopic AEC increases with advanced applications where the AEC assumes control over additional equipment parameters such as pulse length, added filtration and variable aperture setting.

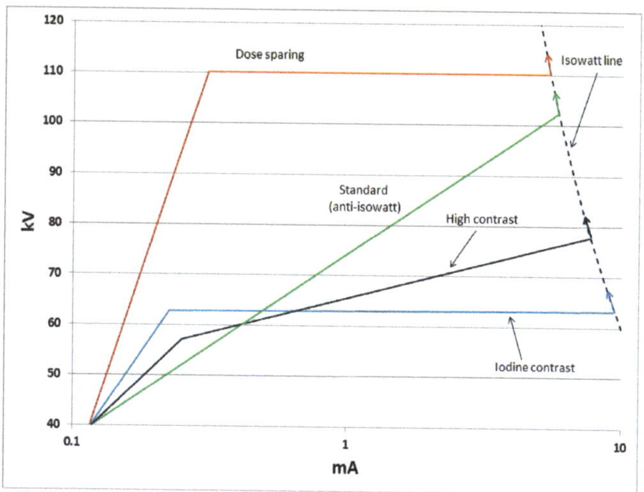

FIG. 8.4. Some typical control curves for different fluoroscopic imaging tasks.

[6] AEC systems may also be known by other names including automatic dose control, automatic dose rate control and automatic brightness control. Automatic brightness control originally described the automatic control of video voltage levels from video cameras, but is now sometimes applied to the general exposure control of fluoroscopic systems.

8.2.3. Electronic magnification

Electronic magnification refers to the use of a focusing electrode to deminify the fluoroscopic image by selecting a smaller portion of the input phosphor to project on to the output phosphor. Electronic magnification improves the image MTF but also decreases minification gain and decreases the sampling pitch of the input phosphor, increasing noise.

In practice, the increased noise in a magnified fluoroscopic image is compensated for by adjusting the technique factors to maintain a constant perceived noise level in the displayed image. In an XRII, the IAKR usually increases as the ratio of the areas of the FOV as the image is magnified. This not only compensates for the decreased photon fluence per image pixel, but also exactly compensates for the decrease in minification gain, and therefore image brightness, in an XRII system. Flat panel based systems also increase the IAKR as the image is magnified in response to changes in the image matrix size.

8.3. IMAGING PERFORMANCE AND EQUIPMENT CONFIGURATION

The factors that must be taken into account when considering image quality (see Chapter 5) in fluoroscopic imaging include contrast, noise, sharpness, temporal resolution and artefacts or image distortions. While each of these quantities is influenced and limited by the design of the fluoroscopic equipment, they are also highly dependent on equipment configuration and use.

8.3.1. Contrast

Subject contrast is inherently poor in fluoroscopic imaging, especially at the high kV values used to maintain patient dose at an acceptable level. Contrast is greatly improved through the use of radiopaque markers on catheters and other instruments, and through the use of exogenous contrast agents. Contrast agents for fluoroscopy are selected on the basis of their chemical properties, toxicities and X ray attenuation properties. Iodine and barium are two contrast agents commonly used in fluoroscopic imaging, with K edges of 33 keV and 37 keV, respectively. Gadolinium or carbon dioxide may be used when iodine contrast is contraindicated owing to allergies or impaired renal function.

8.3.1.1. Spectral shaping

The signal from iodine contrast is highly dependent on the X ray spectrum used to image the contrast agent. The maximal contrast occurs when the polyenergetic X ray spectrum is optimized to be predominantly just above the K edge. However, the use of such low X ray energies may lead to excessive patient dose, requiring careful selection of kV and appropriate filtration, as shown in Fig. 8.5.

The advent of high heat capacity X ray tubes and high power generators has made available another solution, spectral shaping. In its most basic form, spectral shaping involves the use of metal filters to remove, preferentially, low energy X rays from a polyenergetic X ray beam. A commonly used technique is the insertion of small amounts of copper filtration to shape the X ray spectrum. Copper attenuates low energy X rays (below the iodine K edge) that have little chance of penetrating the patient and hence generating contrast (Fig. 8.5). As many low energy X rays that would contribute only to patient dose are removed, a lower kV can be used at the same patient dose rate, resulting in improved iodine contrast.

The energy fluence of the X ray beam is greatly reduced by the addition of Cu filtration, and the tube current must be increased to high levels (50–400 mA) to maintain acceptably short pulse widths (see Section 8.6.2). As patient thickness increases, the additional Cu filtration is gradually reduced to maintain short pulse widths and acceptable tube loading. This is achieved through the programming of the AEC (see Fig. 8.5 and Section 8.2.2).

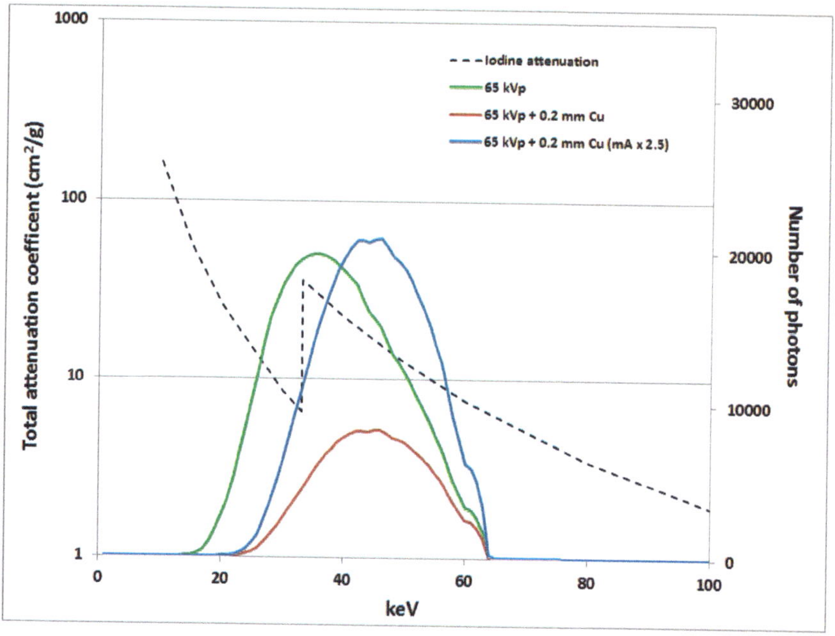

FIG. 8.5. *Impact of spectral shaping on iodine contrast.*

8.3.2. Noise

The noise level in fluoroscopic images is necessarily high, as a low IAKR is typically used to maintain the patient dose at an acceptable level. XRII based fluoroscopic systems are also characterized by low additive electronic noise levels (see Section 7.2.2). Therefore, the system is still quantum limited at low IAKR values. However, flat panel based fluoroscopic systems suffer from high levels of electronic noise (read noise, specifically), and their imaging performance is limited by this noise at low IAKR values. As a consequence, systems utilizing flat panels require a higher IAKR than XRII based systems for fluoroscopic imaging. Conversely, flat panels perform better than XRIIs at high IAKR, such as those used during digital acquisition imaging (see Section 8.4.1). The appearance of image noise in fluoroscopy is also influenced by human perception; for example, an observer will perceive less noise at high frame rates than at lower frame rates.

8.3.3. Sharpness

The sharpness of a fluoroscopic image is influenced by several factors, including the display matrix, FOV, video camera matrix, focal spot size, geometric magnification, image noise and motion. The impacts of both focal spot size and geometric magnification on image sharpness are discussed in Chapter 6. XRII fluoroscopic systems differ from a screen film image receptor in that the limiting resolution varies with operating mode, as described in Section 8.2.3.

Image noise interacts with sharpness, as it can obscure and blur small details in the image that would normally be visible at a higher IAKR. The large number of signal conversions that occur in an XRII also degrade the sharpness of the fluoroscopic image. The sharpness of a fluoroscopic image acquired with a flat panel receptor is affected by the size of the image matrix compared with the display matrix and the pixel size of the receptor, which may vary if pixels are binned at certain FOVs.

8.3.4. Artefacts

Artefacts in fluoroscopic imaging usually stem from image distortions caused by the image chain components. XRIIs suffer from several common image distortions, including veiling glare, vignetting, blooming, pincushion distortion and S distortion, while flat panel image receptors are generally free from image distortions.

Veiling glare is a contrast reducing 'haze', not unlike the effect of X ray scatter, that results from the scattering of information carriers within the XRII, including electrons within the electron optical system and, most importantly, light photons within the glass output window. To address the latter, a thick XRII output window is used that may incorporate dopants to absorb scattered light, and whose sides are coated with a light absorbing material. In some cases, the optical coupling system between the XRII output phosphor and the video camera is replaced by a direct fibre optic linkage, which also reduces veiling glare.

Vignetting is an optical distortion that produces a fall off in light intensity or darkening near the edges of an image. This may be caused by a number of factors, including deterioration of the video camera, and is also inherent to multielement lenses. Vignetting can be reduced in some cases by restricting the aperture size.

Blooming is caused by the input of signals to the video camera that exceed its dynamic range. Such large signals cause lateral charge spreading within the camera target, resulting in a diffuse image that is larger than the original. Blooming can be minimized through the use of tight X ray beam collimation and, as noted in Section 8.2.1, has largely been eliminated in CCD cameras.

Pincushion distortion causes enlargement of the fluoroscopic image near the edges (Fig. 8.6(a)) and results from the curvature of the input phosphor, which is required for proper electronic focusing and structural support. Pincushion distortion is more severe for a large FOV.

S distortion causes straight objects to appear curved (Fig. 8.6(b)) and results from the acceleration of electrons in the electron optical system of the XRII in the presence of an external magnetic field. Common sources of such magnetic fields include the Earth (5×10^{-5} T), fringe fields from nearby magnetic resonance imaging units (0.1–0.5 mT), and steel support structures and reinforcement. S distortion can be minimized by proper site planning and by encasing the XRII in a high susceptibility metal.

(a)
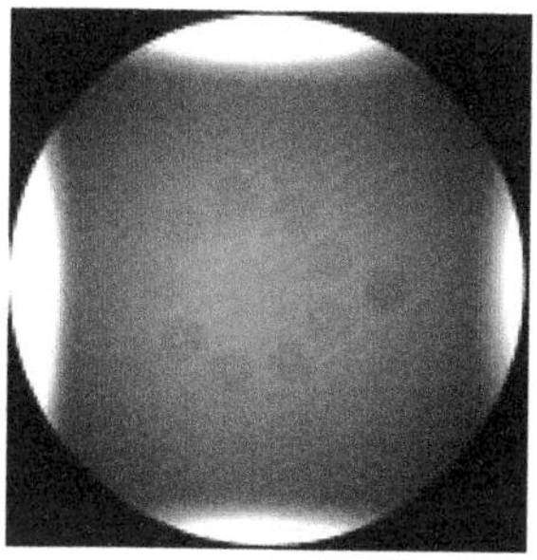

(b)
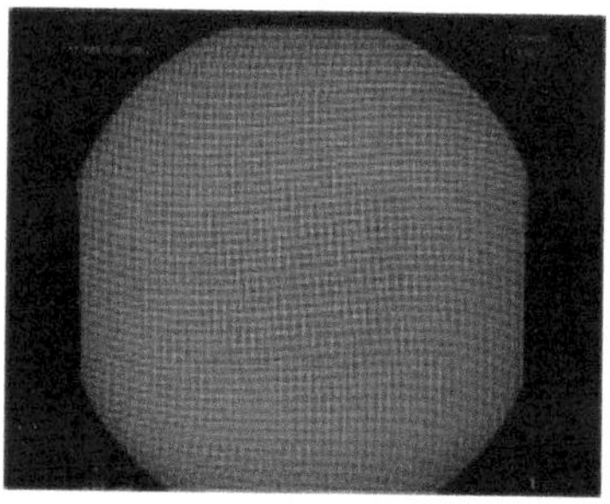

FIG. 8.6. Examples of distortion. (a) Pincushion distortion; (b) S distortion.

8.4. ADJUNCT IMAGING MODES

8.4.1. Digital acquisition imaging

Digital acquisition imaging refers to a mode of operation in which high quality images are recorded and stored for analysis. IAKRs, and therefore patient dose rates, are at least one order of magnitude higher in digital acquisition mode than in the fluoroscopic mode. In order to avoid saturation of the video camera for systems using an XRII, the signal from the image intensifier may be reduced through the use of the variable aperture, as discussed in Section 8.2.1. Digital acquisition images can be acquired at frame rates ranging from 1 to 30 frames/s, or as individual images, which are often referred to as 'spot' or 'single shot' images.

8.4.2. Digital subtraction angiography

Digital subtraction angiography (DSA) is a technique in which sequential 'fill' images that include a contrast agent are subtracted from a 'mask' image that includes only the anatomical background. This subtraction reduces anatomical noise and increases the contrast of the blood vessels (Fig. 8.7) in the subtracted images. Both the mask and fill images undergo a log transform before subtraction. The final result is an image in which the signal in the contrast filled vessels depends only on the amount of contrast in the vessel, and not on the background.

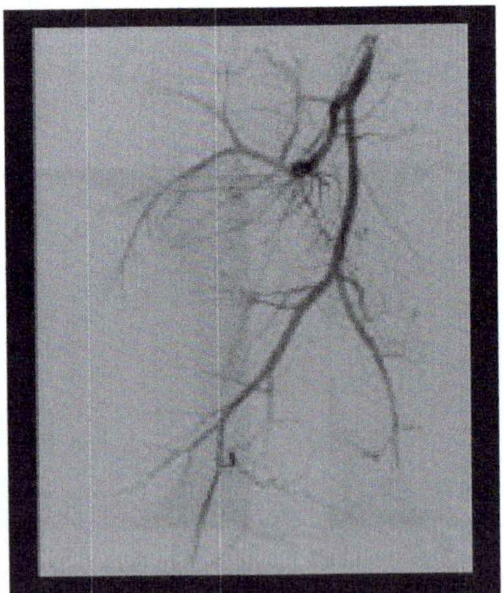

FIG. 8.7. IMAGE of arteries generated using DSA.

As quantum noise sums in quadrature when images are combined, the noise level in the subtracted image is higher by a factor of 1.4 than the noise level in the constituent images. This increase in noise implies that DSA will require higher exposures than digital acquisition imaging if similar image noise levels are to be maintained. However, the reduction in anatomical noise achieved with DSA may offset part or all of the increase in image noise, and advanced techniques such as mask averaging can be used to reduce the exposure requirements for DSA imaging.

The major source of artefacts in DSA is patient motion between the capture of the mask and fill images (Fig. 8.8). These motion artefacts can obscure contrast filled vessels. These types of artefact can be reduced retrospectively in some cases through the use of processing techniques such as manual or automatic pixel shifting of the mask image, or remasking through the selection of a different mask frame for subtraction.

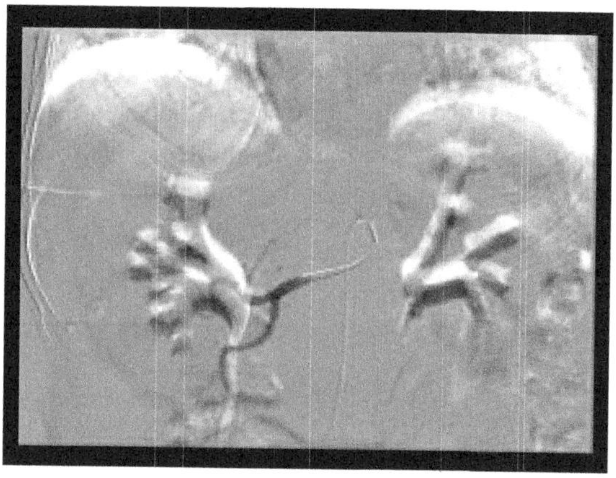

FIG. 8.8. Motion artefact in DSA.

Roadmapping is an adjunct imaging mode used to create a map of vascular anatomy that aids the navigation of catheters within tortuous vessels. A roadmap can be generated very simply by using a stored image of a contrast filled vessel, or in a more complex fashion by using the peak opacification in each image pixel obtained from a series of post-injection images. This is essentially a maximum intensity projection image of the contrast filled vessel and ensures a relatively uniform signal throughout the vessel, as it is less affected by contrast washout. A further improvement on this method is to subtract a fluoroscopic mask image from the fill images. While this is similar to DSA, it uses a lower dose. The roadmap image can either be displayed alongside the live image on another monitor, or overlaid on the live fluoroscopic image. The roadmap image is often greyscale inverted prior to overlaying on the live image.

Peripheral runoff imaging follows a bolus of contrast as it travels from the injection site into the peripheral vasculature, most often in the legs. Many angiographic systems operate in a stepping mode for runoff procedures, sequentially stepping along the patient's body, acquiring images at each step. The images overlap by some amount, often 1/3, to ensure seamless anatomical coverage. This type of study requires the use of compensating filters to equalize image receptor exposure around the patient's legs. Compensating filters can either be external to the system, such as wedges or forms placed around the patient's legs, or internal in the form of wedge shaped metal filters, either attached to the outside of the collimator or contained inside it.

Rotational angiography is an adjunct imaging mode used most often in vascular, interventional and neurointerventional radiology. A series of basis images are acquired as a C-arm (see below) rotates around the patient. A contrast injection can be performed during the scan. The basis images can be viewed as a cine loop and are often used to reconstruct CBCT images (Fig. 8.9). The images can be reconstructed in axial, coronal and sagittal planes, or in arbitrary curved planes. Maximum intensity projection images are often generated to improve visualization of iodine contrast in small vessels. Some manufacturers offer the capability to perform 3-D rendering using the CT images and to perform subtracted rotational angiography.

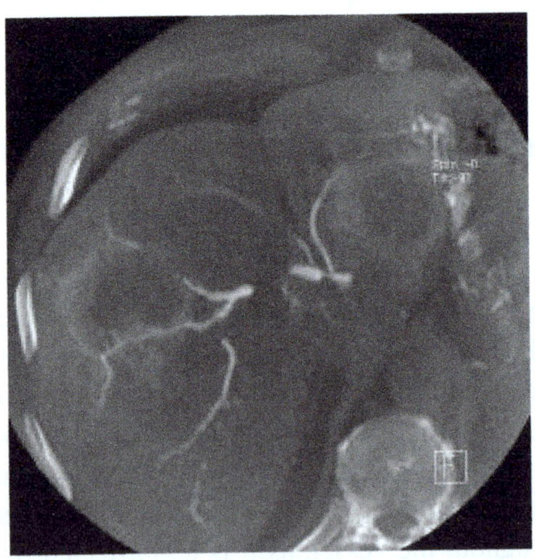

FIG. 8.9. Axial image reconstructed from rotational angiography basis images.

8.5. APPLICATION SPECIFIC DESIGN

Fluoroscopic imaging systems can be configured in several ways. The most common is the configuration in which the X ray tube is located under the patient table and the XRII and auxiliary imaging equipment are placed on a movable 'tower' above the patient table (Fig. 8.10). Lead curtains hang from the XRII tower and shield the operator from stray radiation scattered from the patient. This configuration is commonly used for genitourinary and gastrointestinal imaging.

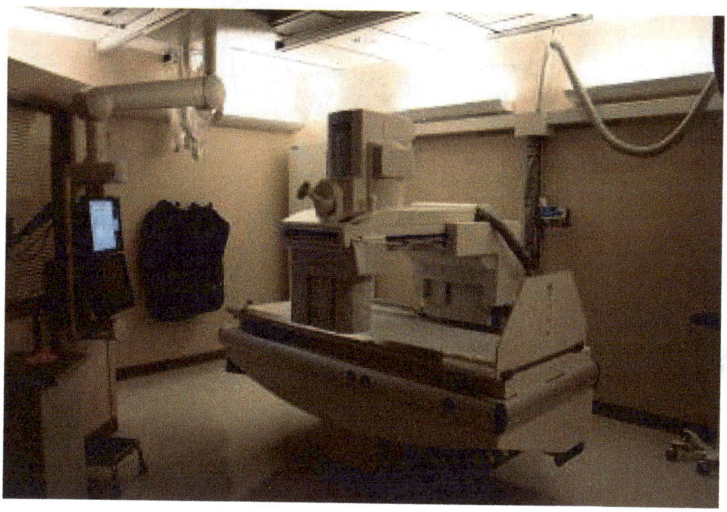

FIG. 8.10. Conventional fluoroscopy system.

8.5.1. Remote fluoroscopy systems

Remote fluoroscopy systems are commonly used for gastrointestinal procedures, including barium swallow and barium enema examinations, and utilize a configuration with the X ray tube located above the table and the XRII assembly below the table (Fig. 8.11(a)). The system can be rotated to achieve other necessary projections or to distribute contrast agents within a patient. It can also be configured vertically for seated examinations, such as the barium swallow (Fig. 8.11(b)). The focus to image distance is usually continuously variable between two extremes, and a remote controlled compression cone may be available for the radiologist to manipulate air and barium contrast within the patient's abdomen. There are distinct advantages in the use of remote fluoroscopy rooms, namely related to radiation safety, as exposure of the operator and technical staff to stray radiation is greatly reduced. By increasing the focus to image distance to its maximum, the patient entrance surface air kerma rate (see Chapters 20 and 21) can also be decreased by 15–20%. However, remote fluoroscopy rooms are more expensive than conventional rooms and are often unsuitable for young patients who require close supervision. The dose to individuals remaining in the room with a difficult patient can be much greater than the dose in a conventional fluoroscopy room, owing to the position of the X ray tube and absence of integrated radiation shielding.

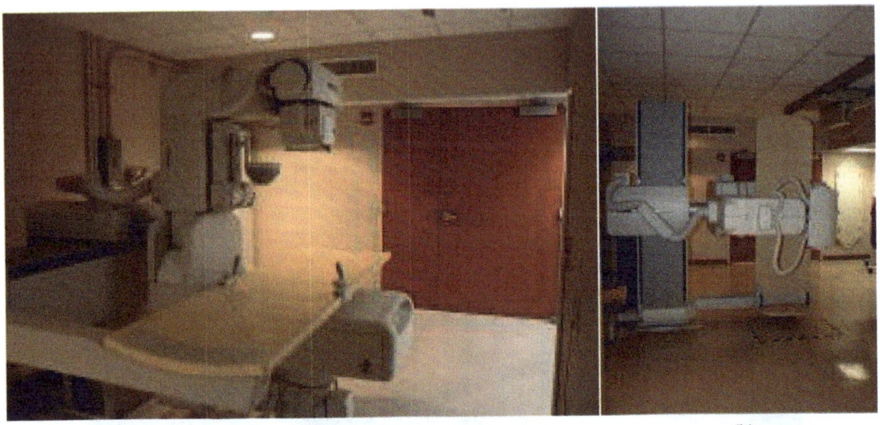

(a) (b)

FIG. 8.11. Remote fluoroscopes configured for (a) supine examination (e.g. barium enema) and (b) seated examination (e.g. barium swallow).

8.5.2. Vascular and interventional radiology

Vascular and interventional radiology procedures are usually performed in angiographic suites equipped with C-arm fluoroscopes. A C-arm fluoroscope comprises a mechanically coupled X ray tube and image receptor. The X ray tube and image receptor rotate in unison about a point called the isocentre that remains at the centre of the FOV when the C-arm is rotated. The table is often cantilevered to allow continuous, unobstructed rotation of the C-arm around the patient during procedures. Vascular and interventional suites are equipped with more powerful generators with high heat capacity and water or oil cooled X ray tubes. Also, variable spectral shaping filters are often included to maximize iodine contrast while maintaining the patient dose at an acceptable level. Typical XRII sizes for vascular and interventional laboratories range from 28 to 40 cm.

8.5.3. Cardiology

Interventional cardiology suites also use C-arm fluoroscopes for ease of positioning at a variety of angles around the patient. Cardiology suites can be either single plane or biplane systems. Biplane systems use two C-arms that can be independently positioned around the patient for simultaneous digital acquisitions during a single contrast injection. This is important because iodinated contrast is nephrotoxic, and the total volume of contrast that can be administered is limited by the patient body mass. This is particularly critical in paediatric catheterization laboratories, owing to the low body mass of paediatric patients, which severely limits the amount of contrast that can be administered during an imaging study, and the small size of blood vessels, which may require higher iodine concentrations for acceptable visualization.

Image receptors used for cardiac imaging are smaller than those used for vascular and interventional radiology, owing to the small size of the heart. A typical XRII size for a cardiac laboratory is 23 cm. Some newer flat panel based cardiac catheterization laboratories incorporate large image receptors (30 cm × 40 cm) for the primary or A plane, which make possible adjunct imaging modes such as runoff imaging or rotational angiography. The lateral or B plane is sized for normal cardiac imaging.

8.5.4. Neuroradiology

Neuroradiology equipment is very similar to cardiology equipment, as the required FOVs are similar.

8.5.5. Mobile fluoroscopes

Mobile fluoroscopes are fluoroscopes mounted on wheels that can be moved between locations. They are useful when the expense of a permanent installation cannot be justified, or when imaging capability is needed briefly in several adjacent rooms, for example, in the operating room (see also Chapter 10). Mobile fluoroscopes often use shorter focus to image distances, and smaller FOVs than other types of fluoroscope.

8.6. AUXILIARY TOPICS

Advanced fluoroscopic applications and equipment are changing with the rapid deployment of digital image acquisition devices. The use of film is decreasing, and in many cases, specialized films are no longer available. In other cases, precision mechanical equipment needed for radiographic screen film cassette changers and high speed large format film changer systems have become obsolete.

8.6.1. Spot film device

A spot film device is used to acquire radiographs during a fluoroscopically guided procedure. While fluoroscopy is activated, a radiographic cassette is retracted into and held in a lead shielded enclosure. When a spot film is desired, a button is pressed and the radiographic cassette is extended in front of the XRII, behind an antiscatter grid. After the cassette is exposed, it is ejected and manually exchanged for an unexposed cassette, which is retracted into the lead shielded enclosure until needed. Most spot film devices offer several framing options, including a single full size image, two or four images per film, etc. Spot film devices are still common in conventional and remote fluoroscopic systems.

8.6.2. Operating modes

8.6.2.1. Continuous fluoroscopy

Continuous fluoroscopy is the most basic form of fluoroscopic imaging. The X ray beam is on constantly and a video refresh rate of 25 or 30 frames/s yields a frame integration time of 40 or 33 ms. This can lead to blurring of moving objects.

8.6.2.2. Pulsed fluoroscopy

Most modern fluoroscopic equipment is capable of operating in pulsed fluoroscopy mode. When configured properly, pulsed mode offers several advantages over continuous mode, including:

- Lower radiation dose when fluoroscopic pulse rates of less than 30 pulses are used.
- Improved image quality due to reduction in motion blur because of the reduced integration time. Pulsed mode operation ‘freezes’ the motion of objects in the image, resulting in sharper images and improved image quality.
- Reduced tube loading at low pulse rates.

While pulsed fluoroscopy produces sharper images, the reduction in temporal resolution at low frame rates may be unacceptable for rapidly moving organs or instruments within the body. Higher frame rates provide superior temporal resolution for these cases.

Grid controlled or grid switched X ray tubes

Pulsed fluoroscopy can be accomplished either by operating the generator in pulsed mode, or by using a grid controlled or grid switched X ray tube. The long high voltage cables used in many fluoroscopic rooms are characterized by significant capacitance. As a result, power continues to be applied to the X ray tube after the generator switches off between pulses. This results in unnecessary patient dose and possibly additional motion blurring. A grid controlled X ray tube uses a negatively biased grid near the filament to stop the flow of electrons from the cathode to the anode, preventing unwanted X ray production between radiation pulses.

Pulsed fluoroscopy and the human visual system

Since the temporal response of the human visual system has a typical integration time of approximately 0.1 s (up to 0.2 s for low light levels), it has the capacity to integrate several frames of pulsed fluoroscopy during a single integration cycle. Consequently, fluoroscopic images appear noisier as the pulse rate decreases for the same IAKR per frame. When changing from one pulse rate to another, the input air kerma per pulse can be adjusted to account for this phenomenon.

8.6.3. Recursive filtering

Fluoroscopic images are inherently noisy, but increasing the IAKR to reduce noise comes at a penalty of increased patient dose. Noise reduction can also be accomplished through image processing, including the averaging of images. Recursive filtering is an image processing technique that combines portions of both the most recent fluoroscopic frame and several previous fluoroscopic frames to reduce noise in the resultant image. The recursive filtering process can be described mathematically as:

$$\text{frame}_{\text{displayed}} = \sum_{i=N-n}^{N} f_i w_i \tag{8.1}$$

where w_i is a prospectively determined weighting coefficient and f_i is the ith frame in the video buffer.

The recursive filter is thus a moving filter that incorporates information from several frames into the current fluoroscopic frame, reducing noise in the final image. Both quantum (X ray) noise and additive noise from the video camera or image receptor are averaged. There are still, however, potential penalties for this reduction in noise. The recursive filter works well if changes in the image from one frame to the next are small. In anatomical regions where motion is rapid, excessive recursive filtering can lead to unacceptable artificial lag. Artificial lag may also be noticed if instruments are moved rapidly or if the patient table is shifted. Most modern fluoroscopic systems use motion detection algorithms or other methods to prevent artificial lag. These algorithms monitor the change in image pixels from one frame to the next, and if the change exceeds a preselected threshold, the strength of the recursive filter is reduced or zeroed until the image stabilizes, at which time the filter, which also reduces fluoroscopic contrast, is reset to its normal level. The strength of the filter, which also reduces fluoroscopic contrast, is related to both the weighting coefficients applied (w_i) and the number of fluoroscopic frames combined ($N - n$).

8.7. DOSIMETRIC CONSIDERATIONS IN FLUOROSCOPY

The dosimetric quantities used to describe the patient dose from fluoroscopic imaging are outlined in Section 22.4.5. It is important to note that fluoroscopy, particularly when it involves interventional procedures, can give rise to both stochastic and deterministic (tissue) effects, primarily radiation induced skin injury that occurs once a certain dose has been exceeded (see Chapter 20). The discussion that follows focuses solely on deterministic effects from fluoroscopic procedures, and any reference to patient dose is assumed to refer to skin dose.

8.7.1. Skin dose indicators

Dosimetric indicators for skin dose can be direct ('real time') or can be determined after the irradiation event. Examples of direct indicators include the kerma area product (KAP) and the reference air kerma, $K_{a,r}$, while non-direct methods include the use of thermoluminescent dosimeters (TLDs), optically stimulated luminescence or semiconductor detectors and radiographic or radiochromic film (see Chapter 21).

8.7.1.1. Fluoroscopic timers

Fluoroscopy time is commonly used as a surrogate for patient dose in fluoroscopy, as it is a widely available function on fluoroscopic equipment. It is, however, far from ideal, as it ignores many large contributions to patient dose, including digital acquisition imaging. Digital acquisition imaging is frequently, but not always, the largest contributor to patient dose during fluoroscopic procedures.

8.7.1.2. KAP

The KAP can be measured directly using a KAP meter, or it can be calculated from known operating parameters, and is discussed in some detail in Chapters 20 and 21. While KAP is an ideal quantity for assessing stochastic risk, it has limited application as an indicator of skin dose. However, when carefully combined with direct skin dose measures, it has been used to determine trigger levels for specific procedures to alert operators to the possible danger of skin damage.

8.7.1.3. Reference point air kerma

Reference point air kerma ($K_{a,r}$) or cumulative dose (CD) refers to the cumulative air kerma at the interventional reference point (IRP) at any time during a fluoroscopically guided procedure. The IRP is a point 15 cm back towards the focal spot from the isocentre (Fig. 8.12). The location of the IRP does not vary with changes in C-arm angle or focus to image distance. The $K_{a,r}$ is the quantity most closely correlated to skin dose in fluoroscopically guided procedures, as all contributions to skin dose (i.e. both fluoroscopic and digital acquisition imaging) are included in the $K_{a,r}$.

8.7.1.4. Peak skin dose

The peak skin dose (PSD) refers to the highest dose to any single area of a patient's skin. In practice, it is difficult to determine PSD with a high degree of accuracy. It must be considered that the CD is measured at a single point in space that may not correlate with the patient's skin surface (Fig. 8.12). Even in the case where the IRP is located exactly on the skin surface, backscatter will increase the PSD beyond the indicated CD by 30–40%.

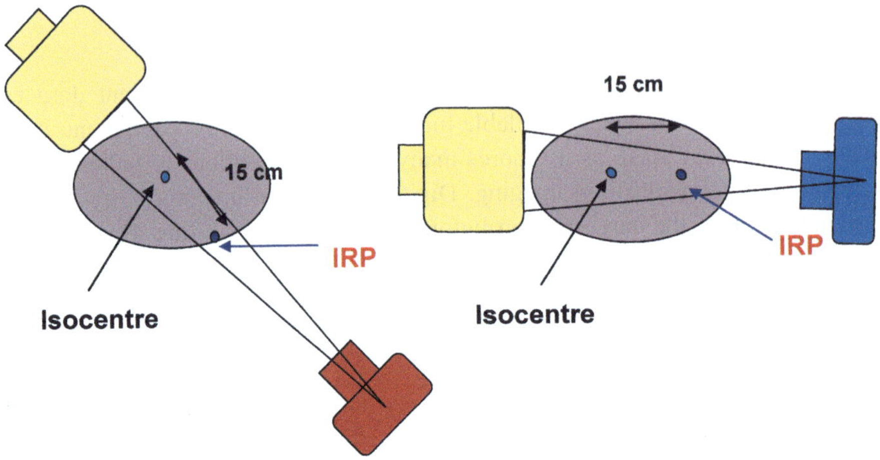

FIG. 8.12. A schematic of the IRP.

PSD can be measured with some degree of accuracy using various dosimeters. While TLD arrays and conventional film have been used, the use of radiochromic film provides the most promising approach.

Finally, it should be noted that the CD or KAP may overestimate PSD when multiple C-arm angles are used. Consider a procedure using two distinct, non-overlapping X ray beam entrance sites on a patient's skin. If the irradiation time were divided equally between the two sites, the PSD would be expected to be one half of the total skin dose.

One other caveat to consider when comparing the CD or KAP with the PSD is the use of a highly attenuating couch. Calibrating CD or KAP to measured skin doses will result in the most accurate estimates (see Section 21.3.1.3)

8.7.2. Radiation safety considerations for patient protection

Chapter 24 outlines general radiation protection concerns, including occupational dose monitoring and shielding requirements. Fluoroscopically guided procedures can result in high patient and operator doses, and radiation safety is a critical component of a fluoroscopic imaging programme. In general, the use of good practice by the operator will result in the minimum patient dose required to complete a fluoroscopically guided procedure safely. Good practice refers to the use of commonly known techniques to deliver the best image quality at the lowest radiation dose. These actions include, but are not limited to:

- Moving the patient as far from the X ray source as practical;
- Placing the image receptor as close to the patient as possible (i.e. no air gap);
- Using the lowest electronic magnification (largest FOV) required to perform the procedure;
- Collimating the X ray beam tightly to the anatomy of interest.

In addition to good practice, all dose reduction tools available on the fluoroscopic equipment should be used. Spacers provided by the manufacturer are used to maintain a minimum distance between the focal spot and the patient. Operators often find them inconvenient and as a consequence spacers are frequently removed and left off the equipment. The reduction in source to skin distance allowed when the spacer is removed can increase the maximum possible patient entrance surface air kerma rate by 100%. Antiscatter grids should be removed when imaging small patients or thin body parts.

Most modern fluoroscopic systems provide additional tools that can be used to reduce the patient and operator doses. Last image hold is a feature that maintains the last fluoroscopic image on the viewing monitor pending fluoroscopy or acquisition being resumed. It allows the physician to contemplate a static image without the use of additional radiation. Many systems allow the operator to archive the last image hold image to permanent storage in lieu of acquiring a digital acquisition image. Some systems extend this further by providing the capability to archive the entire previous sequence of fluoroscopic images instead of acquiring a digital acquisition series.

8.7.3. Radiation safety considerations for operator protection

Occupational radiation protection considerations are often variations on the three cardinal rules of radiation protection: time, distance and shielding. Operators and other personnel remaining in the procedure room during fluoroscopically guided procedures are exposed to scattered radiation and are at risk of developing both stochastic effects, including cancer, and deterministic effects, namely cataracts (see Chapter 20 for more details).

Non-essential personnel should exit the room while the X ray tube is energized, and those persons remaining in the room should wear protective garments made of lead or an acceptable lead free material. Mobile barriers are useful for reducing the radiation dose to persons who remain stationary during procedures, and suspended shields can be used to reduce the dose to the face, eyes and neck of physicians while they are near the patient. It should be noted that the highest scatter radiation fields occur near the patient entrance field;

therefore, standing closer to the image receptor is generally consistent with lower occupational dose levels.

BIBLIOGRAPHY

AMERICAN ASSOCIATION OF PHYSICISTS IN MEDICINE, Functionality and Operation of Fluoroscopic Automatic Brightness Control/Automatic Dose Rate Control Logic in Modern Cardiovascular and Interventional Angiography Systems, AAPM Rep. 125, AAPM, College Park, MD (2012).

AUFRICHTIG, R., XUE, P., THOMAS, C.W., GILMORE, G.C., WILSON, D.L., Perceptual comparison of pulsed and continuous fluoroscopy, Med. Phys. **21** 2 (1994) 245–256.

BALTER, S., Methods for measuring fluoroscopic skin dose, Pediatr. Radiol. **36** Suppl. 2 (2006) 136–140.

BALTER, S., HOPEWELL, J.W., MILLER, D.L., WAGNER, L.K., ZELEFSKY, M.J., Fluoroscopically guided interventional procedures: A review of radiation effects on patients' skin and hair, Radiology **254** 2 (2010) 326–341.

COWEN, A.R., DAVIES, A.G., SIVANATHAN, M.U., "The design and imaging characteristics of dynamic, solid-state, flat-panel X-ray image detectors for digital fluoroscopy and fluorography", Clin. Radiol. 63 (2008) 1073–1085.

GEISE, R.A., Fluoroscopy: Recording of fluoroscopic images and automatic exposure control, Radiographics **21** (2001) 227–236.

INTERNATIONAL COMMISSION ON RADIOLOGICAL PROTECTION, Avoidance of Radiation Injuries from Medical Interventional Procedures, Publication 85, ICRP, Stockholm (2001).

INTERNATIONAL ELECTROTECHNICAL COMMISSION, Medical Electrical Equipment – Part 2-43: Particular Requirements for the Safety of X-Ray Equipment for Interventional procedures, IEC 60601-2-43, IEC, Geneva (2010).

KOCH, A., et al., "Detective quantum efficiency of an X-ray image intensifier chain as a benchmark for amorphous silicon flat panel detectors", Medical Imaging 2001: Physics of Medical Imaging (ANTONUK, L.E., YAFFE, M.J., Eds), SPIE, Bellingham, WA (2001) 115–120.

MAHESH, M., Fluoroscopy: Patient radiation exposure issues, Radiographics **21** (2001) 1033–1045.

NATIONAL COUNCIL ON RADIATION PROTECTION, Radiation Dose Management for Fluoroscopically-guided Interventional Medical Procedures, Rep. 168, NCRP, Bethesda, MD (2010).

POOLEY, R.A., McKINNEY, J.M., MILLER, D.A., The AAPM/RSNA physics tutorial for residents: Digital fluoroscopy, Radiographics **21** (2001) 521–534.

SCHUELER, B.A., The AAPM/RSNA physics tutorial for residents: General overview of fluoroscopic imaging, Radiographics **20** (2000) 1115–1126.

VAN LYSEL, M.S., The AAPM/RSNA physics tutorial for residents: Fluoroscopy — optical coupling and the video system, Radiographics **20** (2000) 1769–1786.

WANG, J., BLACKBURN, T.J., The AAPM/RSNA physics tutorial for residents: X-ray image intensifiers for fluoroscopy, Radiographics **20** (2000) 1471–1477.

Chapter 9

MAMMOGRAPHY

M.J. YAFFE
University of Toronto,
Toronto, Canada

A.D.A. MAIDMENT
University of Pennsylvania,
Philadelphia, United States of America

9.1. INTRODUCTION

Breast cancer is a major killer of women. The International Agency for Research on Cancer estimates that over 1.38 million women were diagnosed with breast cancer internationally in 2008, with over 458 000 deaths. The causes are not currently known; however, it has been demonstrated that mortality can be significantly reduced if disease is detected at an early stage.

9.2. RADIOLOGICAL REQUIREMENTS FOR MAMMOGRAPHY

Mammography is a radiographic procedure optimized for examination of the breast. For many women, mammography is a highly effective means of detecting early stage breast cancer. It is used both for investigating symptomatic patients (diagnostic mammography) and for the screening of asymptomatic women in selected age groups. A typical mammographic screening examination consists of one or two views of each breast. Common views include the cranial–caudal and mediolateral oblique, an example of which is shown in Fig. 9.1 (see also the Appendix to this book). While it is used primarily for the detection and diagnosis of breast cancer, mammography also has value in presurgical localization of suspicious regions and in the guidance of biopsies.

Breast cancer is detected on the basis of four types of sign on the mammogram:

(i) The characteristic morphology of a tumour mass, which can include irregular margins and spiculations;

(ii) Certain presentations of mineral deposits, visualized as specks called microcalcifications;
(iii) Architectural distortion of normal tissue patterns caused by the disease;
(iv) Asymmetry between corresponding regions of the left and right breasts.

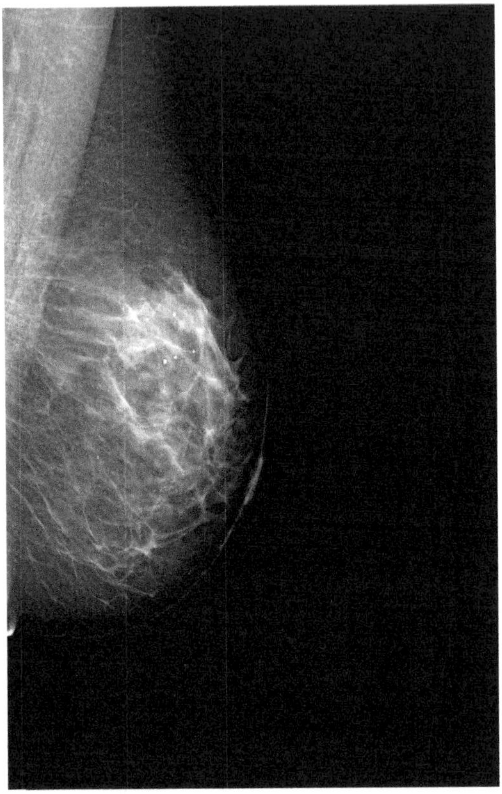

FIG. 9.1. A mediolateral oblique mammogram. In this projection, the pectoralis muscle is visualized down to the level of the nipple. In this mammogram, characteristic benign calcifications are seen.

Figure 9.2 shows X ray attenuation coefficients measured versus energy on samples of three types of material found in the breast: adipose tissue, normal fibroglandular breast tissue and infiltrating ductal carcinoma (one type of breast tumour). Both the attenuation coefficients themselves, and their difference, decrease with increasing energy, resulting not only in a reduction in the radiation dose required to produce an image, but also a decrease in image contrast. As shown in Fig. 9.3, the inherent X ray subject contrast falls as X ray energy increases. Note that the subject contrast of even small calcifications in the breast is similar to that for a tumour mass because of the greater difference in attenuation coefficient between calcium and breast tissue.

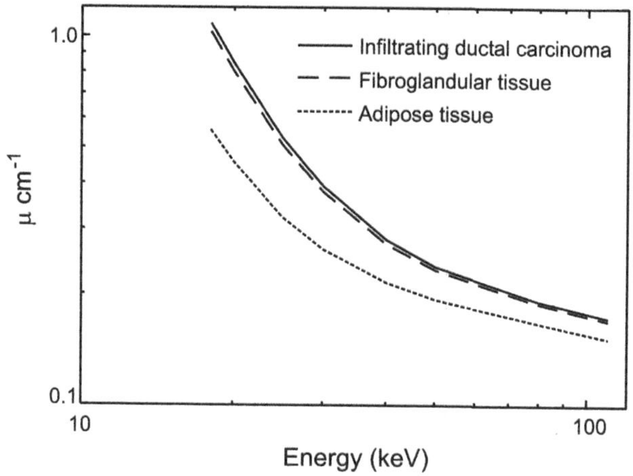

FIG. 9.2. Dependence of the linear X ray attenuation coefficient, μ, on X ray energy.

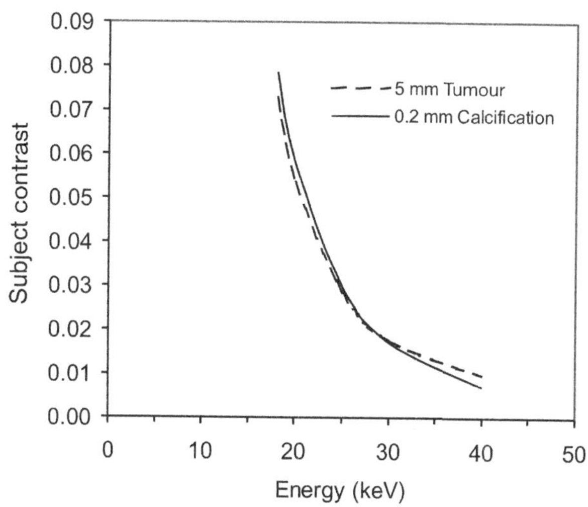

FIG. 9.3. Dependence of image contrast on X ray energy.

The essential requirements for image quality in mammography represent a special case of the general principles of radiography outlined in Chapter 6, adapted to the specific imaging tasks involved in detection of the radiological signs of breast cancer. The physics requirements are summarized here.

The imaging system must have sufficient spatial resolution at high spatial frequencies to delineate the edges of fine structures in the breast. Structural detail, possibly as fine as 50 μm, must be adequately resolved. Variation in X ray attenuation among tissue structures in the breast gives rise to variation in the transmitted X ray signal, and this is the fundamental source of image contrast.

As shown in Figs 9.2 and 9.3, breast tissues intrinsically lack subject contrast, requiring the use of low energy X ray spectra, which emphasize the compositional differences of the breast tissues. Variation in the physical size and internal composition of the breast and age dependent changes in the breast require a broad dynamic range. The use of a radiographic grid and firm breast compression provides some compensation for scatter. The detectability of structures providing subtle contrast is further impaired by random fluctuations in the image, referred to as mottle or noise. However, the breast is sensitive to ionizing radiation, which, at least for high doses, has a small associated risk of inducing breast cancer. It is, therefore, desirable to use the lowest absorbed dose compatible with high diagnostic image quality. In the following sections, the specialized components of the mammographic imaging system will be described, and their design related to the above mentioned imaging performance factors.

9.3. X RAY EQUIPMENT

The mammography unit consists of an X ray tube and an image receptor mounted on opposite sides of a mechanical assembly. Because the breast must be imaged from different aspects, the assembly can be rotated about a horizontal axis, as shown in Fig. 9.4. To accommodate patients of different height, the assembly elevation can be adjusted.

Unlike most general radiography equipment, which is designed such that the image field is centred below the X ray source, in mammography, the system's geometry is arranged as in Fig. 9.5(a). Here, a vertical line from the focal spot of the X ray source grazes the chest wall of the patient and intersects orthogonally with the edge of the image receptor closest to the patient. If the X ray beam were centred over the breast as in Fig. 9.5(b), some of the tissue near the chest wall would not be imaged.

Radiation leaving the X ray tube passes through a metallic spectral shaping filter, a beam defining aperture and a plastic plate, which compresses the breast on to the breast support platform. Those X rays transmitted through the breast and breast support are incident on a specially designed antiscatter grid, and then are incident on the image receptor, where they interact and deposit most of their energy locally. In screen film and cassette based digital mammography systems, a fraction of the X rays passes through the receptor without interaction and these X rays impinge upon the sensor of the automatic exposure control (AEC) mechanism of the mammography unit. In other digital mammography systems, the AEC mechanism is typically integral with the digital image receptor. In all systems, any remaining primary X rays are attenuated by a primary beam stop.

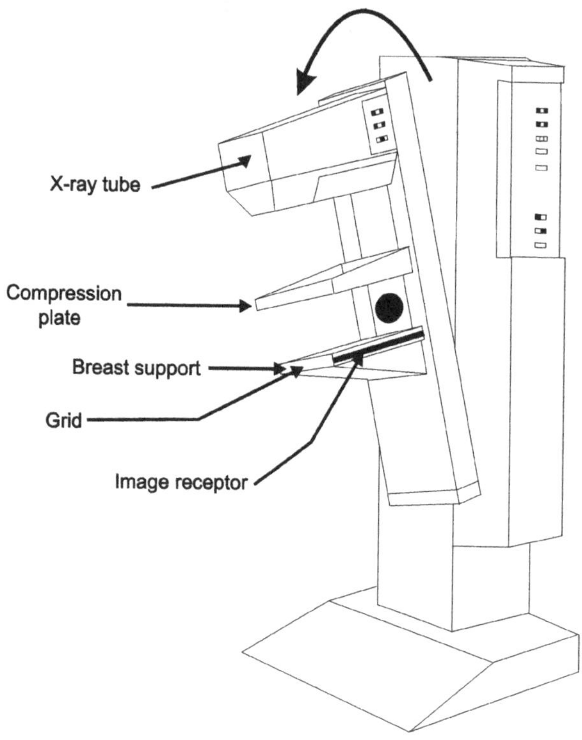

FIG. 9.4. Schematic of a mammography imaging system.

9.3.1. Tubes, filters and spectra

In modern mammography systems, the power supply is typically of the high frequency type (see Section 5.4.2.3) and provides a nearly constant potential waveform during the exposure. The X ray tube employs a rotating anode design in which electrons from the cathode strike the anode target material at a small angle from normal incidence (Fig. 9.6).

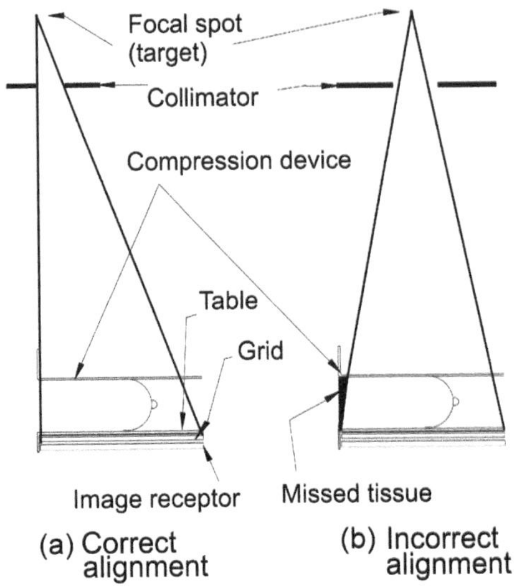

FIG. 9.5. System geometry for image acquisition showing (a) correct alignment and (b) missed tissue associated with incorrect alignment.

On modern equipment, the typical nominal focal spot size for contact mammography is 0.3 mm, while the smaller focal spot used primarily for magnification is 0.1 mm. The nominal focal spot size is defined relative to the effective spot size at a reference axis. As shown in Fig. 9.6, this reference axis, which may vary from manufacturer to manufacturer, is normally specified at some midpoint in the image. The effective size of the focal spot will monotonically increase from the anode side to the cathode side of the imaging field, as illustrated in Fig. 5.7. In mammography, the X ray tube is arranged such that the cathode side of the tube is adjacent to the patient's chest wall, because the highest intensity of X rays is available at the cathode side and the attenuation of X rays by the patient is generally greater near the chest wall. Frequently in breast imaging, there may be different target angles according to the focal spot size. In addition, the angulation of the X ray tube itself may be changed according to the choice of focal spot size and target material.

Most mammography tubes use beryllium exit windows between the evacuated tube and the atmosphere, and no oil is present in the radiation path exiting the tube. Oil, glass or other metals used in general purpose tubes would provide excessive attenuation of the useful energies for mammography.

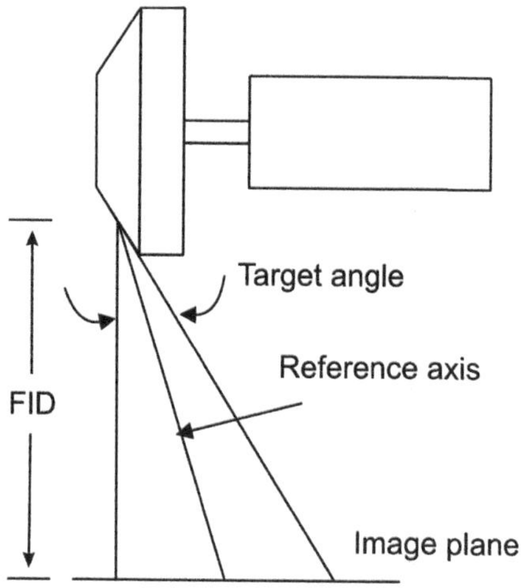

FIG. 9.6. The geometry of an X ray tube (FID: focus to image distance). The perpendicular line abuts the chest wall. The reference axis on a particular system will be specified by the manufacturer.

As in general radiography, one attempts to define a spectrum that provides energies that give an appropriate compromise between radiation dose and image quality. In mammography, the spectral shape is controlled by adjustment of the tube voltage, choice of the target material and the type and thickness of the metallic filter placed between the X ray tube and the breast. The strategies for optimization of the X ray spectrum for screen film mammography and digital mammography are quite different. In screen film mammography, the contrast of the displayed image is constrained by the fixed gradient of the film, while in digital mammography, the quality of the displayed image is constrained by the image signal to noise ratio (SNR).

Using monoenergetic models of mammographic imaging, it has been suggested that the optimum energy for film imaging lies between 18 and 23 keV, depending on breast thickness and composition. It has been found that for a breast of typical thickness and composition, the characteristic X rays from molybdenum (see Fig. 1.3) and rhodium (see Table 9.1) provide good imaging performance for screen film mammography. For this reason, molybdenum and/or rhodium target X ray tubes are available on most mammography machines. Because the contrast of digital images can be controlled during image display, higher energies may be more optimal for digital mammography. For this reason, some digital mammography machines provide tubes equipped with tungsten targets.

TABLE 9.1. CHARACTERISTIC X RAY ENERGY (keV) FOR MOLYBDENUM (Mo) AND RHODIUM (Rh) ANODE X RAY TUBES

Anode	$K_{\alpha 1}$	$K_{\alpha 2}$	$K_{\beta 1}$
Mo	17.48	17.37	19.61
Rh	20.22	20.07	22.72

As in conventional radiology, metallic filters are used in mammography to provide selective removal of low X ray energies from the beam before it is incident upon the patient. In mammography, a molybdenum anode X ray tube is commonly employed with a molybdenum filter that is 30–35 μm thick. This filter acts as an energy window providing greater attenuation of X rays both at low energies and above the K absorption edge at 20 keV, while allowing the molybdenum characteristic X rays from the target and X rays of similar energy produced by bremsstrahlung to pass through the filter with relatively high efficiency. As illustrated by Fig. 9.7(a), the resultant spectra are enriched with X rays in the range 17–20 keV.

Although molybdenum spectra are relatively well suited for imaging a breast of average attenuation, slightly higher energies are desirable for imaging thick, dense breasts. Because the molybdenum target spectrum is so heavily influenced by the characteristic X rays, an increase in the tube voltage alone does not substantially change the shape of the spectrum (see Fig. 9.7(a)). The average energy of the beam can be increased, however, by employing filters of higher atomic number than molybdenum. For example, rhodium (atomic number 45) has a K absorption edge at 23 keV, providing strong attenuation both for X rays above this energy and for those at substantially lower energies. Used with a molybdenum target X ray tube and slightly increased kV, it provides a spectrum with increased penetration (reduced dose) compared with the Mo/Mo combination. A Mo/Rh X ray spectrum is illustrated in Fig. 9.7(b).

Further improvement in imaging performance can be obtained by tuning the effective spectral energy using other target materials in combination with appropriate K edge filters. One example is the use of an X ray tube that incorporates a rhodium target. A 25–35 μm thick rhodium filter is used with this target material. Figure 9.7(c) illustrates the spectrum produced with a Rh target and a Rh filter. Similarly, particularly for digital mammography, K edge filtration of tungsten spectra can be used to advantage in that the lack of pronounced K characteristic peaks provides flexibility in spectral shaping with filters, as illustrated in Fig. 9.7(d). Typically, filters composed of aluminium, rhodium or silver are used to shape the tungsten spectrum.

For screen film mammography, the fixed characteristic curve of the film imposes limitations on the suitable energy range; contrast and noise are limited

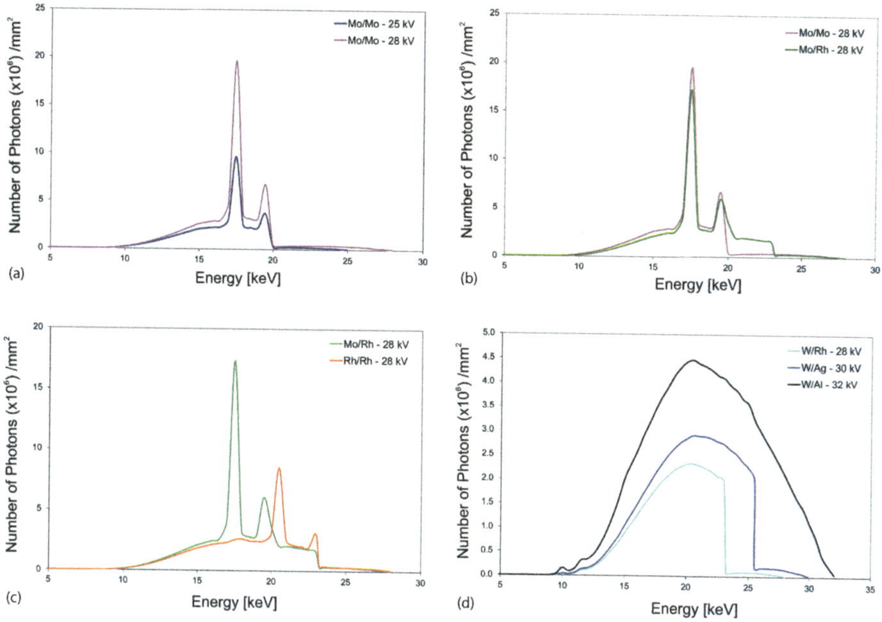

FIG. 9.7. Examples of mammographic X ray spectra.

by the acceptable radiation dose to the breast. For digital mammography, the limitations imposed by the film are removed and the gradient of the image display is freely adjustable at the viewing workstation. This provides an opportunity to use higher energy beams to improve the SNR per unit dose and to provide the potential for dose reductions. For example, whereas a typical exposure technique for an average breast with screen film mammography might be Mo target, Mo filter and 26 kV, with a digital mammography system, either a Mo/Rh or Rh/Rh combination might be used at 28 or 29 kV, or a tungsten target with Ag or Rh filtration operated at similar tube voltage.

9.3.2. Compression

There are several reasons for applying firm (but not painful) compression to the breast during the mammographic examination. Compression causes the various breast tissues to be spread out, minimizing superposition from different planes and thereby improving the conspicuity of structures. This effect may be accentuated by the fact that different tissues (fatty, fibroglandular and cancerous) have different elasticities, resulting in the various tissues being spread out by different amounts and potentially making a cancer easier to see.

As in other areas of radiography, scattered radiation will degrade contrast in the mammogram. The use of compression decreases the ratio of scattered to directly transmitted radiation reaching the image receptor. In Fig. 9.8, the effect of breast thickness on scatter is quantified. Compression also decreases the distance from any plane within the breast to the image receptor, and in this way reduces geometric unsharpness. The compressed breast provides lower overall attenuation to the incident X ray beam, allowing the radiation dose to be reduced. The compressed breast also provides more uniform attenuation over the image. This reduces the exposure range that must be recorded by the imaging system, and in screen film mammography allows a film of higher gradient to be employed. Finally, compression provides a clamping action, which reduces anatomical motion during the exposure, thereby reducing this source of image unsharpness.

It is important that the breast be compressed as uniformly as possible and that the edge of the compression plate at the chest wall be straight and aligned with both the focal spot and image receptor to maximize the amount of breast tissue that is included in the image (see Fig. 9.5). The mechanical properties of the breast are non-linear; after a certain reduction in thickness, application of additional pressure provides little benefit in terms of improved image quality and only contributes to patient discomfort. Specialized mechanisms have been introduced by several manufacturers to try to achieve better compression, while minimizing the risk of overcompression.

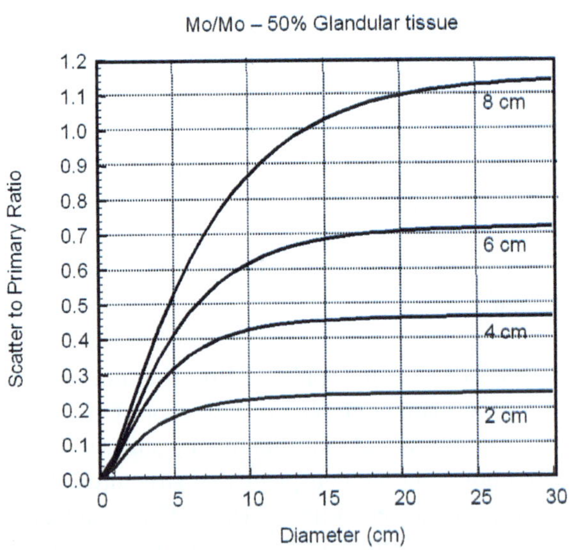

FIG. 9.8. The effect of breast thickness and diameter of the X ray field on the SPR when no antiscatter grid is used.

9.3.3. Grids

In the absence of an antiscatter device, 37–50% of the total radiation incident on the image receptor would have experienced a scattering interaction within the breast. Thus, as seen in Fig. 9.8, the scatter to primary ratio (SPR) will range from 0.3 to 1.2, depending upon the breast size. In addition to contrast reduction, the recording of scattered radiation reduces the useful dynamic range of the image receptor and adds stochastic noise to the image. The actual SPR recorded in the image is determined in part by the detector material; the lower energy and oblique incidence of scattered X rays result in higher attenuation than for primary X rays (see Chapter 6).

It is typical to use focused linear grids in mammography, with grid ratios from 3.5:1 to 5:1. On modern mammography equipment, the grid is an integral part of the system, and during the X ray exposure is moved to blur the image of the grid septa to avoid distracting artefacts in the mammogram. It is important that this motion be uniform and of sufficient amplitude to avoid non-uniformities in the image, particularly for short exposures that occur when the breast is relatively radiolucent. At least one manufacturer provides a crossed grid that consists of septa that run in orthogonal directions. Improved scatter rejection is accomplished at doses comparable to those required with a linear grid, because the interspace material of the crossed grid is air rather than a solid. To avoid artefacts, the crossed grid must be moved in a very precise manner to ensure a uniform blurring.

When a grid is used, the SPR is typically reduced by a factor of about 5, leading in most cases to a substantial improvement in image contrast (Fig. 9.9(a)). As discussed in Chapter 6, to maintain image quality when the grid is used, it is necessary to compensate for losses of X ray fluence at the image receptor that are caused by absorption of primary radiation by the grid septa and interspace material, as well as removal of scatter by the grid. This is reflected in the Bucky factor, which can be as large as 2 to 3 in mammography (Fig. 9.9(b)).

The improvement in image contrast in screen film mammography and SNR in digital mammography is generally considered to justify this increase in dose to the breast. Some differences do exist between digital and screen film mammography. In mammography, the benefit of a grid is clear for thick breasts; however, in the digital mammography of small or thin breasts, the signal difference to noise ratio (SDNR) improvement from scatter reduction may not justify the dose increase from the use of a grid. Also, for digital mammography, it is not necessary to compensate for removal of scattered radiation. This allows a reduction of the Bucky factor in digital imaging and a corresponding dose reduction.

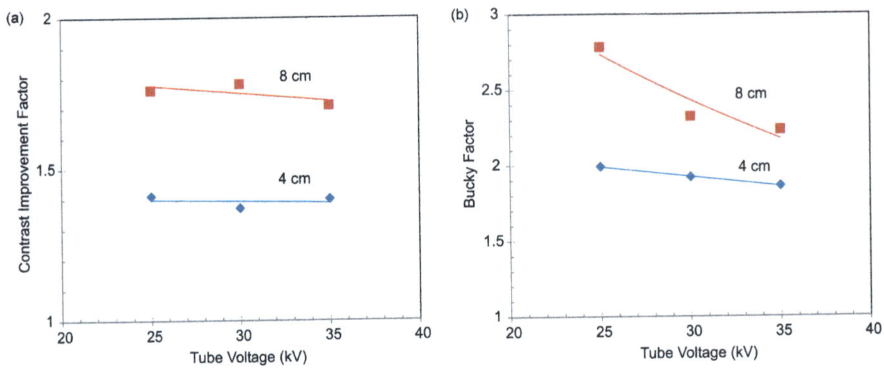

FIG. 9.9. The effect of the use of a grid for screen film mammography on (a) contrast improvement factor and (b) the Bucky factor. Curves are shown as a function of tube voltage for a Mo/Mo target filter combination with 4 cm and 8 cm thick breast equivalent phantoms.

9.3.4. AEC

It is difficult to estimate the attenuation of the breast by visual inspection; therefore, modern mammography units are equipped with AEC (see Sections 5.4.3, 6.2.7 and 19.5.2). For screen film mammography, it is very important for both image brightness and contrast to achieve a target optical density (OD) in the image, while with digital mammography, it is more useful to achieve a target SNR or, preferably, a target SDNR in the image.

For screen film mammography and for cassette based digital systems, the AEC radiation sensor(s) is/are located behind the image receptor to avoid casting a shadow on the image. The sensors measure the X ray fluence transmitted through both the breast and the image receptor and provide a signal to discontinue the exposure when a preset amount of radiation has been received by the image receptor. The location of the sensor is adjustable so that it can be placed behind the appropriate region of the breast to obtain proper exposure. The AEC performance must be independent of variations in breast attenuation, tube voltage or filter settings, and field size. With modern equipment, AEC is generally microprocessor based, so that relatively sophisticated corrections can be made during the exposure for the above effects and for reciprocity law failure of the film (see Section 7.3.5).

Penetration through the breast depends on both breast thickness and composition. For a breast that is thick or dense, it is possible that for a relatively low tube voltage, a very long exposure time would be required to achieve adequate film darkening or digital signal. This would result in a high dose to the breast and possibly blur due to anatomical motion, while a more penetrating beam allows a lower dose to be used but with a loss of image contrast. Thus, many mammography AEC systems also incorporate automatic control of the tube voltage or target/filter/tube voltage combination. These systems sense the compressed breast thickness and the transmitted exposure rate and employ an algorithm to choose automatically the X ray target and/or beam filter as well as the tube voltage. Typically, a short (usually <100 ms) X ray pre-exposure is made first. The recorded X ray signal and the breast thickness measured from the compression plate position are used to infer the composition of the breast and determine the optimal exposure conditions.

To provide greater flexibility, multiple user selectable algorithms can be incorporated into the system to weight the exposure factor selection towards either lower dose or higher image quality, according to the requirements of the examination. All systems build in appropriate constraints to ensure that the operation of the equipment is in compliance with applicable regulatory radiation dose limits and within the functional limitations of the X ray tube, generator and image receptor. These requirements may cause the image quality to be less than optimal under certain circumstances.

For digital mammography, it is typical for the pre-exposure concept described above to be used, but in this case, rather than a single sensor measurement, an entire low dose image is created in the digital detector. This image can be analysed to determine the overall SDNR or the minimum value over a set of small (~1 cm^2) regions of interest in the image. Then, the target material, the filter and the tube voltage can be selected automatically to attempt to ensure a desired SDNR when the main exposure is performed.

As digital detectors can be operated at a range of input dose levels, it is possible to optimize imaging according to a priority of SDNR, lowest dose or a combination. Different manufacturers have approached this challenge in different ways and development in this area is ongoing. For example, the location of the edges of the breast can be determined automatically so that the algorithm is only sensitive to the region of the image within the breast. The algorithm can be ‘trained’ to identify automatically the critical areas that will dominate the selection of exposure parameters. Special modes of operation can be developed for tasks such as imaging breasts containing implants. Again, these systems must be constrained by radiation regulations.

9.3.5. Magnification mammography

Magnification mammography is often used intentionally to improve the diagnostic quality of the image. This is accomplished by elevating the breast above the image receptor, in effect reducing the focus to object distance and increasing the distance from the object to the image receptor. Magnification mammography achieves three key benefits: (i) increased SNR, (ii) improved spatial resolution and (iii) dose efficient scatter rejection. These benefits are illustrated in Fig. 9.10.

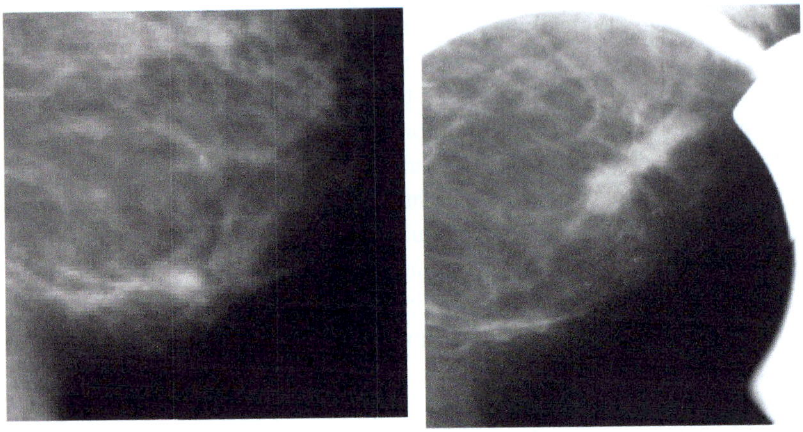

FIG. 9.10. A suspicious region is visible in the lower aspect of the mammogram (left). A magnified image of this region obtained with focal compression shows an obvious mass (right).

Magnification causes structures to appear larger when projected on to the image receptor, thereby increasing the effective modulation transfer function (MTF) of the receptor with respect to structures within the breast (see Sections 6.2.3 and 6.2.4). In screen film mammography, the limiting resolution of the image receptor is already quite high and is rarely a limiting factor. The increase in focal spot unsharpness that occurs as a result of magnification, even with a small focal spot, typically offsets any improvement in the MTF of the image receptor. The main benefit of magnification is to increase the size of the projected anatomical structures compared with the granularity of the image, thereby improving the SNR in the image. This improvement can be valuable, particularly for the visualization of fine calcifications and spiculations.

In digital mammography, where the film grain noise has been eliminated, but where the limiting spatial resolution of the detector is lower than that provided by the screen film image receptor, the benefits of magnification may be different in nature. In this case, the increase in projected size of anatomical features does improve the effective resolution of the detector, which in some cases is a limiting factor.

The spatial resolution of magnification mammography is always limited by focal spot size. As such, the use of a small spot for magnification imaging (typically a nominal size of 0.1 mm) is critical. Loss of spatial resolution can be controlled in part by using the minimum necessary magnification. It is typical to use magnifications of between 1.5 and 1.8. For the small focal spot, the X ray tube current must be reduced, necessitating increased exposure times. As a result, there is an increased likelihood of resolution loss resulting from motion of anatomical structures. It is common to apply focal compression to the breast in magnification imaging (Fig. 9.10, right), reducing the breast thickness and hence exposure time.

By moving the breast closer to the X ray source in magnification mammography, the dose to breast tissue increases compared with that in contact mammography. The increased air gap between the breast and the image receptor provides some scatter rejection; thus, antiscatter grids are not employed for magnification. This partially offsets the increase in dose and the increase in exposure time that occurs from use of the small focal spot.

9.4. IMAGE RECEPTORS

9.4.1. Screen film mammography

In screen film mammography, a high resolution fluorescent intensifying screen is used to absorb the X rays and convert the pattern of X rays transmitted by the breast into an optical image (see Section 7.3). These screens are used in conjunction with single emulsion radiographic film, enclosed within a lightproof cassette. The film is typically available in two sizes: 18 cm × 24 cm and 24 cm × 30 cm. It is customary to use the smallest possible size that ensures complete radiographic coverage of the breast; this results in superior breast positioning and compression. In women with large breasts, multiple films may be required to image the breast fully.

The screen and film are arranged as shown in Fig. 9.11, such that the X rays must pass through the cover of the cassette and the film to impinge upon the screen. Absorption is exponential, so that a larger fraction of the X rays is absorbed and converted to light near the entrance surface of the screen. The lateral spread of the light increases with the distance that light quanta travel through the phosphor. By minimizing the distance that the light must travel before being collected, this geometry reduces blurring due to lateral spreading, thus providing the maximal spatial resolution. To discriminate further against light quanta travelling along long oblique paths, the phosphor material of the screen may be treated with a dye, which absorbs much of this light, giving rise to a sharper image.

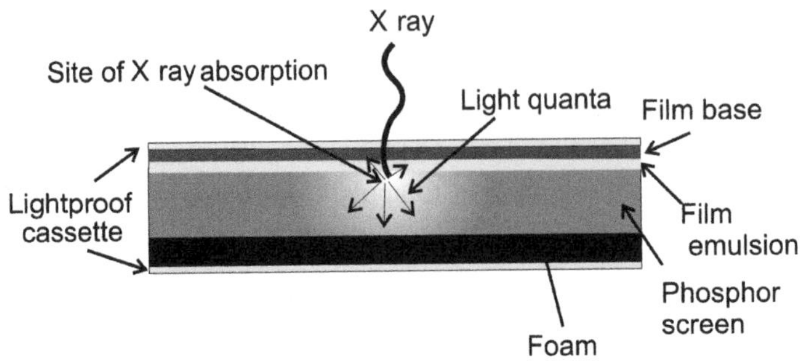

FIG. 9.11. Configuration for a mammographic screen film image receptor. A single emulsion radiographic film is held in close contact with a fluorescent screen in a lightproof cassette.

A typical phosphor used for screen film mammography is gadolinium oxysulphide (Gd_2O_2S:Tb). Although the K absorption edge of gadolinium occurs at too high an energy to be useful in mammography, the phosphor material is dense (7.44 g/cm^3), so that the quantum detection efficiency (QDE) (the fraction of incident X rays that interacts with the screen (see Sections 7.2.1 and 7.3.2)) is reasonably high (approximately 60% for a typical screen thickness and X ray spectrum), and K fluorescence (a potential source of noise) is avoided. Also, the conversion efficiency (fraction of the absorbed X ray energy converted to light) exceeds 10%, which is high for a phosphor. The amount of light emitted from the fluorescent screen is linearly dependent upon the total amount of energy deposited by X rays within the screen.

The photographic film emulsion for mammography is matched to be sensitive to the spectrum of light emitted from the particular phosphor screen, and to the range of X ray fluence exiting the breast. As such, it is important to examine the overall characteristics of the screen and film combination rather than those of the individual components.

In mammography, compression of the breast reduces the overall range of X ray fluence exiting the breast, as compared with a breast that is not uniformly compressed. This allows films with high gradient to be used in an effort to enhance the contrast between subtly varying soft tissue structures (Fig. 9.12(a)). In addition, mammography film has a high maximum OD (D_{max}) (4.0–4.8 OD) to maximize the exposure latitude over which the high gradient exists (Fig. 9.12(b)). This is particularly important near the periphery of the breast where its thickness decreases rapidly. Nevertheless, some regions of the mammogram will generally be underexposed or overexposed, i.e. rendered with suboptimal contrast.

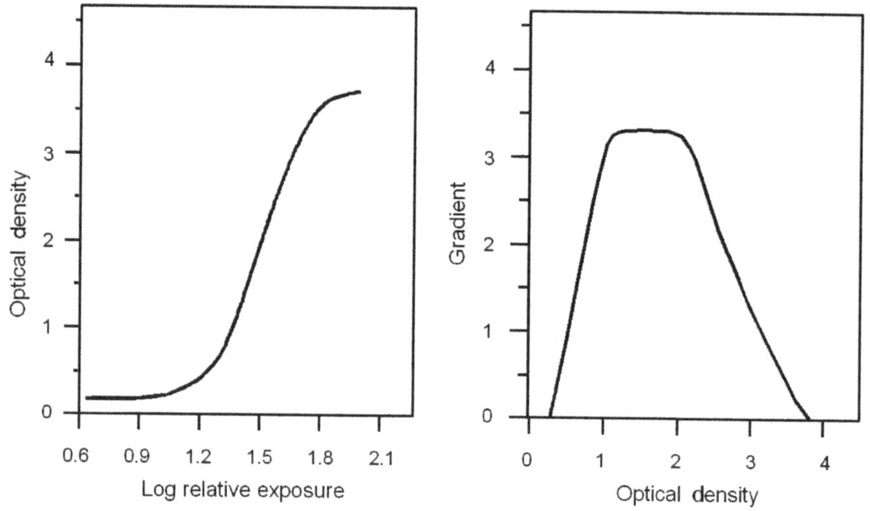

FIG. 9.12. Characteristic curve of a film emulsion used for mammography.

Mammography film is processed in an automatic processor similar to that used for general radiographic films. It is important that the development temperature, time and rate of replenishment of the developer chemistry be compatible with the type of film emulsion used and be designed to maintain good contrast of the film. Daily quality assurance is required in mammography to ensure ongoing optimal performance.

There are several technical factors associated with screen film mammography that limit the ability to display the finest or most subtle details and produce images with the most efficient use of radiation to the patient. In screen film mammography, the film must act as an image acquisition detector as well as a storage and display device. The sigmoidal shape of the characteristic curve results in limited latitude — the range of X ray exposures over which the film display gradient is significant. If a tumour is located in a region of the breast that is either more lucent or more opaque, then the contrast displayed to the radiologist may be inadequate because of the limited gradient of the film. This is of particular concern in patients whose breasts contain large amounts of fibroglandular tissue, the so called 'dense breast'.

Another limitation of screen film mammography is the effect of fixed pattern noise due to the granularity of the phosphor screen and the film emulsion used to record the image. This impairs the detectability of microcalcifications and other fine structures within the breast. Finally, screen film mammography suffers from compromises in spatial resolution versus quantum detection efficiency, which are inherent in the screen film image receptor.

9.4.2. Digital mammography

Digital mammography, introduced commercially in 2000, is able to overcome many of the technical limitations of screen film mammography. In digital mammography, image acquisition, processing, display and storage are performed independently, allowing optimization of each. Acquisition is performed with low noise X ray detectors having a wide dynamic range. As the image is stored digitally, it can be displayed with contrast that is independent of the detector properties and defined by the requirements of the particular imaging task. Whatever image processing techniques are found useful, ranging from simple contrast enhancement to histogram modification and spatial frequency filtering, can conveniently be applied prior to image display.

The challenges in creating a digital mammography system with improved performance are mainly related to the X ray detector and the display device. The detector should have the following characteristics:

- Efficient absorption of the incident radiation beam;
- A linear or logarithmic response over a wide range of incident radiation intensity;
- Low intrinsic noise and little to no fixed pattern noise, to ensure that images are X ray quantum noise limited;
- Limiting spatial resolution of the order of 5–10 cycles/mm (50–100 μm sampling);
- It can accommodate at least an 18 cm × 24 cm and preferably a 24 cm × 30 cm field size;
- It can image immediately adjacent to the chest wall;
- An acceptable imaging time and heat loading of the X ray tube (e.g. in detectors that must be scanned to image the entire breast).

Two main approaches have been taken in detector development — area detectors and scanning detectors. In the former, the entire image is acquired simultaneously, while in the latter only a portion of the image is acquired at one time and the full image is obtained by scanning the X ray beam and detector(s) across the breast. Area detectors offer fast image acquisition and can be used with conventional X ray machines equipped with a grid to reduce scatter. By comparison, scanning systems have longer acquisition times and are mechanically more complex, but use relatively simple detectors and have excellent intrinsic scatter rejection.

Various detector technologies are employed in full field digital mammography systems, i.e. those capable of imaging the entire breast (see also Section 7.4.3). In one approach, the detector consists of an amorphous silicon thin film transistor panel containing a rectangular matrix of 2000–3000 columns by 3000–4000 rows of detector elements (dels). Each del is connected by a thin film transistor switch to electrical lines running along each row and column (see Section 7.4.3). This array is covered by a phosphor or a photoconductor X ray detector.

In so-called 'indirect' detectors, each del includes both a light sensitive photodiode and a thin film transistor switch. The array is covered with a phosphor layer, typically made of thallium activated CsI. X rays transmitted by the breast are absorbed by the phosphor and the light produced is converted in the photodiode to charge, which is stored on its capacitance. After the X ray exposure, readout signals sent sequentially along the lines for each row activate the corresponding switches and the charge is transferred down the columns to readout amplifiers and multiplexers, and digitized to form the image. This readout system allows the signals from all of the dels to be read in a fraction of a second. The needle-like phosphor crystals of CsI (see also Fig. 8.2) behave somewhat like fibre optics, conducting the light to the photodiodes with less lateral spread than would occur with granular phosphors. This allows the thickness of the phosphor to be increased relative to a granular phosphor, to improve the QDE of the detector without excessive loss of spatial resolution.

A second system employs a similar readout strategy but replaces the phosphor with an X ray absorber composed of amorphous selenium, which is a photoconductor. In this so-called 'direct' detector, the energy of the absorbed X rays causes the liberation of electron hole pairs in the selenium. The charged particles are drawn to the opposite faces of the detector by an externally applied electric field. To collect the signal, an array of electrode pads (rather than photodiodes) forms the dels. Unlike the phosphor based detectors, the electric field can be tailored to collect the charge with minimal lateral spread. This allows the use of a relatively thick detector to achieve excellent QDE without significant reduction in resolution at near normal incidence. Other materials in which X ray energy is directly converted to charge are under development and include lead iodide, zinc cadmium telluride and thallium bromide. The use of higher atomic number materials would allow the thickness of the X ray converter to be reduced. This mitigates against the degradation of the MTF resulting from the oblique incidence of the X rays.

Another technology used for digital mammography employs a plate formed of a photostimulable phosphor material, housed in a lightproof cassette, described in more detail in Section 7.4.2. When exposed to X rays, electrons in the crystalline material are excited and subsequently captured by traps in the phosphor. The number of trapped electrons is proportional to the amount of X ray energy absorbed at a particular location in the detector. After exposure, the plate is placed in a reader device and scanned with a red HeNe laser beam.

The energy of the laser light stimulates the traps to release the electrons. The transition of these electrons through energy levels in the phosphor crystal results in the formation of blue light. The light is collected by an efficient optical system, measured with a photomultiplier tube and the signal digitized. By correlating the time of measurement of the signal with the position of the scanned laser beam, the signal can be attributed to a particular pixel in the image. The resolution of the image is determined by the size of the scanning laser beam, the underlying scatter of the readout laser light in the phosphor and the distance between sample measurements.

Mammography photostimulable phosphor systems differ from the general radiography photostimulable phosphor systems in several key areas. In general, the mammography photostimulable phosphor system is designed for higher spatial resolution and thus uses a thinner phosphor material and is scanned with finer sampling pitch (typically 50 μm). The result is less signal per pixel. To overcome this limitation, various innovations have been developed to improve light coupling and reduce readout noise, including the use of dual sided readout of the phosphor plates and needle-like phosphors that permit the use of thicker detectors with superior QDE.

The detector systems discussed thus far acquire the image by integrating the signal from a number of X ray quanta absorbed in the detector and digitizing this signal. The image noise from these systems depends on both the Poisson X ray quantum fluctuations associated with X ray absorption and the additional noise sources associated with the production of the converted electronic signal. As discussed in Chapter 7, these noise sources can arise from the fluctuation in the amount of light produced in a phosphor in response to absorption of an X ray of a particular energy, or from the X ray spectrum itself (different amounts of signal are produced when X ray quanta of different energies interact with the detector material).

As an alternative, it is also possible to count the number of interacting quanta directly, thereby avoiding these additional noise sources. Typically, quantum counting detectors are multiline devices employing a geometry in which the X ray beam is collimated into a slot or multislit format and scanned across the breast to acquire the image. The detector can be based on either a solid state approach, where electron hole pairs are produced in a material such as crystalline silicon, or a pressurized gas, where the signal is in the form of ions formed in the gas. In either case, collection of the charge signal and appropriate amplification produces a pulse for each interacting X ray quantum and these pulses are simply counted to create the signal. An additional feature of these detectors is that as the beam is collimated to irradiate only part of the breast at a time, the SPR is reduced without the need for a grid, and this increases the dose efficiency of the system.

9.5. DISPLAY OF MAMMOGRAMS

9.5.1. Display of film mammograms

To allow visualization of as much of the information recorded in the mammogram as possible, it is essential that the viewing conditions be optimal. Mammograms should be interpreted under conditions that provide good visibility and comfort and incur minimal fatigue. For film, viewing transillumination systems are available that have been specifically designed to produce the appropriate luminance levels for reading mammograms. The illuminator surface should provide diffused light of uniform brightness. The luminance level must be sufficient to illuminate areas of interest in the mammogram. It has been recommended that illuminators for film mammograms be capable of producing a luminance of at least 3000 cd/m^2.

The contrast sensitivity of the eye (the ability to distinguish small differences in luminance) is greatest when the surroundings are of about the same brightness as the area of interest. Therefore, to see detail in a mammogram, it is important to reduce glare to a minimum, to avoid surface reflections, and to reduce the ambient light level to approximately that reaching the eye through the mammogram. Glare and reflections can be reduced by locating illuminators away from bright surroundings such as windows, by switching off surrounding viewboxes when not in use, and by using masks to cover unused portions of a viewbox or to cover low density areas in the mammogram being examined.

Subdued lighting is preferred in the viewing room. It is also important to have a variable brightness high output light source (with appropriate masks) to view high OD areas on the film mammogram, and to ensure that films are properly exposed and processed.

9.5.2. Display of digital mammograms

The display system plays a major role in influencing the overall performance of the digital mammography unit in terms of both the ease of image interpretation and the image quality presented to the radiologist. While some radiologists use ‘hard copy’ systems (laser printed films) for interpretation, the flexibility of adjustment of display brightness and contrast in digital mammography are best realized when the images are viewed on a computer display using either a cathode ray tube or flat panel display monitor. This is often referred to as ‘soft copy’ display.

The display must have a suitable number of high quality monitors (normally two 5 megapixel monitors are recommended) to allow viewing of as much of the mammogram as possible at the required resolution level. Digital mammograms generally contain more pixels than can be displayed at one time on a soft copy device. A 5 megapixel monitor is capable of displaying only a single mammogram with approximately 2000 × 2500 100 µm pixels at full resolution. Larger images must be resampled to reduce temporarily the image size so that the entire mammogram can be viewed. Then, zooming and scrolling operations can be employed to allow inspection of regions of interest in the image at the full acquired spatial resolution.

Hard copy display systems produce a printout of the digital image on transparent film sensitized to laser light. The image brightness and contrast are usually adjusted by the radiographer before printing out the image, making use of the controls provided at the acquisition work station. Hard copy image displays have the disadvantage of not allowing the radiologist to control the image processing operations during viewing. Therefore, it is strongly recommended that images be displayed for interpretation on a high quality soft copy device.

Both wet and dry laser printers are available and produce breast images of similar quality. The spatial sampling (resolution) of laser printers should at least match the del size, so the printing device should not be the limiting factor. Using too low a resolution for printing results in printed films with coarse looking pixels, or in an overly magnified image of the breast. Most of the commercially available printers for digital mammography have two pixel sizes, 100 µm and 50 µm or smaller (around 600 dpi), with pixel bit depth of 12 or 14 bits, and can print on one or more sizes of transparent film (typically 18 cm × 24 cm or 24 cm × 30 cm).

Typically, transparent laser films do not have the same maximum OD capability as mammography film. The dynamic range of laser films varies from approximately 0.2 OD up to around 3.2 OD, depending on the film type. It is recommended that the laser printer characteristic curve be in conformance to the Digital Imaging and Communications in Medicine (DICOM) Grayscale Standard Display Function (GSDF) (see Chapter 16).

9.5.2.1. Image processing of digital mammograms

In addition to resampling, zoom and scrolling operations, several other types of image processing can be employed to improve the presentation of information in digital mammograms. Common image processing steps include: (i) segmentation of the breast from the background (air), so that the greyscale lookup table inversion does not display the air as white, (ii) peripheral enhancement to suppress the effect of changing thickness near the edges of the breast, (iii) image resolution restoration or enhancement to increase the conspicuity of the clinical signs of breast cancer, (iv) lookup table manipulation to allow adjustment of displayed brightness and contrast and to match the performance of the human eye, and (v) noise suppression to enhance the relative conspicuity of small calcific deposits found in the breast. Methods of image processing are discussed in more detail in Chapter 17.

9.6. BREAST TOMOSYNTHESIS

Digital mammography has been demonstrated as providing equal or improved (for women with dense breasts) accuracy compared with screen film mammography. Nevertheless, neither its sensitivity (probability of finding cancers when present) nor its specificity (probability of a negative result when cancer is not present) is 100%. One reason for this is the masking effect of tissue superposition that necessarily occurs in projection radiography. This masking effect can be reduced or avoided using 3-D X ray imaging techniques. The principles of tomosynthesis are discussed in Chapter 10 and of computed tomography (CT) in Chapter 11. Dedicated breast imaging systems have been developed for both of these modalities. Both provide reconstructed planar images of sections of the breast.

Tomosynthesis images are acquired on a modified digital mammography system where the arm supporting the X ray tube pivots about a point, while the compressed breast remains stationary. The detector may also pivot, depending upon the system design. Typically, a small number (9–25) of low dose projection images are obtained over a limited range of angles (±7° to ±30°) about the normal to the desired image plane.

The X ray spectra used in tomosynthesis are generally higher in energy than those used in digital mammography (e.g. the W/Al spectrum shown in Fig. 9.7). The X ray tube may be moved in a continuous or discrete ('step and shoot') fashion; thus, short X ray pulses with higher tube currents are used in tomosynthesis. As with mammography, the total acquisition time must be minimized to avoid image degradation due to patient motion.

Planar cross-sectional images are reconstructed from the projections using filtered back projection or an iterative reconstruction algorithm (see Chapter 11). It should be noted that the spatial resolution of tomosynthesis is anisotropic; tomosynthesis provides the highest resolution in plane and relatively poor resolution between planes. As a result, the reconstructed voxels are generally non-isotropic, with a pixel size approximately equal to the size of the del and a reconstructed slice spacing that is typically 1 mm. Owing to the limited range of acquisition angles, the projection data for tomosynthesis do not form a complete set, so that the reconstructed image is not a true 3-D representation of the breast anatomy. This results in artefacts, which are observed in images.

9.7. BREAST CT

CT provides true tomographic images in which the voxels fairly accurately represent X ray attenuation coefficients of the breast tissue. The breast CT systems that have been developed employ a cone beam geometry and a flat panel area X ray detector. Therefore, the data for all of the CT slices are acquired simultaneously. This allows rapid image acquisition, but causes the SPR to be much higher than would be the case in single slice CT, especially as no grid is used. Voxels tend to be isotropic, but the pixel dimensions in the plane of the tomographic section are substantially larger than those provided with digital mammography or tomosynthesis. The current system designs provide a dedicated prone imaging table, which introduces challenges with respect to imaging tissue adjacent to the chest wall. Owing to the large number of projections, images are generally acquired at a much higher tube voltage (50–80 kV) than for mammography (~30 kV), in order to keep doses at an acceptable level. Nevertheless, the very low dose per projection can result in noisy images. A desirable feature of breast CT is that it can be performed without the need to compress the breast.

9.8. COMPUTER AIDED DIAGNOSIS

The goal of computer aided diagnosis (CAD) is to assist radiologists in detecting breast cancer, principally in screening mammography. CAD has the potential to be a cost effective alternative to independent double reading by two radiologists, in that the CAD algorithm can be used to simulate the second radiologist. Double reading has been shown to increase the cancer detection rate, but it is not widely practised because of costs and logistics. Thus, CAD has the potential to reduce the cancer miss rate, reduce the variability among radiologists, improve the consistency of a single radiologist and make radiologists more productive.

Most CAD schemes are designed using the paradigm shown in Fig. 9.13. A digital mammogram is used as the input for CAD. This can come from a full field digital mammography system or from data acquired by digitizing a screen film mammogram. The first step is to preprocess the image to segment the breast area from the non-breast area and to use image processing to emphasize lesions or certain features of lesions. For example, spatial filters can be used to make microcalcifications more prominent, or specialized non-linear filters can be used to highlight the spiculation associated with malignant masses.

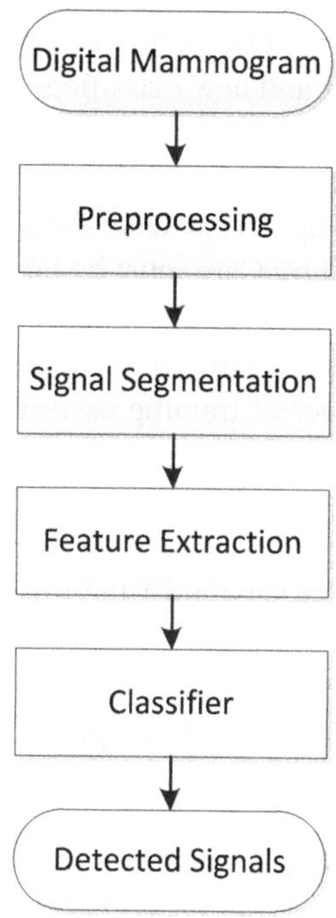

FIG. 9.13. Conceptual outline of a CAD system.

After the image has been preprocessed, potential lesions are identified. The simplest means is to 'threshold' the region containing the lesion (see Chapter 17), as both microcalcifications and masses appear brighter than their surrounding background. Once potential lesions have been identified, they are segmented from the image using various techniques (see Chapter 17). Because the borders of masses are often ill-defined or partially obscured by normal tissues of the breast, gradient based methods are frequently more effective in these situations than grey level based methods.

To reduce the number of false detections, many features of the segmented detections are extracted from the image. Most features fall into one of three categories: intensity based, morphology based and texture based. These features can be extracted from the greyscale image or from the image after it has undergone a mathematical transformation. A subset of these features is chosen for further analysis.

Once the final feature set has been chosen, the features are merged by a statistical classifier to differentiate actual lesions from false detections. Many different types of classifier can be used, such as support vector machines, artificial neural networks, k nearest neighbour and decision trees. Most classifiers have comparable performances and new classifiers are being actively researched.

CAD algorithms must be 'trained' using a set of mammograms for which 'truth' (i.e. the presence or absence of a cancer) is known through biopsy results or subsequent patient follow-up. Truth data are also required to evaluate the performance of a CAD algorithm. Care must be taken in training and evaluating the classifier to avoid bias and to reduce the variance in the measured performance. To avoid a positive bias, cases used to train the classifier are not used to test the classifier. Increasing the number of training cases can improve the performance of the classifier and reduce the variance in the measured performance.

The results of the CAD algorithm are conveyed to the radiologist by means of an image annotated to show the computer detections. For screen film based CAD systems, a low resolution version of the image is either printed on paper or shown on a monitor. The locations of computer detected masses and clustered calcifications are shown using different symbols for the different lesions. For digital mammography systems, the CAD output can be annotated directly on to the radiologist's display workstation.

More recent research in CAD considers the combination of information from multiple images, either from a different view from the same examination, or from the same view from a previous exam. This approach more closely mimics how a radiologist reads a case, and it can improve the performance of a CAD scheme. Other CAD methods combine information from one or more images with clinical findings, as these data are also available and may influence the decisions made by clinicians. CAD may become particularly valuable in 3-D breast imaging, where the amount of image data to be considered is greatly increased. Here, CAD may be useful for automatic detection of microcalcifications in the large image set, allowing the radiologist to focus attention on more sophisticated interpretation tasks.

9.9. STEREOTACTIC BIOPSY SYSTEMS

In mammography facilities that perform diagnostic procedures, digital systems are often used for guidance of stereotactic needle breast biopsy procedures. Stereotactic procedures are used to investigate suspicious mammographic or clinical findings without the need for surgical (excisional) biopsies, resulting in reduced patient risk, discomfort and cost. In stereotactic biopsies, the gantry of a mammography machine is modified to allow angulated views of the breast (typically at ±15° from normal incidence) to be obtained (Fig. 9.14). From measurements obtained from these images, the 3-D location of a suspicious lesion is determined and a needle equipped with a spring loaded cutting device can be accurately placed in the breast to obtain tissue samples. These systems may use small format (5 cm × 5 cm) digital detectors or full field detectors. The image receptor can be located either in a conventional mammography unit or beneath a dedicated table, where the patient lies prone on the table with the breast pendant through an aperture in the table into the imaging region.

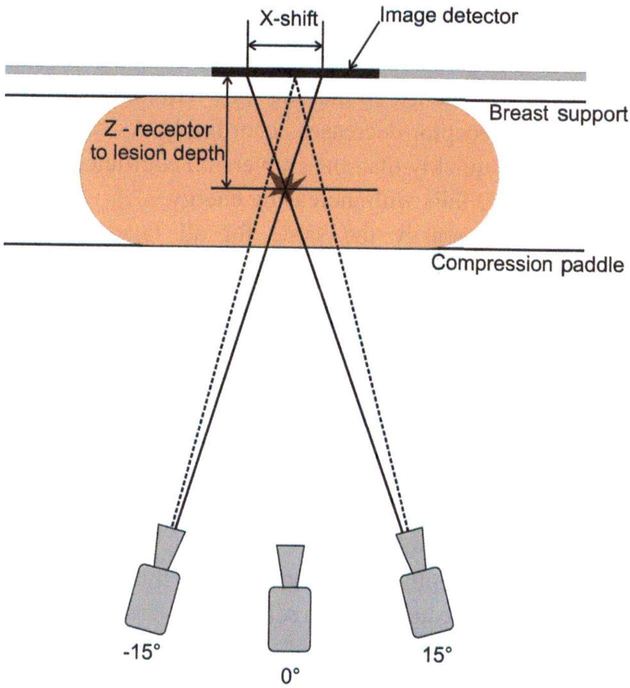

FIG. 9.14. The geometry for stereotactic breast biopsy is shown. The X ray tube is rotated about the breast to produce two views. The Z depth of an object can be determined by the lateral (X) displacement observed between the two views.

9.10. RADIATION DOSE

There are three dosimetric quantities used for mammography: (i) incident air kerma (IAK) (K_i), (ii) entrance surface air kerma (K_e) and (iii) mean dose to the glandular tissue of the breast, known as the mean glandular dose (MGD) (D_G). The MGD is the primary quantity of interest related to the risk of radiation induced cancer in breast imaging.

Details of the calculation of MGD are given in Section 22.5.3.2. The MGD is calculated using factors obtained experimentally or by Monte Carlo radiation transport calculations, which convert from IAK to dose in a breast of specific composition and size. These conversion coefficients are tabulated in various publications, including IAEA Technical Reports Series No. 457 [9.1].

The MGD conversion coefficient increases with the mean energy of the X ray spectrum. To produce an image of appropriate quality, every image receptor requires a specific quantity of X ray energy to be transmitted by the breast and be absorbed by the receptor. As shown in Fig. 9.2, the attenuation of the breast decreases with increasing energy, so that the IAK required to obtain a specified absorbed energy in the receptor decreases accordingly. As energy increases, the required IAK falls more quickly than the conversion coefficient increases, so the net result is that the MGD falls with increasing energy.

Doses are not necessarily the same for all types of film or digital mammography system, especially for the latter where technologies differ quite markedly. In the Ontario (Canada) Breast Screening Program, in measurements made in 2009 and 2010, it was found on standard measurements with phantoms that the dose to each breast from the standard two view examination was 3.2, 2.72 and 2.36 mGy for screen film mammography, digital mammography with photostimulable phosphor cassettes (CR) and digital mammography with captive detectors (DR), respectively. In this survey, a system that used a photon counting detector resulted in a dose that was only 41% of that required on average for imaging with screen film technology.

The dose in mammography is also dependent on the size and composition of the breast, as well as the exposure settings selected. In screen film mammography, where the goal is to maintain a target value for OD on the film, the IAK will increase as the thickness of the breast and the fraction of fibroglandular tissue (often referred to as density) increase. This will cause a corresponding increase in MGD. An increase in beam energy (tube voltage, choice of target material, beam filter) will mitigate against some of the dose increase. However, image contrast will be reduced and at some point this will become unacceptable.

In digital mammography, the goal is to achieve a target SDNR at the detector. Again, the dose will increase with breast thickness and breast density. However, with a digital system where contrast can be adjusted during image display, an acceptable compromise can be achieved at a higher energy than with screen film. This allows the advantage of a greater relative decrease of dose compared with film for large and/or dense breasts. Figure 9.15 shows the effect of breast thickness on the required dose, calculated using a theoretical model. Here, a constant SDNR is achieved at the detector. This requires increased dose for thick breasts and low energies where the breast is more opaque to X rays, but also for higher energies where the signal difference or contrast becomes less.

However, as previously mentioned, the increase with increasing energy is less than would be required for film mammography.

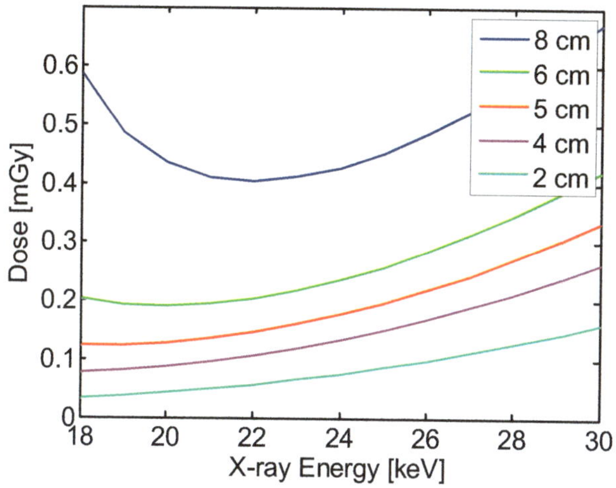

FIG. 9.15. Calculated dose required to achieve a fixed SNR at the detector for breast thicknesses of 2–8 cm. These doses are somewhat low because the effect of scattered radiation on contrast and noise has not been considered and because the beam is assumed to be monoenergetic.

There is a risk of cancer induction associated with the radiation doses received in mammography. Therefore, it is important to understand the magnitude of risk associated with the radiation dose delivered in mammography. The Biological Effects of Ionizing Radiation (BEIR) VII report critically examined data on doses and increased cancer incidence from several studied groups and allowed development of a radiation risk model for breast cancer [9.2]. The report provides a single model for all solid tumours, based on the work of Preston et al. [9.3]. For mammography, this model predicts the excess absolute risk (EAR) of cancer induction as:

$$\mathrm{EAR} = 10^{-3} \mathrm{e}^{-0.05(A_x - 25)} \times \left(\frac{A}{50} \right)^{3.5} \tag{9.1}$$

for women of age, A, less than or equal to 50 and:

$$\mathrm{EAR} = 10^{-3} \mathrm{e}^{-0.05(A_x - 25)} \times \left(\frac{A}{50} \right) \tag{9.2}$$

for women aged over 50 years. Here, EAR is the risk of the radiation induction of a cancer that will ‘surface’ in a 1 year period in a woman of age, A, per Gy of dose to the breasts at some earlier age, A_x, when the exposure occurred. As an example, the risk for a woman aged 60 years from a dose to the breasts of 0.0024 Gy, previously received from mammograms at the age of 45 years is predicted to be 1.06×10^{-6}. Note that in this model, risk is linearly related to the dose received and decreases with age at exposure. No explicit provision for latency is built into this model. However, the predicted risk does increase with attained age following the exposure, the rate of rise being slower after the age of 50 years.

Integrated appropriately, this model can be useful in predicting the lifetime risk following a single mammographic examination or from multiple exposures at different ages, as would occur in periodic screening. For example, in a screening regimen that consists of annual mammography examinations from the ages of 40 to 55 years, and biennial examination thereafter until the age of 74 years (i.e. 25 screenings) with a dose of 2.4 mGy to both breasts, it is estimated that in 100 000 women, 56 radiation induced cancers will be caused, resulting in 6.9 deaths and a loss of 88.5 women-years of life. For this calculation, a latency of ten years was applied, i.e. EAR = 0 for $A < A_x + 10$. For these woman, earlier detection through screening would save 500 lives or 10 670 women-years, resulting in a benefit to risk ratio of 72.5 (in lives) or 120.6 (in women-years). If the same diagnostic accuracy could be achieved at reduced radiation dose, the benefit to risk ratio would be even higher.

REFERENCES

[9.1] INTERNATIONAL ATOMIC ENERGY AGENCY, Dosimetry in Diagnostic Radiology: An International Code of Practice, Technical Reports Series No. 457, IAEA, Vienna (2007).

[9.2] NATIONAL RESEARCH COUNCIL, BEIR VII: Health Risks from Exposure to Low Levels of Ionizing Radiation, National Academies Press, Washington, DC (2006).

[9.3] PRESTON, D.L., et al., Radiation effects on breast cancer risk: A pooled analysis of eight cohorts, Radiat. Res. **158** 2 (2002) 2203–2205.

BIBLIOGRAPHY

AMERICAN COLLEGE OF RADIOLOGY, Mammography Quality Control Manual, American College of Radiology, Reston, VA (1999).

AMERICAN COLLEGE OF RADIOLOGY, Stereotactic Breast Biopsy Quality Control Manual, American College of Radiology, Reston, VA (1999).

BICK, U., DIEKMANN, F., Digital Mammography, Springer, Heidelberg (2010).

EUROPEAN COMMISSION, European Guideline for Quality Assurance in Mammography Screening, Rep. V4.0, Office for Official Publications of the European Communities, Luxembourg (2006).

INTERNATIONAL AGENCY FOR RESEARCH ON CANCER, GLOBOCAN 2008 — Cancer Incidence and Mortality Worldwide in 2008, IARS, Lyon, http://globocan.iarc.fr (accessed on 23 August 2012).

INTERNATIONAL ATOMIC ENERGY AGENCY, Quality Assurance Programme for Screen Film Mammography, IAEA Human Health Series No. 2, IAEA, Vienna (2009).

INTERNATIONAL ATOMIC ENERGY AGENCY, Quality Assurance Programme for Digital Mammography, IAEA Human Health Series No. 17, IAEA, Vienna (2011).

INTERNATIONAL COMMISSION ON RADIATION UNITS AND MEASUREMENTS, ICRU Rep. 82, Mammography – Assessment of Image Quality, J. ICRU **9** 2 (2009).

Chapter 10

SPECIAL TOPICS IN RADIOGRAPHY

I.D. McLEAN
International Atomic Energy Agency

J.A. SHEPHERD
University of California at San Francisco,
San Francisco, United States of America

10.1. INTRODUCTION

Up to this point, this handbook has described the use of X rays to form 2-D medical images of the 3-D patient. This process of reducing patient information by one dimension results in an image of superimposed tissues where important information might be obscured. Chapter 11 begins a section of the book involving the creation of cross-sectional medical images through computed tomography (CT), ultrasound and magnetic resonance imaging (MRI). This Chapter describes a number of special X ray imaging modalities and their associated techniques, and forms a transition between projection and cross-sectional imaging.

The first of these special topics is dental radiography, which is characterized by a diversity of technology and innovation. The common intraoral radiograph of a single tooth has seen little fundamental change since the time of Roentgen and is, today, along with the simple chest radiograph, the most commonly performed radiographic examination. By contrast, the challenge to create an image of all the teeth simultaneously has placed dentistry at the cutting edge of technology, through the development of panographic techniques and, most recently, with the application of cone beam CT (CBCT). Moreover, the small size of the tooth and the consequent reduced need for X ray generation power promotes equipment mobility. The effect of the need for equipment mobility also forms a special topic that is examined in this chapter.

Quantification of the composition of the body is another special X ray imaging technique. Dual energy X ray absorptiometry (DXA) is primarily used to derive the mass of one material in the presence of another, through knowledge of their unique X ray attenuation at different energies. DXA's primary commercial application has been to measure body mineral density as an assessment of fracture risk and to diagnose osteoporosis; thus, the X ray energies used are optimized for bone density assessment. Currently, there are estimated to be over 50 000 whole body DXA systems in use worldwide.

Lastly, this chapter reviews the process of sectional image formation through non-computational methods with the use of relative motion of the X ray source and image receptor.

10.2. DENTAL RADIOGRAPHY

10.2.1. Introduction

The tooth is a low attenuation static object that, when radiographed directly, places very limited demands on X ray generation. The image receptor is placed inside the mouth and irradiated externally. This universal low cost technique is known as an intraoral examination, with bitewing being the most common examination. When radiographs of the entire set of teeth are required, both the image receptor and the X ray source are external to the patient and the X ray beam is transmitted through the head, demanding significant X ray generation power and complex motion control for the X ray tube and image receptor. This procedure, known as an orthopantomograph (OPG), and the intraoral examination produce 2-D images that are captured most commonly on film but increasingly in electronic format. In special cases, when dental diagnosis requires 3-D information, CT units have been specially developed, most recently to include CBCT (also see Chapter 11).

10.2.2. Technology

10.2.2.1. Intraoral radiography

The intraoral X ray tube is a small robust device with a stationary target operating with a tube current of only a few milliamperes (see Fig. 5.9). The X ray generator (see Section 5.4) is typically very simple, often with a fixed tube voltage and tube current allowing output changes only by variations in exposure time. Major concerns with this device are the stability of the tube head and the collimation of the beam. International standards require that the focus to patient surface distance (FSD) be 200 mm. This is assured with the use of a collimating attachment that also restricts the beam to the region of the mouth being radiographed.

While the X ray equipment requires periodic quality control checking (see Chapter 19), the process of film processing (see Section 7.3) requires more diligent attention. The unscreened film is removed from the lightproof moisture protective wrapping and is processed either manually or with varying degrees of automation. Hand processing is probably most common and ideally requires control of temperature and processing time. For higher volume clinics, this can be automated, with film mounted on hangers that progress through the development, stop bath, fixation and rinse processes. Typically, these devices have timing and temperature controls but do not control chemical activity through replenishment. This is achieved in fully automatic processors, although these are typically restricted to major dental hospitals. The uncertainties in film processing are best controlled through sensitometry. While light sensitometers are rare in dentistry, owing to the small film format, adequate results can be achieved by using a simple radiograph of a three step 'wedge' (this can be easily obtained, either through manufacture by folding the lead foil found in the film wrap or by commercial purchase). Increasingly, digital detectors are replacing film.

Digital image capture can be achieved from an intensifying screen that is linked to a charge coupled device (CCD) camera through a tapered fibre optic coupling (see Fig. 7.10). The electronic signal can be transferred to an acquisition computer, either through a direct cable or through 'blue tooth' radiofrequency transmission.

10.2.2.2. OPG

An OPG image (Fig. 10.1) is created by complex equipment where the X ray tube and image receptor assembly move in a horizontal plane around the head of the patient. Importantly, the image receptor itself also moves within the assembly behind a lead aperture (Fig. 10.2). The device uses the principle of tomography (see below) and, more importantly, the principle of panoramic photography. This process can be illustrated through consideration of the panoramic camera used in photography. Here, an acquisition aperture is used to expose an image

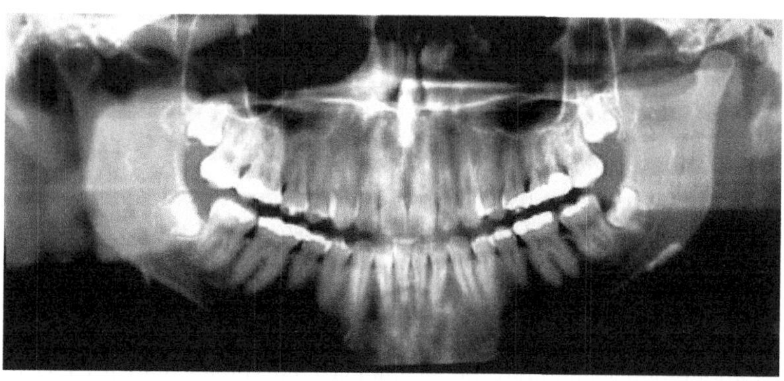

FIG. 10.1. OPG image of the teeth.

plate that is moved behind the aperture slit to capture the image of a 'panorama' while the camera simultaneously slowly rotates to scan a scene. Similarly, with the OPG, the panorama of the teeth is acquired by a narrow vertical fan beam of X rays as the tube rotates around the back of the head. Simultaneously, the image receptor is moved behind the aperture to capture the image. This imaging situation is non-isotropic, with the vertical magnification given by normal projection radiographic principles. The horizontal magnification is determined by the speed of the image receptor behind the acquisition slit and its relationship to the speed of the projected image of the teeth. This is adjusted, however, to give equal magnification in both directions on the resultant image. The tomographic process is described briefly in Section 10.5.

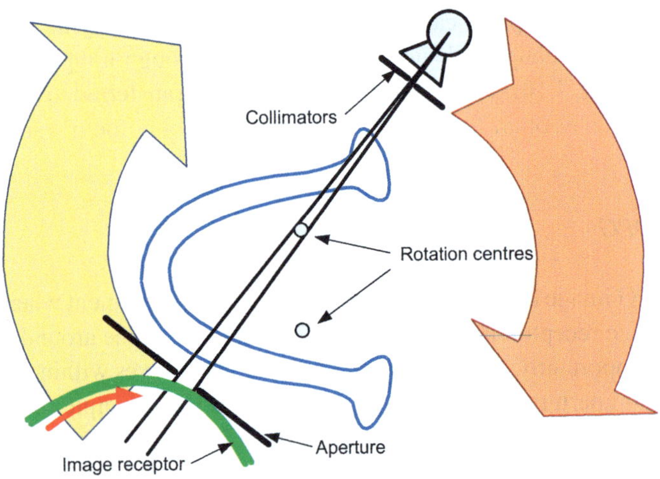

FIG. 10.2. The basic movements of the OPG unit around the mandible are illustrated. A narrow beam of radiation is formed by the tube collimation. Note the image receptor movement behind the image receptor aperture.

10.2.2.3. CBCT

CT imaging has been used for some time in dentistry, including the use of custom designed units for dental applications. However, their use has become more widespread recently with the advent of cone beam technology. There are many CBCT models available, using a variety of acquisition schemes. However, they have in common a flat panel detector for acquisition, typically using either digital radiography technology or an intensifying screen with a CCD camera (see Sections 7.4.4 and 8.2.1). Typically, a CBCT can acquire a full field of view

(FOV) that covers the whole head, although acquisitions that are restricted to the mandible with as little as 10% of full FOV are possible. The use of these lower cost CT units opens up new potential in some areas of dental diagnosis, although their significantly higher dose compared with OPG procedures should be noted.

10.2.3. Dental dosimetry

Since dental examinations are among the most numerous radiological procedures, the dosimetry of these procedures is of great interest. Relevant principles and measurement techniques of dosimetry are covered in Chapter 22 and in IAEA Technical Reports Series No. 457 [10.1]. However, it is useful to discuss the magnitude and possible effect of such doses. While there are large variations in recorded doses between X ray facilities, a recent study in Europe showed that the average incident air kerma (IAK) for an intraoral bitewing projection varied from 1 to 2 mGy, with a corresponding kerma area product (KAP) measurement of 20–40 $mGy \cdot cm^2$. The dose in centres that use slower film would be expected to be significantly higher. Data from OPG examinations, also from Europe, showed KAP values ranging from 40 to 150 $mGy \cdot cm^2$.

The estimation of a population effective dose (see Section 22.3.3), however, is somewhat more difficult, owing to the complex distribution of critical organs. For the main part, there are few radiosensitive organs around the mandible. One exception is the thyroid. However, well collimated X ray units should not directly irradiate this organ, but it will probably receive appreciable scattered radiation.

Other radiosensitive organs include the red bone marrow of the mandible and the brain. The salivary glands, which are subject to considerable irradiation, also need to be considered. They are now included as a remainder organ in the calculation of effective dose in accordance with International Commission on Radiological Protection (ICRP) 103 [10.2].

Some estimates of effective dose have been made for OPG examinations with an average value of about 7 μSv using the earlier weighting factors from ICRP 60. The use of ICRP 103 weighting factors has been variously estimated to increase the effective dose in dentistry by 50 to 400%. Since CBCT units operate with a large FOV, their effective doses are considerably higher than for OPG, with estimates of the dose for full FOV varying from 60 to 550 μSv, still considerably lower than for conventional head CT, which has effective doses of about 2 mSv.

10.3. MOBILE RADIOGRAPHY AND FLUOROSCOPY

10.3.1. Introduction

Mobile X ray equipment ranges from small dental units to CT and magnetic resonance imaging units carried in a large vehicle. This discussion restricts itself to simple radiographic and fluoroscopic equipment. Mobile equipment is needed when the patient cannot be brought to a fixed installation for a radiographic examination. Outside of the clinical centre, this may occur in a resource limited setting, or where needed for a limited time period, such as with a radiographic screening programme for a disease such as tuberculosis. Inside the clinical centre, it may occur when the patient is too ill to be conveniently moved, or if some resources are limited. Limitations of mobile equipment relate to the availability of a suitable electrical power supply, the size and weight of the equipment and the consequent effort required to move it. The design of mobile X ray equipment is varied and innovative, in order to maximize the benefit, given the above constraints.

10.3.2. Technology

As considered in Chapter 5, assuming no loss in the high voltage transformer, the X ray output power in the secondary circuit will equal that of the primary power drawn from the electricity supply. Hence, a domestic single-phase electricity supply may typically be limited to 2.4 kW, in contrast to the needed power for fixed angiographic X ray machines with a capacity to draw up to 100 kW with a high current multiphase supply. While low power is usually not a limitation for fluoroscopic application, it poses a challenge for some radiography. One solution is to charge a capacitor, which is discharged across the X ray tube, the so called 'capacitor discharge' mobile. However, as determined by the physics of the design, the tube voltage will fall rapidly during the discharge of the capacitor, leading to excessive surface kerma for large patient thicknesses. Perhaps the best scenario is to have an integral battery power supply, which is converted to a medium to high frequency alternating current signal, as described in Chapter 5, and leads to substantial reductions in the thickness of the coils needed in the transformer design. There is also the added advantage that it can be used when there is no supply of electrical power available at the examination site.

The variety of possible generator designs leads to the possibility of many types of radiographic waveform being used in the high voltage circuit for X ray generation. As noted in Chapter 5, this leads to varying tube outputs and beam qualities for the same radiographic settings of tube voltage and tube current. Care is therefore needed when determining dosimetric factors for mobile units.

10.3.3. Image quality

Control of image quality and general quality control for mobile X ray units usually follow those used for fixed units. It has been observed that the use of high fluoroscopic image quality can lead to reduced procedural time and hence reduced radiation exposure time. An important part of image quality is the set-up of viewing monitors and the ambient conditions used for operation. Every effort should be made to view monitors in conditions of low ambient light.

10.3.4. Radiation protection

Mobile X ray equipment also raises concerns about occupational and public radiation exposure, as the equipment is not operated in a purpose built shielded environment. Assuming all X ray equipment has been checked for tube leakage, the source of radiation of occupational concern during the procedure is from the input surface of the patient. It is advised that the medical physicist takes field measurements of air kerma levels due to the patient scattered radiation using a patient phantom for typical radiographic and fluoroscopic procedures.

As mobile radiography may take place in environments where other patients or members of the public may be in close proximity, it is essential that good communication exists between the medical physicist and the staff at the location of the radiographic procedure. These staff should attend appropriate radiation safety courses that include information about the radiation risk from mobile radiography. In many cases, such as for mobile chest radiography, the use of good radiographic practice with basic radiation protection allows safe usage in most hospital environments. Simple measurements should be made to demonstrate the safety (or otherwise) of mobile X ray equipment use.

10.4. DXA

The principle of operation for DXA involves two images that are made from the attenuation of a low and a high X ray energy beam, using special imaging equipment comprising special beam filtering and near perfect spatial registration of the two attenuation maps (Fig. 10.3).

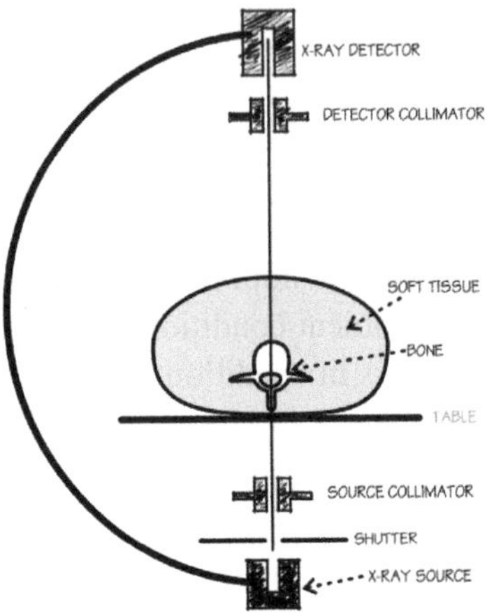

FIG. 10.3. Schematic showing the components of a DXA system. The gantry configuration above, using a pinhole source collimator and single detector, would be referred to as a pencil beam system. To create an image, the pencil beam gantry must physically scan in rectilinear fashion over the region of interest. Other systems with slit source collimators and segmented line detectors are called fan beam systems and have the advantage of higher spatial resolution and shorter scan times.

The process for determining material composition can be outlined from consideration of the total attenuation of an X ray flux passing through a subject, as represented by the following formula:

$$I = I_0 \mathrm{e}^{-\sum_{i=1}^{N} \mu_i t_i} = I_0 \mathrm{e}^{-\sum_{i=1}^{N} \left(\frac{\mu}{\rho}\right)_i \xi_i} \tag{10.1}$$

where

I_0 is the unattenuated X ray intensity before it passes through material N of thickness t_i;
μ_i is the total linear attenuation;
$(\mu/\rho)_i$ is the mass attenuation coefficient of the ith material;

and ξ_i is the ith areal density $= \rho_i t_i$.

DXA can only solve for two materials simultaneously. However, by using the following three fundamental assumptions, three materials may be quantified, which are bone, lean mass and fat mass:

(i) X ray transmission through the body for the two energy windows can be accurately described by an exponential attenuation process (Eq. (10.1)).
(ii) Pixels of the human body image can describe two components, i.e. either soft tissue and bone mineral, or, when bone is not present, lean mass and fat mass. Thus, although DXA can only solve for two compartments within individual pixels, it can describe a three component model for body composition.
(iii) The soft tissue overlying the bone in the image has a composition and X ray properties that can be predicted by the composition and X ray properties of the tissue near to but not overlying the bone.

To further elucidate how DXA measures bone and soft tissue masses, simplified DXA equations will be derived for two monochromatic X ray exposures with different energies (high energy and low energy). The full solution would require integration of the attenuation across the X ray spectrum for each energy. The attenuation equation for each exposure results in the following two equations:

$$I^{\mathrm{L}} = I_0 \mathrm{e}^{-\left[\left(\frac{\mu}{\rho}\right)_{\mathrm{s}}^{\mathrm{L}} \xi_{\mathrm{s}} + \left(\frac{\mu}{\rho}\right)_{\mathrm{b}}^{\mathrm{L}} \xi_{\mathrm{b}}\right]} \tag{10.2}$$

$$I^{\mathrm{H}} = I_0 \mathrm{e}^{-\left[\left(\frac{\mu}{\rho}\right)_{\mathrm{s}}^{\mathrm{H}} \xi_{\mathrm{s}} + \left(\frac{\mu}{\rho}\right)_{\mathrm{b}}^{\mathrm{H}} \xi_{\mathrm{b}}\right]} \tag{10.3}$$

where the H and L superscripts represent the high and low energy X ray beams respectively, and the s and b subscripts represent soft tissue and bone.

The solution of these equations for the areal density of bone (aBMD) is given by:

$$\xi_{\mathrm{b}} = \frac{R_{\mathrm{S}} \ln\left(\frac{I^{\mathrm{H}}}{I_0^{\mathrm{H}}}\right) - \ln\left(\frac{I^{\mathrm{L}}}{I_0^{\mathrm{L}}}\right)}{\left(\frac{\mu}{\rho}\right)_{\mathrm{b}}^{\mathrm{L}} - \left(\frac{\mu}{\rho}\right)_{\mathrm{b}}^{\mathrm{H}} R_{\mathrm{S}}} = \mathrm{aBMD} \tag{10.4}$$

where R_{S} is commonly referred to as the 'ratio value' for soft tissue, measured for tissue surrounding but not containing the bone:

$$R_s = \frac{\left(\frac{\mu}{\rho}\right)_S^L}{\left(\frac{\mu}{\rho}\right)_S^H} \tag{10.5}$$

This process is illustrated graphically in Fig. 10.4.

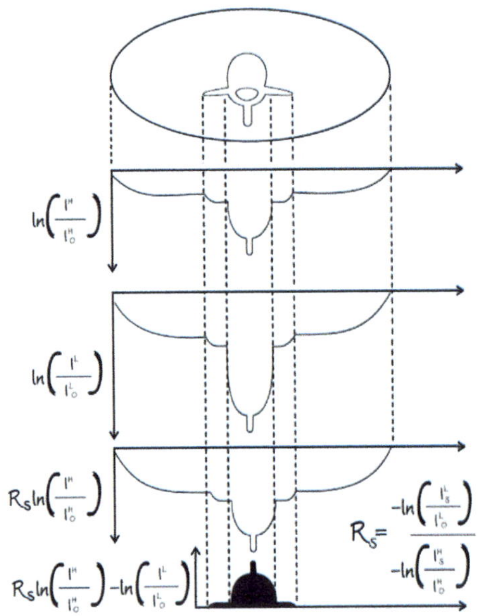

FIG. 10.4. Principle of DXA shown with four intensity profiles as per Fig. 10.3. The high energy absorption profile is multiplied by the soft tissue R value, R_s, which accounts for differences in high and low energy absorption of soft tissue. R_s is measured for pixels that do not contain bone using Eq. (10.5).

In order to use a DXA unit to determine bone mineral density, the DXA unit must be calibrated with a phantom suitable for a particular examination (e.g. spine) and for a particular DXA system type. Universal phantoms that can be used between different types of system have been developed. However, the calibration of DXA units is an important practical subject that is essential for the viability of DXA usage. The T score is the primary diagnostic value used for osteoporosis and is inversely related to fracture risk. The Z score is used to diagnose low bone mass in young adults and children. By international convention, the T score is the difference between the patient's aBMD and a young reference aBMD, in units of the population standard deviation:

$$\text{T score} = \frac{\text{aBMD}_{\text{patient}} - \text{aBMD}_{\text{young adult mean}}}{\text{SD}_{\text{young adult mean}}} \tag{10.6}$$

where SD is the standard deviation of the population of young adults. A patient's bone density can also be expressed as a Z score, the difference between the patient's aBMD and an age matched and, typically, ethnicity matched reference aBMD and standard deviations:

$$\text{Z score} = \frac{\text{aBMD}_{\text{patient}} - \text{aBMD}_{\text{age, ethnicity matched adult mean}}}{\text{SD}_{\text{age, ethnicity matched adult mean}}} \tag{10.7}$$

The reference values used to calculate T and Z scores are derived from normative databases of local populations. More information on the standards used to calculate T and Z scores can be found in the Official Positions of the International Society for Clinical Densitometry [10.3].

10.5. CONVENTIONAL TOMOGRAPHY AND TOMOSYNTHESIS

The usefulness of sectional images, which remove the image of unwanted overlying tissues, has been well understood since the early days of X ray imaging. The formation of such images is through an analogue process known as conventional tomography.

10.5.1. Principles

Conventional tomography uses the principle of image blurring to remove overlying structures from a radiological image while allowing one section of the body to remain in focus. During image acquisition, the X ray tube is in motion, causing the projected image of a given region of the body to move, but in the opposite direction to the tube. To capture this image, the image receptor moves simultaneously in the same direction as the projected image (Fig. 10.5(a)). If the body is considered as being made up of a series of sections parallel to the image receptor, then the images from the different sections will move at different speeds, and only the section whose image is travelling at the same speed as the image receptor will be in focus. This is seen in Fig. 10.5(b), where both the X ray tube and image cassette are travelling at a constant speed. The section in focus is known as the focal plane. Regions of the body above and below the focal plane are increasingly blurred as their distance from this plane increases.

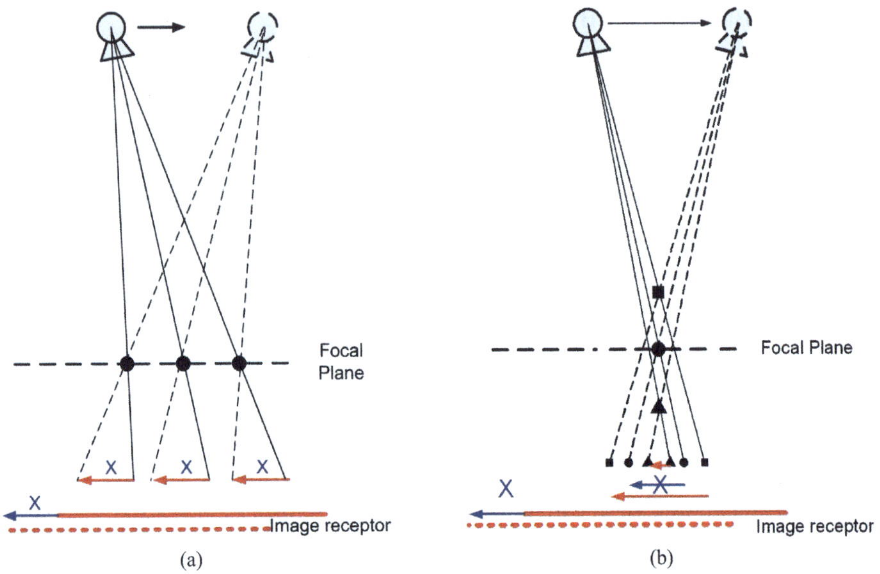

FIG. 10.5. For both (a) and (b), the tube moves at a constant speed to the right. Simultaneously, the image receptor moves in the opposite direction. The focal plane is defined as the surface parallel to the image receptor, where the image movement relative to the image receptor is zero (a). In (b), objects outside of the focal plane are seen to have relative motion between the image and the image receptor.

It can be then understood that the thickness of the focal zone is related to the speed of the image of the areas adjacent to the focal plane relative to the image receptor. For example, when there is a large X ray tube movement, the image speed relative to the image receptor increases rapidly for sections further away from the focal plane; consequently, the volume that is not blurred (the focal zone) will decrease in thickness, and conversely, if the tube motion is small, the focal zone will be large. In the same way, it can be seen that objects above the focal plane, being further away from the image receptor, will be more rapidly blurred than objects closer to the image receptor.

The conventional tomographic example, outlined above, can easily be extended for use in dental radiography by considering the case for varying the speed of the image receptor. If the image receptor speed is increased, it can be deduced that the focal area, known in dentistry as the focal trough, will move away from the image receptor, with the converse happening should the speed reduce. In this way, a curved focal trough can be created, as seen in Fig. 10.6, where it can be seen that the focal trough decreases in thickness as it moves away from the image receptor (see Section 10.2.2.2).

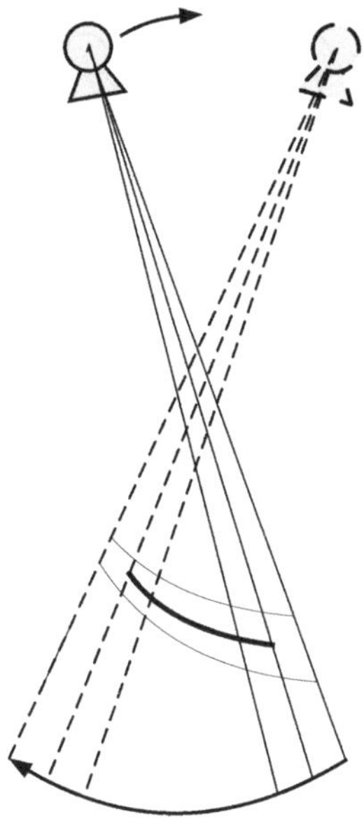

FIG. 10.6. The X ray moves at constant speed to the right while the image receptor below moves to the left. The solid line is the focal plane. In this case, the image receptor accelerates during motion, thus moving the focal plane away from the image receptor.

An interesting development of conventional tomography has been the addition of digital technology to change the speed of the image receptor 'digitally'; this is called tomosynthesis. In this case, one acquisition run might consist of ten individual X ray images each read and erased in sequence throughout the one tube movement. The images are digitally added to reconstruct different focal planes in the body. It can be seen that the focal plane can be altered by advancing or retarding each image in the series by an increasing amount. Figure 10.7 illustrates this for the simplified case of four images.

As discussed in Chapter 9, tomosynthesis can more appropriately be treated as a special case of CT in which data are acquired over a limited angular range. The computed image can then be obtained using the various CT reconstruction methods.

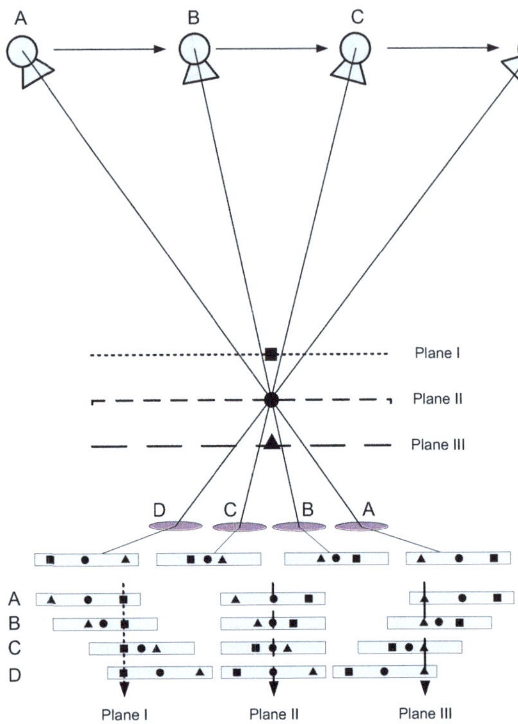

FIG. 10.7. The X ray tube moves at a constant speed to the right, while the image receptor moves at a constant speed to the left. In this figure, four samplings of the image are shown at tube positions A, B, C and D. By combining the four sampled images with appropriate offsets, it is possible to create tomographic images focused on planes I, II and III as indicated.

10.5.2. Tomographic applications

Conventional tomography has been almost completely replaced by CT in the modern radiology department. One area where it may still be in extensive use is for intravenous pyelograms, where contrast in the kidney can be conveniently placed within the focal plane to allow clear visualization of the contrast agent. However, the use of variable speed tomography is very widespread at dental facilities, through the use of pantomographic dental radiography, as described in Section 10.2.

Conventional tomography requires one tube acquisition for each focal plane image or slice. Therefore, examinations requiring many slices are inherently high dose procedures. The use of tomosynthesis, on the other hand, requires only one tube motion to capture enough data to reconstruct multiple slices within the body. It is an emerging technology, with its most notable application so far being in mammography (Chapter 9).

REFERENCES

[10.1] INTERNATIONAL ATOMIC ENERGY AGENCY, Dosimetry in Diagnostic Radiology: An International Code of Practice, Technical Reports Series No. 457, IAEA, Vienna (2007).

[10.2] INTERNATIONAL COMMISSION ON RADIOLOGICAL PROTECTION, The 2007 Recommendations of the International Commission on Radiological Protection, Publication 103, Elsevier (2008).

[10.3] LEWIECKI, E.M., et al., Official positions of the International Society for Clinical Densitometry, J. Clin. Endocrinol. Metab. **89** (2004) 3651–3655.

BIBLIOGRAPHY

CENTRE FOR EVIDENCE-BASED PURCHASING, Digital Cone Beam Tomography (DCBT) Systems, CEP 10048; NHS PASA (2010).

HELMROT, E., ALM CARLSSON, G., Measurement of radiation dose in dental radiology, Radiat. Protect. Dosim. **114** (2005) 168–171.

INTERNATIONAL ATOMIC ENERGY AGENCY, Dual Energy X Ray Absorptiometry for Bone Mineral Density and Body Composition Assessment, IAEA Human Health Series No. 15, IAEA, Vienna (2010).

LANGLAND, O.E., LANGLAIS, R.P., Principles of Dental Imaging, Williams & Wilkins, Baltimore, MD (1997).

LUDLOW, J.B., DAVIES-LUDLOW, L.E., BROOKS, S.L., HOWERTON, W.B., Dosimetry of 3 CBCT devices for oral and maxillofacial radiology: CB Mercuray, NewTom 3G and i-CAT, Dentomaxillofac. Radiol. **35** (2006) 219–226.

WILLIAMS, J.R., MONTGOMERY, A., Measurement of dose in panoramic dental radiology, Br. J. Radiol. **73** (2000) 1002–1006.

Chapter 11

COMPUTED TOMOGRAPHY

J. GELEIJNS
Leiden University Medical Centre,
Leiden, Netherlands

11.1. INTRODUCTION

After its clinical introduction in 1971, computed tomography (CT) developed from an X ray modality that was limited to axial imaging of the brain in neuroradiology into a versatile 3-D whole body imaging modality for a wide range of applications, including oncology, vascular radiology, cardiology, traumatology and interventional radiology. CT is applied for diagnosis and follow-up studies of patients, for planning of radiotherapy, and even for screening of healthy subpopulations with specific risk factors.

11.2. PRINCIPLES OF CT

11.2.1. X ray projection, attenuation and acquisition of transmission profiles

The process of CT image acquisition involves the measurement of X ray transmission profiles through a patient for a large number of views. A profile from each view is achieved primarily by using a detector arc generally consisting of 800–900 detector elements (dels), referred to as a detector row. By rotation of the X ray tube and detector row around the patient, a large number of views can be obtained. The use of tens or even hundreds of detector rows aligned along the axis of rotation allows even more rapid acquisition (Fig. 11.1). The acquired transmission profiles are used to reconstruct the CT image, composed of a matrix of picture elements (pixels) (see Section 11.3).

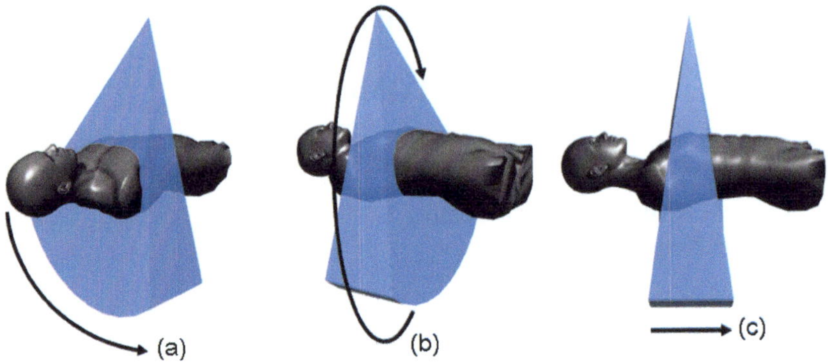

FIG. 11.1. CT image acquisition showing the transmission of X rays through the patient by using a detector row (a), with rotation of the X ray tube and detector (b) and by multiple detector (c).

The values that are assigned to the pixels in a CT image are associated with the attenuation of the corresponding tissue, or, more specifically, to their linear attenuation coefficient μ (m^{-1}) (see Section 2.3.1). The linear attenuation coefficient depends on the composition of the material, the density of the material and the photon energy, as seen in Beer's law:

$$I(x) = I_0 e^{-\mu x} \tag{11.1}$$

where $I(x)$ is the intensity of the attenuated X ray beam, I_0 the unattenuated X ray beam and x the thickness of the material. Note that Beer's law only describes the attenuation of the primary beam and does not take into account the intensity of scattered radiation that is generated. For use in polyenergetic X ray beams, Beer's law should strictly be integrated over all photon energies in the X ray spectrum. However, in the back projection methodologies (see below) developed for CT reconstruction algorithms, this is generally not implemented; instead, typically, a pragmatic solution is to assume where Beer's law can be applied using one value representing the average photon energy of the X ray spectrum. This assumption causes inaccuracies in the reconstruction and leads to the beam hardening artefact.

As an X ray beam is transmitted through the patient, different tissues are encountered with different linear attenuation coefficients. If the pathway through the patient ranges from 0 to d, then the intensity of the attenuated X ray beam, transmitted a distance d, can be expressed as:

$$I(d) = I_0 e^{-\int_0^d \mu(x)\mathrm{d}x} \tag{11.2}$$

Since a CT image is composed of a matrix of pixels, the scanned patient can also be regarded as being made up of a matrix of different linear attenuation coefficient volume elements (voxels). Figure 11.2 shows a simplified 4 × 4 matrix representing the measurement of transmission along one line. For such a discretization, the equation for the attenuation can be expressed as:

$$I(d) = I_0 e^{-\sum_{i=1}^{i=4} \mu_i \Delta x} \tag{11.3}$$

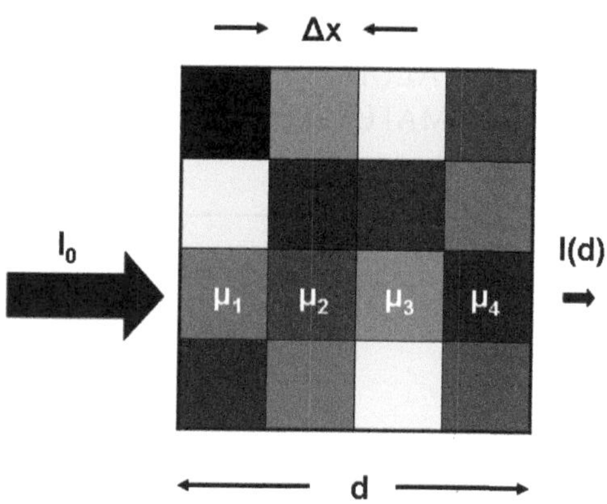

FIG. 11.2. The principle of attenuation of an X ray beam in a simplified 4 × 4 matrix. Each element in the matrix can, in principle, have a different value of the associated linear attenuation coefficient.

From the above, it can be seen that the basic data needed for CT are the intensities of the attenuated and unattenuated X ray beams, respectively $I(d)$ and I_0, and that these can be measured. Image reconstruction techniques can then be applied to derive the matrix of linear attenuation coefficients, which is the basis of the CT image.

11.2.2. Hounsfield units

In the CT image, the matrix of reconstructed linear attenuation coefficients (μ_{material}) is transformed into a corresponding matrix of Hounsfield units ($\text{HU}_{\text{material}}$), where the HU scale is expressed relative to the linear attenuation coefficient of water at room temperature (μ_{water}):

$$\mathrm{HU}_{\text{material}} = \frac{\mu_{\text{material}} - \mu_{\text{water}}}{\mu_{\text{water}}} \times 1000 \tag{11.4}$$

It can be seen that $\mathrm{HU}_{\text{water}} = 0$ ($\mu_{\text{material}} = \mu_{\text{water}}$), $\mathrm{HU}_{\text{air}} = -1000$ ($\mu_{\text{material}} = 0$) and HU = 1 is associated with 0.1% of the linear attenuation coefficient of water. Table 11.1 shows typical values for body tissues. From the definition of the HU, it follows that for all substances except water and air, variations of the HU values occur when they are determined at different tube voltages. The reason is that, as a function of photon energy, different substances exhibit a non-linear relationship of their linear attenuation coefficient relative to that of water. This effect is most notable for substances that have a relatively high effective atomic number, such as contrast enhanced blood and bone.

TABLE 11.1. TYPICAL HU VALUES AND RANGES OF VALUES FOR DIFFERENT TISSUES AND MATERIALS[a]

Substance	HU
Compact bone	+1000 (+300 to +2500)
Liver	+60 (+50 to +70)
Blood	+55 (+50 to +60)
Kidneys	+30 (+20 to +40)
Muscle	+25 (+10 to +40)
Brain, grey matter	+35 (+30 to +40)
Brain, white matter	+25 (+20 to +30)
Water	0
Fat	–90 (–100 to –80)
Lung	–750 (–950 to –600)
Air	–1000

[a] The actual value of the HU depends on the composition of the tissue or material, the tube voltage and the temperature.

The minimum bit depth that should be assigned to a pixel is 12, enabling the creation of a Hounsfield scale that ranges from –1024 HU to +3071 HU, thus covering most clinically relevant tissues. A wider Hounsfield scale with a bit depth of 14 is useful for extending the HU scale upwards to +15 359 HU, thus making it compatible with materials that have a high density and a high linear attenuation coefficient.

CT images are usually visualized on a monitor using an eight bit greyscale offering only 256 grey values. Each pixel HU value then has to undergo a linear mapping to a 'window' 8 bit value. The window width defines the range of HUs that is represented by the mapped values (ranging from white to black) and the window level defines the central HU value within the selected window width. Optimal visualization of the tissues of interest in the image can only be achieved by selecting the most appropriate window width and window level. Consequently, different settings of the window width and window level are used to visualize soft tissue, lung tissue or bone. The greyscale, as defined by window level and window width, is adapted to the diagnostic task and is thus dependent on the clinical question.

In clinical practice, considerable deviations between the expected and the observed HU values may occur. Causes for such inaccuracies may, for example, be the dependence of the HU value on the reconstruction filter, the size of the scanned field of view (FOV), and the position within the scanned FOV. In addition, image artefacts may have an effect on the accuracy of the HU values. When performing longitudinal clinical studies, one should take into account that, even for the same scanner, the HU values for a given tissue type may vary with time. In multicentre studies that involve different CT scanners, there may also be significant variations in the observed HU values. Therefore, quantitative imaging in CT requires special attention and often additional calibrations of the CT scanner.

11.3. THE CT IMAGING SYSTEM

11.3.1. Historical and current acquisition configurations

After preclinical research and development during the early 1970s, CT developed rapidly as an indispensable imaging modality in diagnostic radiology (Table 11.2). It is impressive to realize that most of the modern CT technology that is being used in clinical practice nowadays had already been described by the end of 1983 (Fig. 11.3). The development of multidetector row CT (MDCT) and multisource CT had already been described in a US patent from 1980 [11.1]. The patent describes what the authors call "a multiple purpose high speed tomographic X ray scanner". In the acquisition technique of helical CT (described in Ref. [11.2]), the patent states that "the apparatus enables helical scanning to be effected by the continuous transportation of the table couch". The helix is the pathway of the continuously rotating X ray source seen from the perspective of the patient.

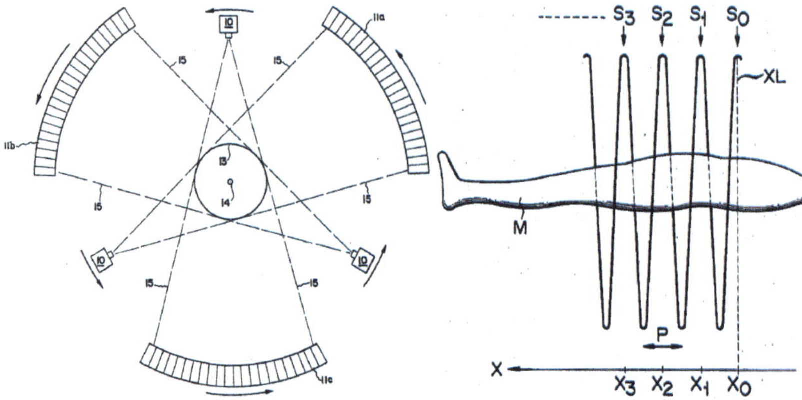

FIG. 11.3. Concepts of multisource and MDCT scanning (left) and of helical CT (right).

Volumetric CT with a scanner that was capable of imaging an entire volume within a fraction of a second was achieved with the installation of the Dynamic Spatial Reconstructor in 1980 at the Mayo Clinic in the USA. This scanner used 14 X ray tubes and 14 image intensifiers and was capable of impressive performance with regard to coverage and temporal resolution, even measured against current standards.

Currently, most scanners are helical MDCT scanners, but the technologies of dual source and volumetric CT scanning have been implemented on a wide scale.

11.3.2. Gantry and table

The gantry contains all the system components that are required to record transmission profiles of the patient. Since transmission profiles have to be recorded at different angles, these components are mounted on a support within the gantry that can be rotated. The X ray tube with high voltage generator and tube cooling system, the collimator, the beam shaping filters, the detector arc and the data acquisition system are all mounted on this support. The engineering of these components is complex, since they need to be able to withstand the strong centrifugal force that occurs during the fast rotation of the gantry. Forces of several tens of g arise for rotation times of the order of 0.25 s.

Electrical power is generally supplied to the rotating gantry by means of slip ring contacts. Recorded projection profiles are generally transmitted from the gantry to a computer by means of wireless communication technologies.

The design and engineering of the table, as with the gantry, are critical to allowing accurate acquisition of data at high rotational speeds. The table must also be able to withstand heavy weights without bending. The position of the patient on the table can be head first or feet first, and supine or prone; this position is usually recorded with the scan data.

TABLE 11.2. OVERVIEW OF DIFFERENT TYPES OF CT TECHNOLOGY

CT technology	Detector configuration	Axial FOV coverage	Acquisition of axial projection angles	Coverage of longitudinal range
First clinical CT scanners, 1974	One single del	Pencil beam, with translation of the X ray tube and del in small discrete steps	Rotation of X ray tube and detector in small discrete angular steps	Translation of the table in small discrete steps
Axial (step and shoot) CT scanners	One single detector row with hundreds of dels	Fan beam, with full coverage of the FOV	One 360° rotation of the X ray tube and detector	As above
Helical CT scanners	As above	As above	Multiple continuous rotations of the X ray tube and detector	Continuous translation of the table
Helical, MDCT scanners, 1998	Multiple detector rows, e.g. 4–64 active channels	As above	As above	As above
Dual source, helical, MDCT scanners	Two detectors with multiple detector rows, e.g. 32–64 active channels	Two fan beams, with at least one fan beam with full coverage of the FOV	Multiple continuous rotations of two X ray tubes and two detectors	As above
Volumetric CT scanners, 2007	Multiple detector rows, currently with up to 320 active channels	Cone beam, with full volumetric FOV coverage	One single continuous rotation of the X ray tube and detector	Coverage (e.g. 160 mm) of the longitudinal range is provided by the cone beam; longitudinal coverage exceeding 160 mm is achieved by step and shoot acquisitions and stitching of the reconstructed volumes

11.3.3. The X ray tube and generator

Owing to the high X ray flux required for CT, the X ray tube uses a tungsten anode designed to withstand and dissipate high heat loads. With long continuous acquisition cycles, a forced cooling system using oil or water circulated through a heat exchanger is often used.

11.3.4. Collimation and filtration

The X ray beam should be collimated to the desired dimensions. The beam width in the longitudinal axis is generally small; therefore, the collimated X ray beam is often referred to as a fan beam. In the plane perpendicular to the table motion, also known as the *x–y* or axial plane, the beam is shaped to reduce the dynamic range of the signal that is recorded by the detectors. Beam shaping (bowtie) filters are used to achieve the desired gradient, with one of a number of mounted bowtie filters moved into the X ray beam during acquisition.

11.3.5. Detectors

The essential physical characteristics of CT detectors are a good detection efficiency and a fast response with little afterglow. Currently, solid state detectors[1] are used, as they have a detection efficiency close to 100% compared with high pressure, xenon filled ionization chambers that were used previously and that had a detection efficiency of about 70%. Solid state detectors are generally scintilators, meaning that the X rays interacting with the detector generate light. This light is converted to an electrical signal, by photodiodes that are attached to the back of the scintillator, which should have good transparency to ensure optimal detection. Typically, an antiscatter grid is mounted at the front of the detector, which consists of small strips of highly attenuating material (e.g. tungsten) aligned along the longitudinal (*z*) axis of the CT scanner, forming a 1-D antiscatter grid.

A detector row consists of thousands of dels that are separated by septa designed to prevent light generated in one del from being detected by neighbouring dels. These septa and the strips of the antiscatter grid should be as small as possible since they reduce the effective area of the detector and thus reduce the detection of X rays. Figure 11.4 shows detector modules for a 4, 16,

---

[1] This is a common term used for imaging detection to cover the use of scintillators and photodiodes and should not be confused with thermoluminescent dosimeters (see Chapter 21 for further discussion).

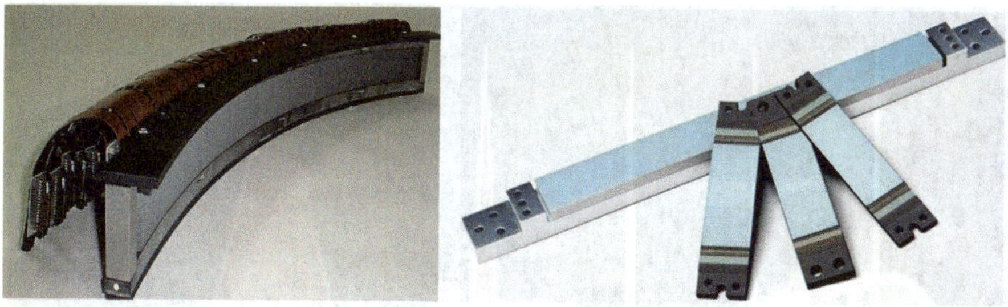

FIG. 11.4. Detector modules for a 4, 16, 64 and 320 slice CT scanner (left). The complete CT detector is composed of many detector modules (right) (courtesy Toshiba Medical Systems).

64 and 320 slice CT scanner. The complete CT detector is composed of many detector modules that are mounted next to each other along an arc.

CT detectors are curved in the axial (x–y) plane and rectangular along the longitudinal (z) axis. While most dels are used to measure transmission profile data (the attenuated intensity $I(d)$), the dels outside the FOV are used to measure the unattenuated intensity of the X ray beam ($I(0)$). Thus, the coefficient $I(d)/I(0)$ from Eq. (11.2) can be easily recorded.

The smallest size of an object (d) within the patient that can be resolved in the reconstructed image depends on the number and size of dels along the detector arc, the size of the dels along the z axis, the number of angles for which projections are recorded during the acquisition, and the focal spot size of the X ray tube. The minimum number of dels in a detector arc covering a specific FOV should be approximately 2FOV/d, to resolve the object, d, in the reconstructed image. About 800 dels are required to achieve a spatial resolution of 1 mm within a reconstructed image for a FOV of 400 mm. Spatial resolution can be improved for an acquisition with a full 360° rotation by a slight geometrical modification of the arrangement of the dels. By shifting the dels by a distance equal to a quarter of their size, the theoretically achievable spatial resolution becomes twice as good. Thus, a quarter detector shift is generally implemented in CT scanners. As a rule of thumb, the number of required projection angles can be approximated by the number of required dels. With the current detector rows of 800–1000 dels, covering an FOV of 400 mm, a spatial resolution of better than one millimetre can be achieved.

Figure 11.5 shows how coverage of MDCT scanners increased when more active detector rows became available. A typical acquisition with a single detector row scanner covered 5 mm. CT scanners with four active detector rows achieved a substantial improvement of the longitudinal resolution. For example, by using four active detector rows in a 4 × 1 mm acquisition configuration, the longitudinal spatial resolution improved from 5 mm to 1.25 mm.

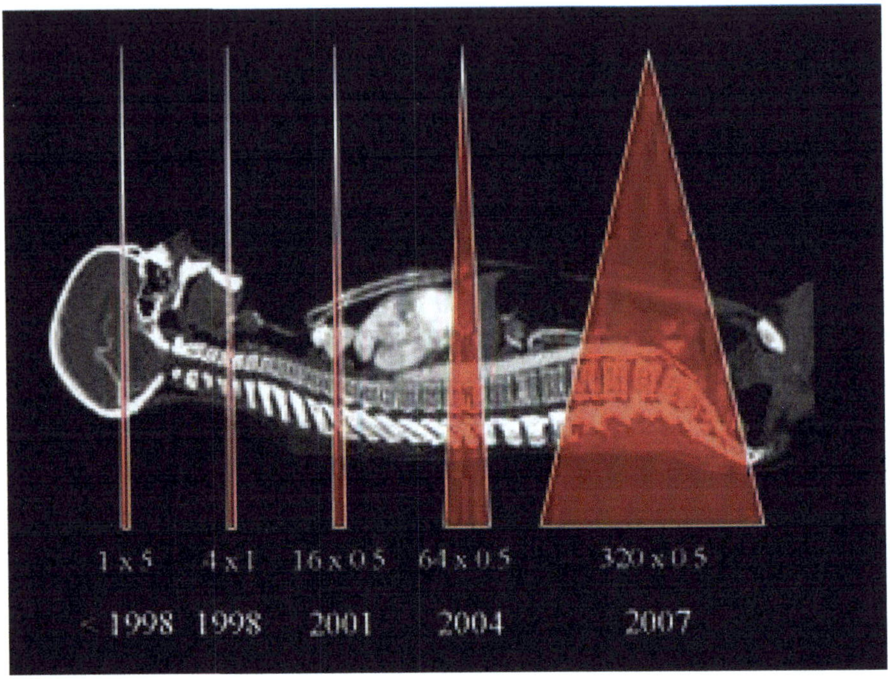

FIG. 11.5. Coverage of the MDCT scanners increased when more active detector rows became available.

In clinical practice, CT scanners with four active detector rows were primarily used to enhance longitudinal resolution, which allowed 3-D visualization of the scanned volume. The CT scanners with four active detector rows could also be used for enhanced longitudinal coverage, for example, by selecting a $4 \times 2 = 8$ mm, or even a $4 \times 4 = 16$ mm coverage. Enhanced longitudinal coverage would allow for shorter scan times but without the benefit of improved longitudinal resolution. The CT scanners with 16 or 64 active detector rows allowed for acquisitions in, for example, $16 \times 0.5 = 8$ mm and $64 \times 0.5 = 32$ mm configurations. These scanners provided excellent longitudinal spatial resolution, high quality 3-D reconstructions and, at the same time, reduced scan times. The MDCT scanners with up to 64 active detector rows do not provide coverage of entire organs, and to cover the prescribed range, the scan is generally a helical acquisition with multiple rotations. With the 320 detector row CT scanner, one single rotation allows for coverage of 160 mm, enough for covering organs such as the brain or the heart within one single rotation.

11.4. IMAGE RECONSTRUCTION AND PROCESSING

11.4.1. General concepts

In order to reconstruct a CT image, numerous measurements of the transmission of X rays through the patient are acquired. This information is the basis for reconstruction of the CT image. Prior to image reconstruction, a logarithm of the measured data is calculated. The logarithm of the (inverse) measured normalized transmission, $\ln(I_0/I(d))$, yields a linear relationship with the products of $\mu_i \Delta x$ (Eqs (11.2, 11.3)).

Intuitively, one might consider that a simple back projection of measured transmission profiles could be used for image reconstruction. This process is visualized in Fig. 11.6, which shows (a) the X ray projection at a certain angle producing a transmission profile (b). The back projection of this profile distributes the measured signal evenly over the area at the same angle as the projection (c). On addition of the back projections of the transmission profiles from all projection angles, it becomes clear that the simple back projection process yields a strongly blurred image (d). A more accurate reconstruction can be obtained by filtering the profiles prior to back projection. This is the method of filtered back projection, which is discussed in the following sections, and is the standard technique used for image reconstruction in CT.

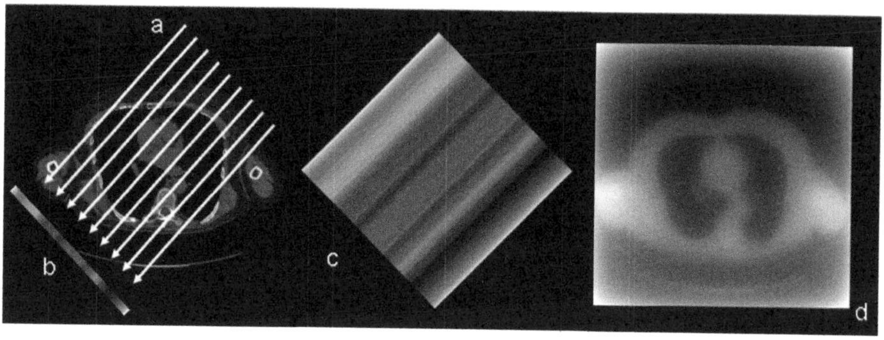

FIG. 11.6. A simple back projection yields a strongly blurred image. The contours of the chest and lungs can still be recognized in the image.

11.4.2. Object space, image space and Radon space

In order to understand the technique of filtered back projection better, it is essential to introduce three interrelated domains: (i) the object space (linear attenuation values), (ii) the Radon space (projection values, this domain is also

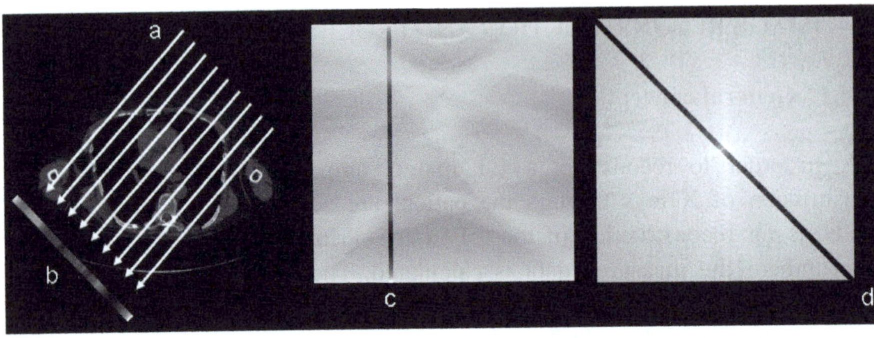

FIG. 11.7. Projection (b) recorded by the CT scanner for (a); one specific projection angle corresponds to one line in Radon space (c) and a 1-D FT of the recorded line in the sinogram yields one line in Fourier space (d) at the same angle.

referred to as sinogram space, in which case Cartesian coordinates are used) and (iii) the Fourier space, which can be derived from the object space by a 2-D (FT).

Figure 11.7 illustrates the interrelations between the three domains for one projection angle with the transmission projection (b) at one specific projection angle; this projection corresponds with one line in Radon space (c). A 1-D FT of the recorded line in the sinogram yields an angulated line in Fourier space (d) (see Section 11.4.3).

The interrelationships between the three domains, object space, Radon space and Fourier space, are illustrated in Fig. 11.8. A 2-D Radon transform converts the object space into Radon space. The 2-D Radon space is actually created during the CT scan: projections are recorded and stored as raw data in 2-D Radon space.

As will be shown in the next section, the combination of 1-D FTs of transmission profiles at many angles allows creation of the Fourier space of the object space. One could intuitively expect that an inverse 2-D FT of Fourier space would be used in CT to reconstruct the object space. However, this does not yield the best result, since the rebinning of the Fourier transformed angulated projections, and the associated interpolations that are required to achieve a Fourier space in Cartesian coordinates, are prone to induce artefacts in the reconstructed images (this will be explained in more detail in the next section). A better technique for CT reconstruction is to use a filtered back projection.

11.4.3. Filtered back projection and other reconstructions

The mathematical operations that are required for a filtered back projection consist of four steps, which are elaborated in the following paragraphs. First, an FT of Radon space should be performed (requiring many 1-D FTs). Then, a high

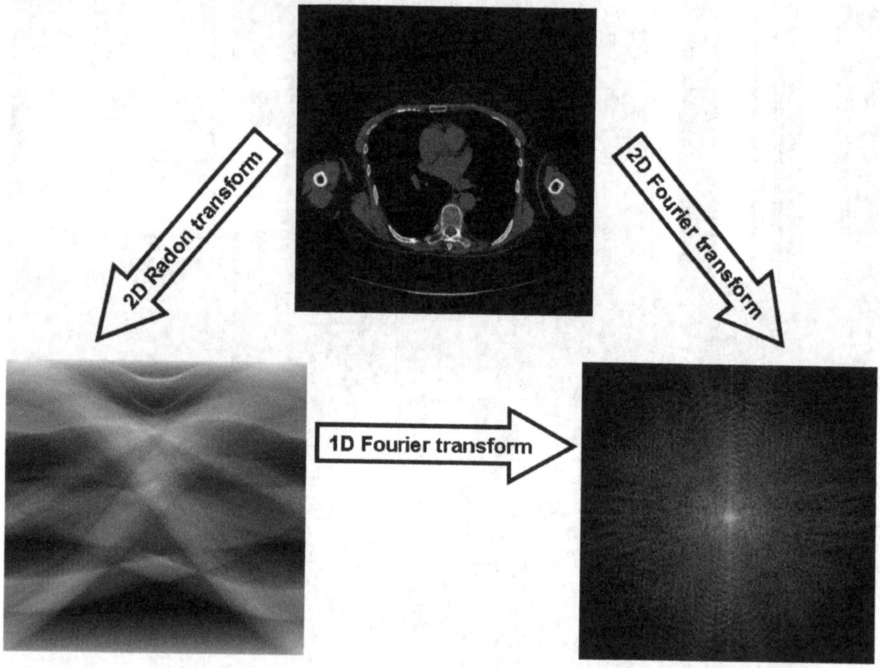

FIG. 11.8. The interrelationships between the three domains, object space, Radon space and Fourier space. Note that multiple 1-D FTs of lines in the Radon space allow the creation of the 2-D Fourier space (the number of 1-D transforms is equal to the number of profiles registered).

pass filter should be applied to each one of the 1-D FTs. Next, an inverse FT should be applied to the high pass filtered FTs, in order to obtain a Radon space with modified projection profiles. Finally, back projection of the filtered profiles yields the reconstruction of the measured object. Figure 11.9 illustrates this by showing how successive filtered back projections at different angles can be used to achieve a good reconstruction of the space domain. It may be noted at this stage that (in accordance with the convolution theorem for FTs) the filter that is applied to the Fourier domain can be substituted by a direct convolution of profiles in the Radon domain with an appropriate kernel.

Image space is generally represented on a regular grid. Let the 2-D image space be defined as $f(x, y)$, where (x, y) are rectangular Cartesian coordinates. A single 1-D projection of the 2-D image space with equidistant and parallel rays yields one line in Radon space, expressed as the projection $p(t, \theta)$, where t is the distance from the projected X ray to the isocentre and θ is the projection angle (Fig. 11.10). The central slice theorem, also referred to as the Fourier slice theorem, states that the FT of such a parallel projection of image space at the projection angle θ yields one line in 2-D Fourier space, $F(u, v)$, angulated at the same angle θ (the 2-D Fourier space is sometimes also referred to as k space).

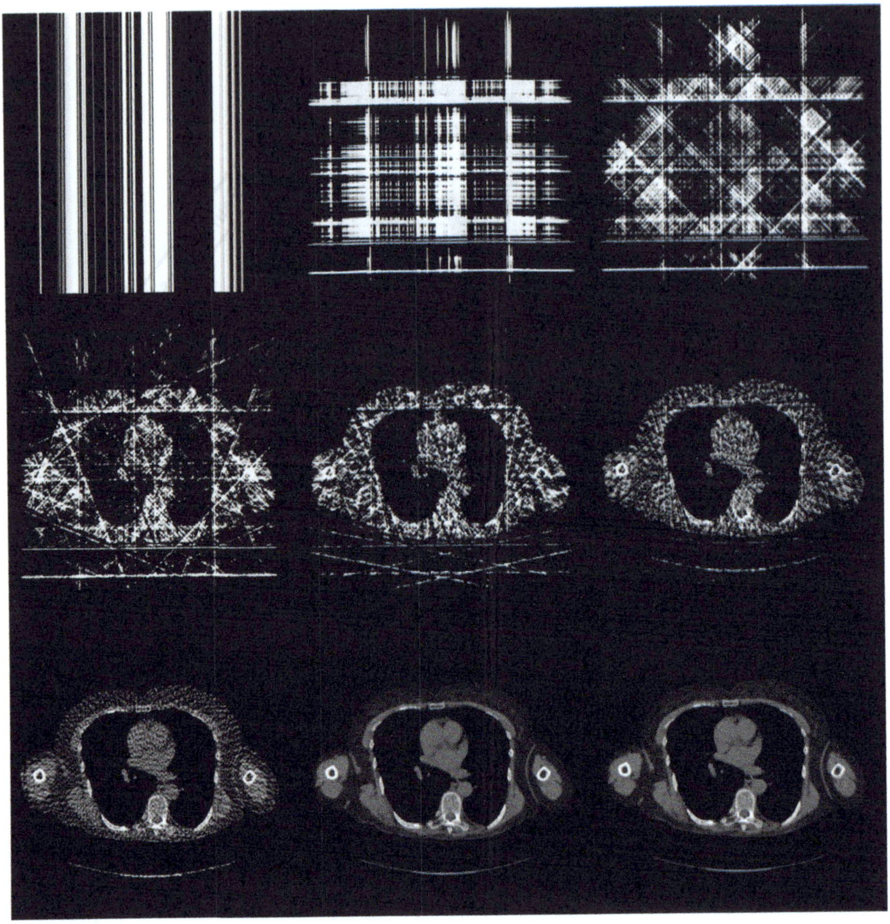

FIG. 11.9. Successive filtered back projections can be used to achieve a good reconstruction of the space domain. The images are associated with, respectively, 1, 2, 4, 8, 16, 32, 64, 256 and 1024 filtered back projections at different angles.

This can be demonstrated as follows. At the projection angle $\theta = 0$, the projection $p(x, 0)$ and the corresponding line in Radon space is described as:

$$p(x,0) = \int_{-\infty}^{+\infty} f(x,y)\,\mathrm{d}y \tag{11.5}$$

The 1-D FT with respect to x, of the projection $p(x, 0)$ at the projection angle $\theta = 0$ is given by:

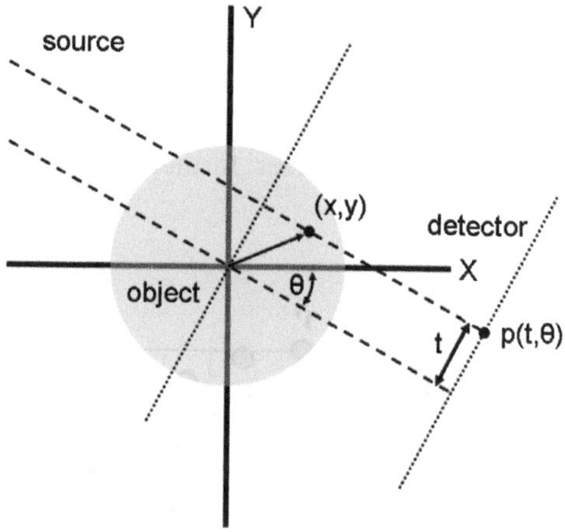

FIG. 11.10. Some geometrical aspects of the generation of transmission profiles. The Cartesian coordinates (x, y) apply to the image space, f. The coordinates that apply to the projection, p, are t, being the distance from the projected X ray to the isocentre, and θ, being the projection angle.

$$P(u) = \int_{-\infty}^{+\infty} p(x,0)\mathrm{e}^{-\mathrm{i}2\pi ux}\,\mathrm{d}x = \int_{-\infty}^{+\infty}\int_{-\infty}^{+\infty} f(x,y)\mathrm{e}^{-\mathrm{i}2\pi ux}\,\mathrm{d}x\,\mathrm{d}y \tag{11.6}$$

and the 2-D FT $F(u, v)$ of the 2-D image space $f(x, y)$ at $v = 0$ is:

$$F(u,v)\big|_{v=0} = \int_{-\infty}^{+\infty}\int_{-\infty}^{+\infty} f(x,y)\mathrm{e}^{-\mathrm{i}2\pi(ux+vy)}\,\mathrm{d}x\,\mathrm{d}y\Big|_{v=0} = \int_{-\infty}^{+\infty}\int_{-\infty}^{+\infty} f(x,y)\mathrm{e}^{-\mathrm{i}2\pi ux}\,\mathrm{d}x\,\mathrm{d}y \tag{11.7}$$

It thus becomes clear that the 1-D FT with respect to x for the projection angle $\theta = 0$ is equal to the 2-D FT $F(u, v)$ of the 2-D image space $f(x, y)$ at $v = 0$:

$$P(u) = F(u,v)\big|_{v=0} \tag{11.8}$$

This conclusion can be generalized for any projection angle θ and it thus provides the proof for the central slice theorem. A reconstruction can thus, at least theoretically, be achieved first by a construction of the 2-D Fourier space $F(u,v)$ by many 1-D FTs of the projection profiles measured under many projection angles, and subsequently by a 2-D inverse FT of the 2-D Fourier space to the 2-D image space. The sampling of the 2-D Fourier space from the 1-D FTs of the projections yields a 2-D Fourier space in regular polar coordinates. Prior to the 2-D inverse FT into image space, the regular distributed points in the polar 2-D Fourier space have to be transformed to regularly distributed points in a

Cartesian 2-D Fourier space. The transformation from a polar coordinate system to a Cartesian coordinate system may lead to artefacts in the reconstructed image, owing to the fact that the sampling of the 2-D Fourier space is denser around the origin (low frequencies), and sparser further away from the origin (high frequencies) (Fig. 11.11).

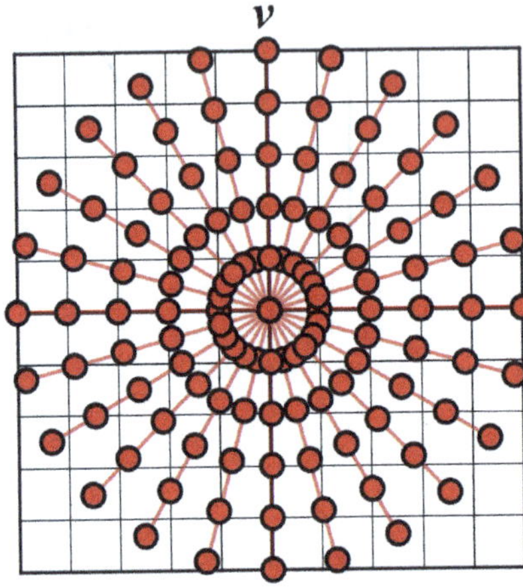

FIG. 11.11. The CT scan yields a regular distributed sampling in polar coordinates of 2-D Fourier space. Transformation into a regular distributed sampling in Cartesian coordinates is complicated, particularly at higher frequencies (further from the origin).

A more accurate and practical reconstruction can be achieved with the formulation known as the filtered back projection. The filtered back projection also starts with 1-D FTs of image space, thus creating the corresponding Fourier space, but the sampling of the 2-D Fourier space $F(u, v)$ is expressed on a polar grid using the coordinate transform:

$$u = \omega\cos\theta, \quad v = \omega\sin\theta \tag{11.9}$$

The image reconstruction — the filtered back projection — is then expressed as:

$$f(x,y) = \int_0^{\pi} d\theta \int_{-\infty}^{+\infty} P(\omega,\theta)|\omega| e^{i2\pi\omega t}\, d\omega \tag{11.10}$$

where $P(\omega, \theta)$ is the 1-D FT of the 1-D projection at angle θ, and $|\omega|$ is known as a ramp filter in the frequency domain.

In practice, different filters can be used in the reconstruction, depending upon the image properties required. The filter (or convolution kernel) in a filtered back projection that theoretically yields an optimal reconstruction is the so-called Ramachandran–Lakshminarayanan filter, also called the Ram–Lak or ramp filter. It provides optimal spatial resolution in the reconstructed images. However, it also yields relatively high noise levels in the reconstructed images. Such a theoretically 'optimal' filter in clinical practice is referred to as a sharp or bone filter. Often, filters are used that reduce the noise level in the reconstructed images; these filters provide some roll-off at higher frequencies.

A modest roll-off is achieved with the Shepp–Logan filter, which provides images that are less noisy and that provide better low contrast resolution and slightly inferior spatial resolution in the reconstructed images; such filters are referred to as normal filters. Even stronger roll-off at higher frequencies leads to further noise reduction, better low contrast resolution, but noticeably poorer spatial resolution. Such filters in clinical applications are referred to as soft tissue filters. CT scanners offer many reconstruction filters that are optimized for specific clinical purposes. It is possible to reconstruct one single CT scan with different reconstruction filters, in order to optimize the visualization of, for example, both bone and soft tissue.

Other reconstruction techniques such as algebraic or iterative reconstruction can also be used in CT. An algebraic reconstruction may seem attractive; however, algebraic reconstruction through equation solving is not feasible in clinical practice, owing to the large (512 × 512) matrices that are used in medical imaging and to inconsistencies in the equations from measurement errors and noise.

Iterative (statistical) reconstructions are now commonly used in CT. The iterative reconstruction is well known in medical imaging, since it is routinely used in nuclear medicine. Iterative techniques provide potential benefits in CT, including the removal of streak artefacts (particularly when fewer projection angles are used), and better performance in low dose CT acquisitions. However, iteratively reconstructed images may be affected by artefacts that are not present in filtered back projection images, such as aliasing patterns and overshoots in the areas of sharp intensity transitions. Iterative reconstruction algorithms are becoming popular in commercial CT scanners and can produce low noise images.

11.5. ACQUISITION

11.5.1. Scan projection radiograph

The CT image acquisition scan sequence is generally preceded by a 2-D scan projection radiograph (SPR) — also termed by manufacturers as scoutview, topogram or scanogram. The SPR is acquired with a static (non-rotating) X ray tube, a narrowly collimated fan beam and a moving table. The X ray tube is fixed, generally, in a position that yields either a frontal or lateral SPR of the patient. Either one or two SPRs are acquired prior to the CT scan. The start position for the SPR is determined by the radiographer during the positioning of the patient on the table prior to the CT scan. This can be achieved with the aid of laser positioning lights that are mounted internally and externally to the gantry. The extent of the SPR is generally predefined for specific CT acquisition protocols and can be adapted for individual patients. The SPR is performed at an intermediate tube voltage (120 kV) and at a low tube current (50–100 mA). The associated radiation exposure for the patient is small compared with the radiation exposure from the CT scan. The image quality, particularly the spatial resolution, of SPRs is modest compared with that for clinical radiographs.

The SPR is used to plan the start and end positions of the CT acquisition sequence (Fig. 11.12). Automatic exposure control (AEC) systems for CT derive information on the X ray transmission through the patient from the SPR and, based on this information, the optimal tube current as a function of longitudinal position of the X ray tube relative to the patient is calculated. This is called *z* axis tube current modulation.

Figure 11.13 shows adaptation of the tube charge (mAs) by AEC at four levels during a helical CT scan. The tube charge is increased in areas with high X ray attenuation and decreased in areas with low attenuation. AEC in CT can also compensate for differences in attenuation at different projection angles. This is referred to as *x–y* axis tube current modulation.

11.5.2. Axial CT scan

An axial CT scan involves an acquisition of transmission profiles with a rotating X ray tube and a static table. An axial acquisition is generally performed

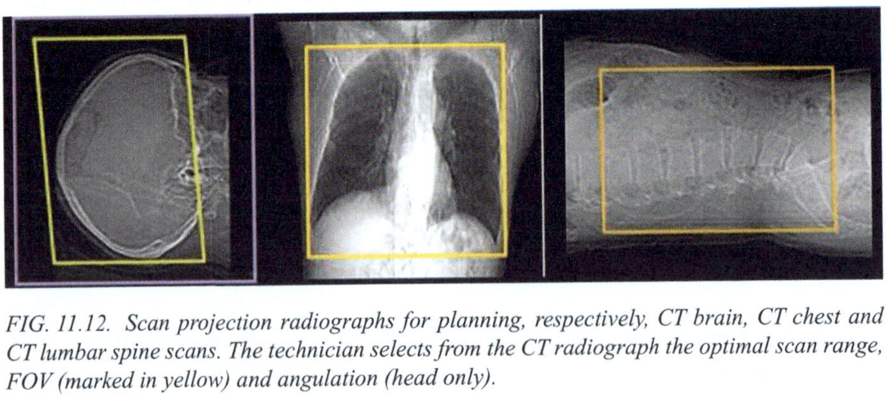

FIG. 11.12. Scan projection radiographs for planning, respectively, CT brain, CT chest and CT lumbar spine scans. The technician selects from the CT radiograph the optimal scan range, FOV (marked in yellow) and angulation (head only).

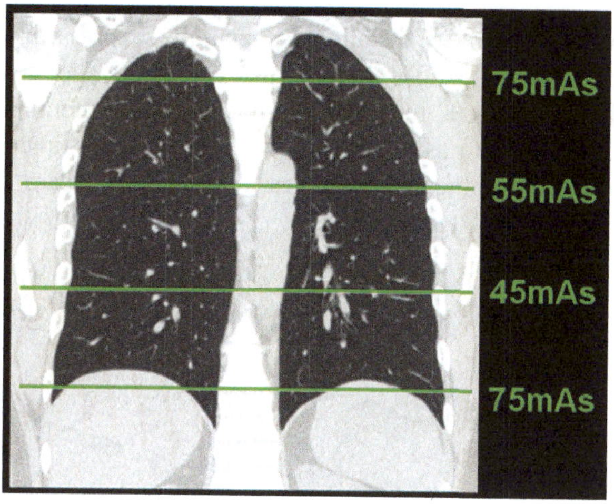

FIG. 11.13. The SPR can be used to achieve AEC during the CT scan. The mAs values are indicated at four levels, but during the helical acquisition, the tube charge is continuously optimized at each level within the scanned range.

with one full 360° rotation of the X ray tube, but to enhance temporal resolution this may be reduced to a shorter '180° + fan angle' acquisition. The rotation angle can be extended to, for example, a 720° acquisition to enhance low contrast resolution by allowing a higher tube charge (mAs). A complete CT scan generally involves subsequent axial acquisitions in order to cover a clinically relevant volume. This is achieved by translation of the table ('step') after each axial acquisition ('shoot'). This is referred to as a step and shoot acquisition. Usually, the table translation is equal to the slice thickness, so that subsequent axial acquisitions can be reconstructed as contiguous axial images. Figure 11.14 (left) shows the geometry of an axial CT acquisition.

11.5.3. Helical CT scan

The helical CT scan was introduced in 1989, whereby the acquisition with a rotating X ray tube was combined with a moving table. The introduction of helical CT scans improved the performance of CT considerably. Advantages of helical CT scans include shorter scan time and more consistent 3-D image information for the scanned volume. Disadvantages of helical CT scans include the introduction of artefacts such as the windmill artefact. Figure 11.14 shows the geometry of a helical CT acquisition (right). The circular trajectory of the X ray tube transforms into a helical course from the perspective of the patient.

Helical scanning allows for the acquisition of a large volume of interest within one breath hold and was a prerequisite for the development of high quality

CT angiography. The table translation is generally expressed relative to the (nominal) beam width (in single slice CT this equals the slice width): the ratio of table translation per 360° tube rotation relative to the nominal beam width in helical CT is referred to as the pitch factor. The rotation time of single slice CT scanners is 1–2 s and the slice thickness (and nominal beam width) in most clinical applications is 5–10 mm.

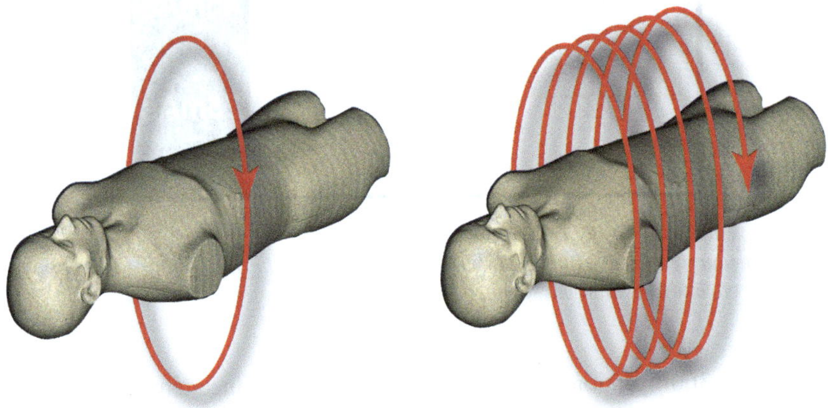

FIG. 11.14. Geometry of an axial CT acquisition (left) and a helical CT acquisition (right).

11.5.4. MDCT scan

Ten years after the introduction of helical CT, the next step in CT technology that provided even more new clinical applications was taken: the introduction of fast rotating MDCT scanners (see Section 11.3) with up to 64 adjacent active arrays of detectors, enabling the simultaneous measurement of a correspondingly large number of transmission profiles. At the same time, the rotation time dropped to 0.3–0.4 s, making it possible to scan almost the entire body of an adult within one breath hold at a slice thickness well below 1 mm. Acquisitions with MDCT scanners are usually obtained in helical mode. Exceptions include high resolution CT of the lungs, and step and shoot cardiac CT for either coronary calcium scoring or coronary CT angiography.

11.5.5. Cardiac CT

Cardiac CT is based on the synchronization of image reconstruction with the electrocardiogram (ECG) and selection of the best cardiac rest phase. Figure 11.15 shows reconstructions of the heart at different cardiac phases.

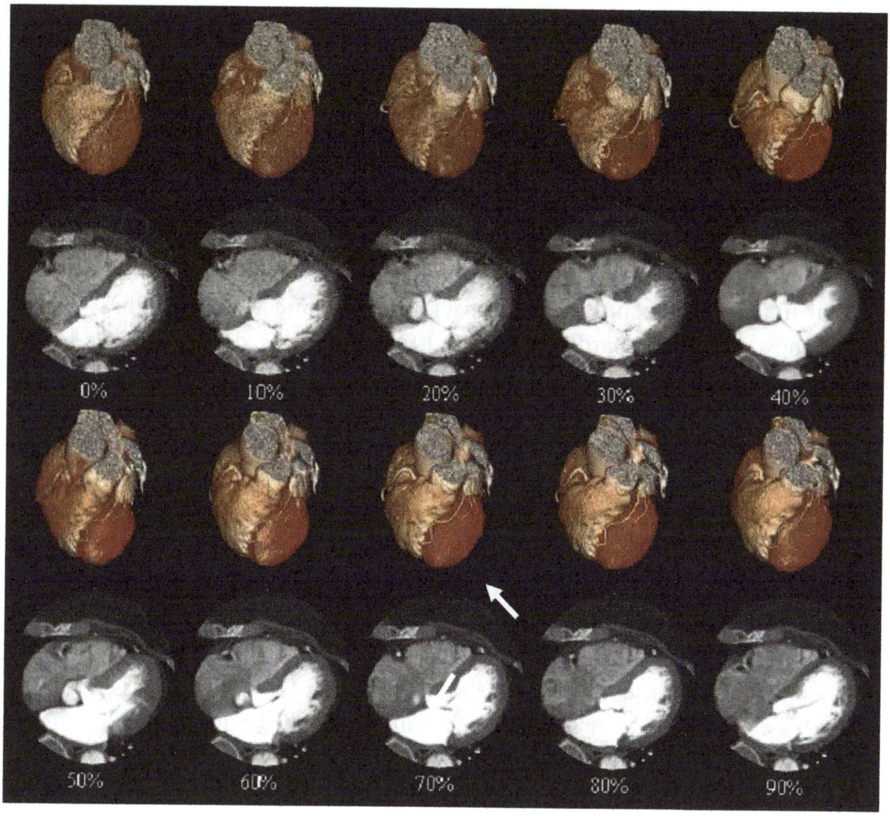

FIG. 11.15. Reconstructions of the heart at different cardiac phases. In this example, the cardiac phase corresponding to 70% of the relative risk interval produces the best motion free result.

Note the difference in blurring of the coronary arteries at different cardiac phases. In this case, the cardiac phase corresponding to 70% of the relative risk interval produces the best motion free result (70% marks the start of the cardiac phase interval). Cardiac reconstruction can be by retrospective ECG gated reconstructions and prospective ECG triggered reconstructions. Reconstructions using retrospective cardiac phase selection are based on registration of the raw data and the ECG during one or more entire cardiac cycles. An alternative to retrospective ECG gated reconstructions is prospectively acquired step and shoot acquisition. An advantage of such acquisitions is the reduction of patient dose. Some CT scanners allow for prospective scanning of the entire heart within one single heartbeat at the preselected cardiac rest phase. Two notable examples include a fast dual source CT scanner capable of performing a helical acquisition of the entire heart, and a wide cone beam CT scanner that performs an acquisition of the entire heart within one single rotation. Such novel 'single heart beat' techniques have the potential for substantial dose reduction.

11.5.6. CT fluoroscopy and interventional procedures

Dynamic CT can be used for image guided interventions, using a technique known as CT fluoroscopy. Technical developments in CT, such as the continuously rotating X ray tube, short rotation time and hardware fast enough for real time image reconstruction, provide the technical preconditions for CT fluoroscopy. Additional hardware required for CT fluoroscopy includes a device to allow operation of the scanner from within the scanner room, and installation of viewing monitors that allow display of images also within the scanner room.

Figure 11.16 shows an axial plan scan that is used to prepare a puncture; the markers on the skin allow planning for the entrance position of the needle and identification of the target for the puncture. Note that noise is much higher in the image of the CT fluoroscopy guided puncture compared with the diagnostic plan scan. During CT fluoroscopy, modest image quality is usually sufficient and the procedure should be performed using a low tube current in order to reduce exposure of the patient and the staff.

The number of clinical indications for MDCT fluoroscopy is growing steadily. The entrance skin dose for the patient should be monitored to ensure that deterministic skin effects do not occur. Staff present in the scanning room during CT fluoroscopy should be protected against exposure to scattered radiation

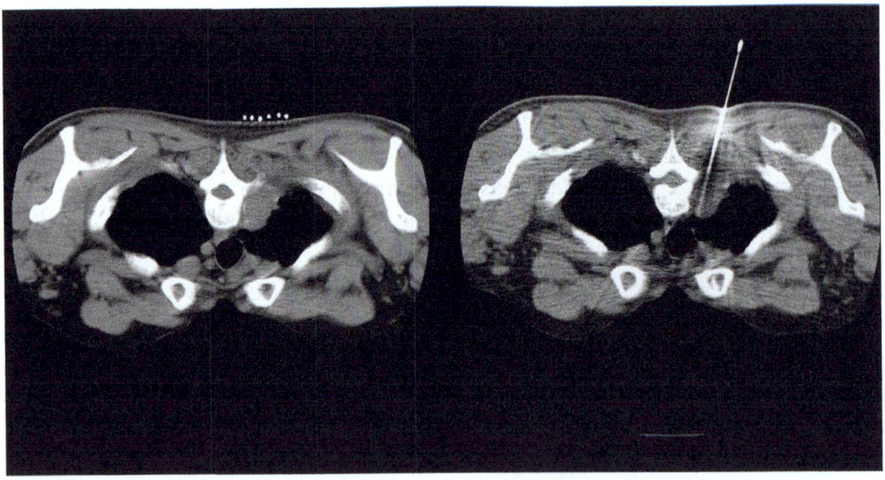

FIG. 11.16. A plan scan for preparing a puncture allows identification of the target (left). During a CT fluoroscopy guided puncture, the position of the needle can be visualized accurately (right).

by wearing a lead apron and by maintaining as great a distance from the scanner as possible. The operator should adhere to the same precautions as in routine fluoroscopy, with the number of CT acquisitions being as few as possible and the duration of acquisition runs as short as possible. These measures reduce radiation exposure of both the patient and the staff. A low dose single axial scan is often sufficient to obtain information about the status of the procedure; dynamic CT fluoroscopy should only be applied when one axial scan does not provide sufficient information. Special care should be taken to avoid direct exposure of the operator's hand. The operator should only manipulate the needle during CT fluoroscopy with a special needle holder that provides greater distance between the hand of the operator and the X ray beam, thereby preventing direct exposure of the hand.

11.5.6.1. Contrast enhanced CT

Contrast can be artificially created within or between structures that would not be visible on non-enhanced scans. For example, in CT angiography, iodine is administered intravenously during the CT scan to enhance the contrast between the vessel and the vessel walls (Fig. 11.17 (left)). In certain studies of the abdomen, a diluted iodine solution is administered orally prior to the scan to enhance contrast within the gastrointestinal tract. In CT colonography, gas may be inflated through the rectum to enhance contrast between the colon and its surrounding tissues (Fig. 11.17 (right)).

11.5.7. Special applications

Special applications of CT include the well established use for radiotherapy treatment planning and for more experimental applications such as dual energy CT imaging and dynamic volumetric CT studies.

For radiotherapy treatment planning applications of CT, the patient is scanned in the treatment position, using a flat table top. Wide bore scanners provide a gantry opening that is large enough to allow the patient to be scanned in this position. Care with patient positioning and alignment to accurate laser systems is essential. The mapping of HU values to electron density requires careful calibration by a medical physicist.

Dual energy CT imaging requires imaging of the volume of interest at two tube voltages. Extra filtration of the beam can be used to optimize the choice of the two X ray spectra used. Dual energy CT promises to give improved discrimination of certain tissues and pathologies, including accurate differentiation between urinary stones that may or may not contain uric acid. Other applications may include improved visualization of tendons of the hand and foot, and support for bone removal from the reconstructions of CT angiography scans.

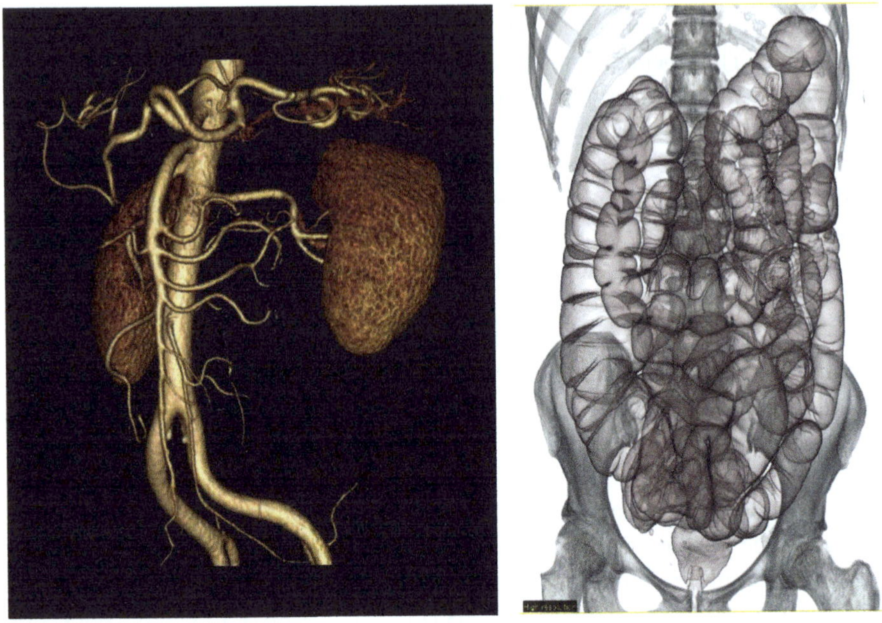

FIG. 11.17. CT angiography. 3-D rendered image of vessels with iodine administered intravenously (left). CT colonography using gas inflated through the rectum (right).

Some scanners allow dynamic CT imaging (also known as 4-D CT), where a volume of interest can be followed as a function of time. Such studies can be used to visualize the movement of joints or the contrast enhancement of organs (perfusion or dynamic CT angiography). Figure 11.18 shows an example of a dynamic CT angiography study of the entire brain with a volumetric CT scanner. In these images, time resolved contrast enhancement of the vessels allows the arterial and venous phases to be followed. Other anatomical sites for CT perfusion studies include the heart and the liver. During dynamic CT studies, as with CT fluoroscopy, the operator should be aware that the skin dose might accumulate rapidly. The patient skin dose should be maintained below 2 Gy to avoid deterministic skin effects such as erythema and epilation (see Section 20.12).

11.6. CT IMAGE QUALITY

The main acquisition parameters in CT are tube voltage, tube current and rotation time. A relatively high tube voltage (120–140 kV) is used to achieve good X ray transmission and sufficient detector signal. For special applications, such as

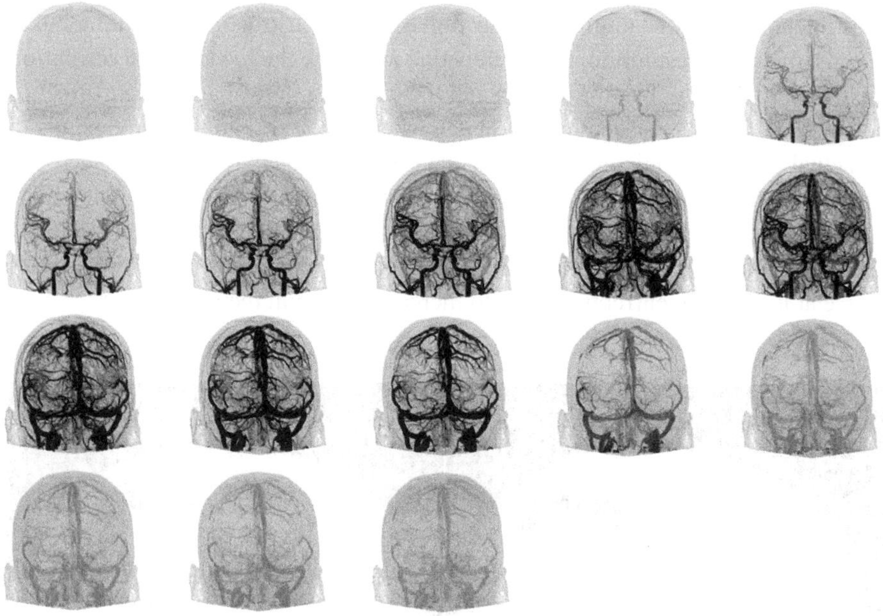

FIG. 11.18. A dynamic CT angiography study of the brain with a volumetric CT scanner that covers the entire brain (Aquilion ONE, Toshiba).

contrast enhanced studies and paediatric examinations, it might be advantageous to use a relatively low tube voltage, in the range 80–100 kV. The tube current used is limited by the long scan time and the heat capacity of the X ray tube, and by patient dose considerations. To avoid motion artefacts, the rotation time needs to be as short as possible. For scans that are less prone to motion artefacts and that require good low contrast resolution (such as scans of the brain), a longer rotation time may be selected to allow appropriate low contrast resolution.

11.6.1. Image quality

The excellent low contrast resolution of CT images is the most prominent characteristic that distinguishes the CT modality from other forms of non-tomographic radiography. Low contrast resolution is the ability to detect structures that offer only a small difference in signal compared with their direct environment. Image noise is the main limitation for low contrast resolution and may be decreased with improved image quality by employing a number of strategies. Most commonly, noise is reduced by increasing photon flux (see Chapter 4), which is achieved by increasing the tube current (mA) at the cost of patient exposure. Alternatively, noise is reduced by increasing the reconstructed slice thickness or by changing the selection of the reconstruction algorithm, but at the cost of spatial resolution. Parameters that influence the low contrast resolution include tube voltage, beam filtration and the use of a contrast agent. The effect of noise in a CT image is seen in Fig. 11.19, where the 100% image corresponds to an actual clinical acquisition. The raw data of the clinical acquisition have been processed with a low dose simulation algorithm that adds noise to simulate image quality for acquisitions that are performed at 75%, 50% and 25% of the clinically used tube current. The appearance of the low contrast lesions in the liver becomes worse at lower tube currents, owing to increased noise in the images.

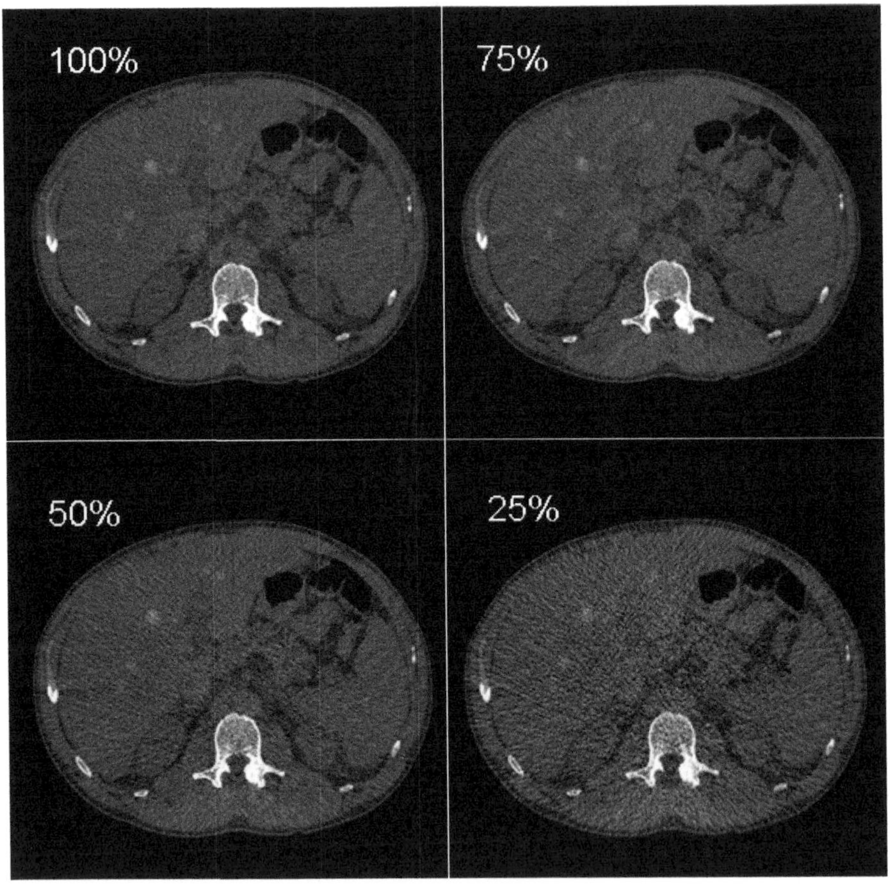

FIG. 11.19. A contrast enhanced CT scan of the liver obtained at normal exposure (100%) and with the addition of extra noise to simulate exposures of 75%, 50% and 25% of the normal exposure.

Physicists usually test low contrast resolution performance using phantoms that contain different sized low contrast inserts. Evaluation of the resultant image can be either subjective, with an observer determining whether an insert is visible or not, or objective, with calculation of the contrast to noise ratio. Determination of the noise power spectrum would provide a more objective measure of scanner performance but is not as yet applied on a large scale.

Spatial resolution, or high contrast resolution, is the ability to observe contours of small objects within the scanned volume. Small objects can only be resolved well when they exhibit a rather large difference in signal. Spatial resolution is limited by the acquisition geometry of the CT scanner, the reconstruction algorithm and the reconstructed slice thickness. Voxel size is often used as an indicator of spatial resolution, although a smaller voxel size does not necessarily imply better spatial resolution. Spatial resolution along the *z* axis is usually determined using a slice sensitivity profile, with the response often quantified as the full width at half maximum, while in the axial plane it is preferably measured as a point spread function (PSF) (see Section 4.4.1). From this, the modulation transfer function (MTF) can be calculated. The MTF does yield useful information on image quality, although clinical assessment of the MTF in the clinical environment may be complex and is usually only performed by medical physicists at acceptance and commissioning of new scanners or after major upgrades. Manufacturers of CT scanners provide information about the MTF, which should be measured according to international standards [11.3]. The performance of current 64 slice scanners with regard to spatial resolution, expressed as the full width at half maximum of the PSF, is within the range 0.6–0.9 mm in all three dimensions.

Figure 11.20 shows images from a CatPhan phantom that is widely used to evaluate the image quality of CT scans. The image on the left allows evaluation of the HU values for four large inserts in the periphery of the phantom for the materials air, low density polyethylene, PMMA and Teflon. Low contrast acrylic inserts of different diameters around the centre are used for determining the effect of object size on low contrast detectability. The image in the middle shows high contrast line pairs that allow the subjective assessment of spatial resolution. The image on the right allows spatial resolution to be measured objectively, as the PSF of a small tungsten bead. The image can also be used to assess the homogeneity of the image.

Temporal resolution is the ability to resolve fast moving objects in the displayed CT image. Good temporal resolution avoids motion artefacts and motion induced blurring of the image. Good temporal resolution in CT is realized by fast data acquisition through the fast rotation of the X ray tube. Reconstruction algorithms that are used for general CT applications provide, in principle, a temporal resolution equal to the time of a 360° rotation with full reconstruction.

The best routinely achievable temporal resolution is slightly longer than 50% of the rotation time using 180° and fan angle rotation reconstruction. Temporal resolution can be improved further by using dedicated reconstruction algorithms, for example, in cardiac CT with a segmented reconstruction, or by using a dual source CT scanner. There are no simple methodologies available yet that allow measurement of temporal resolution in a clinical setting.

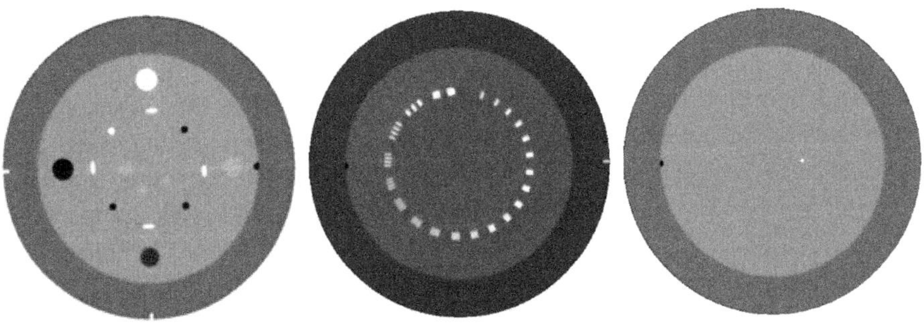

FIG. 11.20. Images of a CatPhan phantom taken at different z coordinates and showing different modules of the phantom.

11.6.2. Clinical observer studies

Fundamental physical image quality testing as described in the previous section yields information on CT scanner performance. Such information can be used for product specifications and for quality control (see Chapter 19). However, it does not provide sufficient information for the development of clinical acquisition protocols. The reason is that it is largely unknown what clinical image quality is required by radiologists for specific clinical tasks, in terms of either fundamental image quality quantities or pragmatic parameters that are derived from test objects. For practical reasons, clinical acquisition protocols are largely based on experience and consensus, but preferably they should be based on clinical observer studies and appropriate scientific evidence (see also Chapter 23). However, observer studies that aim at optimizing acquisition protocols are rare in CT. One reason for this is that the acquisition of additional scans for the same patient is most often considered inappropriate, owing to the extra radiation exposure. As seen in Fig. 11.19 however, lower dose scans can be simulated by mathematical models that add noise to the raw data to determine the image noise levels that can be allowed for specific observer tasks. Appropriately designed observer studies can assist in the determination of the optimal tube current for CT protocols. Optimization of the tube voltage is more difficult to achieve, since no appropriate algorithms have been described for simulating the effect of tube voltage on image quality, and optimization is based mainly on theoretical considerations, phantom studies (for example, aiming at achieving the optimal CNR in iodine enhanced CT angiography studies) and observer consensus.

11.6.3. Effect of acquisition and reconstruction parameters on image quality

Many reconstruction and viewing parameters have an effect on image quality and observer performance. These include the reconstructed slice thickness, the reconstruction filter, the windowing and the image reformats that can be used in addition to the review of axial images. Any CT acquisition can be reconstructed with one or more reconstruction filters. Figure 11.21 shows the same image reconstructed with slice thicknesses of 10, 5 and 0.5 mm. Note that in both the volume rendered and coronal images, the spatial resolution improves considerably at smaller slice thickness. Reconstructions are generally made at a slice thickness of ≤1 mm.

During image reading, the radiologist can choose the appropriate window settings for the specific anatomy and pathology of interest. This is illustrated in Fig. 11.22 for four axial CT head images all created by post-processing and derived from the same acquisition. Images on the left are reconstructed with a soft tissue reconstruction filter; those on the right are reconstructed with a sharp bone reconstruction filter. The images in the upper row are shown in a window setting for brain (window level 50, window width 100); the images in the lower row are shown in a window setting for bone (window level 1000, window width 2500). As can be seen, the image at the top left is processed and windowed appropriately for evaluation of the brain tissue. Likewise, the details in the skull are better presented in the lower right image, owing to appropriate reconstruction and settings. The image at the top right is hampered in its presentation of brain tissue, owing to image noise that results from the inappropriate use of a bone reconstruction filter, while the bone in the image cannot be assessed, owing to the window setting used. Similarly, the image at the bottom left is hampered for bone analysis, owing to blurring of the bone that results from the soft tissue reconstruction filter, while the brain tissue cannot be assessed, owing to the use of the bone window setting.

Many image reformats can be used in addition to the reading of axial images. Figure 11.23 shows an axial image of the brain and three additional reformats: a coronal image, a sagittal image and a volume rendered image.

Figure 11.24 shows two images of the chest, on the left a maximum intensity projection, and on the right a 3-D volume rendered image.

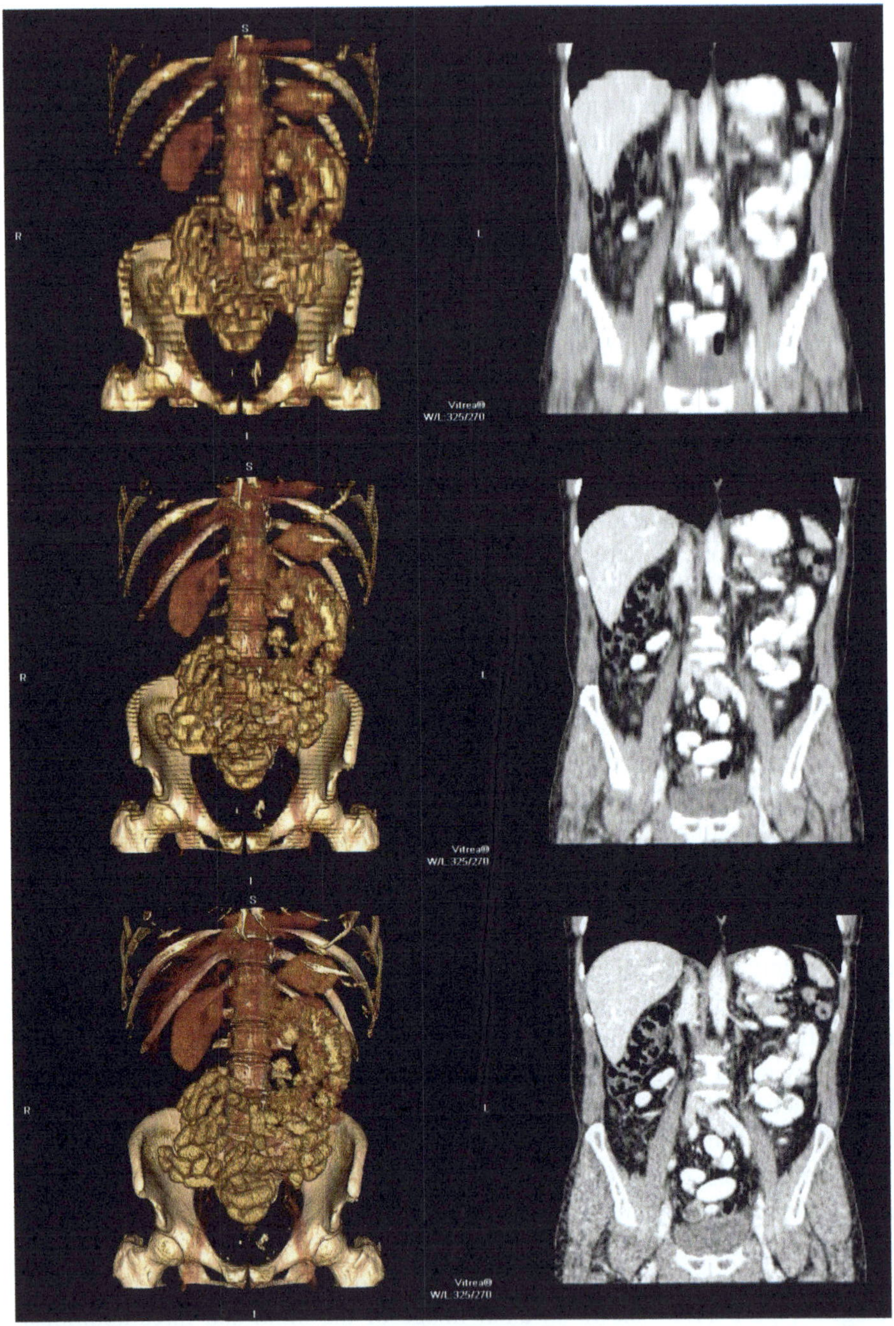

FIG. 11.21. Reconstructions of the same acquisition at slice thicknesses of 10 mm (top), 5 mm (middle) and 0.5 mm (bottom), on the left volume rendered images (3-D representation of the study), on the right coronal views (2-D representation of the study).

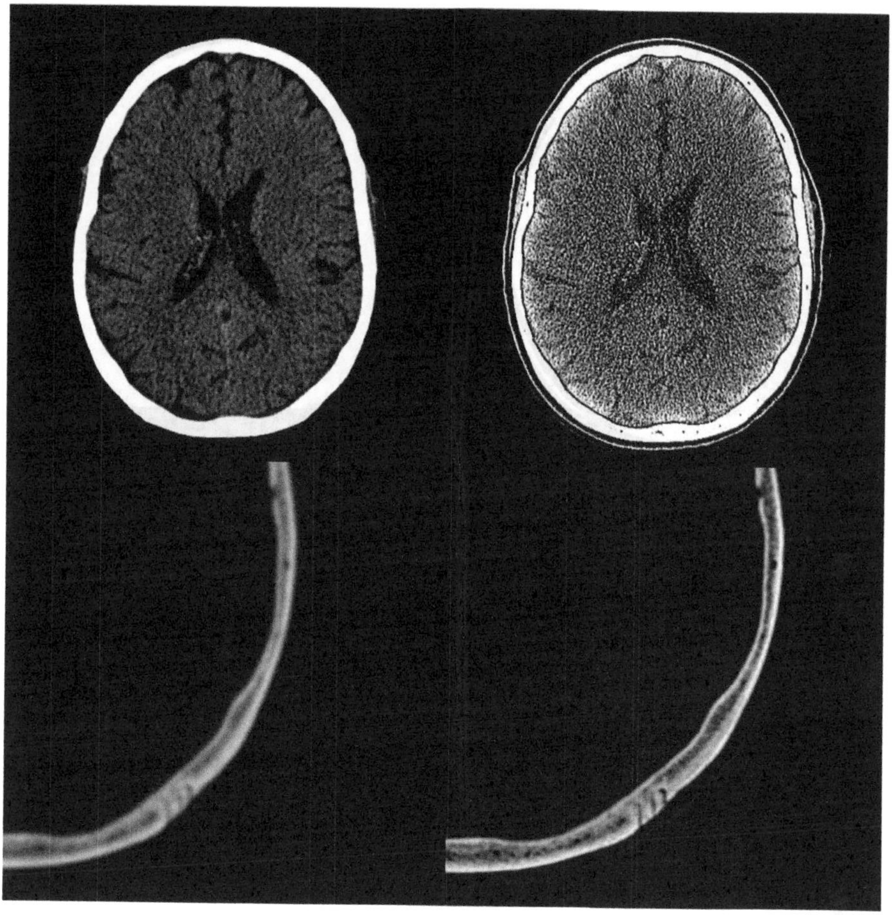

FIG. 11.22. Four CT head images. Images on the left are reconstructed with a soft tissue reconstruction filter; images on the right are reconstructed with a bone reconstruction filter.

11.6.4. Artefacts

Proper image quality in CT is only achieved if calibrations of the scanner are regularly carried out according to the protocols that are prescribed by the manufacturers. These include frequent air calibrations and, less frequently, calibrations with homogeneous water phantoms. Air calibrations provide information about the small differences in the response of individual dels. This is essential since the projections have to be accurate to within 0.5%. Calibrations with phantoms allow some correction of the beam hardening effect.

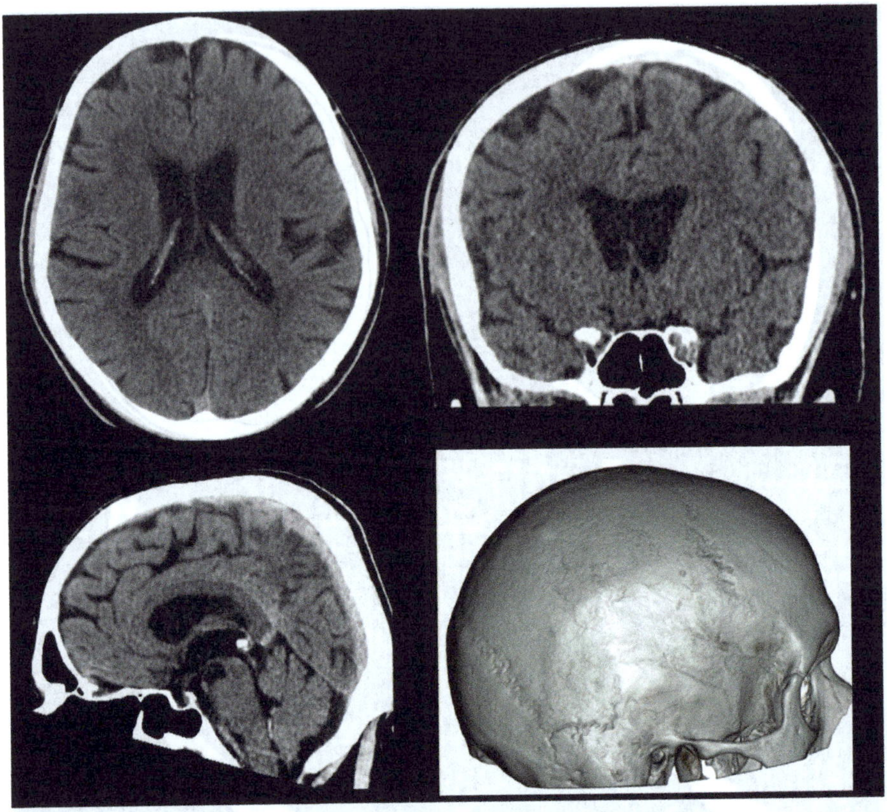

FIG. 11.23. An axial image of the brain (top left) and three additional reformats: a coronal image (top right), a sagital image (bottom left) and a volume rendered image (bottom right).

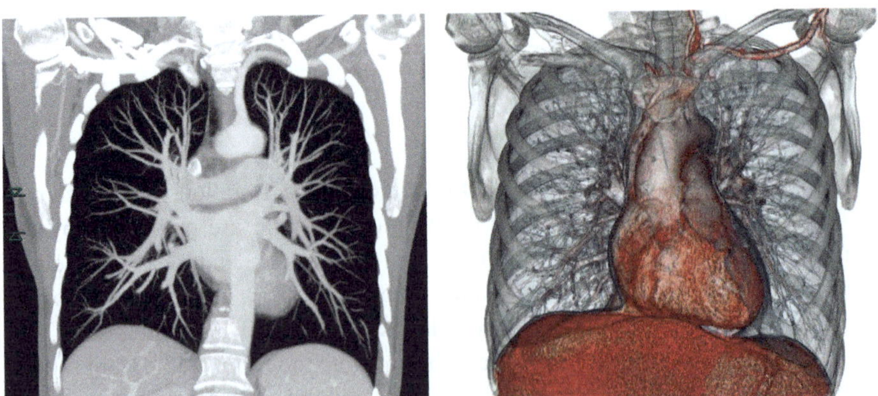

FIG. 11.24. Two images of the chest. On the left a maximum intensity projection and on the right a 3-D volume rendered image.

Artefacts can be acquisition related, reconstruction related, or patient related. Acquisition related artefacts include ring artefacts (Fig. 11.25), which occur with one or more malfunctioning dels, and unusable images, which may result from malfunctioning of the X ray tube during the acquisition. Undersampling the projection data may lead to Moiré patterns, and detector afterglow may induce blurring of the image.

An important artefact that occurs when thick acquisition slices are used results from an averaging of the linear attenuation coefficient along the z axis of a voxel. This artefact is referred to as the partial volume effect; it will make a small high density object appear as a larger lower density object and is seen, for example, when looking at cortical bone in thick CT slices. This can be avoided by the use of thinner slices.

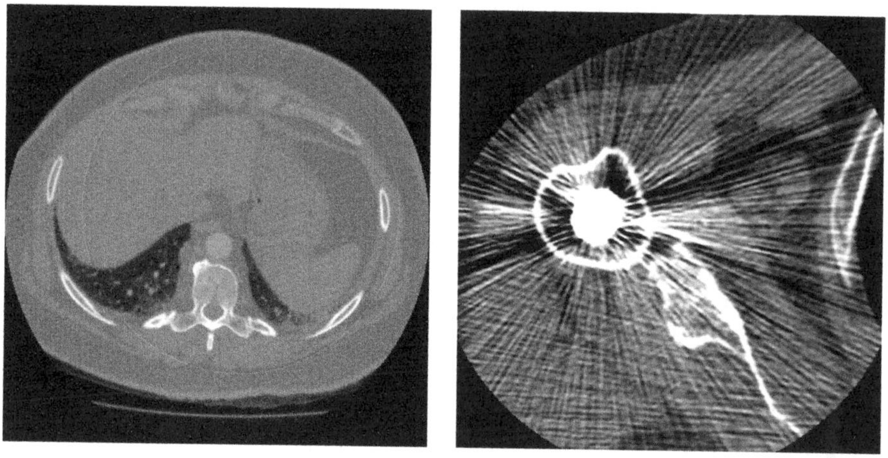

FIG. 11.25. A ring artefact occurs in the case of malfunctioning of one or more dels (left). Metal artefacts occur as a result of beam hardening and low dose on the detector.

Strong attenuation of the X ray beam by compact bone, calcifications or a metal object may lead to a beam hardening artefact. A very severe artefact occurs when a metal prosthesis is scanned; this effect is referred to as a metal artefact and occurs when the prosthesis attenuates the X ray beam almost completely. Figure 11.25 also shows the typical streaks in the axial image that occur when a large metal implant is scanned. In this image, a hip prosthesis was scanned.

Patient related artefacts can sometimes be avoided by properly instructing the patient to refrain from moving during the scan and to maintain the breath holding during the entire scan, particularly during scans of the trunk. Movement of the heart and pulsation of the vessels cannot be avoided. Therefore, it is essential that acquisitions of, for example, the coronary arteries or the aorta are optimized to achieve the best possible temporal resolution. It is well known that pulsation of the aorta may induce artefacts that mimic an aortic dissection. In this case, the artefact may have serious consequences for the patient if it is not recognized as an artefact.

REFERENCES

[11.1] BERNINGER, W.H., REDINGTON, R.W., Multiple Purpose High Speed Tomographic X-Ray Scanner, Patent #4196352, General Electric Company, USA (1980).

[11.2] MORI., I., Computerized Tomographic Apparatus Utilizing a Radiation Source, US Patent #4630202 (1986).

[11.3] INTERNATIONAL ELECTROTECHNICAL COMMISSION, Evaluation and Routine Testing in Medical Imaging Departments – Part 3-5: Acceptance Tests – Imaging Performance of Computed Tomography X ray Equipment, Rep. IEC 61223-3-5, IEC, Geneva (2004).

BIBLIOGRAPHY

BUZUG, T.M., Computed Tomography: From Photon Statistics to Modern Cone-Beam CT, Springer-Verlag, Berlin and Heidelberg (2008).

HSIEH, J., Computed Tomography: Principles, Design, Artefacts, and Recent Advances, 2nd edn, SPIE Press, Bellingham, Washington, DC (2009).

INTERNATIONAL ATOMIC ENERGY AGENCY, Quality Assurance Programme for Computed Tomography: Diagnostic and Therapy Applications, IAEA Human Health Series No. 19, IAEA, Vienna (2012).

KAK, A.C., SLANEY, M., Principles of Computerized Tomographic Imaging, IEEE Press, New York (1988),
http://www.slaney.org/pct/ (accessed on 23 August 2012).

KALENDER, W.A., Computed Tomography: Fundamentals, System Technology, Image Quality, Applications, 2nd edn, Publicis Corporate, Erlangen (2005).

NAGEL, H.D., (Ed.), Radiation Exposure in Computed Tomography: Fundamentals, Influencing Parameters, Dose Assessment, Optimisation, Scanner Data, Terminology, 4th edn, CTB Publications, Hamburg (2002).

Chapter 12

PHYSICS OF ULTRASOUND

J.C. LACEFIELD
University of Western Ontario,
London, Canada

12.1. INTRODUCTION

Ultrasound is the most commonly used diagnostic imaging modality, accounting for approximately 25% of all imaging examinations performed worldwide at the beginning of the 21st century. The success of ultrasound may be attributed to a number of attractive characteristics, including the relatively low cost and portability of an ultrasound scanner, the non-ionizing nature of ultrasound waves, the ability to produce real time images of blood flow and moving structures such as the beating heart, and the intrinsic contrast among soft tissue structures that is achieved without the need for an injected contrast agent. The latter characteristic enables ultrasound to be used for a wide range of medical applications, which historically have primarily included cardiac and vascular imaging, imaging of the abdominal organs and, most famously, in utero imaging of the developing fetus. Ongoing technological improvements continue to expand the use of ultrasound for many applications, including cancer imaging, musculoskeletal imaging, ophthalmology and others.

The term ultrasound refers specifically to acoustic waves at frequencies greater than the maximum frequency audible to humans, which is nominally 20 kHz. Diagnostic imaging is generally performed using ultrasound in the frequency range of 2–15 MHz. The choice of frequency is dictated by a trade off between spatial resolution and penetration depth, since higher frequency waves can be focused more tightly but are attenuated more rapidly by tissue. The information contained in an ultrasonic image is influenced by the physical processes underlying propagation, reflection and attenuation of ultrasound waves in tissue.

12.2. ULTRASONIC PLANE WAVES

12.2.1. One dimensional ultrasonic waves

An acoustic wave is a travelling pressure disturbance that produces alternating compressions and rarefactions (expansions) of the propagation medium. The compressions and rarefactions displace incremental volumes of the medium and the wave propagates via transfer of momentum among incremental volumes. Each incremental volume of the medium undergoes small oscillations about its original position but does not travel with the pressure disturbance.

A mathematical description of a pressure wave, $p(x, t)$, propagating along one spatial dimension, x, through a homogeneous, non-attenuating fluid medium, can be formulated starting from Euler's equation:

$$\frac{\partial}{\partial x}p(x,t)+\rho_{\mathrm{o}}\frac{\partial}{\partial t}u(x,t)=0 \tag{12.1}$$

and the equation of continuity:

$$\frac{\partial}{\partial t}p(x,t)+\frac{1}{\kappa}\frac{\partial}{\partial x}u(x,t)=0 \tag{12.2}$$

In Eqs (12.1, 12.2), ρ_{o} is the undisturbed mass density of the medium, κ is the compressibility of the medium (i.e. the fractional change in volume per unit pressure in units of Pa^{-1}), and $u(x, t)$ is the particle velocity produced by the wave.

Euler's equation, which can be derived starting from Newton's second law of motion, states that the pressure disturbance accelerates the incremental volumes that make up the medium. The equation of continuity, which can be derived by writing a mass balance for an incremental volume of the medium, states that a net flow of material into an incremental volume produces a local change in pressure. These first order equations can be combined to obtain the acoustic wave equation:

$$\frac{\partial^2}{\partial x^2}p(x,t)-\frac{1}{c^2}\frac{\partial^2}{\partial t^2}p(x,t)=0 \tag{12.3}$$

where $c=1/\sqrt{\rho_{\mathrm{o}}\kappa}$ is the speed of sound.

A monochromatic plane wave solution to Eq. (12.3) is given by:

$$p(x,t) = P\cos(\omega t - kx) \tag{12.4}$$

where P is the amplitude of the wave, $\omega = 2\pi f$ is the radian frequency corresponding to the cyclic frequency, f, in hertz, and $k = 2\pi/\lambda = 2\pi f/c$ is the wave number corresponding to the wavelength, λ, which is related to the frequency and sound speed by $c = f\lambda$.

12.2.2. Acoustic pressure and intensity

As implied by Eq. (12.4), an ultrasonic wave can be described by its frequency, f, and its pressure amplitude, P. Diagnostic imaging is typically performed using peak pressures in the range 0.1–4 MPa. The strength of an ultrasound wave can also be characterized by its intensity, I, which is the average power per unit cross-sectional area evaluated over a surface perpendicular to the propagation direction. For plane waves, the time average intensity is related to the amplitude by:

$$I = \frac{P^2}{2\rho_0 c} \tag{12.5}$$

It is often convenient to express acoustic intensity in decibels, dB, which is a relative logarithmic scale defined as:

$$I_{\text{dB}} = 10\lg(I/I_{\text{ref}}) \tag{12.6}$$

where I is the intensity of the wave in linear units and I_{dB} is the intensity of the wave in dB with respect to a reference intensity, I_{ref}.

12.2.3. Reflection and transmission

An ultrasound image displays the magnitude (absolute value of amplitude) of ultrasound echoes, so a physical understanding of acoustic wave reflection is valuable for interpreting the images. Consider the example of shouting within a brick walled gymnasium and hearing one's voice echo from the far wall. As the positive pressure phase of the acoustic wave reaches the wall, the air is compressed against the surface of the wall. The wall is both massive and rigid compared with the air, so the air particles will not possess enough momentum to accelerate the surface of the wall and will, instead, rebound and compress against air particles behind them. The compression of the air as it rebounds from the wall produces an acoustic wave travelling in the opposite direction to the original shout.

The changes in the mass and stiffness of the medium that produce a reflection can be combined into a single property termed the acoustic impedance, Z. For plane wave propagation, the acoustic impedance is given by:

$$Z = \rho_0 c = \sqrt{\rho_0 / \kappa} \tag{12.7}$$

Consider a plane wave travelling in a semi-infinite half space with sound speed c_1 and acoustic impedance Z_1 that is incident upon a planar interface with a second semi-infinite half space with sound speed c_2 and impedance Z_2 (Fig. 12.1). The angles of incidence, reflection and transmission (θ_i, θ_r and θ_t, respectively), which are measured with respect to the normal to the boundary, are related by the acoustics version of Snell's law and where $\theta_r = \theta_i$:

$$\sin\theta_t = \frac{c_2}{c_1}\sin\theta_i \tag{12.8}$$

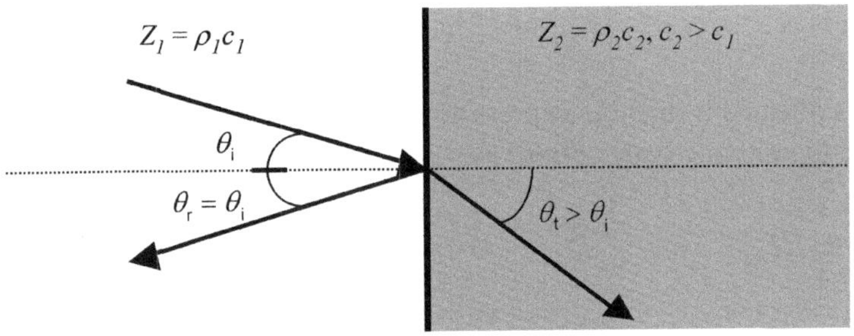

FIG. 12.1. Angles of incidence, reflection and transmission at a planar interface between a material with sound speed c_1 and a second material with a higher sound speed c_2.

Equation (12.8) indicates that the wave transmitted into the second medium is bent towards the normal if $c_1 > c_2$ and away from the normal if $c_1 < c_2$. This change in direction is termed refraction and can be an important source of artefacts in some clinical imaging applications. The limiting case of refraction occurs when $c_2 > c_1$ and $\theta_i > \sin^{-1}(c_1/c_2)$, in which case θ_t is imaginary according to Eq. (12.8) and the wave is totally reflected.

The amplitudes of the incident and reflected waves (P_i and P_r, respectively) are related by the reflection coefficient, R. For plane waves in fluid media, the reflection coefficient is given by:

$$R = \frac{P_r}{P_i} = \frac{Z_2 \cos\theta_i - Z_1 \cos\theta_t}{Z_2 \cos\theta_i + Z_1 \cos\theta_t} \tag{12.9}$$

The numerator in Eq. (12.9) indicates that a reflection is produced when an acoustic wave encounters a difference in acoustic impedance, so an ultrasound image may be thought of as a map of the relative variations in acoustic impedance in the tissues. Since amplitude is a signed quantity, $-1 \le R \le 1$. A negative value of R implies that the reflected wave is inverted with respect to the incident wave.

The amplitudes of the incident wave and the wave transmitted into the second semi-infinite half space, P_t, are related by the transmission coefficient, T, which must be expressed in a form that accounts for the possibility of total reflection:

$$T = \frac{P_t}{P_i} = \begin{cases} \dfrac{2Z_2 \cos\theta_i}{Z_2 \cos\theta_i + Z_1 \cos\theta_t} & c_1 \ge c_2 \text{ or } \theta_i \le \sin^{-1}(c_1/c_2) \\ 0 & c_1 < c_2 \text{ and } \theta_i > \sin^{-1}(c_1/c_2) \end{cases} \tag{12.10}$$

Simpler expressions for R and T for the special case of normal incidence can be obtained by setting $\theta_i = \theta_t = 0$ in Eqs (12.9, 12.10). Comparable expressions for the reflected and transmitted intensities can also be obtained by using Eq. (12.5) to substitute for P_i, P_r and P_t and noting that the $\rho_0 c$ term in the denominator of Eq. (12.5) is the acoustic impedance.

12.2.4. Attenuation

The term attenuation encompasses all mechanisms by which the energy of a wave is dissipated as it propagates, including specular reflections (see Section 12.2.3), divergence, scattering from inhomogeneities in the medium and thermal absorption, in which the mechanical energy of the wave is converted to heat. Absorption is the most significant source of attenuation in diagnostic ultrasound. Inclusion of absorptive terms in the wave equation changes the plane wave solution from the form given in Eq. (12.4) to an exponentially decaying sinusoid:

$$p(x,t) = P e^{-\alpha x} \cos(\omega t - kx) \tag{12.11}$$

where α is the frequency dependent amplitude attenuation coefficient in Nepers (Np)/m. In an ideal viscous fluid, α is proportional to frequency squared, but in soft tissues, the existence of absorption mechanisms in addition to viscous relaxation produces a non-integer power law frequency dependence, i.e. α is proportional to f^m, where $1 < m < 2$. For most applications of diagnostic ultrasound, an approximation of linear frequency dependence of attenuation, i.e. $m \approx 1$, proves sufficient. The primary consequence of frequency dependent attenuation is that higher frequency waves are attenuated more rapidly than lower frequency waves and thus yield shallower penetration depths for imaging.

12.3. ULTRASONIC PROPERTIES OF BIOLOGICAL TISSUE

12.3.1. Sound speed, acoustic impedance and attenuation coefficient

Typical speeds of sound in biological tissues are given in Table 12.1, along with sound speeds for a few other reference materials. As a rule of thumb, solids possess the highest sound speeds and gases possess the lowest sound speeds. Observe that the sound speed in soft tissues is similar to the sound speed in water at body temperature. This similarity between water and soft tissue holds for most acoustic properties and justifies the use of equations for fluid media to analyse wave propagation in biomedical ultrasound.

The acoustic impedances of some representative tissues and other materials are also compared in Table 12.1. Acoustic impedance follows the same general pattern as sound speed — low values in gases, intermediate values in liquids and soft tissues and high values in solids.

Attenuation coefficients of biological tissues are usually reported in dB/(cm·MHz), where the units reflect the fact that acoustic intensities are typically specified in decibels rather than Nepers, as well as the approximation of linear frequency dependence of attenuation. The conversion value for Np and dB is 1 Np ≈ 8.686 dB. Attenuation coefficients for representative tissues are listed in Table 12.1.

12.3.2. Scattering

Similar to the mechanism of specular (mirror-like) reflection described in Section 12.2.3, scattering occurs when an ultrasonic wave encounters a variation in the acoustic impedance of the medium. Scattering occurs when the wave encounters features with dimensions similar to, or smaller than, the wavelength. Scattered echoes are omnidirectional and are significantly weaker than specular reflections. Constructive and destructive interference of echoes scattered backward from cellular scale tissue features to the transducer are the source of the speckle texture that dominates the internal appearance of organs in ultrasound images. Echoes scattered away from the transducer are also a secondary source of attenuation, typically contributing 10–20% of the apparent attenuation in an image.

TABLE 12.1. ACOUSTIC PROPERTIES OF SELECTED MATERIALS
(*acoustic properties taken from Zagzebski [12.1] and Shung [12.2]*)

Material	Sound speed (m/s)	Acoustic impedance (MRayl)	Attenuation coefficient (dB/cm at 1 MHz)
Air	330	0.0004	12
Water	1480	1.48	0.0022
Fat	1450–1460	1.34–1.38	0.52
Liver	1555–1570	1.65	0.96
Blood	1550–1560	1.61–1.65	0.17
Muscle	1550–1600	1.62–1.71	1.2
Skull bone	3360–4080	6.0–7.8	11.3

12.3.3. Non-linear propagation

Non-linearity arises in acoustic propagation because the pressure wave alters the density of the medium and the sound speed depends on density according to $c = 1/\sqrt{\rho\kappa}$. This effect is negligible at low acoustic intensities; however, near the focus of beams used for diagnostic imaging, the density variations produced by the wave become significant, such that the compressive phase of the wave propagates at a higher velocity than the rarefactional phase of the wave. This non-linearity will distort an initially sinusoidal wave into a sawtooth pattern, as the compressive segments of the wave catch up to the rarefactional segments ahead of them. A transformation from a sinusoidal to a sawtooth wave introduces harmonics (i.e. energy at integer multiples of the transmitted fundamental frequency, f_0) into the wave's frequency spectrum. This phenomenon is the physical basis of the tissue harmonic imaging mode discussed in Chapter 13.

12.4. ULTRASONIC TRANSDUCTION

12.4.1. Piezoelectric devices

Ultrasonic transducers were made possible by the discovery of piezoelectricity in quartz by Pierre and Jacques Curie in 1880. Piezoelectricity is a reversible property of certain crystalline materials by which a vibration applied to opposite faces of the crystal produces an alternating net electrical charge across the crystal, whereas an alternating voltage applied across the crystal causes it to vibrate in thickness. The microscopic mechanism of piezoelectricity can be understood by envisioning the material as a collection of randomly oriented electric dipoles. An applied force deforms the crystal, resulting in a rearrangement of the dipoles that induces a net charge across the crystal. Conversely, a voltage difference applied across the crystal will change the arrangement of the dipoles, thereby producing a bulk deformation of the crystal.

The transducers used for diagnostic imaging have conventionally been fabricated using the ferroelectric ceramic lead zirconate titanate, which is commonly referred to by the acronym PZT, from the first letters of the chemical symbols for lead, zirconium and titanium. PZT provides a relatively high electrical to mechanical coupling efficiency at low cost. Many modern transducers are composites of PZT and a non-piezoelectric polymer. The composite materials have a lower acoustic impedance than conventional PZT, which improves acoustic coupling in the tissue and increases the transducer's bandwidth. Therefore, composite materials are now favoured for high end clinical systems, where there is less concern about the cost of the transducer.

The bandwidth and sensitivity of the transducer are improved by sandwiching the piezoelectric crystal between a backing layer and a matching layer. The purpose of the backing layer is to absorb ultrasound radiated from the back face of the crystal and dampen reverberations within the crystal. The matching layer, which is bonded to the front face of the crystal, serves to reduce the reflection coefficient between the transducer and the tissue, thereby increasing the transducer's sensitivity to weak echoes.

12.4.2. Transmitted pulse

The bandwidth of the transducer determines the length of the transmitted pulse and hence the axial (i.e. along the beam direction) resolution of the imaging system. The transmitted pulse takes the form of a gated sinusoid similar to that shown in Fig. 12.2. The full width at half maximum (FWHM) pulse length can be specified most simply as the number of cycles, N_c, of the sinusoid multiplied by the wavelength, in which case the axial resolution, AR, of the imaging system is:

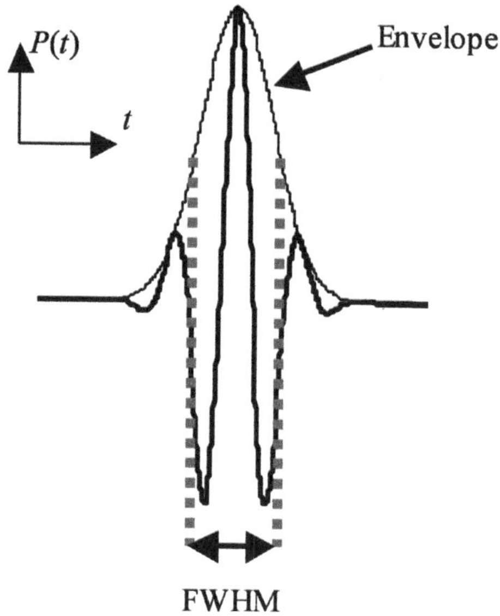

FIG. 12.2. Typical diagnostic ultrasound radiofrequency pulse waveform (thick solid curve) and the corresponding demodulated pulse envelope (thin solid curve). The FWHM of the pulse envelope is also indicated.

$$\mathrm{AR} = \frac{N_c \lambda}{2} \tag{12.12}$$

The division by 2 in Eq. (12.12) arises because the pulse makes a round trip from the transducer to a reflector and back. Ultrasound imaging transducers typically possess high bandwidths yielding transmitted pulses 1.5–2.0 cycles in duration and an axial resolution finer than 1 mm.

12.4.3. Radiation and diffraction

The beam pattern produced by an ultrasound transducer can be analysed in a plane parallel to the face of the aperture using scalar diffraction theory. The Huygens–Fresnel principle is the basis of scalar diffraction theory. This principle states that an aperture can be decomposed into a collection of point sources, such that the field produced by the aperture is a superposition of spherical wavelets radiated from each point source. In the far field of an unfocused aperture, or at and beyond the focus of a focused aperture, the resulting beam is given by the Fraunhofer diffraction integral:

$$U(x,y) = \frac{\exp(jkz)}{j\lambda z}\exp\left(j\frac{k}{2z}(x^2+y^2)\right)\iint_{\Sigma} U(\xi,\eta)\exp\left(-j\frac{k}{z}(x\xi+y\eta)\right)\mathrm{d}\xi\,\mathrm{d}\eta \tag{12.13}$$

where $U(x, y)$ is the field in a plane a distance z from the aperture and $U(\xi, \eta)$ is the field in the aperture plane.

Equation (12.13) can be stated concisely as:

$$U(x,y) = \frac{\exp(jkz)}{j\lambda z} \exp\left(j\frac{k}{2z}\left(x^2 + y^2\right)\right) \Im\{U(\xi,\eta)\} \tag{12.14}$$

where $\Im\{\bullet\}$ is a 2-D spatial Fourier transform (FT) with effective spatial frequencies $\tilde{k}_x = kx/z$ and $\tilde{k}_y = ky/z$.

In medical ultrasound, the lateral (x) dimension within the image plane and the elevation (y) dimension perpendicular to the image plane are typically treated as separable, so the lateral beam pattern can be computed by ignoring the η and y terms in Eq. (12.14). Ultrasound imaging systems typically employ a focused rectangular aperture, so substituting

$$U(\xi) = \mathrm{rect}\left(\frac{\xi}{L}\right) = \begin{cases} 1 & -L/2 \le \xi \le L/2 \\ 0 & \text{otherwise} \end{cases} \tag{12.15}$$

where L is the length of the aperture, into Eq. (12.14), and letting $z = F$, the focal distance, yields the lateral beam pattern:

$$U(x) \propto \mathrm{sinc}\left(\frac{Lx}{\lambda F}\right) \tag{12.16}$$

where $\mathrm{sinc}(a) = \sin(\pi a)/(\pi a)$ is the FT of the rect function (see Section 4.4.2).

Pulse echo imaging is usually performed using the same focused aperture for transmission and reception, in which case the lateral point spread function (Fig. 12.3) is given by the square of Eq. (12.16).

The Rayleigh resolution criterion defines the resolution as the distance from the peak of the beam to the first zero of the beam. Setting the sinc function in Eq. (12.16) equal to zero and solving for x yields an expression for the lateral resolution (LR):

$$\mathrm{LR} = \frac{\lambda F}{L} \tag{12.17}$$

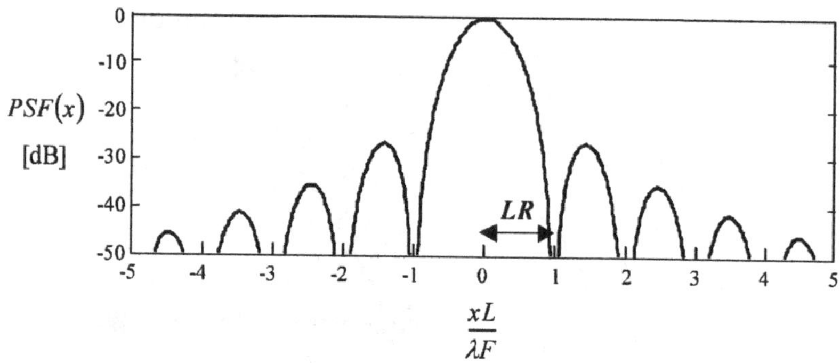

FIG. 12.3. A typical pulse echo lateral point spread function is a sinc² beam with the LR given by Eq. (12.17).

The ratio *F*/*L* in Eq. (12.17) is called the *f* number of the transducer. The *f* number of an ultrasound imaging system is typically between 2 and, at most, 6, so the LR is generally in the 1–2 mm range.

12.5. DOPPLER PHYSICS

12.5.1. The Doppler effect

Doppler ultrasound is a method used to image moving blood and thereby estimate blood velocity by exploiting the Doppler effect, which was studied in 1842 by Christian Doppler. The Doppler effect is explained schematically for a moving continuous wave source in Fig. 12.4. As the source moves to the right, each subsequent cycle of the wave radiates from a point shifted to the right, so the frequency is compressed in the direction of motion and expanded in the opposite direction. This phenomenon is the origin of the change in perceived pitch as a siren or train whistle passes a stationary listener. The Doppler effect applies to echoes from moving reflectors such as red blood cells, as well as to waves radiated from moving sources.

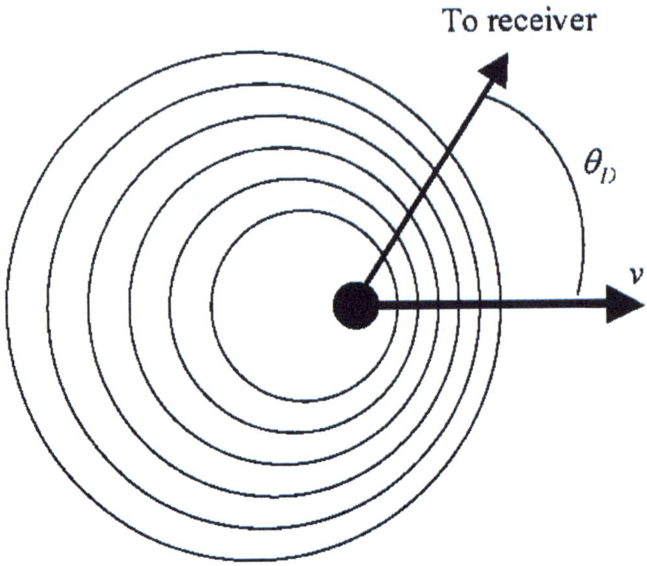

FIG. 12.4. Schematic representation of the Doppler effect for a continuous wave point source (black dot) moving with velocity v in the direction of the thick arrow. The circles represent the relative separation of the maxima of consecutive cycles of the radiated wave as a function of the Doppler angle, θ_D.

The change in frequency produced by motion of a reflector is called the Doppler frequency, f_D, and is given by the Doppler equation:

$$f_D = \frac{2|v| f_0 \cos\theta_D}{c} \tag{12.18}$$

where $|v|$ is the speed of the reflector, f_0 is the frequency of the incident wave and θ_D is the Doppler angle between the direction of motion and a ray pointed from the reflector to the receiver. The cosine term in the numerator indicates that a Doppler system is most sensitive to motion directly towards or away from the transducer, corresponding to $\theta_D = 0$ or $\theta_D = \pi$, respectively.

12.5.2. Continuous wave Doppler

The simplest Doppler systems use continuous wave (CW) Doppler and are usually small handheld devices. A CW Doppler transducer consists of two adjacent piezoelectric elements angled slightly towards one another, as shown in Fig. 12.5. The transmitter emits a continuous sinusoidal wave of the form $\cos(\omega_0 t)$ and the receiver detects echoes returning from the region of overlap between transmitter and receiver beams. If the reflectors are moving, the received echoes are in the form $\cos([\omega_o + \omega_D]t)$, where $\omega_D = 2\pi f_D$, and a Doppler signal of the form $\cos(\omega_D t)$ can be recovered via frequency demodulation of the received signal.

The maximum blood flow velocity under normal conditions is about 1 m/s at the entrance to the aorta, so Eq. (12.18) indicates that the megahertz transmit frequencies used for diagnostic ultrasound will produce Doppler frequencies of a few kilohertz at most, which is conveniently within the range of audible frequencies. Therefore, the simplest CW Doppler devices just direct the demodulated Doppler signal to a speaker, for the physician to interpret audibly.

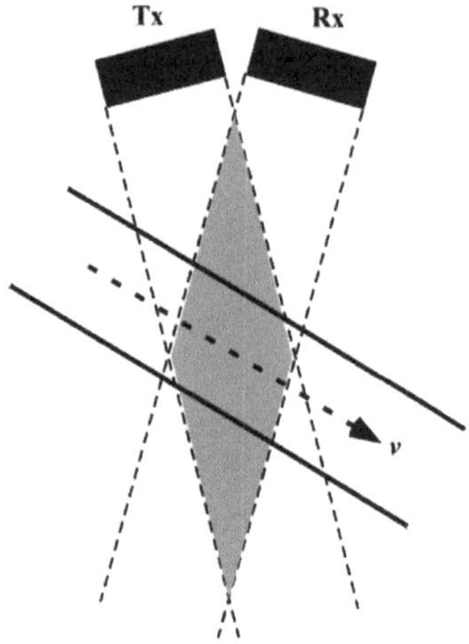

FIG. 12.5. CW Doppler scan geometry showing the orientation of the beams (dashed lines) produced by the transmit (Tx) and receive (Rx) apertures. The measured Doppler spectrum will include contributions from all moving scatterers within the area of intersection of the two beams.

In addition to the audio output, Doppler systems often display the time–frequency spectrum of the demodulated signal (Fig. 12.6). The x axis of this display is time, the y axis is temporal frequency (which can be rescaled to velocity using the Doppler equation) and the pixel greyscale represents the magnitude of the short time FT of the Doppler signal. A pixel in a Doppler spectrum therefore represents the proportion of red blood cells in the field of view that was moving at a particular velocity at a particular time. The spectral display is an effective means of presenting the pulsatile characteristics of intracardiac and vascular flow.

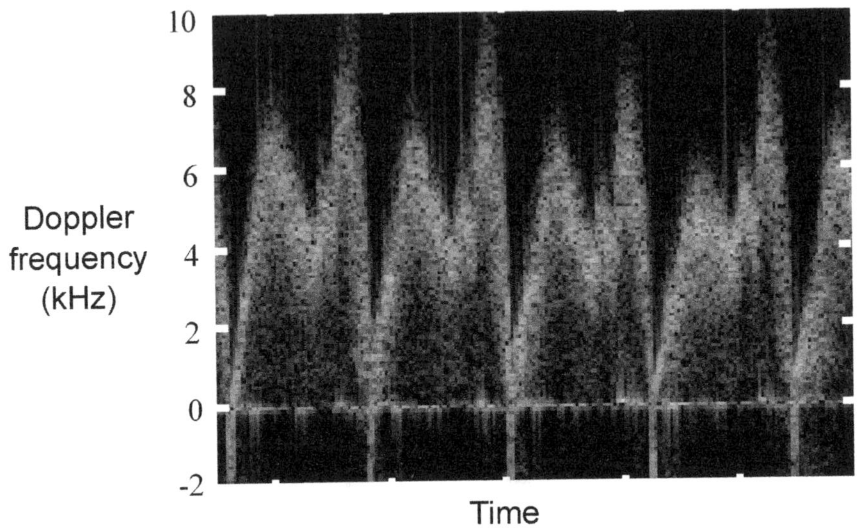

FIG. 12.6. Representative pulsed Doppler spectral display.

12.5.3. Pulsed wave Doppler

The primary shortcoming of CW Doppler is the lack of spatial resolution resulting from the large area of overlap between the transmitter and receiver beams. Pulsed wave Doppler addresses this limitation by transmitting a sequence of short pulses similar to those used for imaging, rather than a continuous sine tone. The user determines the location from which Doppler data will be acquired by positioning a range gate cursor within a B-mode image. As the echo from each successive transmission is received, a single sample at the expected arrival time of echoes from the range gate (the dotted line in Fig. 12.7(a)) is acquired and held until the echo from the next pulse is received. If the scatterers are moving, the signal received from the range gate will change with each subsequent pulse and the sample-and-hold operation will construct a staircase signal as shown in Fig. 12.7(b). The output of the sample-and-hold operation is low pass filtered to obtain a smoothly varying Doppler signal. It can be shown that the frequency of the smoothed signal is equal to the Doppler frequency.

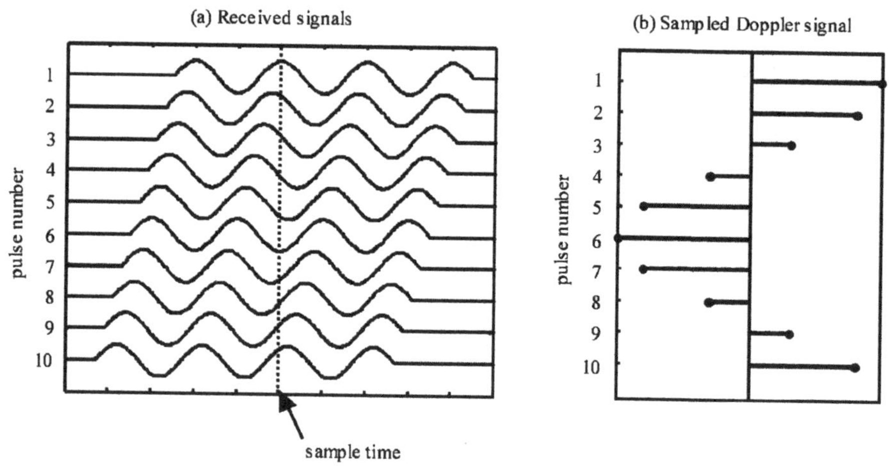

FIG. 12.7. Pulsed wave Doppler sample-and-hold operation. (a) Ten consecutive echo signals received from a scatterer moving towards the transducer. (b) Doppler signal obtained by sampling the 10 echo signals at the sample time indicated by the vertical dotted line in (a).

Since the sample-and-hold step is a sampling operation, it limits the maximum Doppler frequency that can be measured without aliasing. The pulse repetition frequency (PRF) of the transmitted pulses is the effective sampling frequency, so using Shannon's sampling theorem, the maximum unaliased frequency of the smoothed Doppler signal is f_{max} = PRF/2. Substituting f_{max} for f_D in the Doppler equation yields an expression for the maximum velocity, v_{max}, that can be measured by a pulsed wave Doppler system:

$$|v_{max}| = \left|\frac{c(\text{PRF})}{4f_o \cos\theta_D}\right| \tag{12.19}$$

The PRF is, in turn, limited by the depth of the range gate because the second pulse should be transmitted no sooner than the expected arrival time of the echoes from the range gate that arise from the first pulse. If the range gate is positioned at a depth z, the maximum PRF is $c/(2z)$. Substituting this result for PRF in Eq. (12.19) yields:

$$|v_{max}| = \left|\frac{c^2}{8zf_o \cos\theta_D}\right| \tag{12.20}$$

Most scanners offer a high velocity Doppler mode that uses a higher PRF than the range gate depth would ordinarily allow, to increase $|v_{max}|$ according to Eq. (12.19). In high velocity mode, the second (and perhaps the third) pulse is transmitted before echoes produced by the first pulse are received from the range gate, such that the Doppler signal consists of a superposition of echoes from within the range gate owing to the first pulse, and echoes from shallower depths owing to the subsequent pulses.

12.6. BIOLOGICAL EFFECTS OF ULTRASOUND

12.6.1. Bioeffects mechanisms

Ultrasound is generally assumed to be the safest medical imaging modality, but when a high intensity ultrasound pulse is transmitted through tissue, a substantial amount of energy can be transferred from the pulse to the tissue, thereby increasing the risk of adverse effects to the patient. These biological effects can be used beneficially by therapeutic ultrasound devices but are undesirable during diagnostic imaging. The two most important mechanisms for biological effects of ultrasound are thermal absorption and cavitation.

Tissue heating is caused by absorption, the primary mechanism of attenuation. The local temperature rise produced by a single pulse at the intensities used for diagnostic imaging is small. In B-mode imaging, where the beam is continuously steered through the tissue, blood flow typically dissipates the heat deposited by one pulse before the same volume of tissue is insonified again, but in techniques such as pulsed Doppler, where several pulses are transmitted to the same focal point in close succession, local heating can occur at the focus. In therapeutic ultrasound, thermal absorption is exploited for hyperthermia treatment of cancerous tumours by transmitting high intensity pulses that produce more rapid heating than the pulses used for diagnostic imaging.

Cavitation refers to an oscillation in the volume of a gas bubble in response to pressure fluctuations produced by an incident ultrasound wave. Cavitation is most likely to occur in vivo when microbubble contrast agents are employed, or if the lungs are exposed to ultrasound, but most tissues contain small volumes of gas that can coalesce to form cavitation nuclei when exposed to ultrasound. Low intensity ultrasound typically produces harmless stable cavitation, in which gas bubbles are not disrupted. However, higher intensity ultrasound can produce inertial cavitation, in which the rarefactional phase of the pressure wave expands the bubble to greater than its maximum stable volume, resulting in a sudden collapse of the bubble. The sudden collapse produces local heating of the order of 1000–10 000°C.

12.6.2. Acoustic output metrics

Ultrasound exposure has traditionally been quantified by measuring the spatial peak, temporal average intensity, I_{SPTA}, which is the transmitted signal measured at the point with greatest intensity within the radiated field (usually the focus of the transducer) and averaged over a period equal to several pulse repetition intervals. Similarly defined metrics, including the spatial average, temporal average intensity and the spatial average, pulse average intensity, are also commonly reported. The temporal averaging step in the determination of I_{SPTA} results in a greater measured exposure for modalities such as pulsed Doppler, where the same focal point is insonified repeatedly in rapid succession. This characteristic of I_{SPTA} reflects the fact that repeated insonation increases the risk of thermal bioeffects because heat may accumulate more quickly than it can be dissipated by blood flow.

Significant effort has been invested over the past 20 years to develop additional exposure parameters that more accurately reflect the risks of producing thermal and mechanical bioeffects. The definitions of the thermal index (TI) and the mechanical index (MI) were the original result of this effort. Most scanners manufactured since 1992 display real time estimates of the TI and MI.

The TI is the ratio of the acoustic power output of the scanner to the estimated acoustic power needed to raise the temperature of the tissue being imaged by 1°C. Different tissue thermal models, and hence different calculations of the TI, are used for soft tissue, skeletal bone and cranial bone. The tissue thermal model also accounts for the pulse repetition frequency, such that a higher TI will be computed for scanning modes such as pulsed Doppler.

The MI is interpreted as a measure of the relative risk of inducing cavitation and is based on an empirically derived formula:

$$\mathrm{MI} = \frac{\max(p_-)}{\sqrt{f}} \tag{12.21}$$

where $\max(p_-)$ is the peak rarefactional pressure after correction for attenuation. The use of rarefactional pressure in the numerator of the formula reflects the fact that inertial cavitation is triggered by overexpansion of a gas bubble, and the denominator reflects the experimental observation that inertial cavitation is more likely at lower frequencies.

Diagnostic ultrasound scanners manufactured since 1992 and sold in the USA are limited to a maximum I_{SPTA} of 720 mW/cm^2 and a maximum MI of 1.9 [12.3]. The US I_{SPTA} and MI limits are the de facto output limits for most of the world, owing to the relatively large size of the US medical device market.

12.6.3. Patient safety considerations

There is no specific upper limit on TI, but ultrasound operators are encouraged to apply the 'as low as reasonably achievable' principle to the TI. A TI below 2 is generally considered safe exposure for adults. Short exposures at higher TI can also be used safely; a rule of thumb for scanning adults at a TI above 2 is to limit the exposure time, t_e, according to:

$$\mathrm{TI} \leq 6 - \frac{\lg(t_e)}{0.6} \qquad (12.22)$$

where t_e is measured in minutes.

Fetal imaging merits additional caution, especially for scanning at high TI. In the USA, fetal exposure times at TI = 2–6 are restricted to much shorter durations than are suggested by Eq. (12.22), e.g. fetal exposure at TI = 4 is limited to 4 min maximum.

The presence of a significant volume of gas bubbles, as would occur when imaging structures near the lungs or in examinations using microbubble contrast agents, is another circumstance in which additional vigilance is recommended because the presence of excess gas increases the risk of inertial cavitation.

Contrast agents should be avoided in fetal imaging and used with caution in echocardiography of patients with pulmonary hypertension or other unstable cardiopulmonary conditions. When contrast agents are used, inertial cavitation can generally be avoided by maintaining an MI < 0.3, but low MI scanning is not always feasible because some contrast enhanced imaging protocols obtain diagnostic information by intentionally disrupting the microbubbles at MI > 1.

REFERENCES

[12.1] ZAGZEBSKI, J.A., Essentials of Ultrasound Physics, Mosby, St. Louis, MO (1996).

[12.2] SHUNG, K.K., Diagnostic Ultrasound: Imaging and Blood Flow Measurements, CRC Press, Boca Raton, FL (2006).

[12.3] NELSON, T.R., FOWLKES, J.B., ABRAMOWICZ, J.S., CHURCH, C.C., Ultrasound biosafety considerations for the practicing sonographer and sonologist, J. Ultrasound Med. **28** (2009) 139–150.

BIBLIOGRAPHY

BIOEFFECTS COMMITTEE OF THE AIUM, American Institute of Ultrasound in Medicine consensus report on potential bioeffects of diagnostic ultrasound: Executive summary, J. Ultrasound Med. **27** (2008) 503–515.

BLACKSTOCK, D.T., Fundamentals of Physical Acoustics, Wiley, New York (2000).

COBBOLD, R.S.C., Foundations of Biomedical Ultrasound, Oxford University Press, New York (2007).

EVANS, D.H., McDICKEN, W.N., Doppler Ultrasound: Physics, Instrumentation and Signal Processing, Wiley, New York (2000).

HILL, C.R., BAMBER, J.C., TER HAAR, G. (Eds), Physical Principles of Medical Ultrasonics, 2nd edn, Wiley, Chichester (2004).

JENSEN, J.A., Estimation of Blood Velocities Using Ultrasound: A Signal Processing Approach, Cambridge University Press, Cambridge (1996).

KINSLER, L.E., FREY, A.R., COPPENS, A.B., SANDERS, J.V., Fundamentals of Acoustics, 4th edn, Wiley, New York (2000).

KREMKAU, F.W., Diagnostic Ultrasound: Principles and Instruments, 7th edn, Saunders/Elsevier, St Louis, MO (2006).

PIERCE, A.D., Acoustics: An Introduction to its Physical Principles and Applications, Acoustical Society of America, Woodbury, NY (1989).

SZABO, T.L., Diagnostic Ultrasound Imaging: Inside Out, Elsevier Science, Boston (2004).

Chapter 13

ULTRASOUND IMAGING

J.C. LACEFIELD
University of Western Ontario,
London, Canada

13.1. INTRODUCTION

In the conventional method of ultrasonography, images are acquired in reflection, or pulse echo, mode. An array of small piezoelectric elements transmits a focused pulse along a specified line of sight known as a scan line. Echoes returning from the tissue are received by the same array, focused via the delay-and-sum beam forming process reviewed in Section 13.2, and demodulated to obtain the magnitude, or envelope, of the echo signal. The scanner measures the arrival time of the echoes relative to the time the pulse was transmitted and maps the arrival time to the distance from the array, using an assumed speed of sound. The earliest ultrasound systems would display the result of a single pulse acquisition in 1-D A-mode (amplitude mode) format by plotting echo magnitude as a function of distance. A 2-D or 3-D B-mode (brightness mode) image is acquired by performing a large number of pulse echo acquisitions, incrementally increasing the scan line direction between each pulse echo operation, to sweep out a 2-D or 3-D field of view (FOV). The term B-mode imaging reflects the fact that the echo magnitude from each point in the FOV is mapped to the grey level, or brightness, of the corresponding pixel in the image.

13.2. ARRAY SYSTEM PRINCIPLES

13.2.1. Electronic focusing and beam steering

The array transducers employed by modern ultrasound systems enable the use of high speed electronic focusing and beam steering methods, which are the basis of the high frame rates achieved by ultrasound. Beam steering incrementally increases the direction of the scan line to sweep out the B-mode FOV. The details of the beam steering process differ slightly, depending upon which of the three major types of array is being used. The types of array most commonly used for 2-D imaging are reviewed in this section.

A linear array consists of as many as 256 elements aligned in a single row. With a linear array, each pulse echo operation is performed by selecting a small subaperture of an adjacent element, as shown in Fig. 13.1(a). The scan line is always directed perpendicular to the array (i.e. along the axial dimension of the image) from the centre of the active subaperture. The scan line is stepped across the face of the array between each pulse echo acquisition by deactivating an element at one end of the subaperture and activating a new element at the opposite end of the subaperture. This procedure yields a rectangular FOV whose lateral width is comparable to the length of the array.

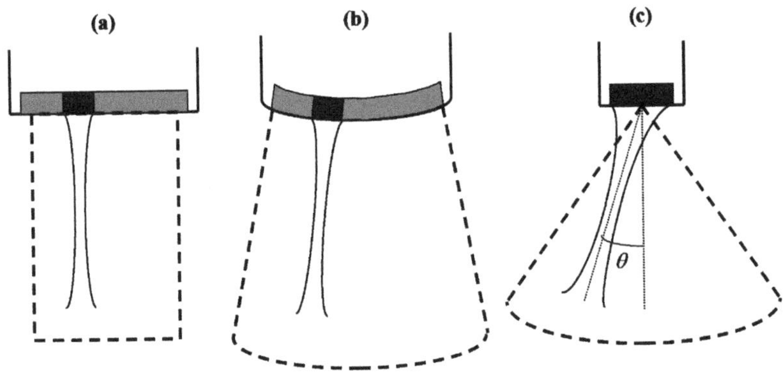

FIG. 13.1. Ultrasound array and B-mode image geometries. (a) Linear array; (b) curvilinear array; (c) phased array with azimuth angle, θ, indicated. In each figure, the dark grey shading represents the active subaperture during acquisition of the scan line indicated by the beam contours. The light grey shaded regions in (a) and (b) represent the inactive portions of the array.

A curvilinear array (Fig. 13.1(b)) functions in a manner analogous to a linear array, except that the face of the array is convex rather than planar. Therefore, as the subaperture is stepped across the array, the scan line both translates laterally and rotates azimuthally. This implementation produces a circular sector image with a wide FOV at all depths.

The elements of a phased array (Fig. 13.1(c)) are also coplanar, but, as discussed below, they are smaller, less numerous and more closely spaced than the elements of a linear array. Each scan line originates from the centre of the array and is acquired using all of the elements. The azimuth angle of the scan line (θ in Fig. 13.1(c)) is incrementally increased by altering the relative timing of the pulses transmitted by each element as described below. This procedure yields a circular sector FOV that spans as much as 90° in azimuth angle and converges to a narrow FOV at shallow depths, near the top of an image.

All array types are laterally focused, and the beam of a phased array is also steered, using delay-and-sum beam formation. The objective of beam formation during transmission is to ensure that the pulses transmitted from each element arrive simultaneously, and therefore constructively interfere, at the focal point. The time at which each element is fired is computed relative to the time the centre of the aperture is fired, by computing the distance from each element to the focal point using elementary trigonometry and assuming the sound speed, c, everywhere in the FOV is equal to 1540 m/s, the average sound speed of soft tissue. During reception, similar delays are applied to align temporally the echo signals received by each array element from the intended focal point. The delayed signals are then summed, such that the echoes received from the focus add constructively, to yield a beam formed receive signal.

13.2.2. Array beam characteristics

The lateral beam patterns produced by a linear or phased array can be modelled by applying linear systems analysis to the Fourier transform (FT) relationship between aperture and beam pattern that arises from the Fraunhofer diffraction integral (see Section 12.4.3). A single element can be viewed as a narrow rectangular aperture of width d, represented mathematically by rect(ξ/d), where ξ is the lateral dimension within the aperture plane and the orthogonal dimension (η in Eqs (12.13, 12.14)) is ignored in the following 1-D treatment. Convolution of that narrow rect function by a sampling, or comb, function, comb(ξ/s), models an infinite array with element spacing s. Multiplication by a broad rect function, rect(ξ/L), produces a linear systems description of a 1-D array extending from $\xi = -L/2$ to $\xi = L/2$:

$$A(\xi) = \left[\text{rect}\left(\frac{\xi}{d}\right) * \text{comb}\left(\frac{\xi}{s}\right)\right]\text{rect}\left(\frac{\xi}{L}\right) \tag{13.1}$$

where $A(\xi)$ is known as the aperture function.

Applying the Fraunhofer diffraction integral (Eq. (12.13)), the lateral beam profile in the focal plane, $U(x)$, is proportional to the FT of $A(\xi)$:

$$U(x) \propto \left[L\,\text{sinc}\left(\frac{Lx}{\lambda F}\right) \times s\,\text{comb}\left(\frac{sx}{\lambda F}\right)\right] d\,\text{sinc}\left(\frac{dx}{\lambda F}\right) \tag{13.2}$$

where sinc(a) = sin(πa)/(πa) and the constant terms in Eq. (12.14) are omitted in the interest of compact notation. The same aperture function is usually employed for transmission and reception, in which case the lateral point spread function (PSF) of the ultrasound system, PSF(x), is equal to the square of $U(x)$:

$$\mathrm{PSF}(x) \propto \left\{ \left[L\,\mathrm{sinc}\left(\frac{Lx}{\lambda F}\right) \times s\,\mathrm{comb}\left(\frac{sx}{\lambda F}\right) \right] d\,\mathrm{sinc}\left(\frac{dx}{\lambda F}\right) \right\}^2 \tag{13.3}$$

The normalized PSF, |PSF(x)|/|PSF(0)|, described by Eq. (13.3) is plotted using a decibel scale in Fig. 13.2(a). The primary beam in Fig. 13.2(a) arises from the sinc($Lx/\lambda F$) term in Eq. (13.3). The largest central peak of the primary beam is called the main lobe of the PSF. Under the Rayleigh resolution criterion, the lateral resolution (LR) of the imaging system is equal to the distance from the maximum to the first zero of the main lobe. Setting sinc($Lx/\lambda F$) = 0 and solving for x reveals that:

$$\mathrm{LR} = \frac{\lambda F}{L} \tag{13.4}$$

which is the same result as obtained for a single element rectangular aperture in Section 12.4.3.

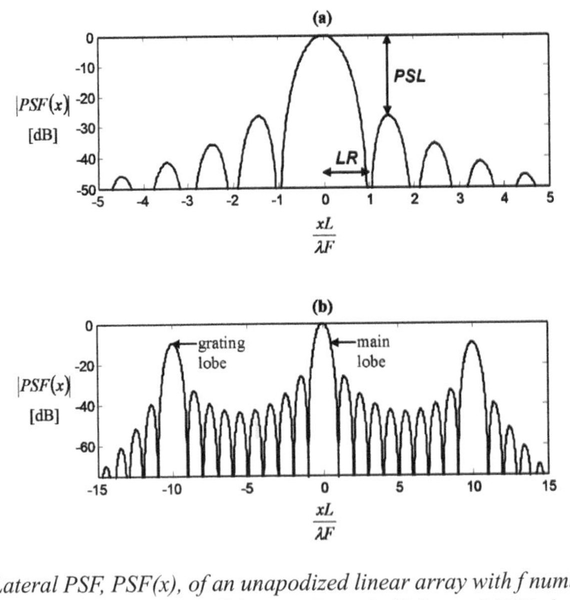

FIG. 13.2. (a) Lateral PSF, PSF(x), of an unapodized linear array with f number, F/L, equal to one. The LR and peak side lobe level (PSL) are indicated. (b) Lateral PSF plotted over a larger range of lateral position, x, to show grating lobes. The element spacing is equal to 2λ in this example.

The peaks flanking the main lobe in the primary beam are called side lobes. The PSL can be expressed in decibels as:

$$\mathrm{PSL} = 20\,\mathrm{lg}\left[\frac{\max\left(\left|\mathrm{PSF}(x)\right|\right)\ \text{outside main lobe}}{\max\left(\left|\mathrm{PSF}(x)\right|\right)\ \text{within main lobe}}\right] \tag{13.5}$$

This may be considered an indication of the contrast resolution of the imaging system, e.g. the image contrast expected for a small anechoic target. For the PSF in Eq. (13.3), the largest side lobe arises from the first off axis peak of the primary beam, which yields PSL = –26.6 dB.

Side lobe levels can be reduced by apodization of the aperture, in which the magnitudes of the transmit and receive signals on the outer elements are weighted during delay-and-sum beam formation, using a function that decreases more gradually than rect(ξ/L). If, for example, a Hamming window is applied for apodization during both transmit and receive beam forming, the resulting primary beam has the form shown in Fig. 13.3. Observe that, with Hamming apodization, the PSL is reduced by about 56 dB compared with the $\mathrm{sinc}^2(Lx/\lambda F)$ primary beam obtained in Eq. (13.3), but the lateral resolution is about twice as coarse as the resolution of the uniformly weighted aperture. Array based ultrasound systems almost always employ some form of apodization because contrast resolution is a crucial design consideration for diagnostic tasks such as lesion detection.

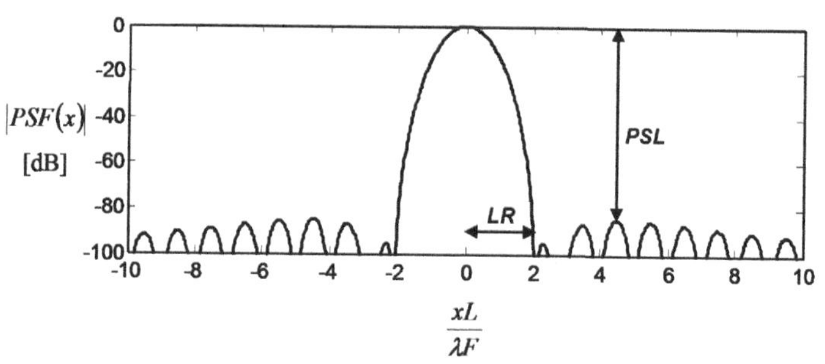

FIG. 13.3. Lateral PSF, PSF(x), of a linear array with f number equal to one and Hamming window apodization. Compare the lateral resolution and PSL of this beam with the lateral resolution and PSL of the unapodized beam in Fig. 13.2.

In Eq. (13.3), the convolution of the sinc($Lx/\lambda F$) term with the sampling function produces copies of the primary beam called grating lobes, spaced by $\lambda F/s$ along the lateral dimension of the focal plane (see Fig. 13.2(b)). Under the paraxial approximation, the azimuth angle, θ_{m}, of the mth grating lobe away from the main lobe peak is:

$$\theta_m \approx \sin^{-1}\left[\frac{m\lambda}{s}\right] \quad (13.6)$$

Grating lobes produce echoes from off axis targets that appear as artefacts in images (see Section 13.6).

Grating lobes are mathematically analogous to the copies of spectral peaks that are observed in the discrete time spectrum of an undersampled signal. In an array system, the aperture function is undersampled if the element spacing, s, is too large. For a linear array, where the beam will be focused but not steered, grating lobes are avoided if the azimuth angle of the first order grating lobe is greater than $\pi/2$ radians. Setting the right hand side of Eq. (13.6) greater than or equal to $\pi/2$ with $m = 1$ and solving for s, yields the linear array design rule for element spacing, $s \leq \lambda$.

For a phased array, which does employ beam steering, the most challenging scenario for grating lobe suppression arises when the beam is steered as far as possible in azimuth angle, i.e. parallel to the face of the array. The corresponding phased array design rule (stated without proof for brevity) is $s \leq \lambda/2$.

The elevation, or out of plane, dimension of the beam is analysed in a similar fashion to the lateral dimension. However, conventional linear and phased arrays possess only one row of elements and therefore electronic focusing and beam steering techniques, including apodization, cannot be applied in the elevation dimension. Elevational focusing is performed by an acoustic lens positioned at the face of the array. Since the height of an element is substantially smaller than the lateral aperture length, the spatial resolution of the imaging system is relatively coarse (of the order of several millimetres) in the elevation dimension. In view of the fixed focal distance of the lens, one benefit of the weaker focusing in elevation is that the depth of field, i.e. the interval over which the beam remains in focus, is greater in the elevation dimension than the lateral dimension. Depth of field is proportional to the wavelength multiplied by the square of the f number.

Electronic focusing in the elevation dimension is possible using multirow linear arrays, a technology that emerged in the late 1990s. The elements of these arrays are divided into a small number of rows (typically 5–9) in the elevation dimension, which permits electronic focusing, but not beam steering, to be performed in the elevation dimension. Compared with a 1-D array with an elevation lens, multirow arrays enable the depth of the elevational focus to be changed, which is a valuable capability for the multifocal imaging methods described in the following section. However, use of multirow arrays also increases the system complexity because arrays with smaller elements are more difficult to manufacture and because the scanner must include additional channels of transmitter and receiver electronics.

13.2.3. Multifocal imaging methods

The control panel of a modern clinical scanner provides the user with some flexibility for selecting the number and depths of transmit foci. Imaging with multiple transmit focal zones is achieved by acquiring each scan line repeatedly, with the transmit beam focused at a different depth along the scan line. A composite scan line is then constructed, consisting, at each point along the line, of the pulse echo sample acquired using the nearest transmit focus. If a multirow array is used, the elevation, as well as the lateral focus position, can be changed for each pulse echo acquisition. Imaging with multiple transmit zones produces an image with more consistent resolution throughout the FOV, but this improvement comes at the expense of a reduced frame rate.

Regardless of whether multiple transmit focal zones are used, all modern scanners also employ dynamic receive focusing. Since the arrival time of echoes at the receive aperture corresponds to the depth of the scatterers that produced the echoes, the receive focusing delays are updated in real time such that the receive focus tracks the pulse as it propagates along each scan line. The receiver delays can be updated essentially continuously as a scan line is acquired, and the lateral width of the receive beam (and its elevation width if a multirow array is used) should be minimized everywhere in the FOV. Use of dynamic receive focusing has no effect on the frame rate, since the receive focus is produced computationally after a pulse echo signal has been acquired.

The frame rate of an ultrasound system is determined by the lateral resolution, the FOV and the number of transmit focal zones. The total time to acquire one scan line can be approximated by the round trip pulse travel time to and from the maximum depth in the image, multiplied by the number of transmit focal zones. The lateral spacing of adjacent scan lines should be no greater than one half of the lateral resolution at the focus, to ensure adequate spatial sampling. The number of scan lines in one frame therefore depends on the lateral FOV and the lateral resolution. The time to acquire one frame is the product of the number of scan lines and the total time to acquire one line. In general, imaging over a larger FOV, imaging with higher spatial resolution and imaging with a greater number of transmit focal zones will each reduce the frame rate.

13.3. B-MODE INSTRUMENTATION AND SIGNAL PROCESSING

Figure 13.4 shows a generic block diagram of an ultrasound imaging system. High voltage excitation pulses are applied to the array by the transmitter electronics to fire a pulse. The transmit/receive switch then disconnects the array from the transmitter electronics and connects it to the receiver electronics, thereby isolating the receiver electronics from the high voltage excitation pulses. The returning echo signals are immediately amplified, digitized and then combined via delay-and-sum beam forming to produce a beam formed radiofrequency (RF) signal. Time gain compensation is applied to the beam formed RF signal to compensate for attenuation of echoes arriving from deeper targets. Envelope detection is performed to obtain the magnitude signal, which is then logarithmically compressed to maximize the dynamic range of the image. Finally, scan conversion is performed to map each sample of the magnitude signal to its 2-D or 3-D position in the image and map the log scaled magnitude values to grey levels. Once the scan lines covering the entire FOV have been acquired, the resulting image frame is displayed on the scanner's video monitor.

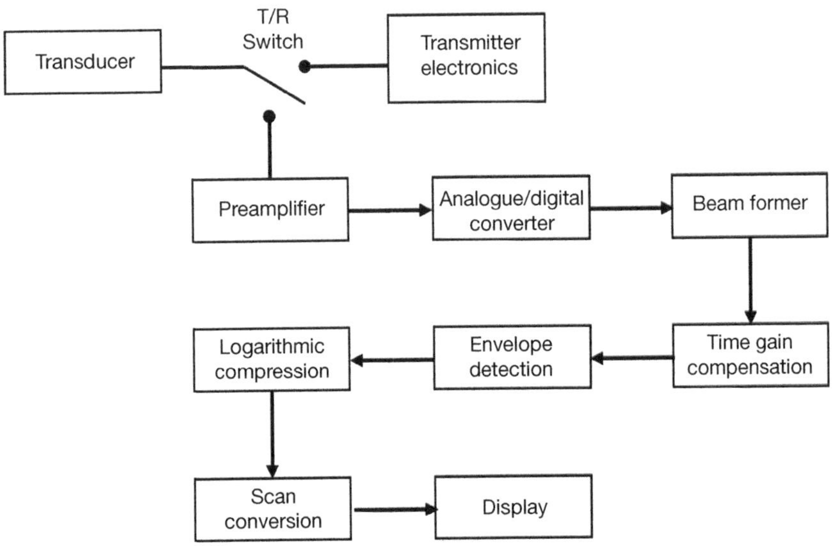

FIG. 13.4. Block diagram of a B-mode ultrasound imaging system.

The design requirements for the transmitter electronics include high bandwidth to enable use of short excitation pulses to obtain high axial resolution and an ability to operate at high power to drive the cable connecting the scanner and transducer with 50–100 V excitation signals. B-mode imaging is conventionally performed using a simple excitation waveform such as a square pulse or a one cycle sinusoid, in which case the spectral characteristics of the transmitted pulse are determined primarily by the frequency response of the transducer. However, some modern imaging methods employ coded transmit pulses (see Section 13.4.3), which require more sophisticated transmit electronics to support the use of programmable excitation waveforms.

Receive signal processing is primarily implemented on modern scanners using digital electronics, so digital techniques are emphasized here. Digital processing requires sampling at sufficiently high frequency (e.g. 25–50 MHz depending on the pulse spectrum) to sample properly the highest frequency components of the RF signal, and quantization using at least 12 bits, and preferably 16 bits, to represent the RF waveform accurately. Analogue to digital conversion is preceded by an anti-aliasing low pass filter. The cut-off frequency of the anti-aliasing filter is often a significant determinant of the ultimate axial resolution of the image.

Beam forming delays must be applied with high precision to obtain a sharp focus. The typical procedure is to interpolate the digitized RF signals to a sampling frequency as high as 1 GHz, which would enable focusing delays to be applied with 1 ns precision, and then downsample the time shifted signals to the original sampling frequency for subsequent processing. Relatively simple interpolation methods must be used to enable this step to be performed in real time for dynamic receive focusing.

Three types of amplification are then applied to the beam formed RF signal. Firstly, a constant gain (often simply labelled 'Gain' on the scanner's control panel) is applied to enable the user to adjust the overall brightness of the image. Secondly, a time dependent amplification called time gain compensation (TGC), is applied to each scan line to offset the effects of attenuation. The TGC should be an approximately exponential function of the echo arrival time (hence the T in TGC), since attenuation causes the signal intensity to decay as an exponential function of propagation distance (Eq. (12.11)). The user sets the TGC slope separately in several depth bands covering the FOV, by adjusting the TGC to obtain visually consistent brightness throughout the image. A third amplification, called lateral gain compensation, is applied to compensate for shadowing or enhancement artefacts (see Section 13.6) that cause image brightness at a given depth to vary as a function of lateral position. The lateral gain is also adjusted manually by the user, in a manner similar to the TGC.

Prior to widespread use of digital processing, envelope detection was performed via analogue amplitude demodulation, e.g. full wave rectification followed by low pass filtering of the beam formed RF signal. Modern scanners typically compute the magnitude signal using the Hilbert transform, which applies an exact $\pi/2$ phase shift to the RF signal, thereby estimating its quadrature component. The instantaneous magnitude at each time sample is then obtained by adding the original and Hilbert transformed signals in quadrature.

The magnitude signal is logarithmically compressed because the dynamic range of the pulse echo data can be greater than 80 dB, which is much greater than the 48 dB (= 20 lg(256)) dynamic range of a standard 256 grey level display. The scanner's electronics can map the data to greater than 256 grey levels, but there would be little benefit to this approach because the human visual system has limited sensitivity to more subtle differences in brightness. An effective algorithm for logarithmic compression is to use the mean of the magnitude signal (averaged over the FOV) as the reference magnitude and convert each of the linearly scaled magnitude samples to decibels with respect the reference value. Then, the decibel values are linearly converted to grey levels, such that a magnitude of $-X/2$ dB maps to grey level 0 (black), 0 dB (i.e. the mean magnitude) maps to grey level 128, and $X/2$ dB maps to grey level 255 (white), where X is the displayed dynamic range and would typically be equal to 60 or 80 dB.

For the rectangular images produced by a linear array system, the spatial sampling of the original echo data is much coarser in the lateral dimension compared with the axial dimension, owing to the anisotropic spatial resolution of an ultrasound scanner (see Section 12.4). If the scan lines are displayed directly, the rectangular aspect ratio of the pixels produces visible banding artefacts in the images. Therefore, the images are laterally interpolated before they are displayed, to obtain a uniform pixel width in all dimensions. A relatively simple 1-D interpolation such as a cubic spline yields a substantial improvement in the appearance of the image.

For the sector images produced by a phased array system, scan lines acquired in a polar format must be displayed on a rectangular pixel grid. Adjacent scan lines will pass through the same pixel, near the origin of the sector, and may be separated by greater than the pixel width near the base of the sector. A single grey level can be assigned to pixels containing overlapping scan lines, by averaging the magnitude signals or by using the maximum of the overlapping magnitude samples. Gaps between scan lines at deeper ranges are typically filled by 2-D interpolation within the image plane. The interpolation process also compensates for the fact that the locations of most of the samples in the original echo data do not correspond exactly to the centre of any pixel in the rectangular display grid.

13.4. MODERN IMAGING METHODS

13.4.1. Contrast enhanced imaging

Gas filled, encapsulated microbubbles with sizes typically ranging from 1 to 4 μm diameter are used as blood pool contrast agents in diagnostic ultrasound. The >3 orders of magnitude difference between the acoustic impedances of a gas and soft tissue (Table 12.1) enables a microbubble to scatter ultrasound strongly despite its small size. Their 1–4 μm diameter means microbubbles are smaller than red blood cells, so they are not trapped in capillary beds but are still too large to extravasate, and this conveniently centres their fundamental resonant frequency in the 2–4 MHz range.

Microbubbles are encapsulated by a shell to prevent the gas from dissolving into the blood. The original 'first generation' microbubble formulation consisted of air bubbles encapsulated in a relatively stiff shell material such as albumin. However, most contrast enhanced examinations are now performed with 'second generation' microbubbles consisting of an inert perfluorocarbon gas core, such as perfluoropropane (C_3F_8), encapsulated in a phospholipid shell. The insolubility of perfluorocarbon gas in blood and the use of a more deformable shell for second generation agents both help to increase the circulating half-life of the agents.

The key to forming contrast enhanced ultrasound images is to differentiate echoes produced by microbubbles from echoes produced by tissue, which is typically accomplished by exploiting the non-linear scattering characteristics of microbubbles. A microbubble's spherical symmetry yields strong scattering resonances at a fundamental frequency determined by the bubble's radius and shell properties as well as resonances at harmonics (integer multiples) and subharmonics (integer fractions) of the fundamental frequency. During the initial development stages of contrast enhanced ultrasound, it was assumed that echo signals containing harmonic spectral characteristics would uniquely identify echoes from microbubbles, but soft tissue was unexpectedly discovered to possess non-linear ultrasonic properties of its own. This discovery led to the development of tissue harmonic imaging (see Section 13.4.2) and also motivated further research into contrast agent detection techniques.

Most scanners perform contrast enhanced imaging using some variation on a multipulse imaging method such as pulse inversion. In pulse inversion imaging, each scan line is acquired twice in close succession, first by transmitting a standard pulse and then by transmitting an inverted pulse (e.g. the original pulse multiplied by −1) and adding the resultant echo signals. The tissue component of the second echo signal is, approximately, an inverted copy of the first echo signal and the echoes from tissue cancel when the two received signals are summed. The harmonic components of the microbubble echoes, on the other hand, are not inverted in the second echo signal, so the summation step cancels the fundamental frequency components of the microbubble echoes but maintains the harmonic components. Contrast enhanced images produced in this fashion reveal abnormalities in vascular function via regional differences in the timing and spatial patterns of contrast enhancement.

Destruction replenishment imaging is another important approach to contrast enhanced imaging. In destruction replenishment imaging, a sequence of high mechanical index (see Section 13.6) transmit pulses are used to fragment all microbubbles in the FOV, then contrast enhancement kinetics are measured as new microbubbles flow into the region of interest. Contrast replenishment can be imaged using a low mechanical index technique, such as pulse inversion, that does not destroy the agents. Measurements of replenishment kinetics are analysed using biophysical models to estimate parameters such as blood volume, transit time through the region of interest and perfusion. Destruction replenishment imaging is arguably the contrast enhanced ultrasound method with the greatest biomedical value because the functional parameters estimated with this technique are most similar to the perfusion parameters measured using modalities such as dynamic contrast enhanced computed tomography or magnetic resonance imaging.

13.4.2. Tissue harmonic imaging

As discussed in Section 12.3.3, harmonic spectral components are generated as an ultrasound pulse propagates through tissue as a result of the pressure wave modulating the sound speed of the tissue. Substantial harmonics are only produced at the transmit focus of a diagnostic imaging system, where the pulse intensity is greatest. If the transducer possesses sufficiently high bandwidth to detect echoes at the second harmonic of the transmit frequency, the received signal can be bandpass filtered to construct an image that depicts only the harmonic component of the echoes. This imaging method is typically called tissue harmonic imaging or native tissue harmonic imaging.

Tissue harmonic imaging partially offsets the trade-off between spatial resolution and penetration depth that is encountered in conventional B-mode imaging. The lower frequency transmit pulse experiences less severe frequency dependent attenuation than the returning second harmonic echoes, whereas the lateral width of the receiver focus is narrower than the transmit beam. The resulting image thus exhibits lateral resolution and penetration that are intermediate between the resolution and attenuation that would be observed in conventional imaging at the fundamental frequency and conventional imaging at twice that frequency. Tissue harmonic imaging has also been shown experimentally to reduce clutter artefacts compared with conventional B-mode imaging. These benefits are achieved at the expense of reduced axial resolution because the passband of the transducer's frequency response must be divided into fundamental and harmonic segments rather than using the entire bandwidth to generate the shortest possible transmit pulse.

13.4.3. Coded excitation imaging

Imaging using coded excitation pulses is a technique that was developed to increase the penetration depth of ultrasound imaging systems. A coded excitation system transmits a relatively long duration signal such as a chirp (a sinusoid with increasing or decreasing instantaneous frequency) or a pulse modulated code, in which a sinusoid is switched on and off in a specific temporal sequence to create a binary code. Both approaches spread the transmitted energy over the pulse duration, so the signal to noise ratio (SNR), and hence the penetration depth, can be increased without exceeding regulatory requirements for patient exposure (see Section 12.6). Matched filtering techniques are used during reception to deconvolve the transmitted code from the echo signal and maintain axial resolution close to that achieved by conventional B-mode imaging.

Coded excitation also enables high frame rate (e.g. several hundred frames per second) imaging by simultaneously transmitting multiple orthogonal pulse codes steered to different azimuth angles. The echoes produced by each of these transmit beams will interfere at the receive aperture, but matched filtering by the receiver discriminates among the echoes produced by each pulse code, thereby enabling multiple scan lines to be acquired simultaneously.

13.4.4. Three and four dimensional imaging

Advances in ultrasound technology have enabled conventional 2-D B-mode imaging to be supplemented, and in some applications supplanted, by 3-D (or volumetric) imaging and so-called 4-D imaging, in which a temporal sequence of 3-D images is presented as a cine loop. There are three primary methods of acquiring 3-D images that present differing combinations of advantages, disadvantages and technical complexity.

The 3-D images can be acquired by mechanically translating or rotating a linear or curvilinear array transducer with the motion directed out of the 2-D image plane. A sequence of 2-D images is acquired at regular linear or rotational increments and a 3-D image volume can readily be reconstructed from the set of 2-D images because the spatial relationship among the images is known in advance. This method of 3-D imaging is relatively uncomplicated to implement and, for stationary structures, provides the highest image quality since large aperture or higher frequency linear arrays can be used and the transducer motion is not operator dependent. One 3-D image volume can require several seconds to acquire by this method since the translation or rotation speed of the transducer should be slow compared with the 2-D frame rate. Therefore, this technique requires extensive respiratory and cardiac gating if it is to be used to image dynamic structures such as the beating heart.

Freehand scanning is a second method of 3-D imaging. In this approach, a conventional linear or curvilinear array is translated or rotated manually by the sonographer. In some implementations of this approach, transducer motion is measured by an external electromagnetic or optical tracking system that is synchronized with the 2-D image acquisition, and the transducer position measurements are used to position each 2-D image plane within the reconstructed 3-D volume. Alternatively, transducer motion can be estimated directly from the image sequence, using cross-correlation to estimate motion within the 2-D plane and using the rate of decorrelation of consecutive frames to estimate out-of-plane motion. The quality of the reconstructed image will also depend on the ability of the sonographer to sweep the transducer in a regular, smooth pattern so that the 3-D image volume is uniformly filled in by the constituent 2-D images.

Real time 3-D imaging using a 2-D phased array is the most technologically sophisticated method of 3-D imaging. A 2-D array consists of a matrix of small (ideally $<\lambda/2$ on each side) square elements to enable phased array beam steering in both the azimuth and elevation angles. The resulting image volume is a 3-D pyramidal sector that can be considered an extension of the circular sector produced by a conventional 1-D phased array. An alternative approach is to construct a 2-D array of slightly larger elements and acquire images in a manner comparable to a conventional linear array system. In this approach, each scan line is acquired using a 2-D subaperture of active elements that are focused straight ahead. The subaperture is stepped in both dimensions across the face of the array, to build up a 3-D rectilinear image volume.

13.5. COLOUR FLOW IMAGING

13.5.1. Flow imaging modes

Doppler based methods for blood flow imaging can be viewed as extensions of the pulsed wave Doppler method introduced in Section 12.5, in which Doppler processing is applied to a large number of sample volumes to produce 2-D or 3-D images of blood flow. There are two principal formats for displaying the resulting flow images. One format, commonly referred to as colour Doppler, uses a red and blue colour scale to represent the mean axial velocity in each sample volume, with flow towards the transducer shown in red, flow away from the transducer shown in blue and the velocity magnitude mapped to the colour intensity. The colour pixels are superimposed on a B-mode image, such that the echo magnitude from tissue volumes containing no detectable flow is displayed in greyscale. Colour Doppler is primarily used when imaging the heart and major blood vessels in applications for which mean flow velocity is a diagnostically useful parameter. Acquisition of the B-mode image must be interleaved with acquisition of the Doppler data, so the colour Doppler frame rate is always less than the frame rate of conventional B-mode imaging. The velocity estimates in a colour Doppler image are therefore prone to aliasing (see Section 12.5.3) because the Doppler pulse repetition frequency is also limited by the need to acquire

B-mode scan lines. The spectral wraparound produced by aliasing in pulsed wave Doppler appears as sudden changes in pixel colour, and thus the apparent direction of flow, in a colour Doppler system.

The second display format, commonly referred to as power Doppler, uses a red-to-orange-to-yellow colour scale to represent the total power in the Doppler spectrum at each sample volume, with the lowest powers displayed in red and the highest powers displayed in yellow. As in colour Doppler, the B-mode signal is displayed in greyscale for tissue volumes with no detectable flow. Doppler power is, in theory, proportional to the concentration of moving blood cells in the sample volume, so power Doppler is typically used for applications such as tumour imaging, where blood volume is a diagnostically useful parameter. Although power Doppler images contain no information about flow direction or velocity, the method provides several advantages over colour Doppler. Doppler power does not depend on the Doppler angle, so power Doppler provides a more continuous display of tortuous vessels. Power Doppler images are not susceptible to aliasing artefacts, since aliasing does not affect the total power in the Doppler spectrum. Power Doppler also performs better than colour Doppler for imaging small, slow flow vessels, since the integrated power is less affected by low SNR than is the mean Doppler frequency.

Colour Doppler and pulsed wave Doppler are sometimes combined in a mode known as triplex Doppler, where the B-mode information included in the colour Doppler image represents the third component of the triplex. In triplex Doppler, acquisition of a pulsed wave Doppler spectrum from a user selected sample volume is interleaved with acquisition of a colour Doppler image. The colour flow image and the Doppler spectrum are displayed side by side. This mode provides the 2-D spatial information of a colour flow image and, for the pulsed wave sample volume, the higher maximum velocity and velocity resolution of a conventional pulsed Doppler examination. Triplex Doppler can be considered an extension of duplex Doppler, a scanning mode that predates colour flow imaging in which a B-mode image and a pulsed Doppler spectrum are acquired and displayed simultaneously.

13.5.2. Tissue Doppler imaging

Conventional Doppler systems, including continuous wave Doppler, pulsed wave Doppler and colour flow imaging, discriminate blood flow from soft tissue motion based on velocity. Specifically, a Doppler system assumes blood flow is concentrated at intermediate and high velocities, while scatterers moving at low velocities correspond to soft tissue. This discrimination is achieved by applying a high pass wall filter, also known as a clutter filter, to eliminate low Doppler frequency components from the Doppler signal.

In applications that require measurements of soft tissue motion, it has proven useful to replace the Doppler wall filter with a low pass filter so that only low velocity motion will be displayed. This imaging mode is called tissue Doppler. Diagnosing regional abnormalities in ventricular wall motion, e.g. following a myocardial infarction or in heart failure patients, is a principal application of tissue Doppler. Tissue Doppler can be performed at a single sample volume in a manner analogous to pulsed wave Doppler, in which case the Doppler spectrogram attributed to tissue motion is displayed. Alternatively, tissue Doppler measurements can be performed over a 2-D region of interest in a manner analogous to colour Doppler, to produce an image of mean Doppler frequency or axial velocity.

13.6. IMAGE ARTEFACTS AND QUALITY ASSURANCE

13.6.1. B-mode image artefacts

Many common B-mode image artefacts can be understood in terms of the fundamental physical concepts introduced in Chapter 12 and also in this chapter. One such example is a reverberation artefact, which can occur when the transmitted pulse bounces back and forth between two strongly reflecting interfaces. Each time the reverberating pulse returns to the top interface, a portion of the acoustic intensity is transmitted through the interface and continues to the transducer. The pulses transmitted back to the transducer appear in the image as copies of the deeper boundary separated by a distance equal to the thickness of the object causing the reverberation. The image intensity of the reverberations decays as a function of depth, owing to the cumulative multiplication of the pulse intensity by the reflection and transmission coefficients at the object's boundaries, so the resulting artefact usually obscures only a limited region below the reverberating object. Reverberation artefacts are often produced by large calcifications or metallic foreign bodies. Sonographers describe the sequence of bright echoes below a reverberating object as a comet tail artefact and use the artefact as an aid to indentifying such hard inclusions.

Refraction artefacts are other B-mode artefacts produced by transmission through specular interfaces. As discussed in Section 12.2.3, refraction occurs when a pulse is obliquely incident on a boundary between two tissues with differing sound speed. The change in propagation direction of the pulse as it crosses the boundary deflects the pulse away from the intended direction of the scan line. If the refracted pulse subsequently encounters a strong reflector, the resulting echo can be refracted back towards the transducer at the overlying specular boundary. The scanner will display the refracted echo at a point along the intended direction of the scan line, so the reflecting object will be displayed at an incorrect lateral position in the image.

A very strongly reflective (i.e. R~1) specular interface can also produce a mirror image artefact. Consider a transmit pulse that is redirected by a specular reflector as described by Snell's law. Echoes from the redirected pulse are scattered back to the specular reflector, which reflects them back to the transducer. The scanner will display those echoes along the direction of the original scan line and hence behind the specular reflector, owing to their longer round-trip travel time. In practice, the most striking mirror image artefacts are produced by the diaphragm when imaging the liver with the scan plane oriented cranially. The redirected echoes will cause features to appear in the image above the diaphragm in the lungs, i.e. at locations from which no signal would be expected, owing to the difficulty of coupling ultrasound with an air filled organ.

Localized variations in the attenuation coefficient can produce shadowing and enhancement artefacts. If, for example, the transmitted pulse traverses a feature that attenuates more strongly than the surrounding tissue, the pulses incident on features located behind that feature will possess a lower intensity than expected. The tissue behind the strongly attenuating feature will thus appear darker than the laterally adjacent tissue in the B-mode image. Enhancement artefacts are produced by the opposite scenario: a weakly attenuating inclusion will cause the tissue behind it to appear brighter than expected in the image. Since the time gain compensation curve is the same along all scan lines and the lateral gain compensation is the same at all depths in the image, neither of these processing steps is effective for eliminating shadowing or enhancement artefacts. In cancer imaging applications, the relative attenuation in a lesion can be correlated with regard to whether the lesion is benign or malignant, so sonographers sometimes consider shadowing or enhancement artefacts to be diagnostically useful observations.

Localized variations in the speed of sound can cause reflectors to be displayed at incorrect depths in the image. For example, a region with elevated sound speed will cause echoes backscattered from behind that inclusion to arrive at the receiver sooner than echoes from the same depth along other scan lines, making scatterers behind the region of high sound speed appear closer to the transducer than their true position. The sound speeds of most soft tissues are clustered near 1540 m/s (see Table 12.1) and thus geometric distortion of the image due to sound speed variations is rarely perceptible. However, sound speed artefacts are a cause for concern when ultrasound is employed in image guided interventional procedures.

As stated in Section 13.2.2, side lobes reduce image contrast. Consider a small anechoic feature such as a fluid filled cyst. When acquiring scan lines passing through the cyst, the side lobes of the beam may extend outside the cyst into the surrounding tissue. Under this circumstance, a portion of the intensity contained in the side lobes can be backscattered towards the transducer. Echoes that are received as a result of scattering from the side lobes will be displayed along the beam axis of the scan line, i.e. within the cyst. The cyst thus appears to be a weakly scattering lesion rather than an anechoic lesion and therefore its image contrast is lower than it would be in the absence of the side lobe artefact. This reasoning is the basis for interpreting the PSL (Eq. (13.5)) as an approximate measure of the contrast resolution of an ultrasound system.

Grating lobes, if present, can produce B-mode artefacts in a manner analogous to a side lobe artefact. A grating lobe artefact will occur if the main lobe of the beam is propagating through tissue with moderate scattering strength while a grating lobe is incident upon a strongly reflective feature. The echo produced by the grating lobe will be displayed along the axis of the scan line. Since the grating lobe pattern is usually symmetrical about the main lobe axis, grating lobe artefacts tend to be visually symmetrical as well. The strong reflector will be displayed brightly in the image at its correct position, and less intense copies of the reflector will be displayed at equal lateral distances to the left and right of the reflector. The artefact depictions of the reflector occur on scan lines in which the grating lobes having $m = 1$ and $m = -1$ (Eq. (13.6)) are directed towards the reflector. However, most array systems are designed to prevent the formation of grating lobes, as discussed in Section 13.2.2, and therefore grating lobe artefacts are relatively uncommon.

13.6.2. Speckle

Speckle, which is the granular or mottled texture observed in B-mode images, is sometimes considered another type of artefact, but speckle is also a fundamental characteristic of a B-mode image and, therefore, merits special consideration. The formation of speckle can be understood using a random walk model that was originally developed in optics. At any instant in time, the ultrasound system receives echoes from multiple unresolved tissue structures located in a resolution volume defined by the 3-D PSF. In the narrowband limit, each of these echoes can be represented by a phasor with a distinct magnitude and phase. The instantaneous value of the received RF signal is equal to the coherent sum of those phasors. If all of the scatterers are similar structures, the magnitude of each phasor can be modelled as a Gaussian random variable; if the scatterers are positioned randomly throughout the resolution volume, the phase of the individual echoes can be modelled as a uniformly distributed random variable from $-\pi$ to π. In this case, the coherent summation can be visualized graphically as a random walk in the complex plane (Fig. 13.5(a)). If there are at least 10 scatterers in the resolution volume, the magnitude of the phasor sum, which corresponds to the envelope detected echo signal, follows the Rayleigh probability density function (Fig. 13.5(b)).

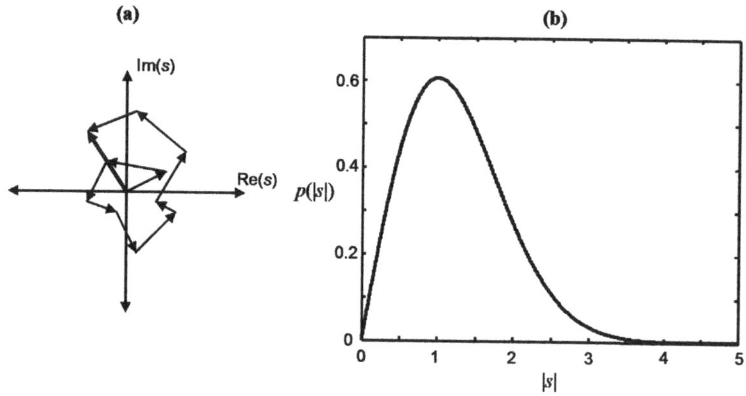

FIG. 13.5. (a) Random walk model for a speckle signal. The thin black phasors represent echoes from individual scatterers and the thick grey phasor is the coherent sum, s, of those echoes. (b) Rayleigh probability density function for the magnitude of a coherently summed speckle signal, $|s|$. *The Rayleigh scale parameter equals one in this example, so the mean value of* $|s|$ *is* $\sqrt{\pi/2} \approx 1.25$.

Speckle in a B-mode image is, thus, a random process that is sometimes alikened to noise. This comparison can be instructive under certain circumstances. For example, if all of the foregoing assumptions are satisfied, the ratio of the mean to the standard deviation of the envelope detected signal, which is termed the point SNR, is a constant ≈ 1.91 and the histogram of the signal provides little information about the tissue beyond its mean scattering strength. The size of the individual speckle grains, i.e. the spatial autocorrelation length of the speckle pattern, is entirely determined by the PSF and so also carries little information about the tissue. In applications such as lesion detection, the grey level fluctuations due to speckle can obscure low contrast lesions, so considerable effort has been devoted to developing speckle reduction methods. Spatial compounding, in which several images of a region of interest are acquired from different angles and averaged, is the most successful speckle reduction method and is now implemented under various trade names on most modern scanners.

However, the comparison of speckle to noise can also be misleading. Since speckle is the coherent sum of all of the echoes scattered from the interior of a tissue structure, the speckle is the signal in most of the image. Furthermore, if the transducer and the tissue are both stationary, the speckle pattern, in contrast to any electronic noise in the image, is constant. More importantly, all tissues possess some degree of spatial organization to be able to perform their biological functions, so the condition of randomly positioned scatterers is only strictly met in simulated images and tissue mimicking phantoms. There may also be fewer than ten scatterers per resolution volume, or two or more populations of scatterers in real tissue. Therefore, the first and second order statistics and spectral characteristics of echoes acquired from tissue do carry more information about the tissue than the random walk model suggests. This observation is the motivation behind ongoing efforts to develop quantitative tissue characterization methods that employ ultrasound imaging.

13.6.3. Quality assurance phantoms and methods

Tissue mimicking phantoms designed for use with ultrasound imaging systems consist principally of a material designed to reproduce the bulk acoustic properties of soft tissue. Simple laboratory phantoms can be made from suspensions of gelatin or agar, but commercially available phantoms typically employ proprietary polymer materials that possess much greater shelf life than gelatin based or agar based phantoms. The sound speed and attenuation coefficients of a phantom are usually carefully controlled to be 1540 m/s and either 0.5 or 0.75 dB/(cm·MHz), respectively. The phantoms also contain suspensions of small scatterers fabricated from materials such as graphite, polystyrene or collagen, whose purpose is to produce visually realistic speckle when the phantom is imaged.

In addition to the background material described above, a phantom will contain several imaging targets that enable verification of the spatial and/or contrast resolution of an ultrasound system. The spatial resolution targets are usually small diameter metal wires or monofilament fibres suspended horizontally through the phantom to produce bright point-like targets when imaged in cross-section. A set of wire targets is positioned at the same depth below the phantom's acoustic window in a regular pattern such that each pair of adjacent wires presents a progressively decreasing separation ranging from several millimetres down to 0.5 mm or less. When this pattern of targets is imaged, the lateral spatial resolution can be estimated by observing which pairs of targets are resolved in the image. Additional wire targets are oriented in a vertical pattern, to enable evaluation of the axial spatial resolution. A phantom may include several laterally and axially oriented groups of wires at different depths, to enable resolution to be assessed throughout the scanner's FOV.

Contrast resolution targets are usually spherical inclusions several millimetres to several centimetres in diameter that are similar to the background material but with a mean backscattering coefficient greater than or less than the background material. The targets are designed to produce a specific backscatter contrast relative to the background material. The contrast of the inclusions can be varied by changing the size or composition of the scattering particles, but it is usually easiest to manipulate contrast by changing the concentration of scattering particles relative to the background material, including a few targets with no scatterers to mimic anechoic lesions. Targets of differing depth will be distributed throughout the phantom to enable image contrast to be evaluated as a function of lesion size and depth.

Flow phantoms provide a useful means of evaluating the performance of spectral, colour and power Doppler systems. In a typical flow phantom design, the tissue mimicking material is moulded to include one or more hollow channels that mimic blood vessels of varying diameter and/or orientation. The channels are connected via tubing to a calibrated pump that is used to pump blood-mimicking fluid through the flow channels at controlled flow rates. Blood-mimicking fluid, as the name implies, is a suspension of small scatterers that is designed to reproduce the acoustic properties of blood.

Phantoms are also available for training sonographers for specific applications and for testing emerging ultrasound imaging methods. Examples of training phantoms include anthropomorphic breast or prostate phantoms designed for practising ultrasound guided biopsy or brachytherapy procedures. Elastography phantoms are an example of a phantom intended for testing emerging imaging methods. Elastography phantoms are similar in design to contrast phantoms, except the emulated lesions differ from the background material in elastic modulus as well as backscattering coefficient.

BIBLIOGRAPHY

EVANS, D.H., McDICKEN, W.N., Doppler Ultrasound: Physics, Instrumentation and Signal Processing, Wiley, New York (2000).

JENSEN, J.A., Estimation of Blood Velocities Using Ultrasound: A Signal Processing Approach, Cambridge University Press, Cambridge (1996).

KREMKAU, F.W., Diagnostic Ultrasound: Principles and Instruments, 7th edn, Saunders/ Elsevier, St. Louis, MO (2006).

QIN, S., CASKEY, C.F., FERRARA, K.W., Ultrasound contrast microbubbles in imaging and therapy: Physical principles and engineering, Phys. Med. Biol. **54** (2009) R27–R57.

SHUNG, K.K., Diagnostic Ultrasound: Imaging and Blood Flow Measurements, CRC Press, Boca Raton, FL (2006).

SZABO, T.L., Diagnostic Ultrasound Imaging: Inside Out, Elsevier Science, Boston (2004).

ZAGZEBSKI, J.A., Essentials of Ultrasound Physics, Mosby, St. Louis, MO (1996).

Chapter 14

PHYSICS OF MAGNETIC RESONANCE

HEE KWON SONG
Hospital of the University of Pennsylvania,
Philadelphia, United States of America

14.1. INTRODUCTION

The discovery of nuclear magnetic resonance (NMR), a property of nuclei in a magnetic field where they are able to absorb applied radiofrequency (RF) energy and subsequently release it at a specific frequency, goes back many decades to the early 1900s. Physicist Isidor I. Rabi, fascinated by the work of Otto Stern and Walther Gerlach which demonstrated that particles have intrinsic quantum properties, delved into the magnetic properties of nuclei, and in 1938 Rabi discovered the phenomenon of NMR. Several years later, in 1946, Felix Bloch and Edward Purcell refined the methods and successfully measured the NMR signal from liquids and solids. For their discoveries, Rabi received the Nobel Prize for physics in 1944 and Bloch and Purcell in 1952.

While Rabi, Bloch, Purcell and other physicists working in this field had laid the foundations, a major discovery that transformed the NMR phenomenon for imaging was not made until 1973, when Paul Lauterbur developed a method for spatially encoding the NMR signal by utilizing linear magnetic field gradients. About the same time, Peter Mansfield had also discovered a means of determining the spatial structure of solids by introducing a linear gradient across the object. The idea of applying magnetic field gradients to induce spatially varying resonance frequencies to resolve the spatial distribution of magnetization was a major milestone and the beginning of magnetic resonance imaging (MRI). For their work, Lauterbur and Mansfield were awarded the Nobel Prize for medicine in 2003.

Since its discovery, MRI has quickly become one of the most important medical imaging devices available to physicians today. Unlike other imaging modalities, such as X ray and computed tomography, MRI does not involve ionizing radiation. MRI also offers superior soft tissue contrast that is not possible with other imaging modalities. Furthermore, in MRI, the desired level of image contrast among different tissues can often be precisely controlled with simple adjustments to the acquisition timing parameters. MRI has become an invaluable tool for the assessment of many types of disease.

This chapter provides a brief introduction to the phenomenon of NMR and describes how it can be used to create magnetic resonance (MR) images. Along with some of the basic concepts of NMR, the technique for spatial encoding is presented, and methods to measure the properties of tissues, namely their intrinsic T_1 and T_2 relaxation constants, are explained. The goal of this chapter is to present to the reader the fundamental concepts of MRI, and in the following chapter some of the more advanced topics in MRI are described.

14.2. NMR

Aside from a few exceptions encountered in research, MRI involves imaging the nucleus of the hydrogen atom, i.e. the proton. Fortunately, the human body has an abundant supply of hydrogen, mostly in the form of water and fat, which, in healthy subjects, comprise 50–70% and 10–20% of the total body weight, respectively.

14.2.1. The nucleus: Spin, angular and magnetic momentum

Among the properties of an atomic nucleus is its angular momentum $\boldsymbol{p}$ (Fig. 14.1):

$$\boldsymbol{p} = \boldsymbol{I}\hbar \tag{14.1}$$

where $\hbar$ is Planck's constant divided by 2π (reduced Planck's constant) and $\boldsymbol{I}$ is the nuclear spin (or quantum number). All vectors are presented in bold. For the hydrogen nucleus, I = ½. As the proton is positively charged, the angular momentum also produces a nuclear magnetic moment, $\boldsymbol{\mu}$, where:

$$\boldsymbol{\mu} = \gamma \boldsymbol{p} \tag{14.2}$$

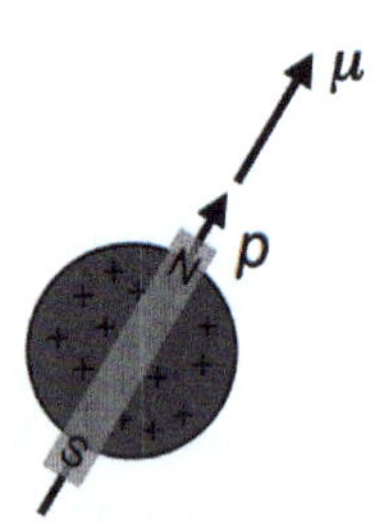

FIG. 14.1. The hydrogen nucleus, the proton, is characterized by an angular momentum, ***p****. Owing to its positive charge, the spin also possesses a magnetic moment,* ***μ****.*

where γ is the gyromagnetic ratio. The gyromagnetic ratio is specific to each type of nucleus, and for protons is roughly equal to 42.58 MHz/T. Aside from hydrogen, other nuclei have also been imaged, including carbon (^{13}C), fluorine (^{19}F), phosphorus (^{31}P), sodium (^{23}Na) and potassium (^{39}K) (Table 14.1). However, because of their relatively low abundance in vivo, MRI of these species is not commonly performed, owing to their limited available signal.

TABLE 14.1. COMMON NUCLEI FOR MR
(adapted from Stark and Bradley [14.1])

Nucleus	Relative abundance (%)	Spin (I)	Gyromagnetic ratio (MHz/T)	Relative sensitivity[a]	Abundance in the human body (% of atoms)
^{1}H	99.98	½	42.58	1	63
^{13}C	1.11	½	10.71	0.016	0.13
^{19}F	100	½	40.05	0.83	0.0012
^{23}Na	100	3/2	11.26	0.093	0.037
^{31}P	100	½	17.23	0.066	0.14
^{39}K	93.1	3/2	1.99	5.08×10^{-4}	0.031

[a] Per equal number of nuclei.

14.2.2. External magnetic field and magnetization

In the absence of an external magnetic field, a collection of these magnetic moments (or spins) will be aligned in random orientations in space, resulting in zero net magnetization. However, if an external magnetic field, $\boldsymbol{B}_0$, is applied, each spin will align either parallel or antiparallel to the direction of the applied field; they become polarized (Fig. 14.2). Owing to the lower energy state of the parallel orientation, there will be a slightly greater number of spins aligned along that direction. The ratio of the number of spins aligned in parallel (N_+) to those aligned in antiparallel (N_-) directions is given by the Boltzmann distribution:

$$\frac{N_+}{N_-} = e^{\Delta E/kT} = e^{\hbar\omega_0/kT} \tag{14.3}$$

where

ΔE is the energy difference between the two states;
k is the Boltzmann constant;
T is the absolute temperature;

and ω_0 is the Larmor or resonance frequency.

At a field strength of 1.5 T, this ratio is approximately 1.000010 for the proton, indicating that in a collection of about 2 million spins, there will only be a net of ten spins aligned with the field! Although in effect only a small fraction of the spins can be detected, the great number of spins available in any given voxel permits the detection of magnetization to generate MR images.

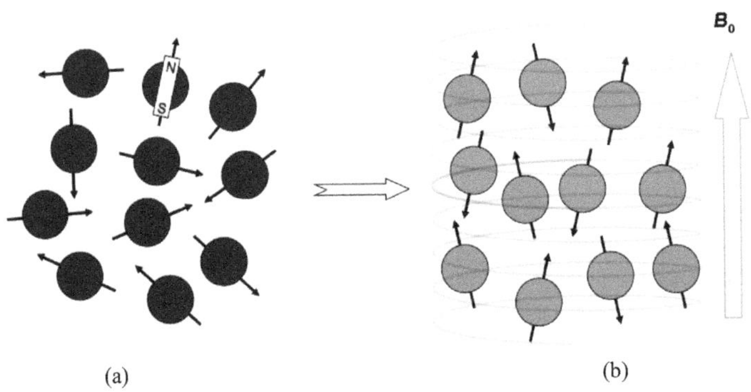

FIG. 14.2. The spins, initially oriented randomly in space (a), become aligned either parallel or antiparallel to an externally applied magnetic field $\boldsymbol{B}_0$ (b).

In addition to the polarization due to the external field $\boldsymbol{B}_0$, the magnetization will experience a torque that will cause the magnetization to rotate, or precess, about the direction of the field. The motion of the spin in a magnetic field can be described by the Bloch equation, which is the fundamental equation describing the behaviour of magnetization:

$$\frac{\mathrm{d}\mu}{\mathrm{d}t} = \gamma \mu \times \boldsymbol{B} \tag{14.4}$$

where $\boldsymbol{B}$ is the magnetic field. According to this equation, the spin will precess at the Larmor frequency, ω_0, which is linearly proportional to the external field strength (Fig. 14.3):

$$\omega_0 = \gamma \boldsymbol{B}_0 \tag{14.5}$$

The motion of the proton in the presence of an external field is analogous to the precession of a spinning top, which possesses an angular momentum owing to its spin, about the direction of a gravitational field resulting from a torque acting on the top.

In a stationary (or laboratory) frame of reference, spins thus precess at the Larmor frequency, ω_0, about the direction of $\boldsymbol{B}_0$ (by convention, along the z axis). It is also useful to consider a different frame of reference — one that rotates at the Larmor frequency. In this rotating frame, the spins that precess exactly at the Larmor frequency will appear to be stationary, while those that precess at a different frequency ω in the stationary frame will appear to precess with frequency ω_r in the rotating frame, where:

$$\omega_r = \omega - \omega_0 \tag{14.6}$$

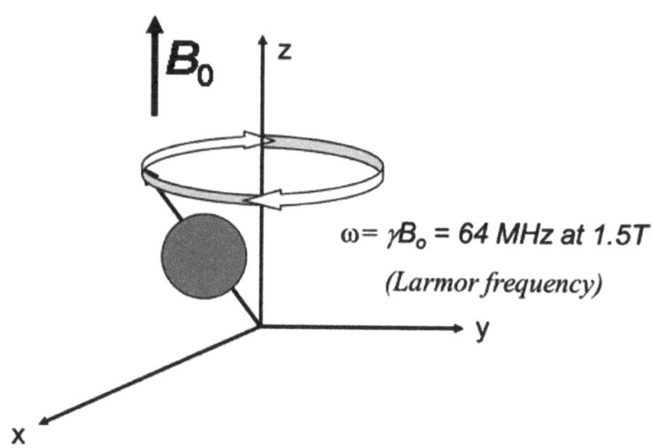

FIG. 14.3. Under the influence of an external magnetic field $\boldsymbol{B}_0$*, the spins precess about the direction of the field at the Larmor frequency, which is proportional to* $\boldsymbol{B}_0$*.*

The total magnetization within a voxel, its net magnetization, is a vector sum of all the spins contained within the voxel and will be aligned along the $+z$ direction, i.e. the direction of $\boldsymbol{B}_0$ (Fig. 14.4). Henceforth, any reference to magnetization will refer to the net magnetization of a collection of spins, such as from an entire voxel, as opposed to magnetization of a single nucleus. The behaviour of the net magnetization, $\boldsymbol{M}$, can be treated similarly to that of a single spin. The Bloch equation can thus be modified to reflect the behaviour of the net magnetization:

$$\frac{\mathrm{d}\boldsymbol{M}}{\mathrm{d}t} = \gamma \boldsymbol{M} \times \boldsymbol{B} \tag{14.7}$$

Note that since the net magnetization is aligned along the *z* axis, it remains stationary and does not precess about any axis. Thus, in the presence of a constant external magnetic field, $\boldsymbol{B}_0$, the net magnetization remains constant and oriented along the direction of the field. The magnetization is said to be at its equilibrium magnetization, $\boldsymbol{M}_0$. However, when additional magnetic fields are applied, including those that are time varying, the magnetization will deviate from its equilibrium position and may begin to precess about an effective magnetic field. Such perturbation of the magnetization is needed for signal detection and is the topic of the following section.

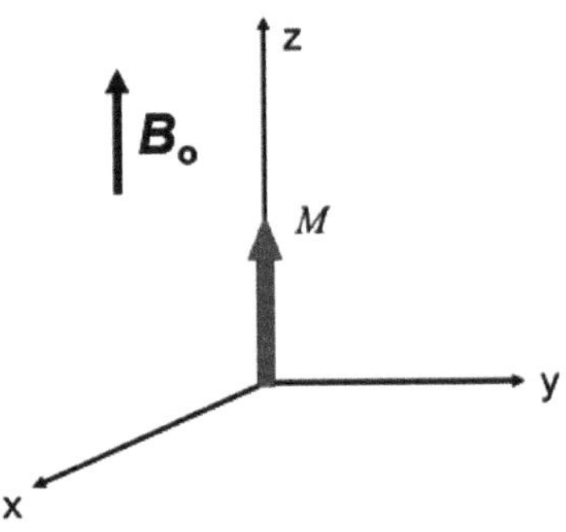

FIG. 14.4. The net magnetization, ***M****, is the aggregate sum of all the spins within a voxel and is aligned along the direction of the applied field,* $\boldsymbol{B}_0$.

14.2.3. Excitation and detection

Consider an external RF field, $\boldsymbol{B}_1(t)$, resonating at the Larmor frequency, applied to the spins in a magnetic field $\boldsymbol{B}_0$. According to the Bloch equation, the magnetization will precess about an effective magnetic field that is the vector sum of the static $\boldsymbol{B}_0$ field and the time varying $\boldsymbol{B}_1$ field. The behaviour of the magnetization can be simplified by considering its motion in the rotating frame, where the $\boldsymbol{B}_1$ field will appear to be stationary, since its frequency is identical to that of the rotating frame. In the rotating frame, the magnetization, initially aligned along the *z* axis, will then simply precess about the direction of the $\boldsymbol{B}_1$ field (Fig. 14.5(a)). If the RF field, $\boldsymbol{B}_1$, lies along the *x* axis in the rotating frame, the magnetization will precess, or nutate, about this axis, and will continue to do so as long as $\boldsymbol{B}_1$ is applied. In order to detect an MR signal, the $\boldsymbol{B}_1$ field is

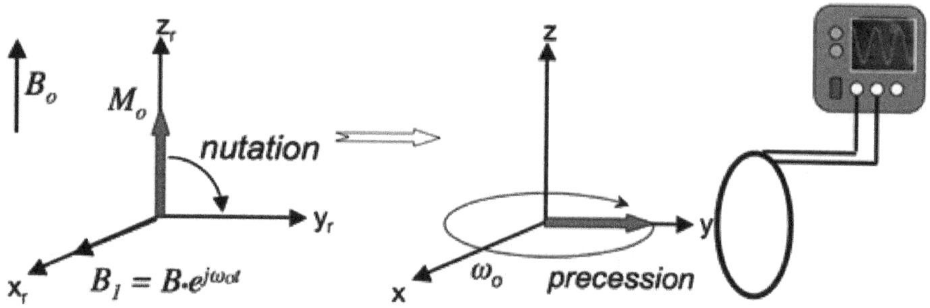

FIG. 14.5. (a) In the rotating frame, indicated by the coordinate axes (x_r, y_r, z_r), the RF field $\boldsymbol{B}_1(t)$ resonating at Larmor frequency will cause the magnetization to precess, or nutate, about its axis. (b) For optimal signal detection, the $\boldsymbol{B}_1$ field is removed once the magnetization reaches the transverse (x, y) plane. The magnetization subsequently processes about the z axis in the stationary reference frame and the signal can be detected with an RF coil.

typically applied just long enough (the order of 2–3 ms) to cause a 90° rotation, so that the magnetization becomes aligned along the y axis at the end of the RF pulse. Once the magnetization is rotated into the transverse (x, y) plane and the RF is removed, the spins will then precess about $\boldsymbol{B}_0$ at the Larmor frequency (in the stationary frame of reference) according to the Bloch equation (Eq. (14.7)). This rotating magnetization could subsequently be detected with an RF coil placed in its vicinity, according to Faraday's law (Fig. 14.5(b)), as the changing magnetic flux through the coil induces voltage changes that can subsequently be detected by a receiver. Among other factors, the type and size of the RF coil used for signal reception, its proximity to the imaging object, and the voxel size determine the overall strength of the received signal. Details of the RF coil and other imaging hardware relevant to MRI are discussed in Chapter 15.

14.3. RELAXATION AND TISSUE CONTRAST

14.3.1. T_1 and T_2 relaxation

Owing to interaction of the spins with their surroundings, the precession does not continue indefinitely following rotation of the magnetization into the transverse plane. Once the magnetization is perturbed by the RF pulse and is tilted away from the equilibrium (minimum energy) position along the z axis, there are two mechanisms that drive the magnetization back to its equilibrium state. Some of the energy absorbed by the spins from the applied RF pulse is lost to its surroundings (or its 'lattice') through the so-called spin–lattice relaxation, and the time constant for this phenomenon is termed T_1. T_1 relaxation depends on the mobility of the lattice and the efficiency of energy transfer from the excited spins to the lattice. Through T_1 relaxation, the longitudinal (or z) component of the magnetization returns to its equilibrium state, $\boldsymbol{M}_0$, in an exponential fashion:

$$M_z(t) = M_z(0)e^{-t/T_1} + M_o\left(1 - e^{-t/T_1}\right) \tag{14.8}$$

where $M_z(0)$ is the longitudinal magnetization immediately following RF excitation. For a 90° excitation, this term is zero and only the second term remains. After a time period of several T_1 s, the magnetization has almost fully returned to its equilibrium state, with amplitude M_o and aligned back along the z axis (Fig. 14.6).

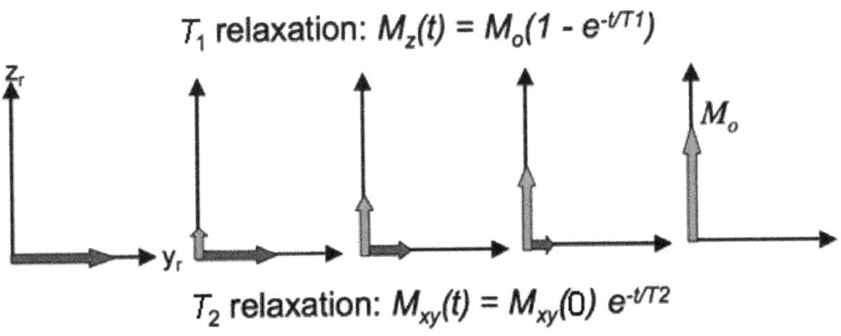

FIG. 14.6. Evolution of the magnetization following RF excitation. Following excitation into the transverse plane, T_1 and T_2 relaxation returns the magnetization to its equilibrium position, M_0. T_2 relaxation reduces the transverse component towards zero, while T_1 relaxation returns the longitudinal component towards M_0.

In addition to interactions with the lattice, the spins also interact with each other, causing spin–spin relaxation. Each spin, which is essentially a magnetic dipole, creates a magnetic field of its own, which slightly alters the field in its surroundings. Any spin that is in close proximity to another will experience this additional field, slightly altering its precessional frequency. As the spins are in constant motion, the precessional frequencies of each spin are in constant flux, subsequently leading to increasing loss of phase coherence, in which different spins accumulate different amounts of phase over time. This loss of coherence leads to an exponential decay of signal in the transverse plane, with a time constant of T_2:

$$M_{xy}(t) = M_{xy}(0)e^{-t/T_2} \tag{14.9}$$

where $M_{xy}(0)$ is the initial transverse magnetization following excitation.

To create an image, the process of excitation and signal detection is typically repeated many times until sufficient data have been acquired for image reconstruction. The time between subsequent excitations is the repetition time, or TR, and from Eq. (14.8) it is clear that TR determines to what extent tissues with various T_1 values have returned to their fully relaxed, equilibrium state. When short TR is prescribed, tissues with short T_1 values, which relax more quickly, will appear brighter than those with longer T_1 values, and differences in T_1 will be emphasized. On the other hand, image contrast due to differences in T_2 can be controlled by adjusting the echo time, the time between excitation and data acquisition. With longer echo times, tissues with short T_2 values will appear darker than those with longer T_2 values. Table 14.2 lists typical T_1 and T_2 values for various human tissues. Note that T_1 and T_2 are dependent on field strength, and their values at 1.5 T are shown in Table 14.2.

TABLE 14.2. TISSUES AND THEIR T_1 AND T_2 VALUES AT 1.5 T
(adapted from Bernstein et al. [14.2])

Tissue	T_1 (ms)	T_2 (ms)
Muscle	870	50
Fat	260	80
Liver	490	40
Blood (oxygenated)	1200	200
Blood (deoxygenated)	1200	50
White matter	790	90
Grey matter	920	100
Cerebrospinal fluid	4000	2000

14.3.2. Bloch equations with relaxation terms

The T_1 and T_2 relaxation constants can be incorporated into the Bloch equation from Eq. (14.7) to obtain a more complete description of the evolution of the MR signal:

$$\frac{\mathrm{d}\boldsymbol{M}}{\mathrm{d}t} = \gamma \boldsymbol{M} \times \boldsymbol{B} + \frac{M_0 - M_z}{T_1}\hat{z} - \frac{M_x\hat{x} + M_y\hat{y}}{T_2} \tag{14.10}$$

This equation incorporates the effects of both static and dynamic (RF) magnetic fields applied to the imaging object, $\boldsymbol{B}_0$ and $\boldsymbol{B}_1$, respectively, as well as the relaxation of spins back towards their equilibrium values due to T_1 and T_2 relaxation. The high level of image contrast between different tissues that MRI is able to provide is mainly because the relaxation time constants T_1 and T_2 are different among different tissue types. One of the main factors that give MRI an advantage over other imaging modalities, such as X ray and computed tomography, is that in MRI, one is able to control precisely to what degree the images are to be influenced by these two time constants, as will be explained in more detail below.

14.3.3. T_2* relaxation

In MRI, a homogeneous $\boldsymbol{B}_0$ field is critical for obtaining high quality images. In a homogeneous field, the transverse signal decays exponentially with an intrinsic T_2 time constant, as described above. However, in the presence of magnetic field inhomogeneities, owing to either imperfect magnet shimming or induced field perturbations (e.g. owing to susceptibility differences at tissue boundaries), transverse relaxation could be enhanced due to additional loss of coherence among the spins. Since the precession frequency of each spin is directly proportional to the local magnetic field, any perturbations in the field within a voxel will cause the spins to precess at different frequencies, causing loss of coherence and enhanced signal loss within the voxel. Although the exact evolution of the signal is dependent on the distribution of the fields within a voxel, it can be well approximated by a decaying exponential function (similar to that of T_2 relaxation) and the time constant is referred to as T_2* ('T-two star'). T_2* must be less than T_2 and represents the total transverse relaxation time, consisting of both the intrinsic T_2 time constant and the component due to the inhomogeneous field, T_2' ('T-two prime'):

$$\frac{1}{T_2^*} = \frac{1}{T_2} + \gamma \Delta B = \frac{1}{T_2} + \frac{1}{T_2'} \tag{14.11}$$

where ΔB is the field inhomogeneity across a voxel.

The most common incidences of these undesired field perturbations in vivo occur near regions of air–tissue boundaries, such as the sinus cavity in the head, or in trabecular (or spongy) bone where the susceptibility difference between bone and bone marrow induces local field inhomogeneities within the marrow. Fortunately, while signal loss due to intrinsic T_2 relaxation cannot be avoided, signal loss due to field inhomogeneities can be reversed by applying a second RF pulse, a 'refocusing' pulse, as is done in spin echo imaging described later in the chapter.

14.3.4. Contrast agents

Although MRI has excellent soft tissue contrast, it is sometimes desirable to enhance contrast among tissues through the introduction of a contrast agent. Contrast agents are highly paramagnetic in nature, enhancing the spin–lattice interaction and shortening the T_1 time constant, resulting in a much more rapid restoration of the longitudinal signal following excitation. Common T_1 contrast agents include those based on the gadolinium (3^+) ion, which contains seven unpaired electrons and is thus strongly paramagnetic. The longitudinal relaxation rate ($R_1 = 1/T_1$) following the introduction of contrast agent into the tissue is directly proportional to the concentration of the T_1 shortening agent:

$$R_1 = \frac{1}{T_1} = \frac{1}{T_{10}} + r[C] \tag{14.12}$$

where T_{10} is the intrinsic tissue T_1 without the contrast agent, and r and $[C]$ are the relaxivity and concentration of the contrast agent, respectively.

Relaxivity is a specific property of each type of contrast agent and can vary significantly from one agent to another. The T_2 time constant is also affected by contrast agents, and the relaxivities for T_1 and T_2 are roughly the same. However, since the intrinsic T_1 of tissues is often much longer than the intrinsic T_2, the relative effect of a given concentration of a contrast agent on the longitudinal time constant is typically much greater than that for the transverse time constant.

In contrast enhanced MRI, imaging sequences with short TRs are utilized to highlight the enhancing structures. Using short TRs prevents the magnetization of most tissues from sufficient recovery after excitation, and only those tissues affected by the contrast agent, whose T_1 values have been drastically shortened, will have recovered sufficiently to produce a high signal.

One of the major clinical applications of contrast enhanced imaging is in tumour detection. Contrast agent is typically injected into bloodstream through a vein in the arm. As most tumours typically have a rich blood supply, these agents can locally reduce the T_1 of these lesions as they diffuse from the vessels into the extravascular space and effectively enhance the lesion signal intensity with respect to the surrounding tissues (Fig. 14.7). The degree to which the signal is enhanced can also be utilized to assess tumour perfusion. Other applications of contrast enhanced MRI include MR angiography for visualization of arteries and veins and for imaging myocardial infarction.

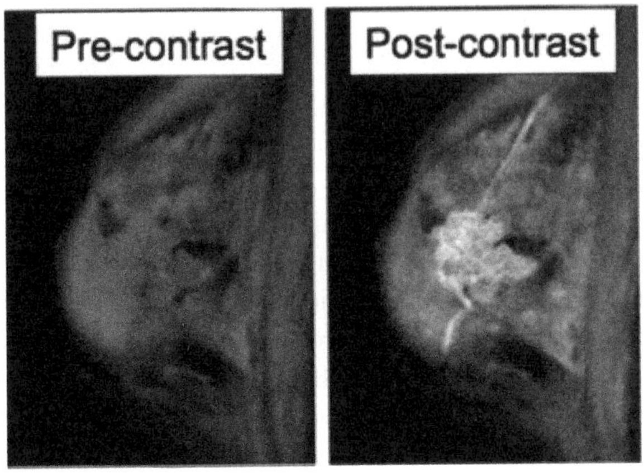

FIG. 14.7. Images of a breast with and without contrast agent. Without the contrast agent, no lesion is visible. Following intravascular contrast administration, the lesion is clearly enhanced, owing to high vascularity within the lesion.

14.3.5. Free induction decay

Following the excitation of the magnetization with an RF pulse, the z component of the magnetization returns to its equilibrium state, M_o, via T_1 relaxation, while the signal in the transverse plane decays exponentially with a T_2 time constant (for the sake of simplicity, additional relaxation due to field inhomogeneities — the T_2* effect — will be ignored). In the laboratory frame of reference, the spins also precess at the Larmor frequency, in addition to the decay of the amplitude from T_2 relaxation. This evolution of the signal in the transverse plane is called the free induction decay (FID) (Fig. 14.8(a)). In a homogeneous magnetic field, the spins all precess at a single Larmor frequency and the detected signal would be a perfect sinusoid, modulated by a decaying exponential function with a T_2 time constant. If one looks at the frequency content of this signal, or its frequency spectrum, which can be obtained simply by taking the Fourier transform (FT) of the free induction decay, the result is a function whose real component is a Lorentzian. The peak of the Lorentzian curve will be centred at the Larmor frequency, and its width, or more specifically the full width at half maximum, will be equal to $1/\pi T_2$ (Fig. 14.8(b)). A more rapidly decaying signal, which has shorter T_2, will have a broader width, while that of a species with longer T_2 is characterized by a sharper peak.

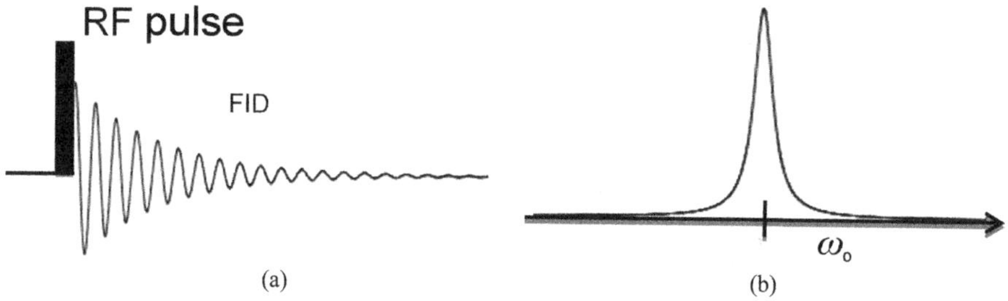

FIG. 14.8. (a) Following excitation with an RF pulse, the transverse magnetization undergoes FID. (b) The real component of the FT of the free induction decay is a Lorentzian function centred at the precession frequency.

14.4. MR SPECTROSCOPY

In vivo, the protons of hydrogen atoms in water and fat are the most abundant source of MR signal. Although the properties of all hydrogen nuclei are identical (spin number, angular moment), those of water and fat precess at slightly different frequencies even in the presence of a homogeneous external field. This is because hydrogen nuclei of different chemical species do not all experience identical local magnetic fields. Slight differences in these local fields are due to differences in magnetic shielding from the electron clouds, which vary between different molecules. Between water and fat, there is a frequency difference of 3.35 ppm, or approximately 215 Hz at 1.5 T.

Thus, the FID of an imaging object that contains both fat and water is no longer a simple decaying sinusoid, but rather the sum of two decaying sinusoids resonating at slightly different frequencies and with different T_2 relaxation time constants. The FT of this FID subsequently yields not a single Lorentzian, but a superposition of two Lorentzians, separated by 3.35 ppm.

In fact, even different hydrogen nuclei of a single molecule, such as fat, can also resonate at different frequencies. For example, the protons bound to a carbon atom containing only single bonds resonate at a slightly different frequency to those bound to carbon atoms containing double bonds. In addition, their T_2 values could vary as well. The resulting frequency spectrum is more complex, containing multiple peaks of different amplitudes and widths. In MR spectroscopy, these differences in frequencies can be utilized to determine the chemical and structural properties of molecules. Since the separations between resonant peaks of different protons are proportional to the external magnetic field strength, high field systems are particularly advantageous in MR spectroscopy as they can provide higher spectral resolution. This phenomenon of frequency shifts due to different nuclei being exposed to different electronic environments is known as chemical shift.

MR spectroscopy can be used to monitor biochemical changes in tumours, stroke and metabolic disorders. Most common nuclei used in MR spectroscopy include proton, phosphorus and carbon. As the same type of nucleus (e.g. hydrogen) associated with different molecules and compounds will often resonate at slightly different frequencies, the relative concentrations of these compounds can be determined by observing the peak heights of the spectrum, with higher peaks corresponding to higher concentrations. For example, in proton MR spectroscopy, an increased ratio of choline to creatine may indicate the presence of malignant disease and high lactate levels may indicate cell death and tissue necrosis. Further details of MR spectroscopic methods are discussed in Chapter 15.

14.5. SPATIAL ENCODING AND BASIC PULSE SEQUENCES

14.5.1. Slice selection

Previously, we have discussed the application of an RF pulse $\boldsymbol{B}_1(t)$ with a constant amplitude and a fixed duration, causing rotation of the magnetization into the transverse plane. The application of the RF pulse in a homogeneous field causes the entire imaging volume to be excited. However, in most applications, an image of a thin slice of the object is desired, and we wish to limit the excitation to the slice while leaving the rest of the object unperturbed. In MRI, this is easily achieved by applying a linear magnetic field gradient across the object during the RF pulse. In addition, by appropriately shaping the RF pulse, a desired slice profile can be achieved.

If an RF pulse is applied simultaneously with a linear magnetic field gradient, the excited slice profile along the same direction as the gradient will closely resemble the FT of the shape of the RF pulse. This Fourier relationship is most accurate for smaller flip angles (e.g. <30°) and is even acceptable for a 90° excitation, but the slice profile will increasingly deviate from the predicted shape at larger flip angles. Nevertheless, the Fourier relationship is often assumed in conventional MR applications. Thus, if a sinc shaped RF pulse is applied in the presence of a linear gradient, a narrow band of spins within the object will be excited, as the FT of a sinc function is a 'top hat' or rectangular function in frequency space (Fig. 14.9). The thickness of the excited slice, Δz, is a function of the bandwidth (BW) of the sinc pulse and the slice selection gradient amplitude, G_z, where:

$$\Delta z = \frac{\text{BW}}{\gamma G_z} \tag{14.13}$$

Since the sinc pulse is infinitely long, the excitation pulse is typically truncated to include only one or two lobes on either side of the main lobe. In addition, a low pass smoothing filter, such as Hanning or Hamming, is often applied to the pulse to create a smoothed pulse that tapers to zero at either end of the truncated pulse. The filtering is desired to improve the slice profile, as abrupt truncations of the RF pulse can lead to Gibbs ringing, undesirably causing some of the magnetization outside the slice to be excited. Although sinc is the most commonly used RF pulse, other RF pulse shapes, such as Gaussian, could also be utilized.

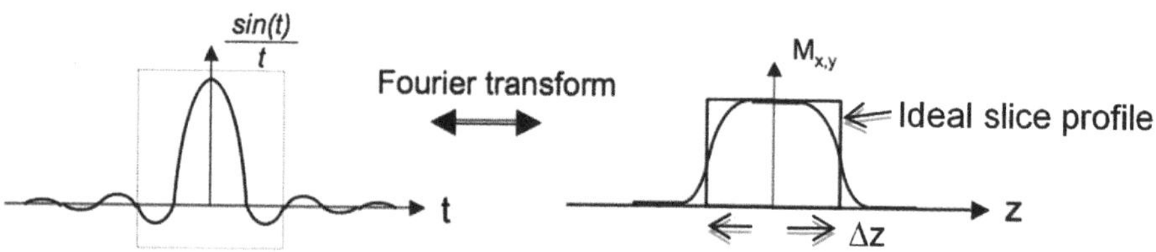

FIG. 14.9. As there is an approximate Fourier relationship between the RF pulse shape and the excited slice profile, a sinc pulse is often used to excite, selectively, a single slice. The sinc pulse is often truncated (dotted box) to limit its width and the ends are often tapered in order to create a smooth transition region in the slice profile.

In MRI, three sets of linear gradient coils are used for spatial encoding: G_x, G_y and G_z, which provide linear gradients along the x, y and z axes, respectively. The slice selection gradient can be any of these three. In addition, a slice plane with any spatial orientation could be selected by turning on more than one gradient during excitation. The excited plane will always be perpendicular to the direction of the net gradient.

14.5.2. Frequency and phase encoding

Subsequent to exciting the desired slice, spatial encoding within the plane of the selected slice follows. This is achieved by utilizing linear gradients along the two perpendicular in-plane directions (by convention assumed to be G_x and G_y following excitation of an axial slice plane with the G_z gradient). A linear gradient field will cause spins at different positions along the direction of the gradient to precess at different frequencies, varying linearly with position:

$$\omega_x = \gamma G_x x \tag{14.14}$$

where G_x is the applied magnetic field gradient along the x axis and ω_x is the resonance frequency of the spins at position x. Note that the precession frequency in the above equation, as well as in all subsequent equations, is with respect to the rotating frame; the additive frequency term due to the static field B_0 is not included. Thus, during MRI data acquisition, a spatial gradient is applied (along the x axis by convention), causing the spins along this axis to resonate at linearly varying frequencies. Encoding spatial information in this manner is referred to as 'frequency encoding', and the gradient that is applied is referred to as the readout gradient, since data points are acquired while the gradient is applied.

The detected signal $S(t)$ is the sum of all the excited spins in the imaging volume, each resonating at a frequency corresponding to its position along the gradient direction:

$$S(t)=\int \rho(x)\mathrm{e}^{j\omega_x t}\mathrm{d}x=\int \rho(x)\mathrm{e}^{j\gamma G_x x t}\mathrm{d}x \tag{14.15}$$

where $\rho(x)$ is the spin density, and T_1 and T_2 dependence have been ignored for simplicity. The term in the exponent is a phase term and represents the relative phase accumulated in the rotating frame due to the frequency encoding gradient. Typically, the readout gradient is constant during data acquisition. More generally, time varying gradients could instead be used, and in this case, the accumulated phase term in the above equation needs to be replaced by a time integral of the readout gradient. If we define:

$$k_x=\gamma\int G_x(t)\,\mathrm{d}t \tag{14.16}$$

it follows that the detected signal $S(t)$ in Eq. (14.15) could be expressed as a function of the variable, k_x, where:

$$S(t)=S(k_x)=\int \rho(x)\mathrm{e}^{jk_x x}\mathrm{d}x \tag{14.17}$$

Equation (14.17) shows that the position variable x and the 'spatial frequency' variable k_x, or, equivalently, image space and 'k-space', are a Fourier pair. This could be extended to 2-D:

$$k_y=\gamma\int G_y(t)\,\mathrm{d}t \tag{14.18}$$

$$S(k_x,k_y)=\int \rho(x,y)\mathrm{e}^{j(k_x x+k_y y)}\mathrm{d}x\,\mathrm{d}y \tag{14.19}$$

Equation (14.19) shows that once a sufficient number of data $S(k_x, k_y)$ are acquired in k-space, the imaging object $\rho(x, y)$ is obtained by a simple FT (Fig. 14.10). The desired k-space data $S(k_x, k_y)$ are acquired by navigating through k-space with the G_x and G_y gradients according to Eqs (14.16, 14.18). While it is possible to collect all the necessary data by filling the entire k-space after a single excitation, signal loss due to T_2 or T_2* during data acquisition often limits the availability of sufficiently high signal level, potentially leading to blurring or other image artefacts. Instead, data are typically acquired in multiple TRs in which a single line of k-space is acquired following each excitation. Details of typical data acquisition strategies are described in the following sections.

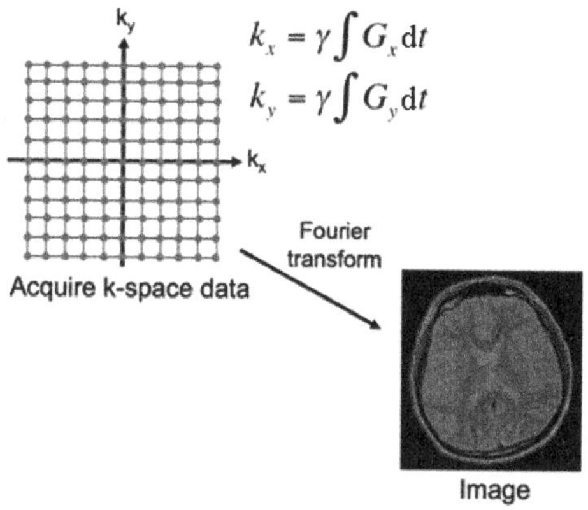

FIG. 14.10. The image acquisition process. Data are acquired in k-space, whose FT yields the desired image.

14.5.3. Field of view and spatial resolution

The field of view (FOV) and spatial resolution of the reconstructed image are determined by the manner in which the k-space is sampled. As the image space (units of distance) and k-space (units of 1/distance) are Fourier pairs and correspond to spatial position and spatial frequency, respectively, there is an inverse relationship between the two. Thus, a larger FOV requires finer sampling in k-space, while a smaller FOV requires coarser sampling. Specifically, the FOV is given by the inverse of the distance between adjacent sampled k-space points and can vary between the two in-plane directions:

$$\mathrm{FOV}_x = \frac{1}{\Delta k_x} \text{ and } \mathrm{FOV}_y = \frac{1}{\Delta k_y} \qquad (14.20)$$

where Δk_x and Δk_y are the spacings between adjacent k-space points along k_x and k_y.

Conversely, the spatial resolution, or the pixel widths Δx and Δy, is determined by the range of the sampled k-space region:

$$\Delta x = \frac{1}{2k_{x,\max}} \text{ and } \Delta y = \frac{1}{2k_{y,\max}} \tag{14.21}$$

where $k_{x,\max}$ and $k_{y,\max}$ are the maximum positions along k_x and k_y that are sampled, respectively. Thus, a high spatial resolution (or small pixel size) requires that a larger region of k-space be sampled.

14.5.4. Gradient echo imaging

In MRI, various pulse sequences are available for acquiring images. A pulse sequence is a set of instructions that control how MRI data are acquired and includes the application of the G_x, G_y and G_z magnetic field gradients, the RF pulse and data acquisition. Over the years, many different types of pulse sequence have been developed, including those optimized for rapid imaging or for estimating diffusion or flow, or those that are robust to motion, to list a few.

One of the simplest imaging pulse sequences is the gradient echo sequence. The main components of the gradient echo sequence are the following (Fig. 14.11):

- Slice selective RF pulse which rotates the magnetization into the transverse plane (conventionally along the *z* axis);
- Frequency encoding, or readout, gradient (*x* axis);
- Phase encoding gradient (y axis).

The slice selection and frequency encoding and phase encoding described in the preceding sections essentially comprise the main components of this basic pulse sequence. There are a few more gradient pulses that must be added to the sequence to make it complete. First, following slice selective excitation, a slice 'rewinder' must be applied. The rewinder gradient, which has opposite polarity as the slice selection gradient amplitude is needed to undo the phase accumulation that occurs during the RF pulse. If a non-selective RF pulse had been used where the slice selection gradient is not concurrently applied, no rewinding would be necessary as all spins are rotated in phase on to the transverse plane as depicted in

Fig. 14.5(a). However, in the presence of a slice select gradient, spins at different positions along the z axis will accumulate different amounts of phase during the RF pulse. By applying a rewinder gradient whose area is approximately half that of the slice select gradient, the undesired phase dispersion can be effectively reversed and cause the spins to be aligned along the same direction in the transverse plane.

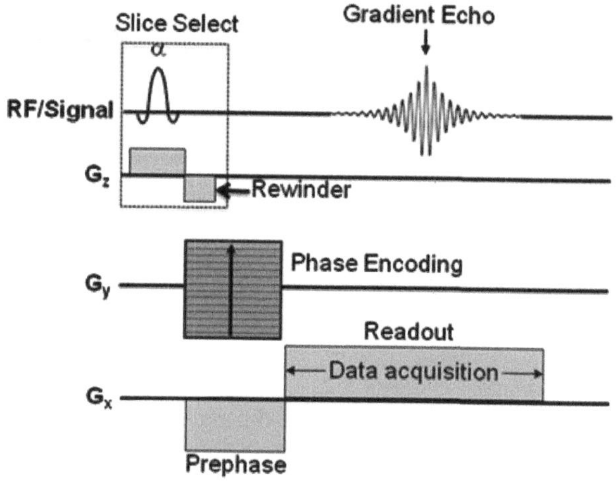

FIG. 14.11. The gradient echo sequence. Following slice excitation, phase encoding and readout prephase gradients are applied. Data acquisition subsequently follows, in which one line of k-space data is acquired. The phase encoding gradient amplitude is linearly increased from one TR to the next, in order to fill different k_y lines of k-space.

The second gradient pulse that must be added to the sequence is the readout prephase gradient. In order to reconstruct an MR image, data should fill the k-space symmetrically about its centre ($k_x = k_y = 0$). This can be accomplished most efficiently by sampling a full line of data in a single readout, and the purpose of the prephase gradient is to move the k-space location to the $-k_{x,\max}$ position prior to data acquisition (Fig. 14.12(a)). Since the position in k-space is simply the integral of the spatial gradients (Eqs (14.16, 14.18)), a negative G_x prephase gradient is used to move the k-space position along the negative k_x direction. Subsequently, the readout gradient is applied while data points are simultaneously acquired, filling one full line of k-space along k_x (Fig. 14.12(b)). Without a prephase gradient, data readout would begin at $k_x = 0$, and a second readout in the opposite direction would be required to fill the other half of k-space.

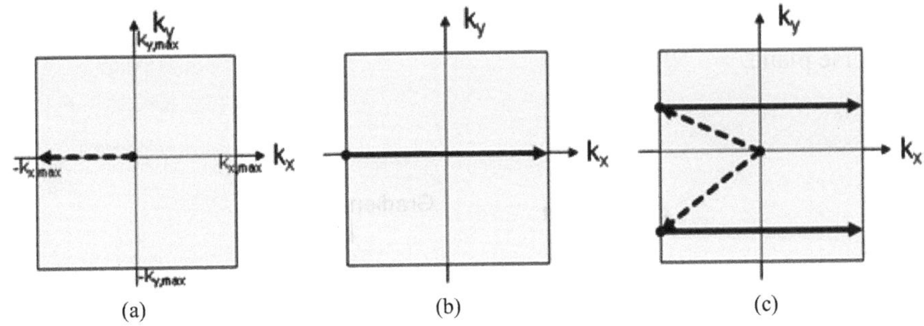

FIG. 14.12. (a) The prephase gradient of the readout axis moves the current k-space position to $-k_{x,max}$. (b) Subsequently, one complete line of data is acquired during readout. (c) By also applying the various phase encoding gradients prior to readout, one is able to move along the k_y axis. Two different phase encoded acquisitions are shown in (c), one positive and one negative. This process continues until the desired k-space region is filled (grey box).

As explained in the previous section, during each repitition time (TR), one line of data is acquired, and from TR to TR, different phase encoding amplitudes are applied to sample different lines along the k_y axis (Fig. 14.12(c)). Note that the phase encoding and prephase gradients, as well as the slice rewinders, could be applied simultaneously. The only requirement is that these gradients be applied after the end of the excitation pulse and before the beginning of the readout window. In terms of the gradient amplitude and duration, what matters is the total area under these pulses, since the k variable is the time integral of the gradient. For example, the area under the readout prephase gradient should be one half of the area of the readout gradient, so that the echo occurs midway through the readout window. While using the same amplitude with opposite polarity and half the pulse width is one option, it is also possible to use the highest possible gradient amplitude with a shorter pulse width. The latter option would be preferable if the shortest possible echo time is desired. Once the desired data are acquired, e.g. 256 phase encoding lines and 256 readout points per line, a simple 2-D discrete fast FT of the k-space data will yield the desired image.

The time between the peak of the excitation pulse and the centre of the readout window ($k_x = 0$) is called the echo time. In MRI, the echo time (TE) can be arbitrarily chosen above a certain minimum to achieve the desired image contrast. During this period, signal decays with a T_2 relaxation time constant, reducing the transverse magnetization (Eq. (14.9)). In addition, as mentioned above, the presence of any field inhomogeneity will cause additional loss of signal within each voxel, owing to the loss of phase coherence among the spins. The sum of these two factors — the intrinsic T_2 decay and signal loss due to field inhomogeneity — results in a T_2* decay during the TE, and the gradient echo image will be weighted by the factor $e^{-\left(\mathrm{TE}/T_2^*\right)}$.

Gradient echo sequences utilizing short TRs are often used when T_1 contrast is desired. Following each excitation, T_1 relaxation takes place, allowing the longitudinal magnetization to recover. By keeping TR short, the longitudinal magnetization of tissues with relatively long T_1 values remains low, owing to insufficient time for recovery, while those with shorter T_1 values will have achieved a greater amount of signal recovery. This leads to higher signal in tissues with short T_1 values and lower signal in those with long T_1 values. In these T_1 weighted images, image contrast results from differences in T_1 among the tissues. The TE is typically kept to a minimum in T_1 weighted imaging, to minimize T_2 or T_2* contrast.

The gradient echo sequence is also often used in applications that require fast data acquisition, such as dynamic imaging, where multiple images are rapidly acquired to detect changes in the image over time. For these applications, the TR needs to be kept short to minimize the scan time. Since short TR reduces the time available for magnetization to recover following excitation, small excitation angles (e.g. 5–30°) are often employed in order to maximize the acquired signal. If the tissue T_1 is known or can be approximated, the Bloch equation can be solved analytically to determine the optimum flip angle that yields the highest signal intensity for a given TR. This optimum flip angle is known as the Ernst angle, θ_{E}, and is given by the following expression:

$$\theta_{\mathrm{E}} = \cos^{-1}\left(\mathrm{e}^{-\mathrm{TR}/T_1}\right) \tag{14.22}$$

14.5.5. Spin echo imaging

One of the drawbacks of the gradient echo sequence is its sensitivity to magnetic field inhomogeneities. In the presence of inhomogeneous fields, either resulting from imperfect shimming of the main $\boldsymbol{B}_0$ field, or along air–tissue boundaries in which the susceptibility difference between the two regions distorts the local magnetic field, additional signal loss may occur owing to dephasing of the spins within the affected voxels (T_2* decay). This leads to reduction in the voxel signal and can even lead to complete loss of signal, particularly at longer TEs.

This reduction of signal due to the magnetic field inhomogeneities can be avoided with the spin echo sequence. In spin echo imaging, a second RF pulse of

180° flip angle is applied at TE/2 following an initial 90° excitation pulse. Termed the 'refocusing pulse', the 180° pulse reverses the phase that a spin may have accumulated because of the inhomogeneous field. Following the refocusing pulse, the magnetization again accumulates phase, but this time the total phase is reduced because of the earlier phase reversal, eventually causing the phase to cancel at TE. Figure 14.13(a) shows the spin echo sequence. Notice that the polarities of the prephasing and phase encoding gradients are opposite to that of the gradient echo sequence, to account for the phase reversal effect of the refocusing pulse. In k-space, the refocusing pulse essentially moves the current k-space location to its conjugate position (Fig. 14.13(b)), or, equivalently, reflects the point about the k-space origin. The acquired signal at TE in spin echo imaging evolves with the intrinsic T_2 time constant and is not affected by inhomogeneous fields, unlike the gradient echo sequence, where it evolves with T_2*.

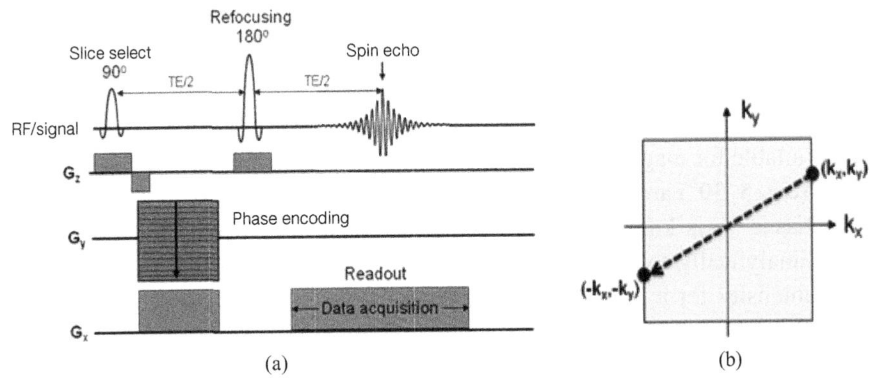

FIG. 14.13. (a) The spin echo sequence. The spin echo sequence is similar to the gradient echo sequence, except that a 180° refocusing pulse is applied to refocus any spin that may have dephased as a result of magnetic field inhomogeneities. Since the refocusing pulse reverses any accumulated phase, the polarities of both the phase encoding gradient and the prephase gradient of the readout are reversed. (b) In k-space, the 180° pulse moves the current k-space location to its conjugate position (reflection about the origin).

While gradient echo imaging can be used for fast data acquisition by employing a short TR and small angle excitation, imaging with a short TR is generally not recommended with the spin echo technique, because of the rapid reduction of available magnetization. The 180° refocusing pulse in a spin echo sequence inverts any positive longitudinal magnetization to the $-z$ direction. As a short TR does not allow sufficient time for the longitudinal magnetization to recover, spin echo imaging typically requires longer TRs, and consequently longer scan times. However, in a modified spin echo technique in which multiple refocusing pulses are applied and multiple lines of data are acquired following a single excitation pulse, the scan time can be drastically reduced. Commonly referred to as 'fast spin echo' or 'turbo spin echo', 16 or 32 or more lines of data are typically acquired within each TR. Single shot acquisitions are also possible, in which all of the k-space data are acquired (e.g. 128 refocusing pulses and 128 readout lines), following a single excitation pulse. Details of this sequence are presented in Chapter 15.

14.5.6. Multislice imaging

In MRI, there are two ways to acquire multiple slice data. The first method is to acquire multiple 2-D slices by repeating the single slice strategy described previously, but with shifted slice positions. In this multiple slice technique, the location of each slice is varied simply by modulating the frequency of the excitation RF pulse. All gradient waveforms, including the slice selection gradients, remain unchanged. Since the RF pulse excites spins within a specified frequency band, the excited slice location can be chosen by modifying the carrier frequency of the RF pulse. The centre location of the excited slice is determined according to the following equation:

$$z_{\text{slice}} = \frac{\Delta f}{\gamma G_{\text{sl}}} \tag{14.23}$$

where z_{slice} is the location of the slice with respect to the scanner isocentre, Δf is the offset frequency of the RF pulse relative to the resonance frequency at the isocentre and G_{sl} is the slice select gradient amplitude.

Acquisition of multiple 2-D slices is typically carried out in an interleaved fashion. In interleaved slice acquisition, one does not need to wait until the end of a TR following data acquisition from one slice before acquiring data from other slices. A different slice is excited and data acquired during the 'dead time' of the previous slice, the time between the end of data acquisition and the next excitation pulse for that slice (Fig. 14.14(a)). With this strategy, the total scan time can be equal to that of a single slice if the dead time is sufficiently long enough to accommodate the acquisition of all the slices. However, if the TR is not sufficiently long enough to allow data collection from all the slices within a single TR period, the scan is repeated until all remaining slices are acquired. Following data acquisition, images are reconstructed by applying separate 2-D FTs for each slice.

14.5.7. 3-D imaging

An alternative way to acquire multiple slices is with 3-D imaging. In 3-D MRI, instead of exciting and acquiring data from individual 2-D slices, a single large volume is excited, followed by additional phase encoding along the slice direction. Thus, phase encoding is applied along both in-plane and slice (or 'through plane') directions (Fig. 14.14(b)). With this scheme, in the same way that spatial information is encoded then separated with phase encoding along the y axis followed by an FT, slice information is phase encoded along z and the individual slices separated with an FT. Thus, following data acquisition, images are obtained by a 3-D FT of the acquired k-space data. Data for all the slices are acquired simultaneously, as opposed to the 2-D scheme in which each data acquisition window receives data only from a single slice. In 3-D imaging, the scan time is increased by a factor equal to the number of slices relative to a single slice acquisition using the same TR. Note that in 3-D imaging, all the slices are contiguous, whereas in 2-D imaging, relative slice positions can be arbitrary. Thus, 3-D imaging is often preferred when multiple contiguous slices are needed. It is also preferred over the multislice 2-D counterpart when thin slices are desired, since excitation of a single large 3-D volume requires much lower gradient amplitude than that for exciting individual thin slices.

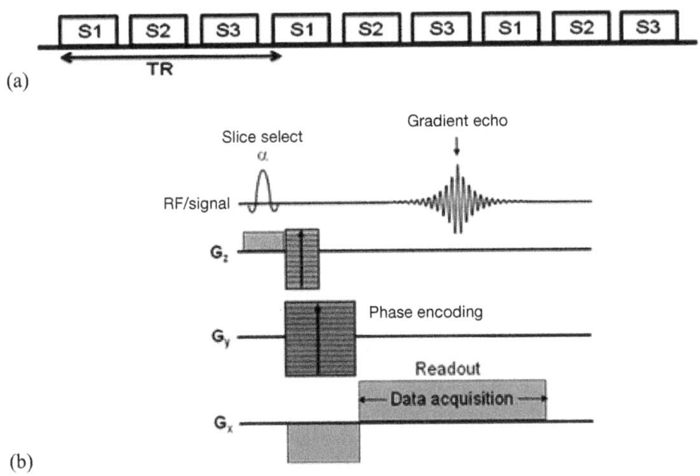

FIG. 14.14. Multislice imaging. (a) Interleaved multislice 2-D imaging. A multislice 2-D sequence is shown in which three slices (S1, S2, S3) are acquired in an interleaved fashion. Each of the numbered boxes can represent any imaging sequence (e.g. gradient or spin echo). The effective TR for each slice is the period between acquisitions of the same slice, as this period corresponds to the time between excitations for any given spin. (b) A 3-D gradient echo sequence with additional phase encoding along the slice encoding direction.

14.5.8. Measurement of relaxation time constants

It is sometimes desirable to quantify the relaxation time constants of tissues, as they may be helpful in determining the status of certain diseases or to track their progress following treatment.

14.5.8.1. Measurement of T_1

There are various ways to measure T_1, but one of the more accurate and widely utilized means is the inversion recovery technique. In this technique, a 180° pulse is initially applied to invert the spins from their equilibrium position into the $-z$ axis. Subsequently, a waiting period, the inversion time, follows before data are acquired (Fig. 14.15). During the inversion time period, the magnetization recovers from its inverted state towards its equilibrium magnetization in an exponential fashion, with a longitudinal relaxation time constant, T_1, according to Eq. (14.8) (with $M_z(0) = -M_0$). A series of images are acquired, each with different inversion times, and the signal intensities are subsequently fitted to Eq. (14.8) to derive T_1. For data acquisition, any sequence can be used, but a rapid sequence, such as a turbo spin echo, is typically preferred to reduce the total scan time. Long TRs (ideally TR ~ $5*T_1$) are often used in inversion recovery sequences to ensure that the spins are fully relaxed prior to subsequent TRs.

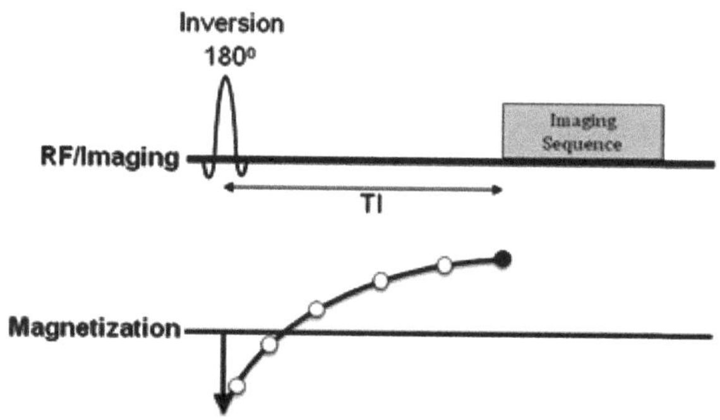

FIG. 14.15. Inversion recovery sequence. In inversion recovery, magnetization is first inverted to the –z axis by an 180° pulse. Data are subsequently acquired following an inversion time. The sequence is repeated with different values of inversion time (open circles) and data subsequently fitted to an equation to determine T_1. Typically, a fast imaging technique, such as a turbo spin echo sequence is used for data acquisition.

*14.5.8.2. Measurement of T_2 and T_2**

Since the signal from a spin echo sequence decays exponentially with a T_2 time constant, one can compute the transverse relaxation time constant by repeating a spin echo scan with increasing TEs and subsequently fitting the signal intensities to a decaying exponential. Although only two TEs are required, four or five or more are typically used to improve measurement accuracy. Ideally, the TR should be sufficiently long enough (TR $\gg T_1$) to ensure complete return of the magnetization to its equilibrium state. This ensures that the magnetization prior to each excitation is identical and not a function of TR or TE.

A more efficient means to measure T_2 is to apply multiple 180° refocusing pulses following a single excitation. For example, a 90° excitation followed by six refocusing pulses and six data acquisition periods, in which data from each TE are used to reconstruct a separate image, would permit acquisition of data for multiple images at different TEs at each TR. The sequence would appear similar to a fast spin echo sequence described above, except instead of all acquired echoes used for reconstruction of one image, different echoes would encode data for different images. Although highly efficient, one drawback of this method is that the measurement accuracy is sensitive to inaccuracies in the 180° refocusing pulses. Imperfect refocusing pulses do not affect the accuracy of T_2 measurement in single echo spin echo sequences, since the signal at different TEs is equally affected by the inaccuracies. This leads to a simple scaling of the signal amplitudes identically for all echoes and does not affect calculation of T_2.

Measurement of T_2* is similar to that of T_2, except a gradient echo sequence is used in lieu of a spin echo. Multiple gradient echoes following a single excitation pulse, with each echo encoding images at different TEs, can be used to measure T_2*. Unlike the spin echo counterpart, the accuracy of T_2* measurement is not affected by the accuracy of the RF pulse in multiple gradient echo imaging, since refocusing pulses are not applied.

REFERENCES

[14.1] STARK, D.D., BRADLEY, W.G., Magnetic resonance imaging, 2nd edn, Mosby, St. Louis, MO (1992).

[14.2] BERNSTEIN, M.A., KING, K.F., ZHOU, X.J., Handbook of MRI pulse sequences, Elsevier, Amsterdam (2004).

BIBLIOGRAPHY

BUSHBERG, J.T., SEIBERT, J.A., LEIDHOLDT, E.M.J., BOONE, J.M., The Essential Physics of Medical Imaging, 2nd edn, Williams & Wilkins, Baltimore, MD (2002).

HAACKE, E.M., BROWN, R.W., THOMPSON, M.R., VENKATESAN, R., Magnetic Resonance Imaging: Physical Principles and Sequence Design, John Wiley & Sons, New York (1999).

PAULY, J., NISHIMURA, D., MACOVSKI, A., A k-space analysis of small-tip-angle excitation, J. Magn. Reson. **81** (1989) 435–436.

SPRAWLS, P., Magnetic Resonance Imaging: Principles, Methods, and Techniques, 2nd edn, Medical Physics Publishing, Madison, WI (2000).

Chapter 15

MAGNETIC RESONANCE IMAGING

M.O. LEACH
The Institute of Cancer Research and The Royal Marsden Hospital,
London, United Kingdom

15.1. INTRODUCTION

In Chapter 14, the principles of nuclear magnetic resonance were presented, along with an introduction to image forming processes. In this chapter, magnetic resonance imaging (MRI) will be reviewed, beginning with the hardware needed and its impact on image quality. The acquisition processes and image reconstruction will be discussed, as well as the artefacts that are possible, with discussion of the important area of safety and bioeffects completing the chapter.

15.2. HARDWARE

MRI systems comprise a number of major hardware components, under the control of digital systems that provide instructions, monitor system performance and acquire and process the signals that are used to create images or spectroscopic signals reporting on a wide range of tissue states. These systems are coordinated by one or more computer workstations or PCs that provide the interface to the MR operator, enabling acquisitions to be planned and executed, and images to be calculated, displayed and stored, often providing sets of measurement and analysis software addressing particular clinical questions. Figure 15.1 shows the major components of an MRI system, and these are described in more detail below.

15.2.1. The static magnetic field subsystem

The primary requirement for an MRI system is to provide a highly uniform and stable magnetic field to ensure that the Larmor resonance condition is met to within about 1 ppm over the imaging volume. This condition may be met by both the main magnet design and by additional static and electronic shims (or adjustments) that correct for imperfections in the magnet design, the effects of nearby static steel structures and the effects of the patient's intrinsic magnetic susceptibility.

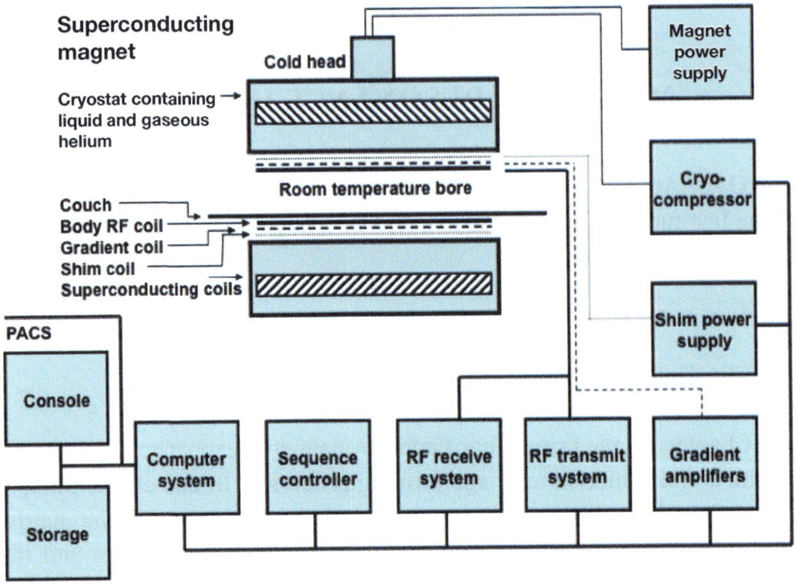

FIG. 15.1. The major components of a clinical MRI system utilizing a superconducting magnet.

15.2.1.1. Common field strengths and magnet designs

There are a number of different types of magnet design, with different geometries that relate directly to the intrinsic orientation of the magnetic field. Some current designs are shown in Fig. 15.2. Perhaps the simplest design is the resistive solenoid, in which a long solenoid generates a uniform magnetic field. In practice, this solenoid design has been reduced to a number of coils, typically four or six, separated by spacers. The coils may have differing radii, but with the relative current density, number of windings and diameter calculated to maximize the uniformity of the field. Both horizontal and vertical configurations have been built. While being relatively cheap, they have the disadvantages that they produce considerable heat, requiring water cooling, and their resistance changes as they heat up. These factors, together with the constraints of power supplies, high current demand and an effective limitation of field strengths to a flux density of about 0.5 T, led to them being superseded by other approaches.

The most common design, based on similar principles but utilizing superconducting cable, has been superconducting magnets. These usually follow the same general horizontal bore geometry, with a number of coils in parallel, but they are wound from superconducting cable, typically niobium–tin encased in copper, and as such they usually have to be maintained at liquid helium temperatures (4 K, –269°C). This requires them to be encased in a vacuum cryostat, where great attention is paid to the design to limit heat conductance into the cryostat and to maintain a number of thermal layers. Typically, in modern magnets, the liquid helium is surrounded by a helium vapour barrier, as well as a vacuum. Cryocompressors reliquefy the helium vapour, maintaining a system with very little cryogen during normal operation, although should the cryocompressor fail because of a fault or loss of chilled water or power, the magnet will have only a limited superconducting life based on the boil-off of the limited reserve of cryogen. Once the cryogen fails to cool the superconducting windings adequately, the windings, which carry a very high current, will become resistive and will rapidly heat up because of the large current, resulting in a magnet quench, whereby the current is rapidly resistively dissipated as heat, leading to a rapid boil-off of any remaining cryogen. The cryostat should be designed to vent safely any explosive release of helium gas to the outside of the building, without it entering the scanner room. Earlier magnet designs used additional layers of liquid and gaseous nitrogen, which had to be replenished relatively frequently, to minimize boil-off of helium.

Superconducting MRI systems in clinical use range from 0.5 to 3.0 T, with experimental clinical systems ranging to 8 T and above. Systems up to 3.0 T (and in some cases 7 T) will usually be self-shielded, generating a second reversed magnetic field that cancels much of the field outside the magnet, but retaining a strong magnetic field in the magnet bore. This considerably reduces the extent of the magnetic field, making site planning and facility design much easier, and reducing the area of restricted access and effects on nearby equipment. Some models provide an accessible central aperture perpendicular to the magnet axis, for interventions, and configurations with a vertically oriented field have been built, offering some signal to noise ratio (SNR) advantages in receiver coil geometries where patients are positioned horizontally. Figure 15.2 provides examples of some superconducting magnets.

Electromagnets reduce the electrical current requirement of resistive magnets by incorporating a ferromagnetic core and also by providing greater stability and minimizing cooling requirements. Such magnets generally operate at lower fields (typically 0.1–0.2 T) with a vertically orientated field between the pole faces. They provide increased access to the patient compared with most superconducting designs. Permanent magnets have predominantly been designed with a vertical field format, with the field constrained between the top and bottom pole faces; this design requires a flux return path. The magnets are very heavy and cannot be switched off, but have a small stray field. A recent innovation has been a permanent magnet with a horizontal field, allowing patients to be positioned in the upright position, which is of value for examining joints.

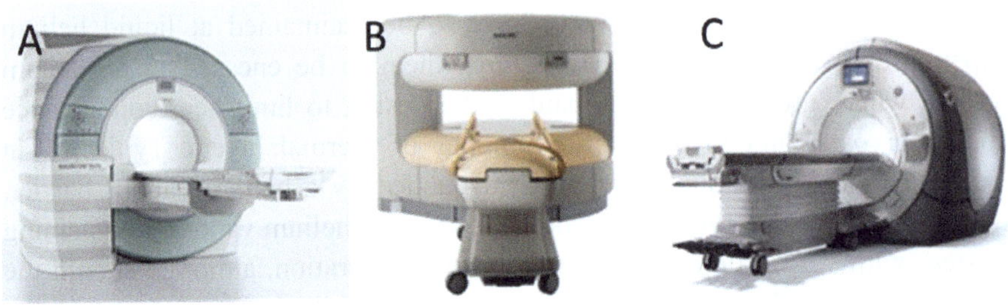

FIG. 15.2. Different superconducting magnets: (a) Siemens 3 T Vario system; (b) Philips 1 T Panorama system; (c) General Electric 1.5 T Discovery MR450 system.

MRI requires a highly homogeneous magnetic field, typically to at least 1 ppm over the field of view (FOV). Homogeneity is limited by the tolerances of magnet design, and is further affected by the environment of the magnet (e.g. structural steel in a building). To compensate for these effects, the magnetic field is usually adjusted at installation by a process termed shimming, whereby a combination of steel shims, and additional magnetic fields generated by adjustable currents in sets of additional shim coils, are employed to adjust the field homogeneity. For some imaging approaches, such as fat suppressed imaging, as well as for MR spectroscopy, where field homogeneities of the order of 0.1 ppm are required, the shim coils have to be adjusted prior to a study to compensate for the distortions in the magnetic field caused by the magnetic susceptibility of the patient.

15.2.1.2. Siting issues

The major concerns in siting a magnet are safety issues, the potential effects of adjacent steel and moving steel objects on the magnet homogeneity, nearby current carrying cables, adequate structural strength and the effects of the magnet on adjacent equipment. Safety is covered in Section 15.6. A site for an MRI system needs to contain the 0.5 mT field contour within a secure zone (see below) with controlled access. The design needs to account for floor loading and equipment access and avoid sources of vibration. The potential effect of the magnetic field on other sensitive equipment, such as gamma cameras, X ray systems, radiotherapy linear accelerators, image intensifiers and electron microscopes, needs to be considered. Nearby structural steel can affect the magnet homogeneity, and while this can be minimized by self-shielding, it may require other steps, such as replacement with non-ferromagnetic stainless steel. Large steel pipes and electrical cables with heavy current load can also present problems. Moving steel from vehicles, trains or subway trains can be particularly difficult to shield. The manufacturer will normally be able to advise on these factors.

15.2.2. The radiofrequency subsystem

The radiofrequency (RF) system comprises the generation of analogue audio frequency pulses modulating the RF Larmor frequency, generated from a digital source and amplified to drive the RF coils that transmit, and in some cases receive, RF to and from the patient. Usually, these are designed to drive a coil that can irradiate the whole body and so the amplifier may be capable of supplying some 15 kW of power. The detected RF signal is weak and so the receive channel(s) must be protected from the transmitted power. There is a wide range of coil designs and these are discussed further below. The RF signal passes through several stages of amplification and is digitally processed, either before or after demodulation from the RF signal. This signal provides the information that builds the images or spectra.

15.2.2.1. Coil designs: Volume, surface, phased array

In conventional systems, the main body coil transmits RF to the patient, and it will typically be circularly polarized (generating a field rotating at the Larmor frequency) and be driven in quadrature. For both transmit and receive, it is important that the coils transmit and detect magnetic fields orthogonal to the main $\boldsymbol{B}_0$ static magnetic field. In some high field machines (3 T and above), parallel transmit systems are being installed to overcome the inhomogeneous $\boldsymbol{B}_1$ transmit fields that arise as the wavelength of the RF approaches the dimensions of the body. These systems have an array of transmit coils with each coil being supplied by a separate amplifier to allow individual amplitude and phase modulation. For some applications, transmit head and transmit surface coils (smaller coils that are positioned over the body — particularly for multinuclear measurements) may be used.

The body coil is also often able to act as a receiver, and allows large volumes of the body and multistation whole body imaging (i.e. images of different sections of the body which are joined together to give a whole body image) to be achieved. Smaller homogeneous volume coils include head coils and knee and extremity coils. There is a range of designs, including birdcage designs (in its simplest form, two circular loops connected by parallel rods). Surface coils have long been used to provide much increased SNR from small volumes, and flexible designs allow many parts of the body to be covered. This approach has been extended to provide phased array coils (a set of closely positioned separate coils each connected to a separate parallel receive channel), allowing signal to be acquired from many such coils simultaneously. These parallel imaging approaches using phased array coils have transformed image quality, with some systems allowing the body to be covered with such coils to enable optimized imaging of many different body areas without the need to move the patient to position additional coils. They have also considerably speeded up many imaging acquisitions. Multinuclear measurements require coils tuned to the particular frequency or, for some applications, coils that will operate at several frequencies. Figure 15.3 illustrates some of the wide range of RF receiver coils available for MRI systems.

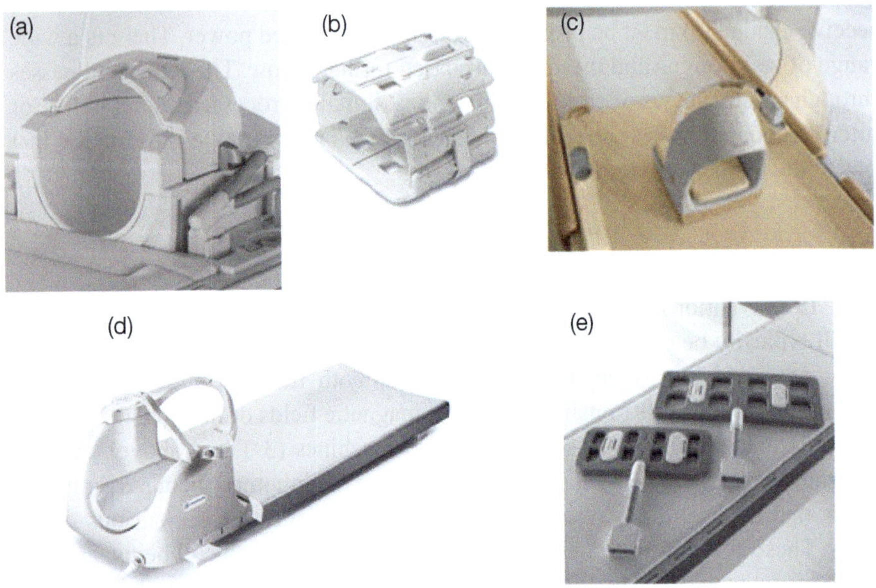

FIG. 15.3. MRI receiver coils: (a) Siemens 32 channel head coil; (b) General Electric torso array coil; (c) Philips Neck coil; (d) General Electric cervical, thoracic, lumbar spine coil; (e) Siemens four channel flex coils.

15.2.2.2. RF shielding requirements: Siting

MRI systems are generally sited in an RF screened room, both to avoid RF interference to other equipment, including radio and television reception, and to prevent external sources interfering with the detection of the very weak RF signal. A screened room will generally be made of copper, aluminium or stainless steel, with RF screened windows and RF doors with knife-edge brushes to maintain the integrity of the screen. The room will be attached to a good earth point, and all services into the room will need to be non-conducting or RF filtered, or incorporate a non-conducting break. It is convenient to have a number of wave guide access channels for non-conducting services.

15.2.3. Gradient coil design and specifications

Spatial localization of image information is governed by three orthogonal sets of coils that can superimpose a field gradient that is added to or subtracted from the main $\boldsymbol{B}_0$ static magnetic field, effectively causing the Larmor frequency to change linearly with position in either the x (usually horizontal: orthogonal to $\boldsymbol{B}_0$), y (usually vertical: orthogonal to $\boldsymbol{B}_0$) or z (orientated along $\boldsymbol{B}_0$) directions. For a standard superconducting magnet, this arrangement is usually effected by two pairs of Golay coils (coils shaped like a saddle positioned on a cylinder) for each of the x and y directions, and a pair of coils coaxial with the magnet windings for the z direction. In practice, these coils are all mounted on a substantial former that is placed within, and concentric with, the room temperature bore of the magnet. The coils are capable of rapid switching and carry heavy currents (of the order of 600 A), as they may generate fields of up to about 60 mT/m. The high power dissipation results in a need for cooling and this is often achieved with water. Particular issues include noise and vibration, which have been addressed by use of a high mass and rigid mountings, with balanced force designs becoming available, and in some cases, the gradients are mounted within a vacuum jacket. Active noise cancellation has also been implemented by some manufacturers.

15.2.3.1. Maximum amplitudes, rise times and slew rates

Large gradient values allow thin slices and high resolution images. They also reduce the impact of $\boldsymbol{B}_0$ field inhomogeneities. Large gradients can also aid diffusion weighted imaging (see below) by allowing larger $\boldsymbol{B}$ values with faster imaging times. Ideally, the gradients will switch very quickly to allow fast imaging sequences. The maximum slew rate describes the maximum rate of change of the gradients, which can also be described as the rise time of the gradients. Slew rates of some 200 $T \cdot m^{-1} \cdot s^{-1}$ are possible in current equipment. In practice, gradient rise times or slew rates are limited by the induction of currents in the body and the possibility of nerve stimulation, discussed under safety below. The slew rate may necessarily be slower for large gradient values to ensure that safe operation is maintained, and manufacturers may offer switching between different modes on a given machine. Often, there may be a choice of gradient options available on a given instrument model.

15.2.3.2. Eddy current effects and compensation techniques

While an imaging pulse sequence may calculate and require a well executed gradient pulse waveform, with a linear ramp up, a maximum value held for a defined period, and a linear ramp down, all within a specified total time, the inductance of the gradient coils, together with the presence of currents in adjacent conducting structures, resulting from the time varying magnetic field generated by the gradient coils, will result in a non-perfect response. Current systems have screened gradient designs that cancel out gradient fields outside the magnet bore, reducing the generation of eddy currents that would themselves generate additional magnetic field gradients. Careful design of the magnet cryostat and heat shields can also minimize the generation of eddy currents. Other imperfections are usually addressed by pre-emphasis corrections in the gradient circuitry that impose additional LC terms on the driving waveform, to compensate for any modulation due to eddy currents. The performance of these will usually be tuned by the manufacturer. Non-corrected eddy currents can lead to time varying magnetic fields that distort the image information.

15.2.4. Computer and control systems

Much of the activity of an MRI system is controlled by the pulse programmer, which is responsible for interpreting the pulse sequences and generates and delivers waveforms to different equipment components, such as the gradient amplifiers and the RF amplifiers, and also receives and digitizes signals at appropriate times. In some cases, the user may have direct access to this programmer, to generate their own sequences. More usually, users will access an interface that allows a limited class of variables to be adjusted for a given sequence. This processor may also be associated with specialized equipment for image calculation, together with the memory to store raw time domain data (the signals that are measured), as well as reconstructed data. Some of these facilities may be provided by one or more host computers or consoles, which may provide for analysis and interface to storage facilities, hard copy output and the picture archiving and communications system. A key feature will be interactive displays, allowing measurements to be planned and image data to be analysed.

15.2.5. Common imaging options

MRI systems have become increasingly complex, with features organized into a range of packages that are marketed for particular clinical applications. In addition, there is a wide range of core equipment configurations, allowing users to match equipment to their clinical requirements, space and budgets. In some areas, the reliability of power supplies, availability of cryogens and quality of engineering support will be important factors in choosing a system. The lifetime cost of adequate engineering and maintenance support should be factored into any decision process. On the basis of these issues, the choice of magnet configuration will be the major determinant. Following this, gradient specification (if an option), clinical packages and RF coils will be the major factors to be investigated. It is helpful to develop a clinical specification against which these needs can be assessed.

15.3. BASIC IMAGE QUALITY ISSUES

Owing to the complexity of MRI systems, there are many features that can contribute to image quality. Here, only some of the major determinants of image quality will be addressed.

15.3.1. B_0 field strength, homogeneity and shimming

The field strength of the B_0 field is a major determinant of SNR, which increases approximately linearly with magnetic field strength. The homogeneity of the B_0 field is fundamental to the quality of an MRI system. This will ultimately depend on the design of the magnet system, and to a first approximation will be governed by the type and size of magnet; generally, the longer it is, the better the homogeneity. This will have been optimized at installation by careful design of the facility to avoid structural steel, etc., and by careful shimming of the magnet (passive and electronic) at set-up, where the engineers should demonstrate that the system meets its specification. Changes in environmental steel or small items of steel that have entered the vicinity of the magnet and have been attracted into the magnet can also cause changes in B_0 inhomogeneity. If there is a sudden deterioration, these potential causes should be investigated.

For some imaging sequences, particularly where frequency dependent fat suppression or water excitation is required, there may be a facility for operator shimming of the B_0 field. This is performed by adjusting the current in a number of the room temperature shim coils incorporated into the system. When spectroscopy is being performed, it is always necessary to shim the system to compensate for the B_0 inhomogeneities arising from the susceptibility of the patient.

Gradient non-linearities, as well as inhomogeneities in the B_0 field, can introduce spatial distortions in the images. Gradient non-linearities are inevitable in the intrinsic design of the system. Eddy currents (as discussed above) and inadequate pre-emphasis correction can also give rise to image distortions and other errors.

15.3.2. B_1 homogeneity and flip angle adjustment

Ideally, MRI sequences are designed with the expectation that a specified flip angle will be delivered uniformly and accurately to the imaging volume. When a new patient is placed in the MRI system, it is usual for the system to perform a calibration to adjust the voltage delivered to the body coil to compensate for the loading of the coil by the patient (which changes coil impedance), the particular coil configurations used and, at higher field, the interaction of the RF field with the subject, where standing wave and dielectric effects can lead to further $\boldsymbol{B}_1$ inhomogeneity. From this calibration, further adjustments can be calculated to allow for particular RF pulse lengths, slice thicknesses and so on.

In practice, a number of factors will lead to inaccurate pulse angles being delivered to all or part of the volume, despite the calculation. These effects include the physical geometry of the transmit coil, the presence of receiver coils within the transmit coil, the effects of standing waves or dielectric effects, and the imperfect shape of slice select pulses resulting from their finite pulse length. In conventional diagnostic imaging, many of these effects may reduce image contrast and SNR, but the image information will remain useable. However, if quantitative imaging is to be performed, these issues need to be addressed or compensated for in the analysis.

15.3.3. Phantoms, equipment assessment and coil loading

The manufacturer will usually provide some standard phantoms that allow basic routine service procedures to be performed. These are often relatively simple and do not fully evaluate the scanner performance. However, they do enable comparison with the manufacturer's specification and maintenance tolerances.

A range of more sophisticated phantoms has been developed and some are available commercially. These will allow measurements of spatial resolution and T_1 and T_2 relaxation times and assessment of spatial distortion. In some cases, purchased or home-made phantoms may not reliably test some equipment features. This is particularly the case with SNR measurements, which depend on the RF coils being appropriately loaded. Sometimes this is achieved by a conducting (ionic) solution in the phantom; in other cases, a separate loading annulus may be provided. If reliance on such measurements is important, users will need to establish that their phantoms are providing accurate measurements and are reflecting the clinical situation. For frequent and routine purposes, a SNR measurement for each imaging coil, under controlled and reproducible conditions, is very useful. This measurement can conveniently be rotated through coils on a regular basis and will demonstrate whether equipment is operating outside its specification. If a local or surface coil measurement is operating outside its specification, it might be appropriate to repeat the measurement on the body coil to rule out or establish whether the fault lies with the surface coil or with the main system. If a fault is found, it may be sufficient to call in the manufacturer, or further tests may be required to identify the nature of the problem. This may require more detailed investigation, for example, to show up distortions or image artefacts, where a phantom with known structures can be used.

To address ghosting or aliasing, 3-D objects with asymmetries can be helpful. There are very many potential sources of artefact and poor image quality, but it is beyond the scope of this book to cover them all in detail. However, a systematic approach to eliminate likely causes can be helpful.

15.3.4. SNR and contrast to noise ratio

With MRI, the signal is highly dependent on the object as well as the imaging sequence and the parameters chosen. Therefore, the SNR needs to be defined for a specific object, coil, geometry and measurement sequence. In a basic MR image, the noise should be distributed by Fourier transform (FT) uniformly throughout the image. However, this can be altered if any temporal frequency filters are applied in the reconstruction or acquisition. With parallel processing and array coils, the distribution of noise can become very complex. These factors, together with the directional propagation of artefacts, suggest that subtraction of two identical images can be a more robust means of determining the noise, taking into account that the noise has a Rician distribution.

15.3.5. Spatial resolution

For 2-D imaging sequences, spatial resolution is governed by slice width and by the in-plane resolution, which is a function of the sampling of the time (k-space) signal, in the two orthogonal directions, together with the FOV, as defined in Section 14.5.3.

While the equations in 14.5.3 provide the theoretical relationship between spatial resolution and sampling, the intrinsic resolution achievable will also be affected by the signal power available at the higher k-space frequencies, which can be affected by the order in which k-space is acquired, and the relaxation properties of the tissues contributing to the signal, together with the duration of the measurement. Reduction in signal due to relaxation, or asymmetrical k-space signal, can result in blurring of the reconstructed images, depending on the nature of the pulse sequence. Images will often be reconstructed to a higher resolution than that originally acquired, effectively interpolating the image. Such interpolation may be greater in the phase encoding direction than in the readout direction, resulting in an asymmetrical resolution. Reduced sampling schemes may also be employed to speed up acquisition, and these can also result in reduced spatial resolution, as well as phase errors and increased noise. Reconstruction may also employ a range of spatial filters that can further affect spatial resolution. Therefore, although the simplistic view is that the in-plane resolution is governed by simple relationships, accurate evaluation of spatial resolution may require experimental verification with phantoms, ensuring that relaxation and susceptibility properties accurately reflect the clinical situation.

Slice resolution will be affected not only by the defined slice thickness, but also by the slice profile, which depends on the RF pulse profiles used to excite the pulse and to generate echoes (where relevant). Slice profile is affected by the relaxation properties of the tissues and by the repetition time (TR), as non-fully relaxed measurements can result in relative suppression of signal in the centre of the slice compared with the tissues at the edges of slices that receive smaller flip angles. Slice profile will vary with pulse sequence and with sequence parameters, and the thickness and profile can be altered by changing the excitation pulse bandwidth, pulse shape and gradient strength. It is best determined experimentally using materials reflecting the relaxation times of the tissues of interest. Slices may be contiguous or have interslice gaps, to reduce overlap of the slice profile, and may be collected serially or in an interleaved fashion.

In 3-D imaging, slice selection is often used to select thick slices, which are then separated into partitions using a phase encoding step. Towards the edges of the slice, the slice profile may affect the accuracy of the flip angle, an effect that is particularly important in quantitative imaging. Partition thickness is defined in the same way as resolution in the readout direction and may also be subject to interpolation, as described above.

15.3.6. Image acquisition time

For simple sequences, image acquisition time is the product of the TR, the number of phase encodes and the number of signal averages used for each image. Multiple slices may be acquired during the TRs, where the sequence has a long TR, at no time cost. With fast imaging, however, the TR may not be long enough to allow additional slices to be obtained, so these will provide a further multiple, extending the measurement time. In practice, however, imaging sequences are much more complex and acquisition time can be extended by preparation pulses, inversion or fat suppression pulses, or acquisition of multiple echoes. Acquisition time can also be reduced by employing different echoes to provide some of the different phase encodes required to encode phase information separately (as in fast spin echo sequences). The 3-D sequences have an acquisition time that is governed by the product of the number of phase steps in each phase encoded direction, but which may also be affected by the other factors detailed above. Reduction in imaging time is now commonly achieved by sharing information from arrays of coils, together with the use of specialized reconstruction algorithms that combine information from each coil (or coil element), allowing for its spatial sensitivity. This process is known as parallel imaging and has led to a significant advance in imaging speed.

15.4. MR IMAGE ACQUISITION AND RECONSTRUCTION

The basic gradient echo and spin echo imaging sequences have been described in Sections 14.5.4 and 14.5.5. These provide the core building blocks of the major families of MR sequences.

15.4.1. Gradient echo sequences

In the gradient echo sequence, the slice select pulse is generally a small angle (typically 5–20°), allowing a very short TR to be employed. The readout can also be shortened by using larger gradients. Spoiling, or destruction of any coherent transverse magnetization at the end of each acquisition, is essential in the widely used spoilt gradient echo sequence. Spoiling can use variations in the phase of RF excitation and acquisition (RF spoiling), or it can use dephasing gradients at the end of the acquisition. There are a number of variants and names for these sequences, with FLASH (fast low angle shot) being frequently used. This approach is commonly used for 3-D sequences, where the small TR allows rapid imaging, enabling many slices to be acquired within a reasonable time. The T_2* weighting is particularly useful in imaging joints.

There are a number of variants based on this sequence. Turbo-FLASH or similar sequences employ preparation pulses, such as an inversion pulse, prior to the acquisition, to condition the contrast, with a whole 2-D image being read out very quickly using very small flip angles and short TRs. A further class of image sequence is based on maintaining a steady state magnetization, by eliminating the spoiling at the end of each repetition, providing image contrast that generally depends upon T_1/T_2*, although the exact contrast dependency depends on the implementation, echo and TRs. One version of this is fast imaging with steady state precession. In order to ensure correct phase encoding, it is necessary to rewind the phase encoding at the end of each repetition, by applying the reverse of the phase encoding gradient pulse.

15.4.2. Spin echo sequence

As the spin echo sequence uses 90° and 180° pulses, it involves relatively high power deposition, which may limit the TR, number of slices or echoes that can be achieved. The sequence is typically used with a long TR of ~2 s, to allow signal recovery and minimize T_1 weighting, and an echo time (TE) of 70–120 ms to obtain T_2 weighted images; a short TE proton density weighted image may also be acquired at the same time by employing two echoes. Usually, a fast spin echo or rapid acquisition with a refocused echo sequence is now used to obtain T_2 weighting, as such sequences are much faster (see below).

15.4.3. Fast spin echo sequence

In the fast spin echo sequence (Fig 15.4), multiple spin echoes are applied, but each receives a separate phase encoding, such that the phase encoding is applied prior to the 180° pulse and is then reversed following readout of the echo. A further phase encoding pulse with encoding for a different line of k-space is then applied prior to a further 180° pulse and read out. Again, that phase encoding is rewound and so on. It will be seen that each line of k-space then has a slightly different TE. The extreme example would acquire sufficient echoes, each differently phase encoded, to sample the entire image. The different echo weighting can affect the quality of the image. Unless the material has a very long T_2, this process will result in very little signal in later k-space lines, affecting spatial resolution. More usual implementations will divide the k-space sampling across multiple repetitions of the sequence, and bands of similar TEs can be grouped together to give echoes that are characterized by a narrow range of TEs. This approach is now very commonly employed for T_2 weighted imaging.

15.4.4. Inversion recovery sequences and applications: Short time inversion recovery and fluid attenuated inversion recovery

Inversion recovery imaging is based on selecting an appropriate point in the recovery of T_1 magnetization following a 180° inversion pulse, to adjust the signal to optimize the contrast between different tissues. The basic principle is described in Fig. 15.5. A frequent application of this technique, using a short inversion time, is to null the signal from fat by selecting image readout at a time when the recovery of the fat signal is passing through the null point. This then provides a fat nulling technique that depends on the T_1 relaxation of fat and provides an alternative to frequency dependent suppression or selection techniques. The imaging readout may be performed using a variety of imaging sequences. This can range from a conventional inversion recovery, with an inversion for each line acquired in the image, to a preparation inversion followed by a very fast gradient echo or fast spin echo readout. Often, sequences will combine these approaches by having suitably spaced inversion pulses and acquiring a number of k-space lines around the inversion time. The fluid attenuated inversion recovery sequence is a variant of this, but designed to null cerebrospinal fluid for central nervous system imaging or, alternatively, fluids

(a)

(b)

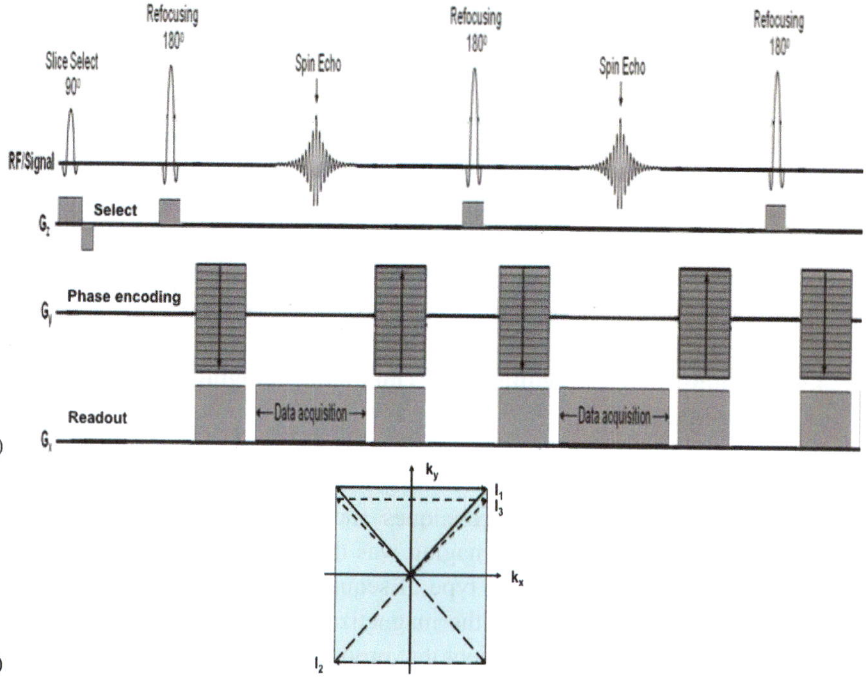

FIG. 15.4. (a) A fast spin echo sequence showing initial slice selection with a 90° pulse, initial slice selective 180° refocusing pulse, followed by phase encoding for the first k-space line, readout of the first k-space line (the spin echo), rewinding of the phase encoding, slice selective 180° refocusing pulse, phase encoding for a second k-space line, readout of the second k-space line (the spin echo), rewinding of the phase encoding, etc. This refocus and readout is continued for as many k-space lines as are required. (b) Lines of k-space acquired by the first three 180° refocusing pulses. The position in the k_y direction depends on the phase of the 180° refocusing pulse and on the amplitude of the phase encoding gradient.

elsewhere in the body. This can facilitate monitoring the structure of tissues that would otherwise be masked by strong signals from adjacent fluids. Further variants of the sequences allow double inversion, so that signal from two tissue groups can be nulled.

15.4.5. Common sequence options: Spatial and chemical saturation techniques

Signal suppression can also be achieved by selecting tissues, using a broad slice defined by a slice select pulse, but with no readout, and immediately followed by spoiler gradients to null the signal in the *x–y* plane. This approach of employing saturation slices or bands is commonly used to reduce an FOV or a null signal that may cause artefacts from moving tissues. The timing of such bands has to consider the T_1 of the tissues concerned and the TR of the sequence. Alternatively, a similar technique, but without the slice select, can be used to saturate a specific type of tissue (usually fat), taking advantage of the fact that it has a different resonant frequency to water. One approach is called CHESS — chemically specific saturation. This excitation is followed by spoilers to dephase the signal prior to imaging. The approach is also used in spectroscopy. There are optimized approaches such as WET, which use a tailored range of pulses to improve suppression of the fat signal. An alternative to this approach is to use chemically specific excitation and to excite just the water signal.

15.4.6. Ultrafast imaging sequences: Echo planar imaging and spiral techniques

While the fast gradient echo techniques can be employed with preparation to obtain very fast readout in techniques such as SNAPSHOT FLASH, an independent form of single shot imaging was developed by P. Mansfield (echo planar imaging technique). In this type of sequence (see example in Fig. 15.5), following a slice selective pulse, the magnetization is continually interrogated by a large alternating readout gradient that produces a string of echoes. A second gradient provides phase encoding, which may be done by a small continual gradient, or repeated blips (blipped echo planar), with the origin of k-space being adjusted by a preparatory phase encoding offset pulse. At one time, this approach required advanced gradient design, but such gradients are now available on most commercial systems. The entire image may be read out in, typically, 50–100 ms, or may be built up from a number of interleaved measurements in a longer measurement time. Such an approach can reduce the T_2* weighting of the measurements and the degree of image distortion. K-space can be sampled by a range of more complex trajectories, including spiral and radial, which can confer time savings and be advantageous for some imaging approaches.

15.4.7. MR angiography sequences

Sequences for measuring vascular structures can make use of the phase encoding that occurs for blood flowing in the presence of a gradient, allowing the direction and velocity of flow to be measured. Techniques that exploit the inflow of unsaturated blood into a saturated slice can also provide high contrast for vascular structures. More often, current techniques exploit the additional contrast and signal that can be generated by making measurements as a bolus of contrast agent passes through the vascular tree, enabling faster imaging and higher SNR than is feasible without contrast agents. Preferably, a blood pool agent would be used, where the contrast agent is bound to a protein such as albumin that

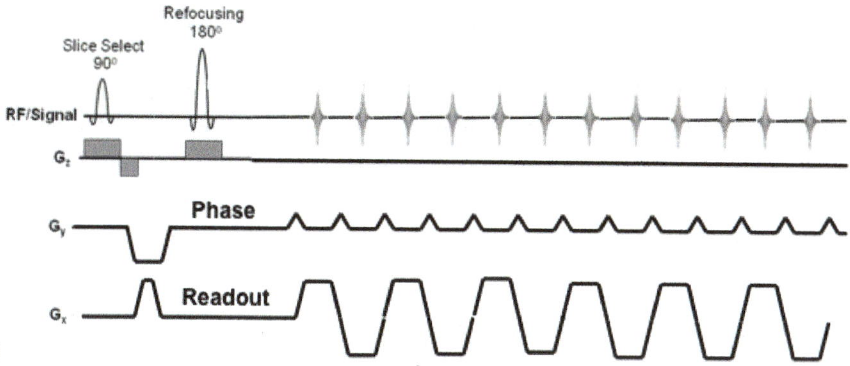

(a)

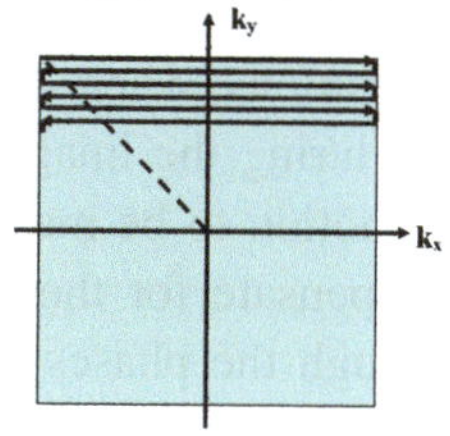

(b)

FIG. 15.5. (a) Blipped echo planar sequence showing single shot spin echo slice selection module, followed by readout of multiple lines of k-space (b) in the presence of an alternating readout gradient, with the initial phase offset and an incremented phase for each line of k-space.

is not rapidly excreted from the vascular system. This avoids the tissue blush that occurs because of smaller molecules leaking from the vasculature into the extracellular space of tissues, which can reduce vascular contrast.

15.4.8. Flow measurements

Bulk flow in blood vessels is most often measured by phase contrast MRI, which allows the direction and velocity of flowing blood to be measured. This uses the phenomenon whereby spins moving along a magnetic field gradient gain (or lose) phase compared with stationary spins. The method usually involves comparing spatial variation in phase between a sequence with no flow encoding gradient with one in which the phase gain due to flow is encoded by a bipolar pair of gradients, enabling the phase gain due to flow to be calculated and presented as an image. This technique is also often used for angiography, as it can be sensitized to pick up flowing spins. Another approach is to set up a slice along the vessel of interest and monitor the distance travelled by labelled blood (either unsaturated or tagged, for instance with an inversion pulse) in a given time. This time-of-flight method also allows profiles to be measured through a particular vessel. Extensions of this approach are also used to measure tissue perfusion by arterial spin labelling. These techniques are based on labelling blood in a slice outside the tissue of interest and then observing the delivery of those labelled spins in blood flowing to the tissue of interest. Tissue in the region of interest may firstly be saturated, or the method may rely on inverting the labelled inflowing signal. To avoid magnetization transfer effects, a set of images using a control labelling slice on the opposite side of the sample may be acquired.

15.4.9. Cardiac measurements

Cardiac measurements have been a major area of development in MRI. Initially, morphological images utilized cardiac triggered acquisitions, but more recently, navigator triggered acquisitions, where a signal profile along a column of tissue is measured frequently during the imaging sequence, have been used to allow the dynamics of tissue motion to be measured and the pulse sequence synchronized or adjusted to compensate for the tissue motion. MR can make excellent anatomical images through the phases of the cardiac cycle, allowing direct visualization of cardiac wall motion. Motion encoding based on the phase techniques described above can allow accurate measurement of tissue motion, as well as providing insight into bulk flow of blood and cardiac valve operation. Tissue tagging studies can impose saturation bands on the images in one phase of motion, allowing the movement of these tagged bands to be mapped in any direction. These techniques, combined with sophisticated display software, can provide highly informative displays and measurements of cardiac function. In addition, real time imaging can also be used to capture and follow irregular cardiac motion. Contrast agents can be used to evaluate cardiac muscle perfusion, helping to identify areas of cardiac damage.

15.4.10. Diffusion measurements

As water molecules move randomly in unimpeded space in the body, their motion can be measured by firstly applying a phase change, and then, after a set time, applying an opposite gradient to rewind the phase gain. Molecules that have not moved will suffer no net phase change and therefore no signal loss. However, spins that have moved will experience a change in phase proportional to the distance moved in the direction of the applied gradients. This loss of phase results in a loss of signal that is dictated by the strength of the gradients, their duration and the interval for movement to have occurred. The signal at TE, $S(T_E)$, compared with that at $t = 0$, $S(0)$, is given by:

$$\ln\{S(T_E)/S(0)\} = -T_E/T_2 - \gamma^2 G^2 D \delta^2 (\Delta - \delta/3) = T_E/T_2 - bD \quad (15.1)$$

where γ is the gyromagnetic ratio, G is the applied gradient, D is the diffusion coefficient of water, and δ and Δ are shown in Fig. 15.6. The effects of the pulse sequence parameters are often incorporated into a term b, known as the b value, largely driven by the strength of the gradients and their duration.

Diffusion weighted imaging exploits the loss of signal resulting from diffusion, which leads to water molecules in fluid spaces losing signal faster than those in more highly cellular tissues, such as tumours, where water can move less freely. The approach is gaining ground in identifying disseminated cancer, where it appears to be sensitive to high cellularity lesions and involved lymph nodes. In this context, whole body imaging is rapidly developing, requiring good fat suppression to maximize lesion contrast. In a number of fields, calculation of the apparent diffusion coefficient, describing the diffusion in a restricted environment, is of interest. Here, a set of images with at least two b values needs to be acquired to calculate the apparent diffusion coefficient. If a whole series of b values are obtained, it is possible to identify different components causing signal loss, one of which may be attributed to tissue perfusion (the intravoxel incoherent motion technique). In structured tissues, the directional properties of the diffusion can be exploited to demonstrate the orientation and connectedness of groups of nerve tissues. Very elegant tractograms of neural interconnections have been developed, which complement other functional and structural neurological examinations. Such measurements rely on diffusion tensor analysis, where diffusion is sensitized in many different directions, requiring multiple acquisitions to build the tractograms. A simpler measurement of value in some applications is to measure diffusion anisotropy.

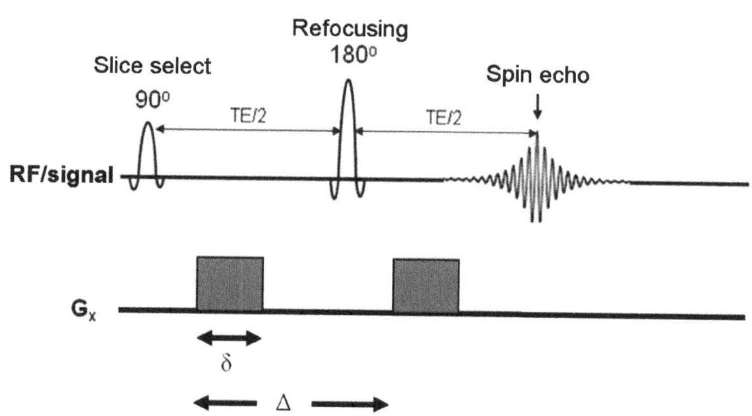

FIG. 15.6. A diffusion weighted spin echo sequence showing the diffusion sensitizing gradient.

15.4.11. Brain activation measurements

Brain activation measurements are based on the blood oxygen level dependent measurement technique. This employs T_2* weighted images of the brain, exploiting the phenomenon that deoxyhaemoglobin is paramagnetic and oxyhaemoglobin is diamagnetic. Areas of the brain with increased function utilize more oxygen, resulting in an increase in deoxyhaemoglobin and a change in magnetic susceptibility, which reduces the signal observed in T_2* weighted images. In parallel, there is an increased perfusion to the area, which can also affect the signal measured. The technique employs paired images with and without a neural stimulus (such as a visual or mechanical paradigm), with the differences between the two images highlighting areas of neural function. Changes in local blood flow as a result of increased demand can also result in changes to these images. The technique has evolved into a very important tool for evaluating and understanding neural function. Advances in field strength are allowing these measurements and specific brain functions to be mapped to increasingly better localized regions of the brain.

15.4.12. Dynamic contrast enhanced MRI

Contrast agents are described in Section 14.3.4, together with their use in improving visualization of disease. Dynamic contrast enhanced imaging utilizes the dynamic behaviour of a contrast agent, usually a chelate labelled with gadolinium, to provide information on tissue function that may inform on pathology. Probably most widely used in cancer, it builds on the success of contrast enhanced imaging to identify suspicious lesions, using the uptake of contrast and the appearance of the lesion. By observing the uptake and washout of the contrast agent, the clinician may be better able to discriminate malignant from benign lesions and to identify the areas of greatest vascular permeability. In tumours, the development of a highly permeable neovasculature is an important feature of tumour development. In some tumours, characteristic types of washout curve serve as diagnostic aids. Building on these approaches, quantitative imaging approaches have been developed that allow the concentration of contrast agent to be calculated. The dynamic change in this concentration can then be used as the input to a pharmacokinetic model that allows parameters describing the vascular properties of tissues and the related exchange of contrast agent to be described.

15.4.13. MR spectoscopy sequences

MR spectroscopy allows chemically specific measurements of a range of nuclei to be made. In the body, these nuclei are most commonly ^{1}H, ^{19}F, ^{31}P and ^{13}C.

By accurately identifying the chemical shift (or resonance frequency) of resonant lines, the molecular origin of the line can be identified. Initially, measurements focused on energy metabolism, in particular, utilizing the measurements of phosphocreatine and adenosine triphosphate that could be performed in muscle using ^{31}P spectroscopy. More recently, the behaviour of phospholipids such as phosphocholine and phosphoethanolamine has been of interest, particularly in tumours. With improvements in technology, ^{1}H spectroscopy has become

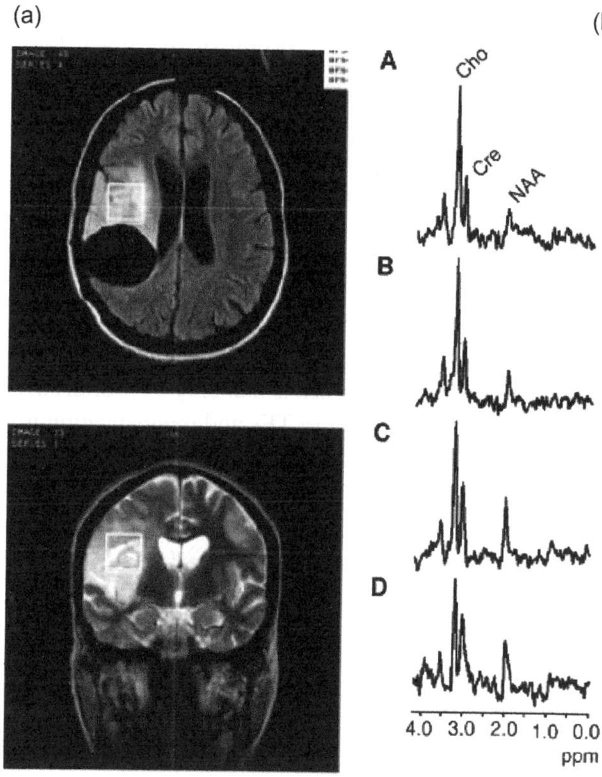

FIG. 15.7. (a) Pretreatment fluid attenuated inversion recovery (top) and T_2 weighted FSE (bottom) pretreatment images from a patient with low grade glioma receiving treatment for recurrent disease, showing the position of voxel selected for spectroscopy and (b) serial ^{1}H spectroscopy measurements from the same patient. Panel showing long TE (TE= 135 ms) STEAM spectra obtained before (A), 3 months (B), 6 months (C) and 9 months (D) after initiation of temozolomide treatment. A progressive decrease in the choline/creatine (Cho/Cre) ratio was observed, suggesting reduced membrane metabolism and diminishing cellular density. Also note the increasing conspicuity of the N-acetyl aspartate peak, a specific neuronal marker whose level may reflect the regression of tumoural tissue and repopulation of normal brain matter (from Murphy, P.S., et al. [15.1] (reprinted with permission)).

practicable and can be provided on many MRI systems without the need for a broadband RF capability and additional amplifiers. The principal signals include total creatine, total choline, *N*-acetyl aspartate in the brain, citrate in the prostate, lipids and lactate. An example is shown in Fig. 15.7. Spectroscopy requires the magnet to be shimmed to optimize field homogeneity over the region of interest, ideally to the order of 0.1 ppm. This is often achieved with automated routines that adjust currents in a number of shim coils. Proton (^{1}H) spectroscopy generally requires suppression of the water signal to avoid saturation of the analogue to digital converter, although some systems have a sufficient dynamic range to avoid the need for water suppression. While much spectroscopy has been performed using a small receive coil to localize the origin of the signal (a surface coil), improved spatial definition is achieved by using a localization sequence.

15.4.13.1. Single voxel localization

Single voxel localization is designed to sample signal from a well defined region, often cubic. For ^{1}H spectroscopy, two different techniques are in common use. Stimulated echo acquisition mode (STEAM) spectroscopy utilizes stimulated echoes, employing a sequence of three orthogonal slice selective 90° pulses. The good slice profile of 90° pulses results in a sharp definition of the volume of interest. An intrinsic feature of this approach is that only 50% of the available signal is sampled. The technique can deliver short TEs and magnetization is stored along the z axis between the last two pulses, reducing T_2 signal decay. Point resolved spectroscopy employs one slice select 90° pulse, followed by two orthogonal 180° pulses. This approach delivers all of the magnetization and is less prone to motion artefacts than is STEAM. With improved gradient performance, short TEs can be achieved. Either sequence can be preceded by water suppression pulses such as CHESS or WET.

Although these techniques can, in principle, be applied to ^{31}P, the short T_2 relaxation times result in considerable signal decay and it is more usual to collect free induction decays, avoiding T_2 losses. The preferred approach for this is to use the image selected in vivo spectroscopy technique, which employs eight separate permutations of three preparation slice selective inversions, each permutation being followed by a 90° readout pulse, generating a free induction decay. By appropriate combination of these eight free induction decays, a localized signal from the space representing the intersection of the planes is generated with no loss of signal.

15.4.13.2. Spectroscopic imaging

The above techniques require careful positioning of the volume of interest prior to acquisition. If further voxels are required, the sequence will need to be repeated. Spectroscopic imaging or chemical shift imaging employs imaging approaches to obtain an array of voxels in 2-D or 3-D. For proton spectroscopic imaging, a stimulated (STEAM) or spin echo (point resolved spectroscopy) based technique is employed to select the FOV, allowing potential contaminating signals such as lipid to be excluded. Phase encoding in two or three dimensions is then employed to sample the large FOV in an array of voxels. While taking longer than single voxel spectroscopy, signal from many voxels is obtained simultaneously. These voxels are not as sharply defined as is the case for single voxel spectroscopy, with a point spread function defined by the number of sampling points in each dimension. For ^{31}P spectroscopy, a preselected volume of interest is less commonly used; the volume of interest covering the entire object space (see Fig. 15.8). However, a smaller FOV can be established using saturation slabs surrounding the desired volume of interest.

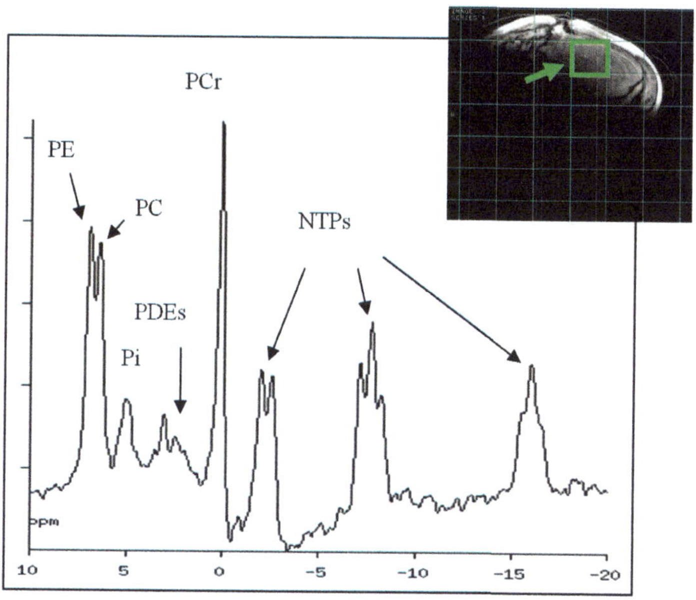

FIG. 15.8. A ^{31}P MR spectrum acquired on a 1.5 T scanner using ^{1}H decoupled ^{31}P chemical shift imaging from an abdominal mass in a patient. The inset image shows the grid used to acquire the array of spectra, and the highlighted voxel shows the position of the selected spectrum. PE: phosphoethanolamine, PC: phosphocholine, Pi: inorganic phosphate, PCr: phosphocreatine, NTPs: nucleotide triphosphates.

15.5. ARTEFACTS

Image artefacts in MRI can be generated by a wide variety of factors resulting from the behaviour of the sample, imperfections in the equipment or its operation, or poorly optimized measurement sequences. Some of the more important artefacts are described below. With careful attention to quality assurance and to tuning and optimizing sequences, many artefacts can be eliminated or substantially reduced. Those intrinsic to the measurement approach, or due to patient characteristics or motion, are harder to avoid, but can often be ameliorated with a judicious choice of imaging sequence, orientation of phase encode direction and application of saturation bands.

15.5.1. Motion

Tissue motion during MR image acquisition results in additional phase gains and incorrect phase encoding, which can lead to characteristic artefacts propagated in the phase encode direction. Common effects are ghosted, displaced, lower intensity images of fat in the body, particularly from the anterior abdomen, and multiple repeat copies of major vessels aligned in the phase direction with major arteries or veins. The former can be reduced by signal averaging, respiratory gating or navigator triggered acquisitions, or rotation of the phase encode direction. Alternatively, fat suppression can be employed, or a saturation band placed over the source tissue. Motion from vessels can be suppressed by out of plane suppression of inflowing blood, and sequences can include motion rephasing gradient lobes.

15.5.2. Aliasing or 'wrap around'

Aliasing occurs where the sample extends outside of the imaging FOV. The reconstruction process cannot then discriminate between tissues positioned within one edge of the FOV and those on the other side that are an equal distance outside the FOV. A common example is areas at the side of the body folded into the FOV. The problem can be increased in areas where the gradients are non-linear, which may result in several areas having the same gradient strength despite being at different positions. These effects can be mitigated by altering the phase encode direction, or by using saturation slabs or RF suppression blankets.

15.5.3. Metal objects

Ferromagnetic materials can cause major distortions on the local magnetic field, resulting in displacement of signals in position and also loss of signal, an effect that can be minimized by utilizing spin echo sequences with short TEs. Other metallic materials can cause susceptibility artefacts, leading to some local distortion and loss of signal, and can also conduct currents induced by switched gradients, although such effects are usually small.

15.5.4. Chemical shift

Water and fat resonate at different frequencies, with a difference of some 3.4 ppm. MRI systems usually adjust their reference frequency to that of water, with the result that where localization depends on frequency, as in slice selection or frequency encoded readout, the fat signal will be spatially displaced from the water signal, effectively producing a shifted image. This can result in areas of signal void or brighter signal, owing to overlap of water and fat signals in the readout direction. In frequency terms, this effect becomes greater the higher the field and thus the larger the frequency separation (in hertz) of fat and water. The effect is generally minimized by ensuring that the bandwidth per pixel is of the order of the frequency separation between fat and water. However, increasing the bandwidth per pixel to this value can result in increased noise per pixel compared with a narrower frequency range per pixel. Alternative strategies include fat suppression and water excitation.

15.5.5. Truncation

Signal digitization involves sampling the echo with a predetermined number of samples (typically 128 or 256), each taking a specified time. The evolution of the echo may mean that at the beginning and end of sampling there is a finite signal and this may be asymmetrical. A similar effect can be seen in the phase encoding direction, where the signal has not reduced to zero at the maximum gradient values. This, effectively, is equivalent to multiplying the signal with a square function, the FT pair of a sinc function. This results in the FT image being convolved with a sinc function that produces ringing at sharp edges in the image, usually seen as parallel bands in the frequency or phase, or in both directions.

15.5.6. System related artefacts

A number of artefacts are intrinsic to the MR hardware and may not be adjustable without major hardware changes or tuning.

15.5.6.1. Distortions

Spatial distortions in images primarily result from the design of the gradient coils, where it is difficult to maintain the linearity of the gradients towards the edges of the FOV, which are physically close to the gradient windings. This problem becomes worse the shorter the magnet bore, as the gradient coils must necessarily be close to the edge of the FOV. The problem can be reduced by increasing the bandwidth per pixel, which, although reducing distortion, will increase the noise in the image. Software correction may be incorporated within the image calculation process.

15.5.6.2. RF coil problems and RF interference

The transmit RF coil dictates the uniformity of the $\boldsymbol{B}_1$ irradiation field and, therefore, how consistent the pulse angle is over the volume of interest. Largely, this will be governed by the design of the coil, but it can also be affected by dielectric interactions between the RF field and the patient, which can result in a further sample dependent variation in $\boldsymbol{B}_1$. This latter effect is reduced by using a circularly polarized transmit coil, but it also increases with increasing RF frequency and thus $\boldsymbol{B}_0$ field strength. These effects will result in flip angle, and therefore excitation or inversion efficiency, variations over the sample. For some pulse sequences, the effects can be reduced by using adiabatic pulses, which, above a certain threshold voltage, are power independent. However, the range of applications of such pulses is limited.

15.6. SAFETY AND BIOEFFECTS

The safety issues in MR can be conveniently divided into acute hazards, with an immediate risk of injury because of interactions between the patient, staff, or extrinsic objects with the fields produced during an MR examination, and biological effects that can arise from interactions between the MR fields and biological processes, which might cause longer term effects. Both of these types of hazard will be considered. For most purposes, the acute fields are of most concern and importance in the day to day operation of MR facilities and their design. The operation of MR equipment and the exposure of patients, staff and the general public are covered by standards, guidance and regulations that are being continually updated. These vary from country to country, as does the degree of regulation. While some examples of current guidelines and regulations are given below, the requirements in any given country should be ascertained. If no local guidance is available, it would be prudent to follow the guidance and regulation in force elsewhere to ensure safe practice. In the UK, the Medicines and Healthcare Products Regulatory Agency publishes a useful guide that summarizes the recommendations of different bodies and, in addition, provides much practical advice on operating clinical MR facilities [15.2]. An American College of Radiology guidance document also provides a range of practical advice on procedures to ensure patient safety [15.3]. For patient exposures, several recommendations adopt three levels of exposure (Health Protection Agency (HPA) [15.4], International Commission on Non-Ionizing Radiation Protection (ICNIRP) [15.5] and the International Electrotechnical Commission (IEC) [15.6]), which are: (i) normal operating mode for routine scanning of patients, (ii) controlled operating mode for specific examinations above the normal operating mode output level, carried out under medical supervision, based on a clinical decision balancing potential adverse effects against foreseen benefits and (iii) experimental operating mode carried out at levels above the controlled operating mode and for which local ethical approval has been obtained.

15.6.1. Static field considerations: Projectile, effects on implants, physiological effects

The static magnetic field affects ferromagnetic materials, magnetic storage devices, magnetically operated switches, sensitive sound receivers and sensitive electronic equipment (particularly that using accelerated electrons such as linear accelerators or photomultiplier tubes). From a safety point of view, the two major areas of concern are (i) the potential for projectiles, equipment and objects containing ferromagnetic materials that can be attracted into the magnet and (ii) implants that are sensitive to magnetic fields.

Equipment and objects to be used close to the magnet need to be tested and ferromagnetic materials must be rigorously excluded from the room containing the magnet, or an otherwise defined inner controlled zone, which should only be accessible to trained and approved staff and to appropriately screened subjects.

Certain implants may also be or become ferromagnetic, or may be damaged or malfunction in a strong magnetic field. These include cardiac pacemakers, which should not be allowed into a field greater than 0.5 mT, hearing aids, other magnetically controlled or programmable implants and some steel clips where some grades of stainless steel can become magnetic when worked or adjusted. A valuable web site maintained by F. Shellock [15.7] provides details on many implanted devices. Details regarding the exact make and model of an implant should be obtained and the safety checked with the manufacturer before scanning. While the attractive effect of the field is strong and, depending on the type of magnet, can extend over a considerable distance, the torque that can be produced on implanted objects, producing a force that seeks to align the object with the field, can also pose considerable hazards if there are ferromagnetic implanted objects. At higher fields (>2 T), the static field can cause a degree of dizziness or disorientation if the head moves quickly in the field. This is believed to be due to currents in the vestibular system interfering with normal sensory perceptions. It is important at high fields to move slowly in the magnetic field and to avoid abrupt motion of the head.

The static field causes few verified biological effects at the fields used in clinical practice. A moving conductor, such as blood, can result in an induced current in the presence of a magnetic field. The levels of current produced by magnetic fields up to 8 T in the body are below levels of concern. No serious or permanent health effects have been found from human exposures up to 8 T, but scientific investigation has been limited. For patient exposure, ICNIRP guidance limits exposure to 4 T and 8 T for normal and controlled operating modes, respectively.

15.6.2. RF field considerations: Tissue heating, specific absorption rate, burn injuries

The RF electromagnetic field used to manipulate the sample magnetization is usually applied by the body coil. Owing to the large volume of the body and thus of the coil, this is driven by high voltages generated by a powerful amplifier, which may typically have an output of 15 kW. The RF field can give rise to heating in the body, and the equipment is designed to limit the specific absorption rate, and thus the heating effects of the RF power, to 0.5–1°C, depending on the condition of the patient, the length of the measurement and the clinical need. Receiver coils used to receive signals should be decoupled from the transmit field, but fault conditions in coils, or incorrect set-up or positioning of connection leads can give rise to local coupling and the deposition of excessive local RF power, resulting in heating.

RF burns are the most common form of accident seen in MRI. Operators need to be continually in communication with patients and volunteers, and should stop an examination immediately if there is any expression of discomfort from the patient. Great care has to be taken over other conducting materials within the magnet, such as electrocardiogram leads, where there are special requirements, and the manufacturer's guidance needs to be followed carefully. While the effects of RF power in producing heating are well established, there is no clear evidence for non-heat induced long term effects at exposures of up to 4 W/kg.

Patient and volunteer exposure to RF irradiation is designed to limit temperature rise in the body, as detailed in Table 15.1, based on HPA [15.4], ICNIRP [15.5] and IEC [15.6] guidelines. Table 15.2 provides simplified derived specific absorption rate limits. More detailed information is provided in the references. As for the static fields, exposure limits are defined in the three modes of operation:

(i) *Normal operating mode*: Exposure of extended volumes of the body should be such as to avoid a rise of more than 0.5°C in the body temperature of patients and volunteers, including those compromised with respect to their thermoregulatory ability.

(ii) *Controlled operating mode*: A relaxation of the basic restrictions on the rise in body temperature to 1°C can be envisaged if the patient or volunteer is monitored medically and with appropriate physiological monitoring. HPA guidelines (2008) [15.4] suggest that particular consideration should be given to restricting the use of the controlled mode for imaging infants, pregnant women, febrile patients and others with reduced thermoregulatory ability or with compromised peripheral circulation.

(iii) *Experimental operating mode*: Any scanning in this mode, which may result in a whole body temperature rise above 1°C, requires ethics committee approval. Medical, thermal and physiological monitoring is required. In addition to the limit on increase in core temperature due to the MR investigation, absolute limits in temperature rise in core and regional temperatures are defined. The 3rd edition IEC (2010) standard differs slightly from the preceding edition of the standard, and from the ICNIRP (2004) guidance [15.5], allowing a 40°C temperature for the core at the controlled level of exposure and >40°C in the research mode. HPA (2008) guidelines [15.4] provide fixed values for the research mode.

TABLE 15.1. BASIC RESTRICTIONS ON THE MAXIMUM TEMPERATURE FOR THE BODY[a]

Operating mode	Spatially localized temperature limits[b] (°C)			Body core temperature (°C)		Local tissue temperature (°C)
	Head	Trunk	Extremities	Maximum core temperature rise[b,c]	Maximum core temperature[c]	Maximum local temperature[c]
Normal	38	39	40	0.5	39	39
Controlled	38	39	40	1.0	40	40
Research	>38	>39	>40	>1.0	>40	>40
Research[d]	39	40	41	2		

[a] From these temperature rise limits, specific absorption rates can be derived. These are usually implemented in the software and hardware of clinical scanners.
[b] ICNIRP (2004) [15.5].
[c] IEC (2010) [15.6].
[d] HPA (2008) [15.4].

TABLE 15.2. PATIENT AND VOLUNTEER SPECIFIC ABSORPTION RATE LIMITS (W/kg) FOR RF FIELD EXPOSURE

Mode ↓	Volume transmit coil				Local transmit coil					
	Whole body	Head		Not head	Head		Trunk		Extremities	
Reference	a, b	a	b	a, b	a	b	a	b	a	b
Normal	2	3	3.2	2–10	10	10	10	10	20	20
Controlled	4	3	3.2	4–10	10	20	10	20	20	40
Restricted	>4	>3	>3.2	>(4–10)	10	>20	>10	>20	>20	>40

a: ICNIRP (2004) [15.5].
b: IEC (2010) [15.6].

15.6.3. Gradient field considerations: Peripheral nerve stimulation, sound pressure levels

The switched gradient fields used for localization give rise to two major effects — induction of currents in conducting materials in the body and vibration — both of which can produce acute effects. Induced currents can give rise to involuntary nerve stimulation, causing muscles to twitch and causing pain at higher levels of stimulation. The likelihood of nerve stimulation depends on the local strength of the gradient, the rate of change of the gradient, the gradient waveform and the individual's susceptibility to stimulation. As the strength of the gradient depends on the position in the bore of the magnet and the design of the gradients, the probability of stimulation in a given individual for a particular sequence can be hard to predict. Gradient levels are usually set to avoid stimulation, but mild stimulation can be acceptable. In principle, much higher gradients can give rise to sufficiently high cellular potentials to induce cardiac fibrillation, but the threshold for this is a factor of approximately ten above that for nerve stimulation.

The HPA (2008) [15.4] has summarized the literature on the effects of low frequency time varying fields typical of those used in MRI. No consistent evidence of harm following short term exposures was found, although some subtle biological effects have been reported. Subjects with epilepsy or who are taking drugs that reduce seizure activity may exhibit increased sensitivity to stimulation by electric fields induced in the cortex and should be imaged with caution.

Both the ICNIRP (2004) [15.5] and IEC (2010) [15.6] provide guidance on the restrictions for time varying magnetic field gradients to be incorporated into MR scanners. They state that the system must not provide a gradient output that exceeds a defined percentage of the limit for peripheral nerve stimulation for a given operating mode:

- *Normal operating mode*: The gradient system shall operate at a level that does not exceed 80% of the directly determined mean threshold peripheral nerve stimulation, where the threshold value is defined as the onset of sensation.
- *Controlled operating mode*: The gradient system shall operate at a level that does not exceed 100% of the directly determined mean threshold peripheral nerve stimulation.

Additionally, the IEC has set a limit to prevent cardiac stimulation.

15.6.4. Common MR contrast agents

Paramagnetic contrast agents are the most common type of contrast used in MRI, where a strong local magnetic field due to unpaired electrons gives rise to a strong net electronic magnetic moment. Such agents are highly reactive, and therefore potentially toxic, and the paramagnetic nucleus has to be bound within a chelate to prevent direct interactions between the ions and biological molecules. The contrast agent works by relaxing nearby water molecules, with the T_1 relaxation process being the predominant mode of relaxation. As well as the strength of the paramagnetic effect, the access of water molecules, governed by the structure of the molecule, also governs the degree of relaxation produced.

The most common paramagnetic contrast agent ion is gadolinium, which can be bound to a number of different chelates. These are grouped as charged and uncharged linear chelates, and charged and uncharged macrocyclic chelates. Recently, a rare condition, nephrogenic systemic fibrosis has been reported, which is associated with exposure to certain gadolinium based contrast agents. It only occurs in patients who have had prolonged exposure to contrast agent, owing to severely impaired renal function. Guidelines now limit the use of such contrast agents in patients with severely impaired renal function (e.g. Ref. [15.3]).

REFERENCES

[15.1] MURPHY, P.S., et al., Monitoring temozolomide treatment of low-grade glioma with proton magnetic resonance spectroscopy, Br. J. Cancer **90** (2004) 781–786.

[15.2] MEDICINES AND HEALTHCARE PRODUCTS REGULATORY AGENCY, Safety Guidelines for Magnetic Resonance Imaging Equipment in Clinical Use, MHRA Device Bulletin DB2007(03), Department of Health, London (2007), http://www.mhra.gov.uk/home/groups/dts-iac/documents/publication/con2033065.pdf (accessed on 12 September 2012).

[15.3] AMERICAN COLLEGE OF RADIOLOGY, ACR Guidance Document for Safe MR Practices: 2007, AJR Am. J. Roentgenol. **188** (2007) 1447–1474, http://www.ajronline.org/content/188/6/1447.full (accessed on 12 September 2012).

[15.4] HEALTH PROTECTION AGENCY, Static Magnetic Fields — Report of the Independent Advisory Group on Non-ionising Radiation, HPA Rep. Docs HPA, RCE-6, HPA, Chilton, UK (2008), http://www.hpa.org.uk/web/HPAwebFile/HPAweb_C/1211184025757 (accessed on 23 August 2012).

[15.5] INTERNATIONAL COMMISSION ON NON-IONIZING RADIATION PROTECTION, Medical magnetic resonance (MR) procedures: Protection of patients, Health Phys. **87** (2004) 197–216.

[15.6] INTERNATIONAL ELECTROTECHNICAL COMMISSION, Medical Electrical Equipment — Particular Requirements for the Basic Safety and Essential Performance of Magnetic Resonance Equipment for Medical Diagnosis, IEC 60601-2-33 Edn 3.0, IEC, Geneva (2010).

[15.7] MRIsafety.com, www.mrisafety.com (accessed on 13 September 2012).

BIBLIOGRAPHY

BERRY, E., BULPITT, A.J., Fundamentals of MRI: An Interactive Learning Approach, CRC Press, Boca Raton, FL (2009).

BROWN, M.A., SEMELKA, R.C., MRI: Basic Principles and Applications, John Wiley & Sons, Hoboken, NJ (2010).

DORAN, S.J., LEACH, M.O., "Spatially localised magnetic resonance", Webb's Physics of Medical Imaging (FLOWER, M.A., Ed.), CRC Press, Boca Raton, FL (2012).

HAACKE, E.M., BROWN, R.W., THOMPSON, M.R., VENKATESAN, R., Magnetic Resonance Imaging: Physical Principles and Sequence Design, John Wiley & Sons, New York (1999).

HASHEMI, R.H., BRADLEY, W.G., LISANTI, C.J., MRI: The Basics, Lippincott Williams & Wilkins, Philadelphia (2010).

INTERNATIONAL COMMISSION ON NON-IONIZING RADIATION PROTECTION, Amendment to the ICNIRP statement on medical magnetic resonance (MR) procedures: Protection of patients, Health Phys. **97** 3 (2009) 259–261.

INTERNATIONAL COMMISSION ON NON-IONIZING RADIATION PROTECTION, Guidelines on limits of exposure to static magnetic fields, Health Phys. **96** 4 (2009) 504–514.

MAMOURIAN, A.C., Practical MR Physics, Oxford University Press, Oxford (2010).

WEISHAUPT, D., KOECHLI, V.D., MARINCEK, B., FROEHLICH, J.M., How does MRI Work?: An Introduction to the Physics and Function of Magnetic Resonance Imaging, Springer, Berlin (2006).

Chapter 16

DIGITAL IMAGING

D. CLUNIE
CoreLab Partners, Princeton,
United States of America

16.1. INTRODUCTION

The original means of recording X ray images was a photographic plate. Nowadays, all medical imaging modalities provide for digital acquisition, though globally, the use of radiographic film is still widespread. Many modalities are fundamentally digital in that they require image reconstruction from quantified digital signals, such as computed tomography (CT) and magnetic resonance imaging (MRI).

16.2. IMAGE ENCODING AND DISPLAY

16.2.1. Characteristics of digital data

16.2.1.1. Pixels and voxels

Individual digital images typically consist of a 2-D rectangular array of regularly sampled picture elements or 'pixels'. When multiple images are grouped together for some purpose, such as to provide multiple samples through time or space, then each image may be referred to as a 'frame' and the entire set of frames may be referred to as a 'multiframe image'.

Pixels are typically square, in that the two physical dimensions in space that they represent are equally sampled along the horizontal and vertical axes. These dimensions are usually expressed as 'pixel spacing', i.e. the distance between the centres of each pixel. The meaning of pixel spacing depends on the type of acquisition. For cross-sectional modalities such as CT and MRI, the pixel spacing within the patient is known at the time of image reconstruction. For projection radiography with a diverging X ray beam, the pixel spacing is dependent on the position along the beam and subject to geometric magnification, and hence can only be defined at a known reference point, typically the central ray at the front face of the detector. For photographic and scanned images, a reference of known size within the image can be used for calibration.

When an image or frame consists of a cross-sectional 'slice' through the body, then an additional dimension exists, normal to the plane of the slice. A pixel is then referred to as a volume element, or 'voxel'. Voxels in a single slice are typically all of the same thickness, and have the same spacing as the previous and following slices. Slice thickness and spacing are not necessarily the same, in that slices may be acquired with a gap between them, or overlapping.

16.2.1.2. Photometric representation and dynamic range: Channels, bits and pixel depth

In consumer applications, images from digital cameras and for the web are usually 8 bits per channel and encoded with three channels of red, green and blue 'true colour' information, at least in their decompressed state.

Medical imaging applications often involve single channels of information representing a single physical quantity, such as X ray density, and are visualized as greyscale images without colour. These may be encoded such that the lower numerical values are displayed as either dark or light.

If colour is applied to such single channel images, it is an artificial 'pseudo-colour' palette, intended to make subtle differences in intensity more visually apparent. The colour may be applied to the greyscale single component image in the display application, the colour palette to be applied may be included in the images, or the greyscale image may be transformed into a colour representation.

Medical images may also contain signal intensities sampled with a dynamic range that exceeds the 256 values that may be encoded in an 8 bit channel and which are typically 10, 12 or 16 bits in depth. Usually, any data that are between 9 and 16 bits in dynamic range are encoded in a single 16 bit (two byte) word in its uncompressed form, and the remaining unused high bits are left empty, or used for other purposes, such as graphic overlays. Medical images may also contain signed data with negative values (e.g. Hounsfield units (HU) in CT may be negative).

16.2.1.3. Image data versus image metadata

The encoded pixel values (or 'pixel data') are distinct from the information that describes the characteristics of the pixel data, sometimes referred to as 'metadata' (data about data). To interpret the pixel data, the recipient needs metadata that describe the dimensions of the pixel data array (rows and columns and number of frames), the number of channels (samples) per pixel, the physical spacing of the pixels, the number of bits used to store each sample, the number of bits within each stored word that contain significant bits and where within the word they are positioned, and the photometric interpretation of each pixel (whether it is greyscale or colour, zero is black, etc.).

These metadata may be stored separately from the pixel data, or, more commonly, in a 'header' preceding the pixel data. The header may be of a fixed length binary format, as variable length binary name–value pairs, or as some sort of structured text format, such as an extensible markup language.

16.2.2. Display of digital images

Acquired medical images typically differ in size and dynamic range from the devices on which they are displayed. Some processing is needed in order to convert them into the matrix of values that is sent to the display device. This processing may involve single point operations applied repeatedly for each pixel, such as conversion of pixel intensity, or multipoint operations, such as resampling and interpolating all or part of the image to fit the display, and other spatial operations, such as rotation, flipping, zooming and panning. Descriptive annotations derived from the metadata are also typically applied to the image during display.

16.2.2.1. Window and level

The number of sample values within an image may exceed the number of discrete displayable intensities available. The human visual system of the observer may not be able to discern small differences in displayed signal intensity (luminance), and only a subset of the sample values may be of interest for a particular diagnostic task. For example, in CT, although the acquired and encoded values typically span a range of 4096 values, they are unevenly distributed over the range of densities corresponding to air, fat, soft tissue, iodinated contrast media and bone. Accordingly, when viewing soft tissue, the user wants to see displayed a narrow range of intensities centred around water density, whereas when viewing bone, a wider range of intensities needs to be displayed, with the subtle variations in soft tissue density sacrificed. Thus, the user selects a 'window' of signal intensities for display, specified by a pair of centre and width values, and this window selects the range of input values that are then linearly mapped to the full dynamic range of the display device. Preset pairs of window values and interactive adjustment are both routinely provided.

For some applications, such as projection radiography, a linear mapping, with its implied clipping to white or black at the upper and lower limits of the window respectively, may not produce a satisfactory result. Application of a non-linear function, such as a sigmoid function, may produce a more satisfactory 'roll-off' of intensity in the light and dark regions, comparable to the behaviour of X ray film (Hurter and Driffield characteristic curve). If provided by the acquisition equipment vendor, this may be encoded in the metadata as parameters of a defined function, or as a 'lookup table' that encodes the mapping of each input to each output value.

16.2.2.2. Display consistency

Display devices translate digital driving levels into light intensity (luminance) that is perceived by the human visual system in combination with a contribution from any ambient lighting. The human visual system is non-linear in its response to luminance, such that the minimum change in luminance that is detectable is not the same at different absolute luminance levels. The net effect is that the same displayed image will be perceived differently with different levels of luminance (displayed and ambient). The change in luminance required for any particular luminance level to be perceptible can be measured or predicted and is referred to as a 'just noticeable difference' (JND).

Efforts to compensate for this effect can be made by applying 'perceptual linearization'. This involves arranging that each change in digital driving level results in a luminance change that has the same perceptual impact (same number of JNDs). In practice, this can be achieved by using the Barten model of the human visual system as described in the Digital Imaging and Communications in Medicine (DICOM) Grayscale Standard Display Function (GSDF) to provide a calibration function for displays. The net effects of such calibration are: (i) that all images intended to be displayed with the GSDF will appear to be of similar contrast on displayed devices of different luminance ranges (a consistency objective) and (ii) that the most effective application of the available luminance range will be used (a quality objective).

To avoid sacrificing JNDs through quantization, application of such calibration should be performed using lookup tables of sufficient bit depth in the display hardware. For greater perceived consistency of appearance, it is also recommended that the same maximum luminance be used on all devices within a site.

Regular recalibration is also required to account for degrading back light intensity or other display technology factors, as well as changes in ambient lighting conditions, if not dynamically accounted for by the display hardware.

16.3. DIGITAL IMAGE MANAGEMENT

16.3.1. Picture archiving and communications systems

16.3.1.1. Overview

A film image, of which there is only one original, is physically located in one place and requires manual handling, storage and transportation. A digital image, which may be replicated as many times as necessary with full fidelity, requires a means of distribution, electronic archival and an electronic method of display. Individual devices can be constructed to perform each of these tasks. Grouping such devices together to form a system that manages these tasks produces a picture archiving and communications system (PACS).

16.3.1.2. Scope

A PACS may be a small system that addresses the needs of one department, perhaps involving a single modality, with a modest number of workstations for a limited number of specialized users; this is a mini-PACS.

More commonly, the term PACS refers to a system that serves an entire enterprise, such as a hospital or a large outpatient (ambulatory) facility. A PACS is responsible not only for acquisition, archival and distribution, but also management of the workflow and integration with other departmental or enterprise wide information systems such as the radiology information system, cardiology information system, laboratory information system or hospital information system. An enterprise PACS will have both local users, accessing information from within the organization, and remote users, accessing information from off-site.

Patients are often investigated and treated at multiple locations. Images and related information need to be interchanged between different sites and organizations. This process of cross-enterprise interchange may involve the use of physical recording media or electronic networks. Individual systems and an entire PACS may be federated (grouped together and accessible in a single view) to provide for such interchange, or data may be pushed over point-to-point connections, or stored (centralized) in regional archives that provide for broader access.

PACS may be linked to electronic medical record systems that provide access to a broader range of information about a patient in one facility, or to a 'longitudinal' patient health record system that provides for information to span facilities.

16.3.1.3. Functions, components and workflow

Ordering and scheduling

Digital images may be acquired in response to a request or order (scheduled), or in the course of investigation or treatment of the patient by a clinician (unscheduled). In either case, there is a source of demographic information about the patient, including their identity and characteristics, as well as other information about their condition and the identity of those placing the orders. This information may come from admission, discharge and transfer systems. Orders may be placed in a computerized physician order entry system or other form of order placer, such as a hospital information system. Scheduling may occur in a separate department system scheduling and order filler system, such as a radiology information system.

The use of an electronic source of identity, ordering and scheduling information, and the integration of disparate sources of information, reduces the opportunity for operator error caused by re-entry of data and allows images to be associated with information in other systems.

Acquisition

A digital acquisition modality is responsible not only for the acquisition of the raw data from which to reconstruct an image suitable for display to a human observer, but also for associating that image with the order and demographic metadata. The human operator of the device may provide this information through direct data entry at the modality console (which is error prone), or by scanning previously printed barcode information, or by selection from a predefined schedule on a worklist or other source of demographic information provided electronically to the modality. The modality includes this metadata in the image 'header'.

The modality then transmits the images and related information to another device on the local network, such as a quality control (QC) or interpretation workstation, an analysis system, an image manager or archive, or a digital film printer. Typically, a modality will send the images once to a single preconfigured location, which will then take responsibility for interpretation, archival and distribution, then purge any local copies after a predetermined interval. Alternatively, some modalities provide for local archival, for example, on removable media such as compact disc (CD) or magneto-optical disc, of the reconstructed image data, the raw data or both.

Older modalities that produce digital images, but which predate the widespread use of standard means of network communication, may require an acquisition gateway or converter 'box' of some type to perform these functions. Similarly, non-medical devices used to capture images from other sources, for example, a digital camera or digital video recorder attached to an endoscope or a microscope, may require such a gateway to convert the consumer format images (e.g. Joint Photographic Experts Group (JPEG) or tagged image file format (TIFF) to a standard medical format containing the additional metadata for transmission.

Other information, including images from previous examinations, is often vital for correct interpretation of the current examination. Digitizers may be used to capture physical material, such as radiographic film, printed images and other paper documents, such as request forms and operators' notes and worksheets. Digital images and reports previously recorded on interchange media such as CD can be imported. These devices are integrated in a similar manner to any other acquisition device, though the workflow may be specialized, for example, to reconcile the identifiers on externally provided material with local record numbers.

Film digitizers designed for scanning radiographic film, or laser printed medical film, differ significantly from their consumer counterparts. The mechanism of scanning is transmissive rather than reflective, the range of optical density (OD) of the film is high, and the requirements for spatial and contrast resolution are demanding. Specialized physical handling of the film transport is also required. Both charge coupled devices (CCDs) and laser devices are available that provide for at least a spatial resolution of 2.5 lp/mm, a contrast resolution of 10 bits and a maximum OD of 4.0. Mechanisms of calibration are provided that allow for predictable values of OD or a standard perceptually linear space.

QC

After acquisition, QC by a human operator is usually required to confirm positioning, technique, absence of motion or other artefact, and correct labelling and identification. A display device may be present at the modality console itself, or a separate QC workstation may be provided to which the images have been automatically routed, or the images may be available from a central location via an ordinary PACS workstation but sequestered from general availability until QC has been completed. The choice of mechanism is dictated by both the type of modality and the local optimization of workflow. A direct digital X ray system may have an integrated display, whereas a computed radiography cassette reader may require a separate workstation. Separation of acquisition and QC responsibility to different operators may improve efficiency. Whatever workstation is used for QC should have a display adequate for the purpose, in terms of size, functionality, calibration and viewing environment.

Analysis and post-processing

Some types of digital image are amenable to automated image processing and analysis to provide for computer aided detection (CAD) (see Section 9.8). Cancer detection on X ray mammography is the primary application for CAD, but CAD is also applied for cancer detection to dynamic contrast enhanced MRI, chest X rays and chest CT, as well as CT virtual colonoscopy.

Distinct from CAD, other types of post-processing may be appropriate for some applications and modalities. A human operator may create additional reconstructions of the acquired image data, for example, 3-D reconstruction, or perform quantitative measurements on the images, prior to transmission to the physician for interpretation. A typical example would be quantitative analysis of coronary arteries on CT angiograms, which requires segmentation of the vessels, presentation in specialized 3-D software and semi-automated quantification of stenosis. Depending on the preferred workflow, specialization and expertise of staff, reimbursement pattern and capabilities of the PACS, this may be performed as a separate workflow step or by the physician during reporting.

CAD and other post-processing devices receive digital images from the modality, either directly or via the PACS, and produce additional information to be used during human interpretation of the examination. The images required for processing may be different from those required for presentation to a human observer, and a modality may be required to send raw data to the CAD system rather than processed images.

The output of any post-processing step may be in the form of derived and annotated images, or structured information that can be rendered and analysed by a user's workstation. The CAD or analysis workstation transmits this information to the PACS for the subsequent workflow steps.

Display and reporting

The end product of a radiological examination includes not only the images, but also the report by the radiologist. Digital images need to be displayed to the radiologist and interpreted by them (soft copy reading).

An appropriate image display needs to be able to provide to the user either a worklist of unreported studies to interpret, or a list of all available studies for them to choose from, preferably with their reported state. The former is more efficient but requires greater integration with the information systems. Images need to be made available in a timely manner, including those from relevant prior examinations. These may be obtained on demand or prefetched, depending on the architecture and performance of the system.

When a study is selected, the full available screen 'real estate' should be automatically populated with images in the most efficient manner for interpretation according to the user's preferences, by the application of hanging (default display) protocols, which recognize the type and content of current and prior images and which lay out the images for comparison accordingly, flipping and rotating into the correct orientation as necessary.

The display software needs to support adequate display of the different image types from the different modalities being supported, including support for:

- Projection (e.g. X ray) and cross-sectional (e.g. CT, MR) images;
- True colour (e.g. ultrasound) and pseudo-colour (e.g. nuclear medicine) images;
- Multiframe and cine images (e.g. cardiac ultrasound and angiography).

Support for 3-D multiplanar reconstruction, maximum intensity projection and volume rendering is expected for CT and MR, especially for musculoskeletal and angiographic interpretation and tumour size measurement. Interpretation of hybrid positron emission tomography (PET)-CT examinations requires not only multiplanar reconstruction support, but also display of fused images (pseudo-coloured PET superimposed on CT).

Basic image review features required include:

- Visual navigation of the available series of images through the use of thumbnails or a hierarchical browser;
- Side-by-side comparison of at least two sets of images, whether they be series from the same study, or different studies, with synchronized scrolling, panning and zooming in the case of cross-sectional modalities;
- Annotation of laterality with orientation as well as spatial localization of cross-sectional images for anatomical reference;
- Annotation of demographics, management and basic technique information to provide for safe identification and usage;
- Simple measurements of linear distance and angle as used for change detection and treatment planning.

Reporting styles vary and some radiologists use conventional dictation (recorded digital audio), some use speech recognition and others fill in predefined structured report templates on the screen. Integration of reporting technology with image display and other information systems for optimal user efficiency and workflow remains challenging.

Management

No system is perfect and human error is possible at each stage of the process. Images may be inadvertently assigned to the wrong patient or order. The wrong study may be performed or images may be of poor quality and need recapturing, which may require rescheduling and recall of the patient. The wrong side may be inadvertently recorded in the image header or pixel data.

These problems require that the systems provide a management function that allows corrections to be made by authorized personnel and for a record of these corrections (audit trail) to be reliably maintained. These corrections may involve changes to various databases that need to be accessed wherever the images may be used.

Over time, the PACS archive will fill with images that may no longer need to be immediately available, for example, when statutory retention periods expire. Prior examinations for reporting rapidly decrease in usefulness over time. The ability to purge selected studies may be desirable, both manually and in a rule based automated manner.

Patients frequently need to be referred elsewhere for further treatment and their records must be available to accompany them, so export capabilities, both over the network to remote facilities and via standard interchange media, are required.

Archival, retention and business continuity

Images must be stored, both in the short term for immediate use, for use as relevant priors for subsequent examinations, for referral for subsequent treatment and for statutory retention purposes.

Many types of storage technology have been used for PACS archives, including hierarchical storage management systems that attempt to provide fast access to current images and slower access to older images, prefetching them from slower storage (tapes or optical discs) in anticipation of demand. With the advent of redundant arrays of inexpensive discs, network attached storage, storage area networks, fixed content storage and the plummeting cost of high capacity hard drives, the use of so-called 'all spinning' media, i.e. all images being available on-line on hard drives with the same level of service, is now routine. A variant on the hierarchical storage management concept involves the use of hard drives of differing performance to minimize cost for less frequently accessed content.

Nowadays, the cost of archival may not be dominated by the cost of the drives, but rather the total cost of ownership, i.e. the infrastructure required to support them, including the cost of power and cooling, and network bandwidth for off-site replication.

Access to images has become mission critical, so inaccessibility of the PACS is unacceptable, either for scheduled (maintenance) or unscheduled (failure) reasons. This requirement drives the need for high availability in the entire system design, but particularly with respect to recent examinations of current inpatients. Satisfying high availability requirements may require the use of an off-site replica of the archive maintained to a near real time consistent state with the local primary archive, and a means of redirecting requests for images in the event of failure.

A separate requirement from high availability is the need for backup in the event of significant local data loss. Since images are now stored exclusively electronically, the loss of the only copy results in total loss, and this is not acceptable either for clinical care or for meeting statutory retention requirements. Backups must be performed and must be stored off-site, in a sufficiently distant and robust facility designed to protect them. High availability and backup requirements can be satisfied with a single solution, but the needs are distinct.

The design and procedures associated with high availability and backup solutions are defined in an organization's disaster recovery or business continuity plan, which analyses failure scenarios and provides predefined procedures to follow for likely contingencies. These plans must include estimates of the time to recover and be regularly tested.

The prospect of using regional and national archives, in addition to, or in place of, local archives, is being considered. These provide an opportunity to provide access across enterprises and to community physicians, and could potentially be reused for backup purposes or the primary archive. Commercial services also offer off-site storage, not just for high availability and backup, but as the primary archive.

Local regulations, site policies, standard of care, risk management and cost dictate what type of information needs to be archived for patient care and medico-legal purposes. Not only the radiological report but also the digital images themselves may form a critical component of the medical record. When images are acquired in multiple forms, for example, thin CT slices intended for post-processing and thicker reconstructions intended for distribution and viewing, policy will dictate whether these are retained or not. Retention of raw data for CT and MR is not typical, but retention of images for processing for screening mammography CAD may be of benefit to the patient, as may the retention of CAD results, to improve accuracy at the next round of screening.

Distribution

Though some referring providers are satisfied with the report alone, many require the images, whether to make their own clinical and diagnostic decisions for surgical or other treatment planning, or for patient education or teaching.

Digital images have an advantage over film in that, subject to the appropriate authorization and access controls, they can be made available at multiple locations simultaneously, both inside and outside a facility, via a secure network or via interchange media. If necessary, for teaching or research purposes, the digital images may be de-identified to protect the patient's confidentiality.

Many PACS systems provide for local and remote network access by authorized users other than the primary interpreting radiologists, often utilizing a web browser component or a software application that has significantly constrained functionality or with images of limited quality, and on displays that may not be calibrated or of medical grade. For many sophisticated users, this may be unsatisfactory, and a complete set of diagnostic quality images must be made available, preferably routinely, but certainly on request. If necessary, the providers of clinical care must be able to import the images into their own systems.

Different users require different functions to be available in their image display software. Though an assumption is often made that only rudimentary image manipulation tools are required, counter examples include the dedicated software needed for orthopaedic prosthesis template application, neurosurgical robotic surgery planning and radiotherapy planning.

Patient access is often neglected. Though the authorization and access control issues of providing external network access to patients are non-trivial, institutions deploying electronic medical record systems are increasingly providing patient portals. More commonly though, a set of images on interchange media is routinely provided to all patients, either immediately after examination or after the final report is issued.

16.3.2. DICOM

16.3.2.1. Background

The earliest digital medical imaging devices were both proprietary in nature and provided output in the form of printed film; the users had no expectation that digital images would be extracted from such devices, nor that they could be exchanged between devices or software from different manufacturers. As use-cases were developed for transferring, storing and remotely displaying images electronically, manufacturers initially provided proprietary solutions that were not interoperable. One could, for example, equip an entire hospital with X ray, CT and MR acquisition devices as well as a PACS and review workstations, but only if everything were purchased from one vendor, or if custom interfaces were developed for each acquisition device. This approach was neither scaleable nor affordable, and the need to develop open standards to promote interoperability between equipment from different manufacturers quickly became apparent.

The first open standard effort for medical imaging was the ACR-NEMA standard published in 1985, jointly sponsored by the American College of Radiology, representing the users, and the National Electrical Manufacturers Association, representing the producers. This standard defined a mechanism for encoding the pixel data of the images themselves, together with information about the images in the form of a list of data elements, and a set of commands and a means of exchanging these data over a point-to-point connection between two devices using a 50 pin parallel interface. There was little adoption of this standard at the time, and it was not until 1993 that an extensively revised version of the standard was produced, renamed DICOM, that significant progress was made. A key feature of DICOM that distinguished it from its predecessor was the use of evolving computer networks and Internet technology and protocols. Nowadays, the use of DICOM is ubiquitous and no manufacturer would be able to market a device that did not conform to the standard. The standard is not static, but rather evolves through extension with additional features, as new imaging and communication technology is developed.

Though initially targeted towards radiology applications, nowadays, the DICOM standard is not so restricted in scope and includes support for many other medical specialties, such as cardiology, dentistry, endoscopy, dermatology and pathology. DICOM has also been extended beyond the scope of medicine to include non-destructive testing of aircraft parts (Digital Imaging and Communications in Non-Destructive Evaluation — DICONDE) as well as baggage screening and other security applications (Digital Imaging and Communications in Security — DICOS).

16.3.2.2. Composite information model and information objects

A primary purpose of DICOM is the interchange of images and their accompanying information. The standard describes information object definitions (IODs), each of which is specific to a type of image produced by a particular modality, but which shares a common structure. For example, there is an IOD for CT and another IOD for ultrasound. These share common information about the patient and the management of the study, but different information about the acquisition technique, spatial and temporal relationships, and encoding of the pixel data. DICOM describes this information in modules that are either general or modality specific. The patient module, for instance, includes the patient's name, their date of birth and their identifier, i.e. characteristics of the patient that are fixed. In addition to information about the patient, additional information is required to manage the study, such as the date and time that the study was started, the identifiers of the request and the study itself, and descriptors of the type of procedure; these may be found in the general study module.

CT images will contain additional modules that are applicable to all forms of cross-sectional imaging, or may be specific to CT. For example, CT, MR and PET images all share the concept of an image as a slice in a well defined 3-D space. The frame of reference module defines the patient relative coordinate system shared by a set of slices acquired in the same procedure, and the image plane module defines the position and orientation of an individual slice. Ultrasound images, on the other hand, traditionally acquired with a freehand transducer and not having a Cartesian geometry, do not contain these modules.

Since CT images are acquired using an X ray beam, they contain specific attributes that describe the characteristics of that beam and its production, including the voltage, tube current, exposure time and filtration. Ultrasound images, on the other hand, include information about the type of transducer used, the transducer frequency, and so on. Accordingly, there are CT image and ultrasound image modules defined to record this modality specific information.

Modules not only describe information that is either general or modality specific, but that is also shared between multiple images during the same procedure. This commonality is defined in a DICOM information model, which describes entities, such as patients, studies, equipment, series and images, and the relationships between them. So, all images that are acquired as part of the same procedure will contain exactly the same information about the patient and study. If the procedure is performed on the same device, then the information about the equipment will be identical in all such images.

Multiple images may be grouped into the same series if they have something in common, for example, if they were acquired in a single run of the CT gantry. When images are encoded, however, all of this common information is replicated into each instance, that is, every image contains a full set of information, and for this reason they are referred to as composite instances (as opposed to normalized instances in which the information about each entity would be managed and transmitted separately). The intent is that a single image may be separated from other images or the system on which it is produced or stored, yet still contain a full set of information necessary to identify and interpret it.

16.3.2.3. *Attributes, data elements, encoding and transfer syntaxes*

Modules are defined as a list of attributes, each of which encodes a specific piece of information such as a name or a numerical value. For transmission and storage, these attributes are encoded as data elements in a single binary dataset. Each data element in the standard is assigned a unique 32 bit numerical tag, usually described as a pair of 16 bit hexadecimal group and element numbers. For example, the patient's name attribute is assigned a data element tag of (0010,0010). A textual description of the name of each data element is not included in the encoding, hence, the format is not 'self-describing' and the recipient needs to have prior knowledge of what each element means.

Each data element is of a predefined type, or value representation, and the standard defines a variety of such types. There are binary types for signed and unsigned integers of 16 and 32 bit lengths, IEEE floating point binary types of 32 and 64 bit lengths, as well as specific and general string types, such as those for names, integers and decimal values, dates and times, and codes, as opposed to free text descriptions. The value representation may either be encoded explicitly, or implied and looked up in a dictionary by the recipient.

The value length of each data element is always explicitly encoded. Value lengths are always even and, where necessary, strings are padded to even length.

The pixel data themselves are encoded as just another data element (7FE0,0010), albeit a very large one with some specific encoding rules. Hence, a DICOM dataset does not consist of a fixed length 'header' that may be skipped to reach the pixel data, nor are the pixel data necessarily at the end of the dataset. Full parsing of the successive data elements, including recursion into any variable length sequences, is necessary to recover the pixel data reliably.

The transfer syntax defines the actual encoding. It is also used to distinguish images whose entire dataset or pixel data may be compressed, as discussed later. In addition to the standard transfer syntaxes defined by DICOM, a manufacturer may define its own private transfer syntax, which can be used as long as both the sender and recipient agree to support them.

16.3.2.4. *Service classes, service–object pair classes, associations and conformance statements*

Once an instance of an image information object has been assembled and encoded in a particular transfer syntax, it is transmitted on the network using one of the many network services that DICOM defines, the storage service class.

Not all devices support all types of image from different modalities. Accordingly, DICOM defines the combination of a service class and an IOD as a service–object pair (SOP) class. For example, the combination of the storage service class and the CT image IOD is the CT image storage SOP class.

The purpose of defining SOP classes is to allow the sender and receiver to negotiate their mutual capabilities when establishing a connection on the network, or what DICOM refers to as an 'association'. This negotiation mechanism allows the sender to offer, for example, CT and MR images each in uncompressed or compressed form, and the receiver to accept either those that it supports, or prefers, and for the sender to then select from among the accepted choices which one to use.

In some cases, a sender may have images to send of more than one SOP class, say CT and ultrasound, yet the receiver, a 3-D workstation for example, may not support ultrasound images and would then reject the corresponding SOP class.

In general, such limitations are known beforehand, at purchase and at installation time, and are determined by comparing DICOM conformance statements. Each manufacturer is required by the standard to document their product's capabilities in a conformance statement. Among other things, these statements contain a tabulation of which SOP classes and transfer syntaxes are supported. A simple statement by a vendor that a device is 'DICOM compliant' is *not* sufficient to describe interoperability; rather, specific review of every pair of devices' conformance statements by an 'educated' purchaser with a critical eye is required to ensure compatibility.

The devices at either end of the association are referred to as application entities. This term is used since there is no requirement that there be a one to one correspondence between physical devices or software applications and application entities.

16.3.2.5. Interchange media and archival storage

In addition to providing for transfer of DICOM instances on a network, the standard also includes rules for the use of interchange media, such as recordable CDs, DVDs, MODs and USB (universal serial bus) devices. These are thought to be sufficiently robust to allow the information to be preserved for physical transfer from one location to another, such as by mail or courier. DICOM has chosen conventional consumer format media and file systems whenever possible, in order to maximize reuse of affordable technology and ensure that they are readable with conventional operating systems on ordinary computers without the need for special hardware or software.

For the file format, a short 'meta' information header is required, preceding the encoded DICOM dataset, which provides a recognition string ('magic number') by which DICOM files can be distinguished from other files, as well as a description of the transfer syntax actually used to encode the dataset that follows.

All DICOM media also need to contain, in the root directory, a DICOMDIR file, which encodes a summary of the content of the media, listing every patient, study, series and instance present, together with a summary of the characteristics of each of those entities. An application can read this file and quickly summarize the contents in a browser for the user, without having to read every file on the media.

16.3.2.6. Composite instances other than images

The initial focus of DICOM was the interchange of images themselves. However, there are other types of bulk data that can be handled in a similar manner to images, such as time based waveforms (e.g. electrocardiograms), spectroscopic data (e.g. MR spectroscopy), documents of various types (e.g. PDFs) and even the raw data that are acquired prior to image reconstruction. These different types of data need description in a similar manner to the information model for images, and hence can share the composite information model that is used for images. Each can be described as a composite IOD with the addition of the appropriate modules, attributes, data elements and encoding mechanisms. The same storage service class can be used for transfer of these objects on the network and they can be encoded on interchange media in the same manner as images.

Additionally, other types of information may be acquired that do not consist of bulk data that need to be described, but which can adequately be described as a set of individual attributes. For example, a radiotherapy plan can be described in such a manner, as distinct from a radiotherapy image (portal image) or a radiotherapy dose map, which are encoded as images. There is an entire family of radiotherapy related objects to support both external beam therapy and brachytherapy.

The need to encode data in an extensible structured form is common to many use-cases, including the recording of quantitative and categorical data from acquisition devices (such as obstetric or cardiac ultrasound measurement), CAD of abnormalities on images such as mammograms, as well as the encoding of human generated reports. This is the province of the DICOM structured report family of objects, which use nested recursive sequence attributes to encode an extensible tree of categorical, coded, numerical and free text information using templates defined for specific applications.

A distinguishing feature of the DICOM structured report, compared with other structured document formats, is a mechanism for making reference not only to DICOM images (and waveforms) in their entirety, but also to specific coordinate referencing locations in space (or time). These are used, for example, to illustrate findings or to define the locations at which measurements were made. DICOM structured report objects are also used to encode radiation dose structured reports, as well as lists of relevant images for some purpose (key image notes or key object selection documents).

When displayed, medical images are often manipulated by the user to zoom or pan to a specific location, or adjusted in contrast and brightness (window width and centre) and may be annotated with text or graphics. These manipulations can be captured as a presentation state, and such states can be stored as composite instances for retrieval and application to the same images at a later time. References to images in structured reports may also contain an accompanying reference to a presentation state, for example, to capture the appearance of the display of a particular region when a measurement was made.

As with the non-image bulk data composite instances, the structured report and presentation state instances are exchanged using the normal storage service class and interchange media profiles.

16.3.2.7. Service classes other than storage

Though the major use of DICOM network services is to transfer (store) images and other composite instances from one application entity to another, there are many other service classes defined. Some of these exist primarily to support storage activity. The storage commitment service allows the sender of a set of instances to ask the receiver if it will take responsibility for the persistence of the stored objects. This service is used by an acquisition modality prior to deleting its local copies of images.

The query/retrieve service class allows one to query a remote device for patients, studies, series and instances, using identifiers and other matching attributes, and then select an item for retrieval.

Other service classes are defined for use-cases that are not directly related to storage. The worklist management service class provides demographic and request and scheduling information via a modality worklist. The responses obtained and scheduled procedure steps provide the modality with the necessary information to choose the correct patient, perform the work of image acquisition and populate the attributes in the resulting images. Once the image acquisition is complete, feedback is provided to the management system in a modality performed procedure step.

16.3.3. Radiology information system–hospital information system interfacing, Health Level 7

16.3.3.1. Background

Just as DICOM is ubiquitous and unchallenged as the single standard for interchange of medical images, other information systems in a health care enterprise depend upon the Health Level 7 (HL7) standard for communication. The first version dates back to 1987, but varieties of version 2.x are the most common in use nowadays, particularly since version 2.3 (1997). An almost completely different and much more complex standard, HL7 version 3, has been defined, but has yet to supplant the dominance of the version 2.x in the field.

HL7 defines a clinical document architecture, a means of encoding and managing structured documents with consistent metadata. Clinical architecture documents may be exchanged using version 2.x or 3.x or other mechanisms and persist independently of the communication mechanism. They can even be exchanged and stored using DICOM services.

The HL7 organization has also grown to absorb, embrace or define several other standards. This includes the clinical context object workgroup, which defines a means of loosely coupling different desktop applications to share the same context (e.g. to enable recognizing that the patient being viewed has changed).

16.3.3.2. Version 2.x

Unlike DICOM, HL7 version 2.x messages are encoded as text messages, rather than as binary. The format and meaning of messages are defined in detail, as are the circumstances under which they will be sent (trigger events).

Most commonly, HL7 devices communicate over a network using transmission control protocol/Internet protocol, using the minimal lower level protocol. In addition, third party ‘interface engines’ can be used to centralize the messages produced by individual sources, transform them and propagate them to other devices that need the information.

HL7 messages are composed of ‘segments’ separated by carriage returns. The first segment is always the message header segment, which, among other things, assigns an ID to the message, specifies the separator (delimiter) characters used and specifies the ‘trigger event’ that stimulated the message to be sent. Subsequent segments carry the payload of a message. Many segments are common to several different types of message. HL7 segments are composed of ‘fields’ that have a ‘data type’ associated with them. There are no explicitly conveyed tags to identify a field in a segment and no explicit data type conveyed. The meaning of a field is conveyed by its position in a segment alone.

The scope of HL7 version 2.x message types and trigger events is broad and only a few are relevant to imaging applications. Of specific interest are those related to the management of patient identification, which includes the admission, discharge and transfer messages, and those related to order entry, such as the general order message. Both are commonly used to construct the set of information required to respond as a DICOM modality worklist query provider. More complex use-cases, such as managing changes to patient identity, are also supported by HL7 messages but are addressed under the subject of integrating the health care enterprise (IHE) (see Section 16.3.4).

16.3.4. IHE

16.3.4.1. Background

The DICOM and HL7 standards define rules for very specific services or messages, but neither defines an overall architecture for building a complete system to support an entire enterprise. Significant gains in interoperability were achieved using both standards, and large and complex systems were built without dependence on proprietary interfaces. However, further progress towards producing turnkey devices that could 'plug and play' required definition of specific use-cases and specific architectures to support them. In 1997, the Radiological Society of North America, an organization that had been instrumental in the promotion and adoption of DICOM, began to engage key stakeholders to establish momentum and direction, and in 1998, allied with the Healthcare Information and Management Systems Society, to initiate the IHE effort.

The premise was that an annual cycle of release of technical specifications and testing of implementations at 'connectathons', followed by public demonstrations, would quickly demonstrate value to product marketeers and customers. Year one focused on one problem; that of scheduling radiology workflow from patient registration through ordering and scheduling to image acquisition, transfer, archival and distribution. This problem involved two standards (DICOM and HL7) and multiple types of device manufacturer (hospital information system, radiology information system and PACS), and resulted in 24 vendors demonstrating 47 systems at the first connectathon, followed by a public demonstration at the Radiological Society of North America annual meeting in 1999. Initially conceived as a 3–5 year project, the project is ongoing and in its 15th year, and the IHE is now a global organization spanning multiple domains well beyond radiology.

16.3.4.2. Profiles, actors and transactions

The IHE approach is to identify a set of use-cases that require a common infrastructure and then to define an integration profile composed of actors and transactions sufficient to support those use-cases. The resulting profile may not be the only way to solve the problem, but it is designed to be sufficient as well as consistent with other integration profiles and, where possible, the installed base of equipment in the field.

For each profile, a family of actors is defined, which abstract devices that, in the 'real world', serve different purposes and are often, but not necessarily, provided by different manufacturers. For example, that part of a hospital information system or radiology information system that performs the scheduling function is referred to as a department system scheduling and order filler actor, and is distinct from the actor that performs the ordering function, the order placer actor. Yet, in reality, these may be grouped together in a single implementation. Similarly, the management and archival functions of a PACS are grouped as the image manager/image archive actor, distinct from the image display functions of a PACS workstation, the image display actor. The various actors are common between various profiles where appropriate.

The behaviour of an actor is not defined generically but, rather, is specified in the context of transactions between actors in the context of a profile. IHE profiles do not define new standards to implement transactions if possible, but use existing messaging standards such as DICOM or HL7; if necessary specializing or constraining particular HL7 messages or DICOM SOP classes to achieve the objective.

For example, in scheduled workflow profile, HL7 messages are specified for patient registration and order entry, and the protocol to be used, the version of HL7 and the content of certain segments and fields are explicitly defined. Further, since other transactions in the same profile use DICOM SOP classes, such as the provision of modality worklist, the mapping from the HL7 messages, segments and fields to the DICOM query return attributes is defined, providing a deterministic bridge between two similar, but not identical, standards. The degree of specificity in the definition of the profiles serves to eliminate uncertainty on the part of the implementers and purchasers.

The profiles are, in general, an all or nothing proposition, rather than having to match the specific capabilities in DICOM conformance statements of two different devices, for example, the purchaser can compare the IHE integration statements of two devices. For example, a CT scanner claiming to be an acquisition modality actor supporting the IHE scheduled workflow profile and a radiology information system claiming to be a department system scheduling and order filler actor for the same profile should interoperate, without the need to evaluate the specifics of the DICOM modality worklist implementation.

IHE integration profiles are not exclusively concerned with workflow. Another type of profile addresses matters concerned with interchange media. The IHE portable data for imaging (PDI) profile requires the use of the DICOM standards for media, but selects and limits the choices provided to those in common use, specifically to the use of uncompressed images on CD. Subsequent extensions to the PDI profile adopt more DICOM features and allow for the use of DVD and USB media, as well as the selective use of compression, and encryption for privacy protection. Encrypted media are required in IHE PDI to provide accompanying on-media decryption software.

Another type of integration profile addresses the behaviour of single actors in terms of the features available to the user. The image display actor, which describes the functions expected in a workstation or viewer, is included in several profiles that specify detailed application specific behaviour. The mammography image profile describes a list of detailed requirements that a display must implement, such as achieving the proper orientation, comparable size of current and prior images, justification to the chest wall, consistency of greyscale contrast (window) settings, completeness of annotations and the display of CAD marks. This profile requires that specific DICOM attributes be used to implement specific behaviour, rather than leaving this to the discretion of the implementer, and also burdens the acquisition modality with requirements to populate these attributes. The mammography image profile and the more general consistent presentation of images profile, also require the display to implement and conform to the DICOM GSDF, in order to facilitate consistency of perceived contrast of displayed images.

Another profile related to the image display is the basic image review profile, which enumerates the minimal features that a clinical review user requires, down to the level of detail that will provide a similar user experience, regardless of manufacturer, going so far as to define standardized icons and minimum performance requirements. The goal of this profile is to improve the consistency of viewers included on PDI media, though it is not limited to that application.

The radiation exposure monitoring profile describes the production of radiation dose structured reports and their transfer to the PACS and their use for local monitoring as well as de-identification and submission to dose index reference level registries.

16.3.4.3. Cross-enterprise document sharing

Though IHE was originally conceived to address issues within a single enterprise, the spread of electronic records and the need for interoperability between enterprises has driven the development of a new family of profiles that may be described under the category of cross-enterprise document sharing. The first of these profiles was cross-enterprise document sharing for imaging, which provides for a central document registry actor that keeps track of metadata about documents (including images), which actually reside in multiple document repositories, originally intended to be collocated with the sites that originated the documents, perhaps implemented as a gateway to the local PACS. The sequence of operations is to query the registry, identify the documents required, then retrieve them from the appropriate repository. For imaging, the document retrieved is a manifest encoded as a DICOM SR, and an additional level of retrieval is then used to return the images themselves.

Mechanisms are continuously evolving in order to address additional complexities encountered with loosely coupled, externally accessible systems, including security and privacy and access control, incorporation of centralized (regional or national) rather than local repositories, and support for distributed consistency after local corrections and updates.

16.4. NETWORKING

Medical image networking applications are all based on conventional underlying wired, optical and wireless network technology using standard Internet protocols such as transmission control protocol/Internet protocol (TCP/IP) for communication. Ordinary network hardware (such as routers, switches, cables) and software (such as firewalls, network protocol stacks in operating systems and network management utilities) are used. Both local area networks within a single site as well as wide area networks are assembled from this conventional technology.

Though image data transfers are very large, switched high speed Ethernet is used for this routinely in local area networks, and networks are often shared with other, lower volume traffic. If necessary, quality of service concerns can be addressed with virtual local area networks, through configuration rather than physically separate networks.

When sites are connected using public facilities such as the Internet, privacy and performance concerns must be addressed. The use of conventional encryption hardware or software to establish either virtual private networks or session based secure connections (using transport layer security as used for electronic commerce) is typical. The greater latency on high speed wide area networks that results in delayed acknowledgement can impact performance if not specifically addressed in the choice of protocol or configuration of the protocol stack or application, or through the use of additional devices such as wide area network accelerators.

The increasing using of mobile computing devices allows for additional channels of distribution of full or partial sets of image data, particularly if forms of compression can be selected that are appropriate to the task.

16.5. IMAGE COMPRESSION

16.5.1. Purpose

When stored or transmitted in uncompressed form, digital images occupy an amount of space proportional to the matrix size of the image, i.e. each pixel occupies a fixed number of bytes. However, images typically contain a significant amount of redundant information that can be represented more compactly. For example, there is often a large amount of black space around the 'important' parts of the image. Reducing the amount of space occupied by an image is a priority, since the cost of storage and bandwidth is significant, and sometimes there is insufficient time to transmit a large set of images over a slow connection to meet the clinical need.

16.5.2. Transformation and coding

Compression schemes typically consist of a series of steps that first transform the original data into an alternative representation that exposes redundancy and then encode the information in a more compact form.

If one is compressing a series of symbols, such as plain text represented as successive characters, one byte per character, it may be directly encoded using a mechanism that substitutes a short encoding for more frequently occurring characters, and a long encoding for less frequently occurring characters. For example, the letter 'e' might be encoded with a single bit, whereas the letter 'z' might be encoded with a large number of bits. This is referred to as 'variable length' coding, in contrast to the one byte per character 'fixed length' coding.

One very common approach to encoding data this way is Huffman encoding. More complex analysis of the plain text might consist of storing sequences of characters in a dictionary, computing their frequency and representing entire sequences with single variable length symbols, an approach referred to as 'dictionary' coding.

Another approach is to encode the entire message as a very long binary fraction, which can produce near optimal output for a given set of symbols and probabilities, given an accurate model of the frequencies of each symbol. A special case that is often considered separately is that of the same symbol occurring multiple times, in which case it may be encoded as the symbol and the number of occurrences, rather than repeating the symbol, referred to as 'run length' encoding.

These approaches are equally applicable to encoding images, except that they do not account for the additional redundancy that is present in two (or more) dimensions in an image.

Transformation steps may be used prior to coding, to expose such redundancy in an image. For example, where successive pixels differ little from their predecessors most of the time, encoding the difference between the current and previous pixel may result in a more compact value to code, or a more compact frequency distribution. For some types of image, more regional 'context' can be considered, for example, the difference between the pixel above as well as the pixel to the left may be included in the difference signal, or yet more complex models that consider the rate of change in the local region can be constructed.

Other types of transformation that are specific to images may expose redundancy in multiple colour channels. For example, colour images are typically encoded uncompressed as red, green and blue channels, yet all three channels carry the same luminance information. Transformation of the colour space into a luminance and a pair of chrominance channels may expose this redundancy. Similarly, transformation from the spatial to the frequency or wavelet domain may also allow for a more compact representation.

16.5.3. 'Lossless' compression

Some forms of compression allow complete and exact recovery of the original data from the compressed data, and these are referred to as lossless or 'reversible' compression schemes. Schemes typically used in consumer applications to compress text documents, for example, would be unacceptable if characters changed when decompressed, regardless of how infrequently this occurred. Likewise, medical imaging applications may require perfect reproduction of the input, and hence lossless compression is widely used.

Lossless compression schemes for images may utilize such transformation steps as described previously, with the proviso that the transformation be implemented in a mathematically reversible form. Difference transformations must have sufficient depth to preserve the maximum possible difference value and its direction (sign) with full fidelity; this requires one more bit to encode than the original values. Similarly, frequency domain or colour space transformations require the use of equations and implementations that use fixed, not floating point, arithmetic of sufficient precision.

In practice, lossless compression of medical images produces compression ratios (relative to the original number of bytes occupied by an unpacked image) ranging from approximately 2.5:1 to 5:1, depending on the modality and type of image. Typically, medical images contain a significant amount of noise, texture and other high frequency content, and this interferes with the ability to achieve higher lossless compression. Even background air or collimated regions contain noise. Despite this, cost and transmission time savings of this order are often sufficient to satisfy the use-case. For example, lossless compression of 512 by 512 by 8 bit cardiac angiograms using a relatively simple scheme is commonly used to fit an entire examination on a single CD.

16.5.4. 'Lossy' compression

Lossy, or irreversible, compression occurs when the decompressed result is not identical to the original, although the amount and type of loss is acceptable for some purposes. For example, lossy compression is routinely applied to colour photographic images obtained from consumer digital cameras and is widely used in consumer Internet web browser pages.

The process involved is similar to that used for lossless compression, with transformation and encoding steps, except that since a finite amount and type of loss is permitted, neither of these steps is required to be entirely reversible. For example, the colour space and frequency or wavelet domain transformations can involve a certain amount of loss, and the use of continuous functions with finite precision can expose more redundancy. Additional, deliberate steps may be applied to discard specific types of information. For example, higher frequency coefficients may be represented with fewer bits of precision than more important lower frequency coefficients, and this can be controlled in a 'quantization' step applied after transformation and before coding. In some compression schemes, this is the step at which the quality of the result is explicitly controlled.

Lossy compression results in artefacts becoming visible in the reconstructed image. For images that are compressed as an entire frame, these artefacts may involve subtle smoothing of the entire image, alterations in the perceived texture of complex regions, or the introduction of distinct small structures that were not present in the original (for example, in the case of wavelet transformation, reflections of the wavelet basis functions may appear, and in the case of truncated high frequency coefficients, ringing may appear around sharp edges such as text). For schemes that involve splitting the image into small tiles before compression, block artefacts may appear where the boundaries of such tiles become distinctly visible.

Despite these artefacts, lossy compressed images may be sufficient for many clinical purposes. Two levels of compression may be defined for medical imaging: 'visually lossless' and 'diagnostically lossless'. Visually lossless compression is that type and level of compression at which a human observer is unable to distinguish visually the original from the reconstructed image, even though mathematical loss has occurred. This can be determined by relatively simple experiments and is known to vary considerably depending on the modality and body part of the image. However, whether such images, or images compressed more or perhaps less, are sufficient for primary interpretation, i.e. are diagnostically lossless, depends on the diagnostic task. As an extreme example, a chest X ray for the purpose of locating the tip of a catheter could undergo extreme degradation and still be sufficient, yet detection of the pneumothorax caused by the insertion of the catheter might require an image with much less compression.

Establishing the appropriate levels of compression for each such modality, body part and diagnostic task may require observer performance studies with sufficient statistical power to detect that any lack of difference found is due to a genuine absence of a clinically significant difference caused by compression, as opposed to a study that is too small to detect any difference. Such studies are expensive and few in number. Despite this, there are emerging guidelines from various professional societies in a few countries on the matter of appropriate use of lossy compression.

A confounding factor is whether or not to perform lossy compression before or after primary interpretation, if one is going to compress for archiving or subsequent distribution. As long as there is the potential for a misdiagnosis, some advocate that optimum patient care requires interpretation of uncompressed images; indeed, the US Food and Drug Administration requires this by regulation for digital mammography. Lawyers argue, however, that exactly what was interpreted should be archived. The most conservative strategy is to avoid lossy compression, regardless of the attractiveness of the potential infrastructure cost savings.

16.5.5. Standard and common compression schemes

The International Standards Organization (ISO) and the International Electrotechnical Commission (IEC) joint technical committee has established subcommittees responsible for both still image and video compression standards, resulting in the JPEG and the Moving Picture Experts Group (MPEG) families of standards, respectively.

In consumer digital camera and web applications, the widely used 8 bit greyscale or colour lossy discrete cosine transform JPEG format is actually only a subset of a large number of schemes defined in several standards. Additional JPEG schemes are used for other types of medical image, including:

(i) 12 bit discrete cosine transform, applicable to greyscale images of greater bit depth;
(ii) 16 bit lossless compression with difference and Huffman coding;
(iii) JPEG 2000, with wavelet transformation and arithmetic coding, in both lossless and lossy variants.

For medical images that involve multiple frames that contain redundancy between frames (e.g. cine images acquired over time or successive 3-D cross-sections), though each frame may be encoded separately using JPEG or JPEG 2000, other schemes, such as multicomponent or 3-D JPEG 2000 or MPEG, may be applied.

Though there are several other common proprietary or de facto standard file formats with inherent compression schemes widely used on the web and for other professional applications, such as GIF (graphics interchange format), PNG (portable network graphics) and TIFF, these are rarely used for medical imaging, owing to bit depth constraints or lack of sufficient compression support. Though file formats with compression schemes that are commonly used for text and data files, such as ZIP, may also be applied to images, they are not generally as effective as methods that take advantage of the image structure.

16.5.6. Compression in DICOM

DICOM makes use almost exclusively of ISO–IEC standard compression schemes and defines transfer syntaxes for each of the appropriate JPEG and MPEG lossless and lossy schemes.

This approach allows devices communicating on the network to negotiate the most appropriate compression scheme to use depending on the circumstance. Association negotiation allows the sender to propose combinations of the type of image (SOP class) and the type of compression (transfer syntax) and the receiver to accept or reject each combination, depending on its capabilities. For lossless compressed images, in every case there is a fallback mechanism to allow the default, uncompressed, transfer syntax to be used in case the recipient does not support any proposed compression scheme. This is not required for lossy compressed images, the principle being that the sender may not have access to the original uncompressed image.

As technology evolves and new compression schemes are standardized, DICOM adds them as new transfer syntaxes, potentially making them available for any type of image, since the transfer syntax is independent of the SOP class.

DICOM is a technology standard and does not address the appropriate use of any particular compression scheme for any particular purpose. The inclusion of a scheme in DICOM is not an endorsement of such a scheme.

BIBLIOGRAPHY

AMERICAN ASSOCIATION OF PHYSICISTS IN MEDICINE, Assessment of Display Performance for Medical Imaging Systems, AAPM On-line Rep. 03, AAPM, College Park, MD (2005),
http://www.aapm.org/pubs/reports/OR_03.pdf supplemental files available at
http://www.aapm.org/pubs/reports/OR_03_Supplemental/ (accessed on 23 August 2012).

AMERICAN COLLEGE OF RADIOLOGY, ACR Technical Standard for Electronic Practice of Medical Imaging (2007), http://www.acr.org/~/media/AF1480B0F95842E7B163F09F1CE00977.pdf (accessed on 23 August 2012).

CENTRE FOR EVIDENCE-BASED PURCHASING, A Beginner's Guide to PACS, MDA Evaluation Rep. 02044, The Stationery Office, Norwich (2002).

CENTRE FOR EVIDENCE-BASED PURCHASING, A Beginner's Guide to Virtual Private Networks in a Picture Archiving and Communication System Environment, CEP 05094, Centre for Evidence-based Purchasing, London (2006).

CLUNIE, D., DICOM, http://www.dclunie.com/ (accessed on 23 August 2012).

DREYER, K.J., MEHTA, A., THRALL, J.H., PACS: A Guide to the Digital Revolution, Springer, New York (2002).

INTEGRATING THE HEALTHCARE ENTERPRISE, http://www.IHE.net

HEALTH LEVEL SEVEN INTERNATIONAL, http://www.HL7.org (accessed on 23 August 2012).

HUANG, H.K., PACS: Basic Principles and Applications, 2nd edn, John Wiley & Sons, New York (2004).

Chapter 17

IMAGE POST-PROCESSING AND ANALYSIS

P.A. YUSHKEVICH
University of Pennsylvania,
Philadelphia, United States of America

17.1. INTRODUCTION

For decades, scientists have used computers to enhance and analyse medical images. At first, they developed simple computer algorithms to enhance the appearance of interesting features in images, helping humans read and interpret them better. Later, they created more advanced algorithms, where the computer would not only enhance images but also participate in facilitating understanding of their content. Segmentation algorithms were developed to detect and extract specific anatomical objects in images, such as malignant lesions in mammograms. Registration algorithms were developed to align images of different modalities and to find corresponding anatomical locations in images from different subjects. These algorithms have made computer aided detection and diagnosis, computer guided surgery and other highly complex medical technologies possible. Nowadays, the field of image processing and analysis is a complex branch of science that lies at the intersection of applied mathematics, computer science, physics, statistics and biomedical sciences. This chapter will give a general overview of the most common problems in this field and the algorithms that address them.

This chapter is divided into two main sections. The first addresses classical image processing algorithms. Under image processing, we discuss image filtering, noise reduction and edge/feature extraction from images. The second section touches on more modern image analysis approaches, including segmentation and registration. The main feature that distinguishes 'image analysis' from 'image processing' is the use of external knowledge about the objects appearing in the image. This external knowledge can be based on heuristic knowledge, physical models or data obtained from previous analysis of similar images. Image analysis algorithms use this external knowledge to fill in the information that is otherwise missing or ambiguous in the images. For instance, a biomechanical model of the heart may be used by an image analysis algorithm to help find the boundaries of the heart in a computed tomography (CT) or magnetic resonance (MR) image.

This model can help the algorithm distinguish true heart boundaries from various other anatomical boundaries that have similar appearance in the image.

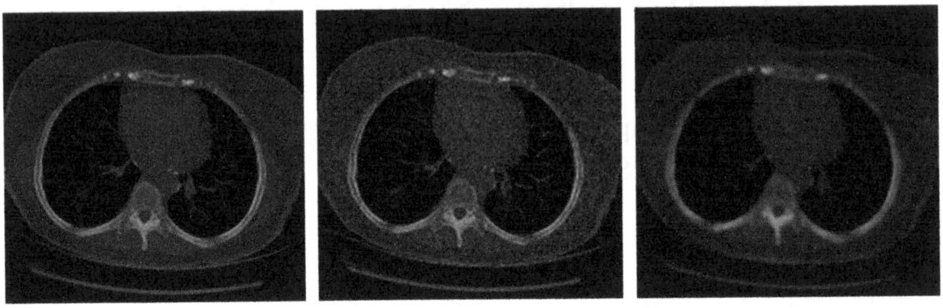

(a) (b) (c)

FIG. 17.1. Example of image denoising. (a) Chest CT slice; (b) same slice with added noise; (c) same slice processed with an edge preserving noise removal algorithm. Note that although noise removal eliminates the noise, it also degrades anatomical features (reproduced from the Lung Cancer Alliance Give a Scan database (www.giveascan.org)).

Before proceeding any further, we must point out the most important limitation of image processing: it cannot increase the amount of information available in the input image. Applying mathematical operations to images can only remove information present in an image. Sometimes, removing information that is not relevant can make it easier for humans to understand images. However, image processing is always limited by the quality of the input data. For example, image noise cannot be eliminated without degrading the contrast between small details in the image. Likewise, the fundamental resolution of the input image (i.e. the ability to separate a pair of nearby structures) is limited by the imaging system and cannot be improved by image processing. These examples are illustrated in Figs 17.1 and 17.2, respectively. In general, when faced with an

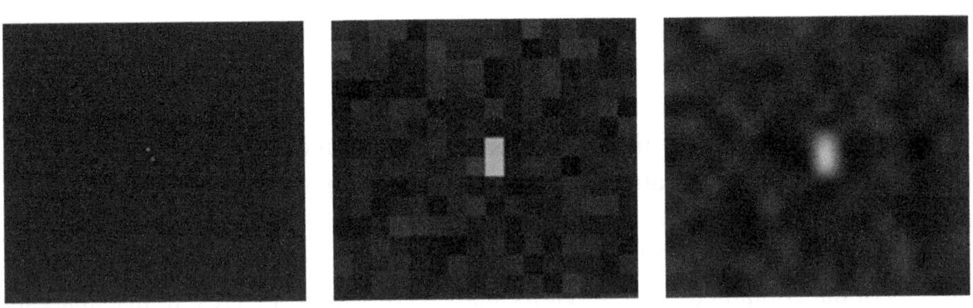

(a) (b) (c)

FIG. 17.2. Changing the resolution of an image. (a) The input to an imaging system consisting of two nearby point impulses. (b) A 16 × 16 image produced by the imaging system. The system's resolution is less than the distance between the impulses. We cannot tell from the image that there were two impulses in the data. (c) Image resampled to 128 × 128 resolution using cubic interpolation. We still cannot tell that there were two impulses in the input data.

imaging system that provides data of unacceptable quality, a wise reader would try to improve the imaging system, rather than hope that the 'magic' of image processing would compensate for poor imaging.

17.2. DETERMINISTIC IMAGE PROCESSING AND FEATURE ENHANCEMENT

In this section, we review the most important image processing operations. We will cover three areas of image processing: filtering, resampling and edge detection. In this section we only discuss 2-D images. In most cases, the same ideas extend to 3-D.

17.2.1. Spatial filtering and noise removal

Filtering is an operation that changes the observable quality of an image in terms of resolution, contrast and noise. Typically, filtering involves applying the same or a similar mathematical operation at every pixel in an image. For example, spatial filtering modifies the intensity of each pixel in an image, using some function of the neighbouring pixels. Filtering is one of the most elementary image processing operations.

17.2.1.1. Mean filtering

A very simple example of a spatial filter is the mean filter. This filter replaces each pixel in an image with the mean of the $N \times N$ neighbourhood around the pixel. The output of the filter is an image that appears 'smoother' and less 'noisy' than the input image (Figs 17.3(a) and (b)). Averaging over the small neighbourhood reduces the magnitude of the intensity discontinuities in the image.

Mathematically, the mean filter is defined as a convolution between the image and a constant valued $N \times N$ matrix:

$$I_{\text{filtered}} = I \circ K; \quad K = \frac{1}{N^2}\begin{bmatrix} 1 & 1 & \cdots & 1 \\ 1 & 1 & \cdots & 1 \\ \vdots & \vdots & \ddots & \vdots \\ 1 & 1 & \cdots & 1 \end{bmatrix} \tag{17.1}$$

The $N \times N$ mean filter is a low pass filter. A low pass filter reduces high frequency components in the Fourier transform (FT) of the image. To understand why the mean filter is a low pass filter, let us recall the relationship between the FT and convolution:

$$F\{A \circ B\} = F\{A\}F\{B\} \quad (17.2)$$

Convolution of a digital image with a matrix of constant values is the discrete equivalent of the convolution of a continuous image function with the rect (boxcar) function (see Section 4.4.2). The FT of the rect function is the sinc function. Therefore, mean filtering is equivalent to multiplying the FT of the image by the sinc, which mostly preserves the low frequency components of the image and diminishes the high frequency components. Figure 17.3 shows a chest X ray image and its FT before and after processing it with the mean filter.

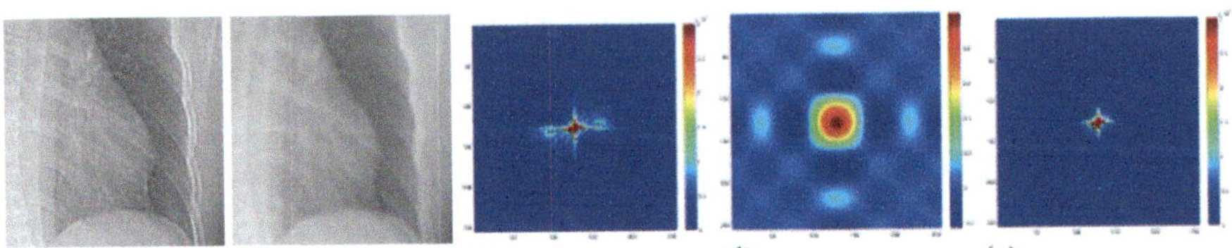

(a) (b) (c) (d) (e)

FIG. 17.3. Mean filtering in the image domain and the Fourier domain. (a) Input X ray image; (b) input image convolved with a 7 × 7 mean filter; (c) FT of the input image (magnitude); (d) FT of the 7 × 7 mean filter, i.e. a product of sinc functions in x and y; (e) FT of the filtered image.

Mean filtering is an example of an image smoothing operation. Smoothing and removal of high frequency noise can help human observers understand medical images. Smoothing is also an important intermediate step for advanced image analysis algorithms. Modern image analysis algorithms involve numerical optimization and require computation of derivatives of functions derived from image data. Smoothing helps make derivative computation numerically stable.

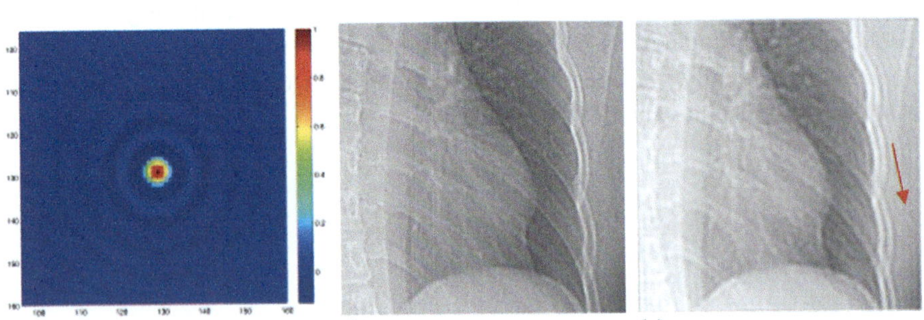

(a) (b) (c)

FIG. 17.4. Ringing artefact illustration. (a) The ideal low pass filter, i.e. a sinc function rotated around the centre of the image. (b) The original image. (c) The image after convolution with the low pass filter. Notice how the bright intensity of the rib bones on the right of the image is replicated in the soft tissue to the right.

17.2.1.2. 'Ideal' low pass filter

The so-called ideal low pass filter cuts off all frequencies above a certain threshold in the FT of the image. In the Fourier domain, this is achieved by multiplying the FT of the image by a cylinder shaped filter generated by rotating a 1-D rect function around the origin. Theoretically, the same effect is accomplished in the image domain by convolution with a 1-D sinc function rotated around the origin. However, the theory assumes that images are periodic functions on an infinite domain. In practice, most images are not periodic and convolution with the rotated sinc function results in an artefact called ringing, illustrated in Fig. 17.4. Another drawback of the ideal low pass filter is the computational cost, which is very high in comparison to mean filtering.

17.2.1.3. Gaussian filtering

The Gaussian filter is a low pass filter that is not affected by the ringing artefact. In the continuous domain, the Gaussian filter is defined as the normal probability density function with standard deviation σ, which has been rotated about the origin in x–y space. Formally, the Gaussian filter is defined as:

$$G_\sigma(x,y)=\frac{1}{2\pi\sigma^2}\,\mathrm{e}^{-\frac{x^2+y^2}{2\sigma^2}} \tag{17.3}$$

The value σ is called the width of the Gaussian filter. The FT of the Gaussian filter is also a Gaussian filter with reciprocal width:

$$F\left(G_\sigma(x,y)\right)=G_{1/\sigma}(\eta,\nu) \tag{17.4}$$

The discrete Gaussian filter is a $(2N+1)\times(2N+1)$ matrix. Its elements are given by:

$$G_{ij}=G_\sigma(i-N-1,j-N-1) \tag{17.5}$$

The size of the matrix, $2N + 1$, determines how accurately the discrete Gaussian approximates the continuous Gaussian. A common choice is $N \geq 3\sigma$. An example is shown in Fig. 17.5.

To apply low pass filtering to a digital image, we perform convolution between the image and the Gaussian filter. This is equivalent to multiplying the FT of the image by a Gaussian filter with a width of $1/\sigma$. As we see in Fig. 17.5, the Gaussian function vanishes rapidly as we move away from the peak. At a

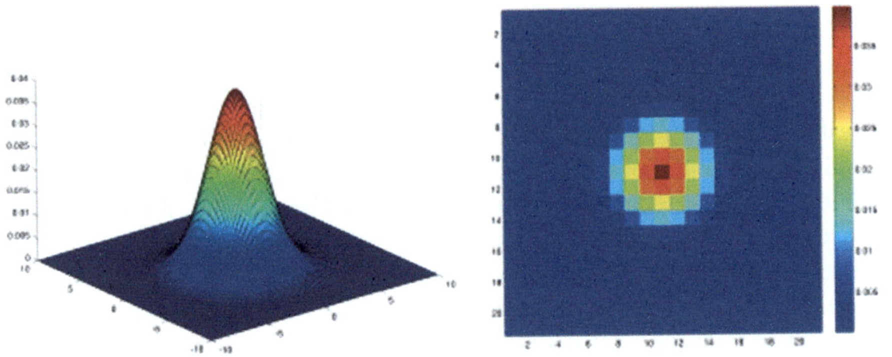

FIG. 17.5. A continuous 2-D Gaussian with $\sigma = 2$, and a discrete 21×21 Gaussian filter with $\sigma = 2$.

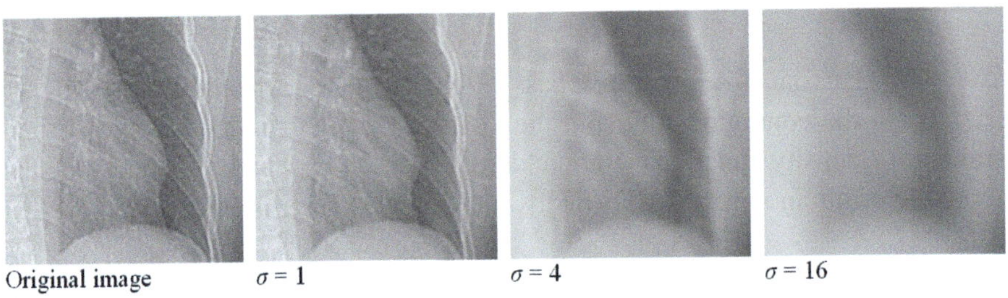

FIG. 17.6. An image convolved with Gaussian filters of different widths.

distance 4σ from the peak, the value of the Gaussian is only 0.0003 of the value at the peak. Therefore, effectively, convolution with the Gaussian filter removes high frequencies in the image. Low frequencies are mostly retained. The larger the standard deviation of the Gaussian filter, the smoother the result of the filtering. Figure 17.6 shows an image convolved with different Gaussian filters.

17.2.1.4. Median filtering

The median filter replaces each pixel in the image with the median of the pixel values in an $N \times N$ neighbourhood. Taking the median of a set of numbers is a non-linear operation. Therefore, median filtering cannot be represented as convolution. The median filter is useful for removing impulse noise, a type of noise where some isolated pixels in the image have very high or very low intensity values (see example in Fig. 17.7). The disadvantage of median filtering is that it can remove important features, such as thin edges.

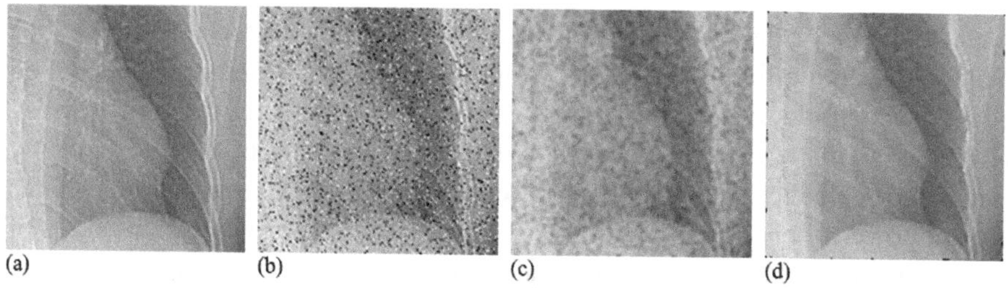

FIG. 17.7. (a) Original image. (b) Image degraded by adding 'salt and pepper' noise. The intensity of a tenth of the pixels has been replaced by 0 or 255. (c) The result of filtering the degraded image with a 5 × 5 mean filter. (d) The result of filtering with a 5 × 5 median filter. Notice how much of the salt and pepper noise has been removed. However, some of the fine lines in the image have also been removed by the filtering.

17.2.1.5. Edge preserving smoothing and denoising

When we smooth an image, we remove high frequency components. This helps reduce noise in the image, but it can also remove important high frequency features such as edges. An edge in image processing is a discontinuity in the intensity function. For example, in an X ray image, the intensity is discontinuous along the boundaries between bone and soft tissue. Some advanced filtering algorithms try to remove noise in images without smoothing edges. An example of these is the anisotropic diffusion algorithm [17.1]. Mathematically, smoothing an image with a Gaussian filter is analogous to simulating heat diffusion in a homogeneous body. In anisotropic diffusion, the image is treated as an inhomogeneous body, with different heat conductance at different places in the image. Near edges, the conductance is lower, so heat diffuses more slowly, preventing the edge from being smoothed away. Away from edges, the conductance is faster. The result is that less smoothing is applied near image edges. However, the approach is only as good as our ability to detect image edges, which is the topic of the next section.

17.2.2. Edge, ridge and simple shape detection

One of the main applications of image processing and image analysis is to detect structures of interest in images. In many situations, the structure of interest and the surrounding structures have different image intensities. By searching for discontinuities in the image intensity function, we can find the boundaries of structures of interest. These discontinuities are called edges. For example, in an X ray image, there is an edge at the boundary between bone and soft tissue.

Edge detection algorithms search for edges in images automatically. As medical images are complex, they have very many discontinuities in the image intensity. Most of these are not related to the structure of interest. Instead, they may be discontinuities due to noise, imaging artefacts or other structures. Good edge detection algorithms identify edges that are more likely to be of interest. However, no matter how good an edge detection algorithm is, it will frequently find irrelevant edges. It is important to remember that edge detection algorithms are not powerful enough to identify, completely and automatically, structures of interest in most medical images. Instead, they are helpful tools for more complex segmentation algorithms, as well as useful visualization tools.

Some structures in medical images have very characteristic shapes. For example, blood vessels are tube-like structures of gradually varying widths. Tube-like structures have two edges that are roughly parallel to each other. This property can be exploited by special tube detection algorithms.

Properties of image intensity at edges and tubes are illustrated in Fig. 17.8.

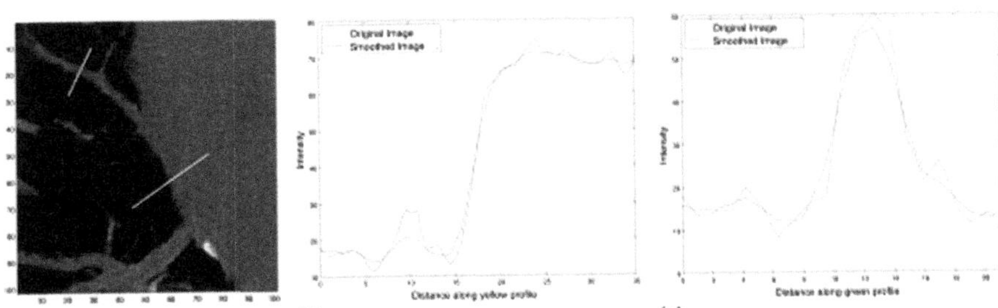

(a) (b) (c)

FIG. 17.8. Illustration of edges and tubes in an image. (a) A detail from the chest CT image is shown. The yellow profile crosses an edge and the green profile crosses a tube-like structure. (b) A plot of image intensity along the yellow profile (in blue) and a plot of image intensity after smoothing the input image with a Gaussian filter with $\sigma = 1$ (in red). (c) A plot of image intensity along the green profile. Edge and tube detectors use properties of the image derivative to detect edges and tubes.

17.2.2.1. How image derivatives are computed

An edge is a discontinuity in the image intensity. Therefore, the directional derivative of the image intensity in the direction orthogonal to the edge must be large, as seen in Fig. 17.8. Edge detection algorithms exploit this property.

How do we compute the derivatives of image intensity? In order to compute derivatives, we require a continuous function. However, an image is just an array of numbers. One solution is to use the finite difference approximation of the derivative. From the Taylor series expansion, it is easy to derive the following approximation of the derivative. For a 1-D function, $f(x)$, we have:

$$f'(x) = \frac{f(x+\delta) - f(x-\delta)}{2\delta} + O(\delta^2) \tag{17.6}$$

where δ is a real number and $O(\delta^2)$ is the error term, involving δ to the power of two or greater. When $\delta << 1$, these error terms are very small and can be ignored for the purpose of approximation.

Likewise, the partial derivatives of a function of two variables can be approximated as:

$$\begin{aligned}\frac{\partial f(x,y)}{\partial x} &= \frac{f(x+\delta_x,y)-f(x-\delta_x,y)}{2\delta_x}+O(\delta_x^2)\\ \frac{\partial f(x,y)}{\partial y} &= \frac{f(x,y+\delta_y)-f(x,y-\delta_y)}{2\delta_y}+O(\delta_y^2)\end{aligned} \tag{17.7}$$

If we treat a digital image as a set of samples from a continuous image function and we set δ_x and δ_y to be equal to the pixel spacing, we can compute approximate image derivatives using these formulas. However, the error term is relatively high, of the order of one pixel width. In practice, derivatives computed using finite difference formulas are dominated by noise.

There is another, often more effective approach to computing image derivatives. We can reconstruct a continuous signal from an image by convolution with a smooth kernel (such as a Gaussian), which allows us to take the derivative of the continuous signal:

$$\begin{aligned}f(x,y) &= (I \circ G)(x,y)\\ D_{\boldsymbol{v}}(f)(x,y) &= D_{\boldsymbol{v}}(I \circ G)(x,y)\end{aligned} \tag{17.8}$$

where D_v denotes the directional derivative of a function in the direction $\boldsymbol{v}$. One of the most elegant ways to compute image derivatives arises from the fact that differentiation and convolution are commutable operations. Both are linear operations, and the order in which they are applied does not matter. Therefore, we can achieve the same effect by computing the convolution of the image with the derivative of the smooth kernel:

$$D_{\boldsymbol{v}}f(x,y) = I \circ (D_{\boldsymbol{v}}G)(x,y) \tag{17.9}$$

This leads to a very practical and efficient way of computing derivatives. We create a filter, which is just a matrix that approximates D_vG. Such filters are illustrated in Fig. 17.9. Then, we compute the numerical convolution between this filter and the image. This is just another example of filtering as described earlier in the chapter.

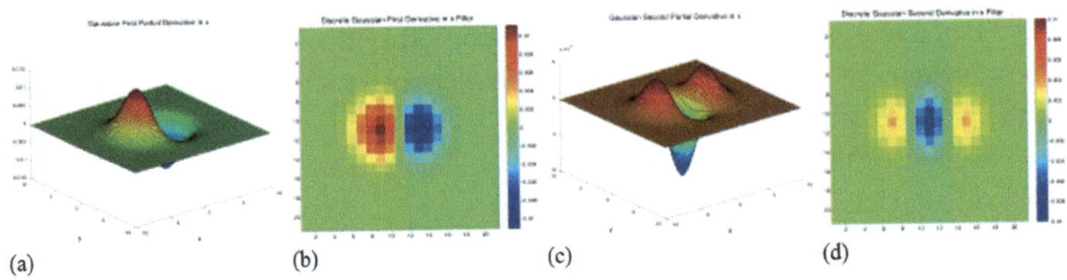

FIG. 17.9. (a) and (c) First and second partial derivatives in x of the Gaussian with $\sigma = 1$. (b) and (d) Corresponding 21 × 21 discrete Gaussian derivative filters.

Most frequently, we let G be a Gaussian filter. The Gaussian is infinitely differentiable, so it is possible to take an image derivative of any order using this approach. The width of the Gaussian is chosen empirically. It determines how smooth the interpolation of the digital image is. The more smoothing is applied, the less sensitive the derivative function will be to small local changes in image intensity. As we will see shortly, this can help us distinguish between the more prominent and less prominent edges.

17.2.2.2. Edge detectors based on the first derivative

A popular and simple edge detector is the Sobel operator. To apply this operator, the image is convolved with a pair of filters:

$$S_x = \begin{pmatrix} 1 & 0 & -1 \\ 2 & 0 & -2 \\ 1 & 0 & -1 \end{pmatrix}; \; S_y = \begin{pmatrix} 1 & 2 & 1 \\ 0 & 0 & 0 \\ -1 & -2 & -1 \end{pmatrix} \tag{17.10}$$

The reader can check that this convolution is quite similar to the finite difference approximation of the partial derivatives of the image described above. In fact, the Sobel operator approximates the derivative at the given pixel and at the two neighbouring pixels and then computes a weighted average of these three values with weights (1, 2, –1). This averaging makes the output of the Sobel operator slightly less sensitive to noise than simple finite differences.

Figure 17.10 shows the results of convolution with S_x and S_y. The last image is the so-called gradient magnitude image, given by $\sqrt{(S_x^2 + S_y^2)}$. Large values of the gradient magnitude correspond to edges, and low values are regions where intensity is nearly constant. However, there is no absolute value of the gradient magnitude that distinguishes edge from non-edge. For each image, one has to establish, empirically, a threshold to apply to the gradient magnitude image in order to separate the edges of interest from spurious edges caused by noise and image artefact. This is one of the greatest limitations of edge detection based on first derivatives.

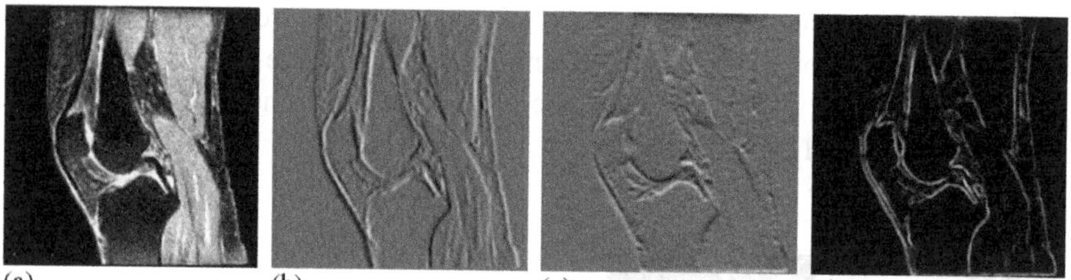

(a) (b) (c) (d)

FIG. 17.10. Illustration of the Sobel operator. (a) MR image of the knee. From the US National Cancer Institute National Biomedical Imaging Archive Osteoarthritis Initiative (https://imaging.nci.nih.gov/ncia). (b) Convolution of the image with the Sobel x derivative filter S_x. (c) Convolution with S_y. (d) Gradient magnitude image. The gradient magnitude is high at image edges, but also at isolated pixels where image intensity varies due to noise. (See also Fig. 17.13.)

Often, the small amount of smoothing performed by the Sobel operator is not enough to eliminate the edges associated with image noise. If we are only interested in very strong edges in the image, we may want to perform additional smoothing. A common alternative to the Sobel filter is to compute the partial derivatives of the image intensity using convolution of the image with Gaussian derivative operators (D_xG) and (D_yG), as shown in Fig. 17.11. The resulting gradient magnitude image should be compared with the gradient magnitude image obtained with the Sobel operator in Fig. 17.10.

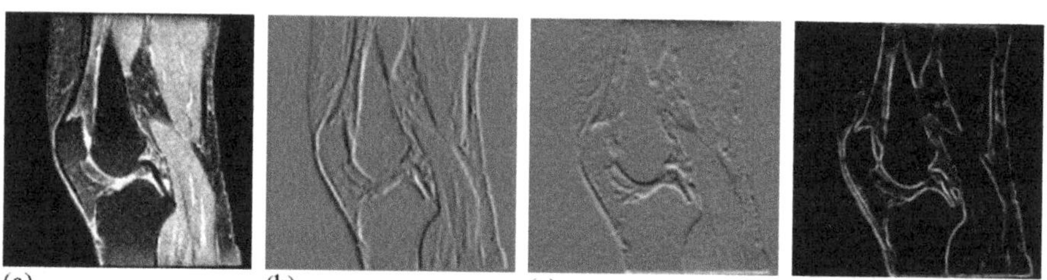

(a) (b) (c) (d)

FIG. 17.11. Illustration of convolution with Gaussian derivative filters. (a) Input image. (b) and (c) Convolution with Gaussian derivative filters with $\sigma = 2$. (d) Gradient magnitude image. Compare with Fig. 17.10.

Of course, too much smoothing can remove important edges too. Finding the right amount of smoothing is a difficult and often ill posed problem.

17.2.2.3. Detectors based on zero crossing of the second derivative

Imagine a particle crossing an edge in a continuously smooth image, F, moving in a direction orthogonal to the edge (i.e. in the direction of the image gradient). If we plot the gradient magnitude of the image along the path of the particle, we see that at the edge there is a local maximum of the gradient magnitude (Fig. 17.12).

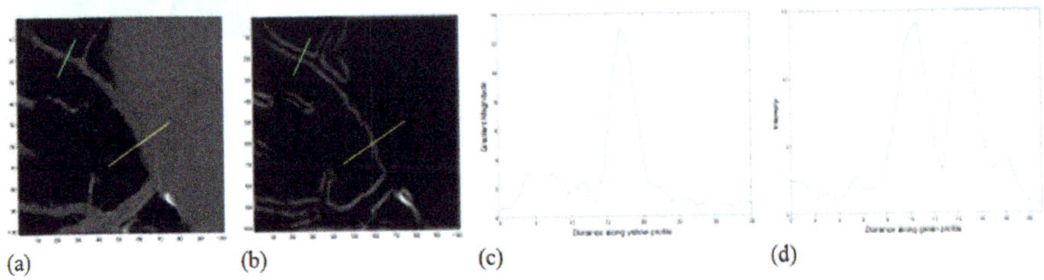

FIG. 17.12. *Gradient magnitude at image edges. (a) A detail from the chest CT image from Fig. 17.8. (b) Corresponding gradient magnitude image. (c) A plot of gradient magnitude image across the edge (yellow profile). (d) A plot of gradient magnitude image across the tube-like structure (green profile). The gradient magnitude reaches its maximum at the points where the profiles cross the image edge.*

Let us denote the unit vector in the particle's direction as $\boldsymbol{v}$, and the point where the particle crosses the edge as $\boldsymbol{x}$. The gradient magnitude of the image F at $\boldsymbol{x}$ is simply:

$$\nabla F(\boldsymbol{x}) = D_{\boldsymbol{v}} F \mid \boldsymbol{x} \tag{17.11}$$

The gradient magnitude reaches a local maximum at $\boldsymbol{x}$ in the direction $\boldsymbol{v}$ if, and only if:

$$D_{\boldsymbol{vv}} F = 0 \text{ and } D_{\boldsymbol{vvv}} F \leq 0 \tag{17.12}$$

Several edge detectors leverage this property. The earliest of these operators is the Marr–Hildreth edge detector. It is based on the fact that the necessary (but not sufficient) condition for Eq. (17.12) is:

$$D_{xx} F + D_{yy} F = 0 \tag{17.13}$$

The operator ($D_{xx} + D_{yy}$) is the Laplacian operator. By finding the set of all points in the image where the Laplacian of the image is zero, we find the superset of all the points that satisfy Eq. (17.12).

When dealing with discrete images, we must, of course, use convolution with a smooth filter, as with the Gaussian, when computing the second derivatives and the Laplacian. That is precisely what the Marr–Hildreth edge detector does. It convolves the discrete image I with the Laplacian of Gaussian filter, given by:

$$J = I \circ (D_{xx}G + D_{yy}G) = D_{xx}(I \circ G) + D_{yy}(I \circ G) \tag{17.14}$$

Next, the Marr–Hildreth edge detector finds contours in the image where $J = 0$. These contours are closed and form the superset of edges in the image. The last step is to eliminate the parts of the contour where the gradient magnitude of the input image is below a user specified threshold. These two steps are illustrated in Fig. 17.13.

The Canny edge detector is also rooted in the fact that the second derivative of the image in the edge direction is zero. This detector first applies Gaussian smoothing to the image. It then finds the pixels in the image with high gradient magnitude, using the Sobel operator and thresholding. Next, it eliminates pixels that do not satisfy the maximum condition of Eq. (17.12). Lastly, it uses a procedure called hysteresis to eliminate very short edges that are most likely the product of noise in the image. The Canny edge detector has very good performance characteristics compared with other edge detectors and is very popular in practice. An example of Canny edge detection results is shown in Fig. 17.13.

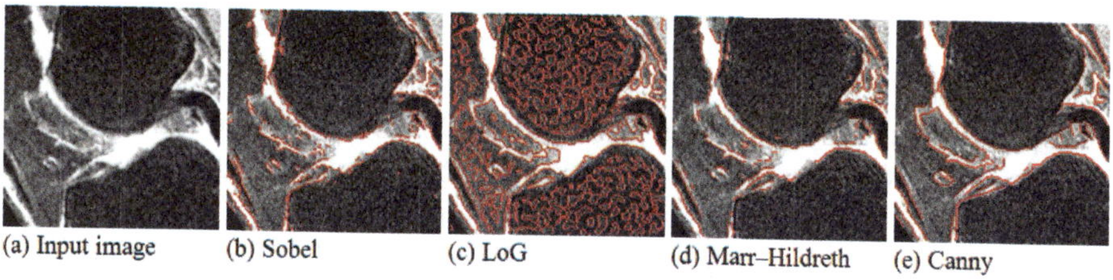

FIG. 17.13. A comparison of edge detection approaches. (a) The input image (a region of interest from the knee MRI in Fig. 17.11); (b) edges produced by the Sobel detector; (c) zero crossings of the convolution of the image with the Laplacian of Gaussian operator; (d) edges produced by the Marr–Hildreth detector, i.e. a subset of the zero crossings that have gradient magnitude above a threshold; (e) edges produced by the Canny detector.

17.2.2.4. Hough transform

So far, we have discussed image processing techniques that search for edges. Edges are important for detecting objects in images. However, sometimes the objects that we are interested in detecting have a very characteristic shape: circles, tubes and lines. In these cases, we are better off using detectors that search for these shapes directly, rather than looking at edges. The Hough transform is one such detector. To understand how it works, let us consider a slightly simpler problem. Given a set of points $(x_1, y_1)...(x_N, y_N)$ in the plane, we want to find lines, circles or ellipses approximately formed by these points (Fig. 17.14).

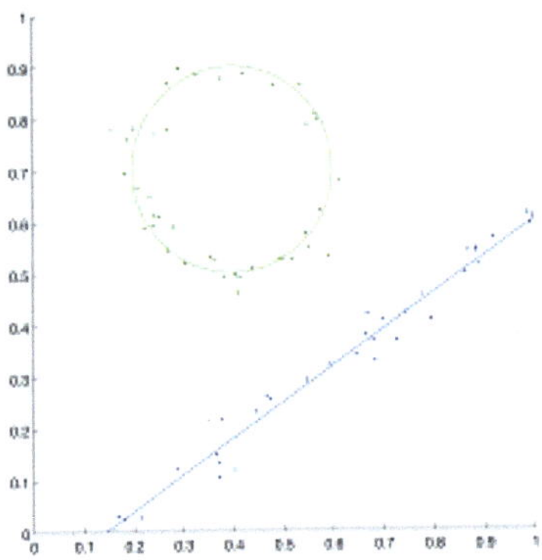

FIG. 17.14. Hough transform (simplified problem). Given roughly collinear data (blue points), find a line that best fits the points. Given roughly circularly distributed data (green points), fit a circle to the points.

Simple shapes, such as lines, circles and ellipses, can be described by a small number of parameters. For example, circles are parameterized by the centre (two parameters) and radius (one parameter). Ellipses are parameterized by four parameters. Lines are naturally parameterized by the slope and intercept (two parameters). However, this parameterization is asymptotic for vertical lines. An alternative parameterization by Duda and Hart (1972) [17.2] uses the distance from the line to the origin and the slope of the normal to the line as the two parameters describing a line.

Each line, circle or ellipse corresponds to a single point in the corresponding 2-D, 3-D or 4-D parameter space. The set of all lines, circles or ellipses passing through a certain point (x, y) in the image space corresponds to an infinite set of points in the parameter space. These points in the parameter space form a manifold. For example, all lines passing through (x, y) form a sinusoid in the Duda and Hart 2-D line parameter space. All circles passing through (x, y) form a cone in the 3-D circle parameter space. This is the meaning of the Hough transform: it transforms points in the image domain into curves, surfaces or hypersurfaces in the parameter domain.

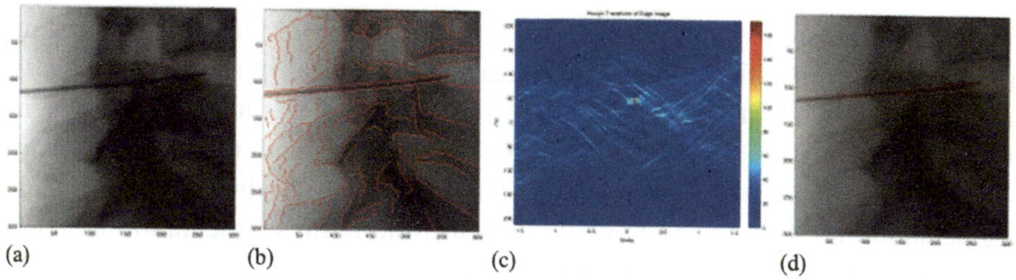

FIG. 17.15. *Hough transform example. (a) An input fluoroscopy image of a surgical catheter. The catheter is almost straight, making it a good candidate for detection with the Hough transform. (b) Edge map produced by the Canny edge detector. (c) Superimposed Hough transforms of the edge points. The Hough transform of a point in image space is a sinusoid in Hough transform space. The plot shows the number of sinusoids that pass through every bin in the Hough transform space. There are two bins through which many sinusoids pass. These bins correspond to the lines in (d).*

If several points in the image domain belong to a single line, circle or ellipse, then their corresponding manifolds in the parameter space intersect at a single point $(p_1,...,p_k)$. This gives rise to the shape detection algorithm. The 2-D, 3-D or 4-D parameter space is divided into a finite set of bins. Every bin, j, is assigned a variable, q_j, that is initialized to zero. For every point (x_1, y_1) in the image domain, we compute the corresponding curve, surface or hypersurface in the parameter space, and find all the bins in the parameter space through which the manifold passes. Every time that the curve, surface or hypersurface passes through the bin j, we increment the corresponding variable q_j by 1. Once we complete this procedure for all N points, we identify the bins where q_j is large. These bins correspond to a set of q_j points that approximately form a line, circle or ellipse.

The Hough transform, combined with edge detection, can be used to search for simple shapes in digital images. The edge detector is used to find candidate boundary points $(x_1, y_1)...(x_N, y_N)$. Then, the Hough transform is used to find simple shapes. In Fig. 17.15, we illustrate this in an intraoperative fluoroscopy image.

The Hough transform is an elegant and efficient approach, but it scales poorly to more complex objects. Objects more complex than lines, circles and ellipses require a large number of parameters to describe them. The higher the dimensionality of the parameter space, the more memory and computationally intensive the Hough transform will become.

17.3. IMAGE SEGMENTATION

The problem of finding objects in images, known as segmentation, is the central problem in the field of image analysis. It is also a highly complex problem. There are many types of segmentation problem. Many involve finding and outlining a specific anatomical structure in a medical image. In addition to looking for specific structures, we are often interested in finding pathology in medical images. These problems are very different, depending on the anatomy and imaging modality. Heart segmentation in CT is very different from heart segmentation in MRI, which is very different from brain segmentation in MRI. Some structures move during imaging, while other structures are almost static. Some structures have a simple shape that varies little from subject to subject, while others have complex, unpredictable shapes. Some structures have good contrast with surrounding tissues; others do not. More often than not, a given combination of anatomical structure and imaging modality requires a custom segmentation algorithm.

This section provides a very brief summary of image segmentation. The aim is to provide the reader with a flavour of the segmentation methods available and to provide an insight into the main ideas used in these methods. Readers interested in learning about image segmentation in more detail should consult the references at the end of this chapter.

17.3.1. Object representation

Before beginning to discuss segmentation methods, we must cover different ways to represent objects in images. Some common approaches are presented next.

- *Binary image or label image*: These are very simple ways of representing an object or a collection of objects in an image. Given an image, I, that contains some object, O, we can construct another image, S, of the same dimensions as I, whose pixels have values 0 and 1, according to:

$$S(\boldsymbol{x}) = \begin{cases} 1 & \text{if } \boldsymbol{x} \in \text{Object} \\ 0 & \text{otherwise} \end{cases} \tag{17.15}$$

Such an image is called the binary image of O. When I contains multiple objects of interest, we can represent them as separate binary images (although this would not be very memory efficient), or as a single label image, L:

$$L(\boldsymbol{x}) = \begin{cases} 1 & \text{if } \boldsymbol{x} \in \text{Object 1} \\ 2 & \text{if } \boldsymbol{x} \in \text{Object 2} \\ \vdots & \\ 0 & \text{otherwise} \end{cases} \tag{17.16}$$

Binary and label images have limitations. Their accuracy is limited by the resolution of the image, I. Furthermore, they represent the boundaries of objects as very non-smooth (piecewise linear) curves or surfaces, whereas the actual anatomical objects typically have smooth boundaries.

- *Geometric boundary representations*: Objects can be described by their boundaries. This representation is more compact than the binary image representation and allows subpixel accuracy. Smoothness can also be ensured. The simplest geometric boundary representation is defined by a set of points on the boundary of an object, called vertices, and a set of line segments called edges (or in 3-D, a set of polygons, called faces) connecting the vertices. Such geometric constructs are called meshes. It is also possible to connect points using smooth cubic or higher order curves and surfaces. In either case, the object representation is defined by the coordinates of the points and the connectivity between the points.
- *Level sets of real valued images*: This representation combines attractive properties of the two representations above. As with binary images, this representation uses an image, F, of the same dimensions as I to represent an object, O, in the image, I. However, unlike the binary representation, the 'level set' representation can achieve subpixel accuracy and smooth object boundaries. Every pixel (or voxel) in F has intensity values in the range $-M$ to M, where M is some real number. The boundary of O is given by the 'zero level set' of the function F:

$$B(O)=\left\{\boldsymbol{x}\in\mathbb{R}^N : F(\boldsymbol{x})=0\right\} \tag{17.17}$$

Of course, we must be a little careful here; F is a discrete image and the definition above requires a continuous function. In practice, linear interpolation is applied to the image F when using the above definition.

Binary and level set representations can be converted to geometric boundary representations using contour extraction algorithms, such as the marching cubes algorithm. A binary or geometric boundary representation can be converted to a level set representation using the distance map algorithm.

Of course, there are quite a few other representations available, but we will not discuss them in this chapter. Examples of the representations covered above are shown in Fig. 17.16.

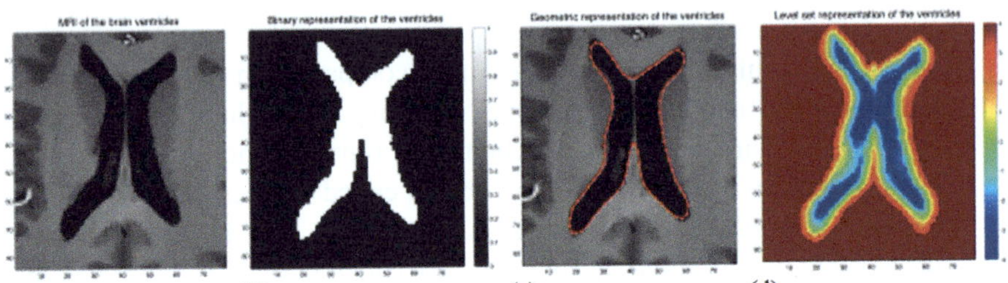

(a) (b) (c) (d)

FIG. 17.16. (a) An axial slice from a brain MRI; (b) binary; (c) geometric; (d) level set representations of the lateral ventricles in this image.

17.3.2. Thresholding

Thresholding is the simplest segmentation technique possible. It is applicable in situations where the structure of interest has excellent contrast with all other structures in the image. For example, in CT images, thresholding can be used to identify bone, muscle, water, fat and air because these tissue classes have different attenuation levels. Thresholding produces a binary image using the following simple rule:

$$S(\boldsymbol{x}) = \begin{cases} 1 & \mathrm{T}_{\mathrm{lower}} \leq \mathrm{I}(\boldsymbol{x}) < T_{\mathrm{upper}} \\ 0 & \text{otherwise} \end{cases} \tag{17.18}$$

Here, T_{lower} is a value called the lower threshold and T_{upper} is the upper threshold. For example, for bone in CT, $T_{\mathrm{lower}} = 400$ and $T_{\mathrm{upper}} = \infty$. The segmentation is simply the set of pixels that have intensity between the upper and lower thresholds. Thresholding is illustrated in Fig. 17.17.

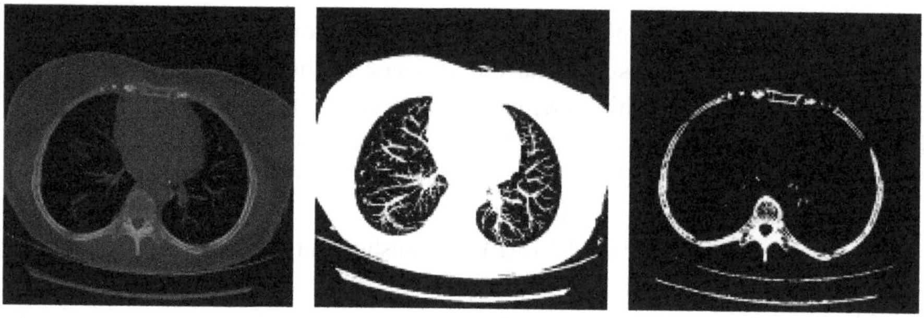

FIG. 17.17. CT image thresholded at different intensity levels.

Of course, in most medical image segmentation problems, thresholding does not produce satisfactory results. There may be several reasons for this. In noisy images, there are likely to be pixels inside the structure of interest that are incorrectly labelled because their intensity is below or above the threshold. In MR images, intensity is usually inhomogeneous across the image, so that a pair of thresholds that works in one region of the image is not going to work in a different region. Lastly, the structure of interest may be adjacent to other structures with very similar intensity. In all of these situations, more advanced techniques are required.

In some images, the results of thresholding are satisfactory, but the values of the upper and lower thresholds are not known, a priori. For example, in brain MR images, it is possible to apply intensity inhomogeneity correction to the image to reduce the effects of inhomogeneity. However, in order to segment the grey matter or white matter in these images, we would typically need a different pair of thresholds for every scan. In these situations, we require automatic threshold detection, which is discussed in the next section.

17.3.3. Automatic tissue classification

Consider the problem where we have to partition an image into regions corresponding to a fixed number of tissue classes, k. In the brain, for example, there are three important tissue classes: white matter, grey matter and cerebrospinal fluid (CSF). In T_1 weighted MRI, these tissue classes produce different image intensities: the white matter is brightest and the CSF is darkest (Fig. 17.18). Unlike CT, the range of intensity values produced by each tissue class in MRI is not known, a priori. Furthermore, because of MRI inhomogeneity artefacts, noise and partial volume effects, there is a great deal of variability in the intensity of each tissue class. Automatic tissue classification is a term used to describe various computational algorithms that try to partition an image into tissue classes based on statistical inference.

The simplest automatic tissue classification algorithms are closely related to thresholding. Let us assume that the variance in the intensity of each tissue class is not too large. In this case, we can expect the histogram of the image to have k peaks, corresponding to the k tissue classes. Such peaks are seen in the histogram of a phantom MR image shown in Fig. 17.18(b).

Tissue classification simply involves finding thresholds that separate these peaks. For example, in Fig. 17.18(b), a threshold at the intensity value 100 would separate the CSF class from the grey matter class, and a threshold of 200 would separate the grey matter from the white matter. Of course, in a real MR image, such as the one in Fig. 17.18(c), the peaks in the histogram are not as well separated (Fig. 17.18(d)) and it is not obvious from just looking at the histogram what the correct threshold values ought to be.

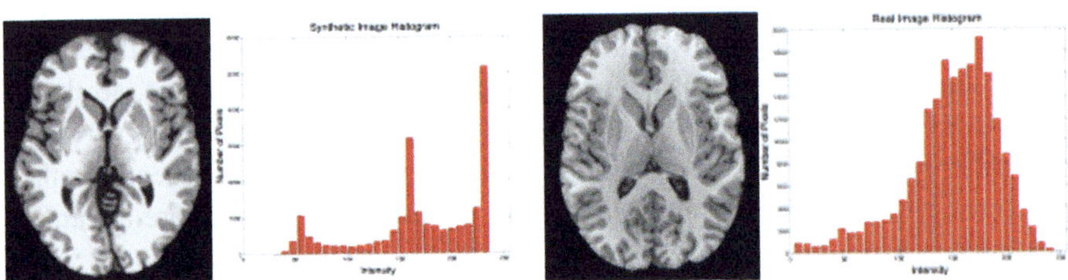

(a) (b) (c) (d)

FIG. 17.18. Histograms of brain MR images. (a) A slice from the digital brain MRI phantom from BrainWeb [17.3]. This synthetic image has very little noise and intensity inhomogeneity, so the intensity values of all pixels in each tissue class are very similar. (b) The histogram of the synthetic image, with clearly visible peaks corresponding to CSF, grey matter and white matter. (c) A slice from a real brain MR image, with the skull removed. (d) The histogram of the real MR image. Peaks in the histogram are much less obvious.

There are several automatic tissue classification methods that examine the image histogram and determine thresholds that are optimal, according to a certain criterion. The simplest of these is k-means clustering. This approach groups intensity values in the image histogram into k clusters. The algorithm seeks to minimize the variability of the intensity within each cluster. Formally, k-means clustering is defined as an energy minimization problem:

$$\alpha_1^* \ldots \alpha_N^* = \arg\min_{\{\alpha_1 \ldots \alpha_N \in [1 \ldots k]\}} \sum_{j=1}^{k} \sum_{\{q:\alpha_q = j\}} (I_q - \mu_j)^2 \quad (17.19)$$

where

α_i is the cluster to which the pixel i is assigned;
N is the number of pixels in the image;
I_q is the intensity of the pixel q;

and μ_j is the mean of the cluster j, i.e. the average intensity of all pixels assigned the label j.

The notation $\underset{x\in\Omega}{\operatorname{argmin}} f(x)$ is read as 'the point x in the domain Ω where the function $f(x)$ attains its minimum'. Theoretically, the optimization problem above is intractable, but in practice, a simple iterative approach yields good approximations of the global minimum. This iterative approach requires the initial means of the clusters to be specified, and one of the drawbacks of k-means clustering is that it can be sensitive to initialization. An example of brain MRI segmentation using k-means clustering is shown in Fig. 17.19.

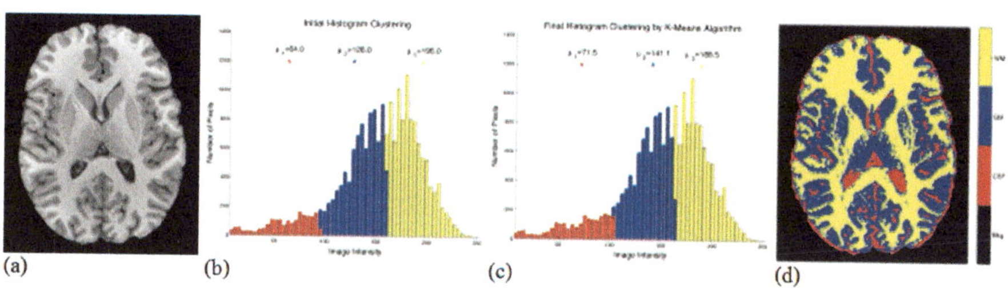

FIG. 17.19. Example of k-means segmentation. (a) A slice from a brain MRI. (b) Partitioning of the image histogram into clusters based on initial cluster means. (c) Partitioning of the histogram into clusters after 10 iterations. (d) Segmentation of the image into grey matter (GM), white matter (WM) and CSF.

Alternatives to k-means are fuzzy c-means clustering and Gaussian mixture modelling. In fuzzy c-means clustering, cluster membership is not absolute. Instead, fuzzy set theory is used to describe partial cluster membership. This results in segmentations where uncertainty can be adequately represented.

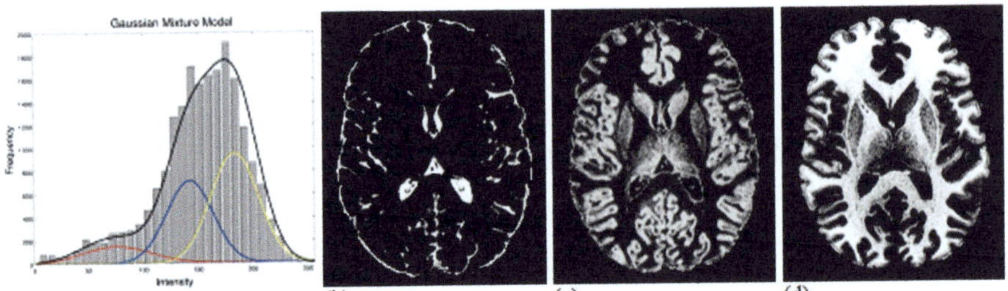

(a) (b) (c) (d)

FIG. 17.20. Example of segmentation using Gaussian mixture models. (a) A mixture model fitted to the histogram of the MR image from Fig. 17.19. The mixture model (black curve) is a weighted sum of three Gaussian probability densities, one for each tissue type (red, blue and yellow curves). (b)–(d) CSF, grey matter and white matter probability maps generated by the method. One advantage of Gaussian mixture modelling is that it produces a probabilistic segmentation, rather than a binary segmentation.

Gaussian mixture modelling assumes that the pixel intensities $x_i,\ldots,x_N$ in an image are samples from a random variable, *X*, with a probability density function *f*(*x*) that is a weighted sum of *n* Gaussian probability densities, with respective weights $\alpha_1,\ldots,\alpha_n$:

$$f(x;\{\alpha_m,\mu_m,\sigma_m\}) = \sum_{m=1}^{k} \alpha_m z(\frac{x-\mu_m}{\sigma_m}) \qquad (17.20)$$

where *z* is the standard normal distribution and parameters α_m, μ_m, σ_m are unknown. The expectation minimization algorithm is used to find the maximum likelihood estimate of these parameters:

$$\{\alpha_m^*,\mu_m^*,\sigma_m^*\} = \underset{\{\alpha_m,\mu_m,\sigma_m\}}{\arg\max}\, P(x_1,\ldots,x_N;\{\alpha_m,\mu_m,\sigma_m\}) = \underset{\{\alpha_m,\mu_m,\sigma_m\}}{\arg\max} \prod_{i=1}^{n} f(x_i;\{\alpha_m,\mu_m,\sigma_m\}) \qquad (17.21)$$

Intuitively, Gaussian mixture modelling fits the image histogram with a weighted sum of Gaussian densities, as shown in Fig. 17.20(a). Once the optimal parameters $\{\alpha_m, \mu_m, \sigma_m\}$ have been found, the probability that pixel *j* belongs to a tissue class *m* is found from:

$$P(L_j = m \mid x_j) = z(\frac{x_j-\mu_m}{\sigma_m}) \qquad (17.22)$$

Like fuzzy *c*-means, Gaussian mixture modelling can describe uncertainty. For each tissue class, a probability image is generated, estimating the probability that a given pixel belongs to a given tissue class. This is an advantage because it is possible to model partial volume effects. For example, a pixel may be assigned 0.5 probability of being white matter, 0.4 probability of being grey matter and 0.1 probability of being CSF. This result can be interpreted as a partial volume effect, i.e. both white matter and grey matter tissues are present in the pixel. An example of segmentation using Gaussian mixture modelling is shown in Figs 17.20(b)–(d). The performance of Gaussian mixture modelling can be further improved by introducing constraints on the consistency of the segmentation between neighbouring pixels. Indeed, methods that combine Gaussian mixture modelling with such spatial regularization constraints are among the most widely used in brain tissue segmentation from MRI.

17.3.4. Active contour segmentation methods

The term 'active contours' is used to describe a family of image segmentation algorithms that emerged from the seminal work of Kass et al. (1988) [17.4] on active snakes. Before active snakes, the mainstream approach to object segmentation involved edge detection, followed by linking edges to form object boundaries. As was pointed out at the beginning of this chapter, such a deterministic approach is limited to simple segmentation problems. Active snakes represented a radical shift from the deterministic paradigm. Snakes are an early example of knowledge based image analysis, where prior knowledge about the shape and smoothness of object boundaries is used to guide segmentation.

Unlike automatic tissue classification, active contour methods address the problem of object segmentation. The goal is to identify a specific anatomical structure, or small set of structures, in a biomedical image. The structure is represented by a contour (a closed curve in 2-D, or a closed surface in 3-D). The goal is to find a contour, *C*, that minimizes an energy function *E*(*C*). This energy function typically comprises two terms: a term that measures how well the contour coincides with the boundaries of objects in the image *I* and a term that measures how simple the contour *C* is. As an example, consider the 2-D contour energy function proposed by Caselles et al. (1997) [17.5]:

$$E(C)=\int_0^1 g_I(C(t))|C'(t)|\,\mathrm{d}t \quad \text{where} \quad g_I(x,y)=\frac{1}{1+|\nabla(G_\sigma \circ I)(x,y)|} \tag{17.23}$$

Here, the contour C is parameterized by the variable t, where $0 \le t \le 1$. The function g_I is called the speed function. It is a monotonically decreasing function of the image gradient magnitude and has very small values along the edges of I, and is close to 1 away from the edges of I. It is easy to verify that the energy, $E(C)$, is decreased by making C fall on edges in I, where g_I is reduced, and by making C shorter, which reduces $|C'(t)|$.

Active contour methods are usually described not in terms of the energy function, $E(C)$, but instead in terms of an evolution equation. This equation has the form:

$$\frac{\partial C}{\partial T} = F\vec{N} \tag{17.24}$$

where F is a scalar function of the image and the contour, and $\vec{N}$ is the unit normal vector to C. This equation describes how to evolve the contour over time, T, such that the energy, $E(C)$, decreases.

The evolution equation and the function F can be derived from the energy function using the calculus of variations. Different active contour methods use different functions, F, which correspond to different energy functions. For example, the evolution equation for the 2-D Caselles energy function has the form:

$$\frac{\partial C}{\partial T} = \left(g_I \kappa - \nabla g_I \cdot \vec{N}\right)\vec{N} \tag{17.25}$$

where κ is the curvature of the contour C. Interestingly, the same equation is used to describe contour evolution in 3-D, except that κ is used to describe the mean curvature of the surface C.

In early active contour methods, the segmentation was represented using a geometric boundary representation, i.e. a piecewise cubic curve for 2-D segmentation, or a surface for 3-D segmentation. In modern active contour methods, the level set representation is used instead, because of its numerical stability and simplicity. The evolution equation can be adapted to the level set representation. If ϕ is a function on $\mathbb{R}^n$, such that $C = \{\boldsymbol{x} \in \mathbb{R}^n : \phi(\boldsymbol{x}) = 0\}$, i.e. C is the zero level set of ϕ, then the evolution equation, Eq. (17.25), can be rewritten in terms of ϕ as $\partial\phi / \partial T = F|\nabla\phi|$. For instance, for the Caselles energy, the level set evolution equation has the form:

$$\frac{\partial \phi}{\partial T} = \left[g_I \operatorname{div}\left(\frac{\nabla\phi}{|\nabla\phi|}\right) + \nabla g_I \frac{\nabla\phi}{|\nabla\phi|}\right]|\nabla\phi| \tag{17.26}$$

The level set representation of the active contour has a number of advantages. As already mentioned, level set methods are numerically robust and simple to implement. With the level set representation, the topology of the segmentation can change; multiple contours can merge into a single contour. As the active contour is represented as a level set, the contour is always a closed manifold.

Active contour segmentation can be used in conjunction with automatic tissue classification using a method developed by Zhu and Yuille (1996) [17.6]. This method uses the following definition for F:

$$F(\boldsymbol{x}) = \alpha\left[\log P(\boldsymbol{x} \in \text{Object}) - \log P(\boldsymbol{x} \in \text{Background})\right] - \beta\,\kappa \tag{17.27}$$

where $P(\boldsymbol{x} \in \text{Object})$ and $P(\boldsymbol{x} \in \text{Background})$ are the probabilities that a pixel at position $\boldsymbol{x}$ belongs to the object of interest or to the background, respectively. These probabilities can be estimated from the image, I, using automatic tissue classification or manual thresholding. The constants α and β are user specified weights that provide a trade-off between the terms in the speed function. The evolution has a very intuitive interpretation. The component of the force weighted by α pushes the contour outwards if it lies inside of the object (i.e. $P(\boldsymbol{x} \in \text{Object}) > P(\boldsymbol{x} \in \text{Background})$) and pushes the contour inwards if it lies outside of the object. The component $-\beta\kappa$ pushes the contour inward at points with large negative curvature, and pushes it outward at points with large positive curvature. The effect is to smooth out the sharp corners in the contour, keeping the shape of the contour simple.

Regardless of the ‘flavour’ of the active contour method, the segmentation proceeds as follows. The user provides an initial segmentation, for example, a circle or sphere placed inside the object of interest. Contour evolution is then simulated by repeatedly applying the evolution equation. Evolution is repeated until convergence, or until the user interrupts it. An example of kidney segmentation in a CT image is shown in Fig. 17.21.

Active contour segmentation is an area of active research. Numerous extensions to the methods have been proposed in recent years, including more general shape priors, constraints on the topology of the segmentation, and various application specific image based criteria.

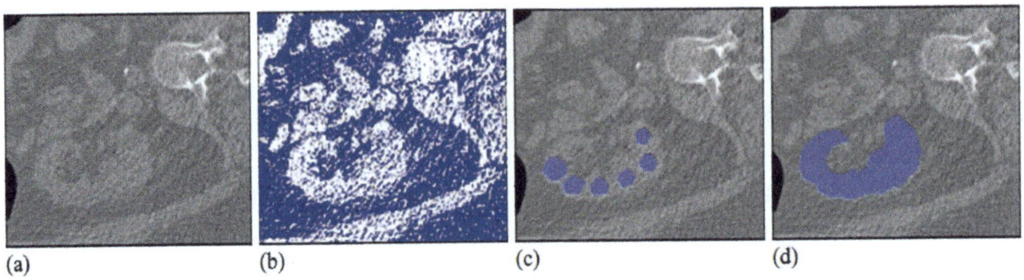

FIG. 17.21. An example of kidney segmentation in a low contrast CT volume. (a) An axial slice from the CT volume. (b) A map of $P(\mathbf{x} \in Object) - P(\mathbf{x} \in Background)$, computed using tissue classification. White pixels have high object probability; blue points have high background probability. Note that simple thresholding of this map will lead to a very noisy segmentation. (c) Initialization of the active contour method using five spherical seeds. (d) Segmentation after 1000 iterations of evolution. Note that the contours have merged to form a single surface.

17.3.5. Atlas based segmentation

Deformable image registration, described in the next section, is a technique that automatically finds correspondences between pairs of images. In recent years, it has also become a popular tool for automatic image segmentation. The idea is simple and is illustrated in Fig. 17.22. Suppose we have one image, called the atlas, in which the structure of interest has been segmented, e.g. manually. Suppose we have another image, I, in which we want to segment the structure of interest. By performing registration between this image and the atlas, we obtain a mapping, $\phi(x)$, that maps every point in the image into a corresponding point in the atlas. This mapping can be used to transform the segmentation from the atlas into image I. The quality of the segmentation is limited only by the quality of the registration.

Several authors have extended this simple idea to using multiple atlases. Each atlas is registered to the image, I, and segmentation from each atlas is mapped into I. Owing to registration errors, these warped segmentations $S_1 \ldots S_K$ do not overlap perfectly. A voting scheme is used to derive a consensus segmentation from $S_1 \ldots S_K$.

The appeal of atlas based segmentation is that it is very easy to implement. Several image registration software applications are available in the public domain. All that the user needs to perform atlas based segmentation is an image, or several images, where the object of interest has been manually segmented. Atlas based segmentation can be applied in various imaging modalities, but its quality may not be as high as with methods that use shape priors.

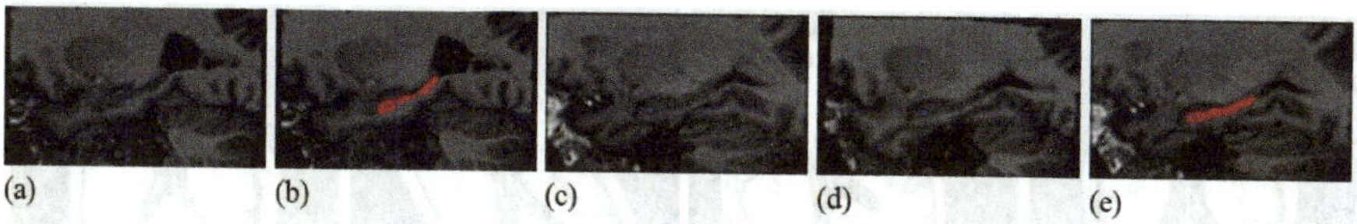

FIG. 17.22. *Example of atlas based segmentation of the hippocampus in brain MRI. (a) An MR image used as the atlas. (b) Manual segmentation of the hippocampus in the atlas. (c) The target image in which we want to segment the hippocampus. (d) Atlas warped to the target image using deformable registration. (e) The atlas based segmentation of the target image, overlain on the target image.*

17.4. IMAGE REGISTRATION

Often in medical image analysis, we have to process information from multiple images. We may have images with different modalities (CT, positron emission tomography (PET), MRI) from the same subject, or we may want to compare images acquired at different time points from a single subject, or we may have images of the same anatomical regions from multiple subjects that we want to compare. In all these, and many other situations, we need a way to find and align corresponding locations in multiple images. Image registration is a field that studies optimal ways to align and normalize images.

Image registration is the problem of finding transformations between images. Given an image $I:\Omega \in \mathbb{R}^n \to \mathbb{R}$ and an image $J:\Omega \in \mathbb{R}^n \to \mathbb{R}$, we seek a transformation $\phi:\Omega \to \Omega$, such that $I(\boldsymbol{x})$ and $J(\phi(\boldsymbol{x}))$ are 'similar' for all $\boldsymbol{x}$ in Ω. The meaning of 'similar' depends on the application. In the context of medical image analysis, 'similar' usually means 'describing the same anatomical location'. However, in practice, such anatomical similarity cannot be quantified and 'similar' means 'having similar image intensity features'.

There are many different types of image registration problem. They can be characterized by two main components: the transformation model and the similarity metric. In the next two sections, we describe each of these components.

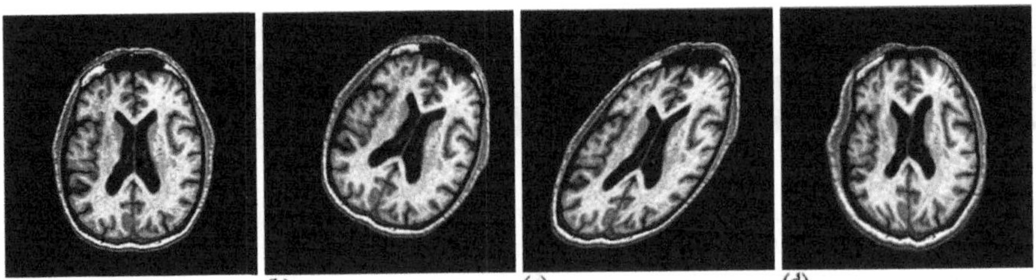

(a) (b) (c) (d)

FIG. 17.23. Examples of spatial transformation. (a) Original image. (b) Image transformed by rigid transformation (rotation and translation). (c) Image transformed by linear affine transformation. (d) Image transformed by non-linear deformable transformation.

17.4.1. Transformation models

The transformation ϕ can take many forms. When it has the form:

$$\phi(\boldsymbol{x}) = \mathbf{A}\mathbf{x} + \boldsymbol{b} \tag{17.28}$$

where $\boldsymbol{A}$ is an $n \times n$ matrix, and $\boldsymbol{b}$ is an $n \times 1$ vector, the transformation is called linear, otherwise, the transformation is non-linear.

A special case of linear transformations is rigid transformations (Fig. 17.23(b)). The matrix $\boldsymbol{A}$ in rigid transformations is a rotation matrix. Rigid transformations describe rigid motions. They are used in applications when the object being imaged moves without being deformed. Non-rigid linear transformations (Fig. 17.23(c)), as well as non-linear transformations (Fig. 17.23(d)), are called deformable transformations.

Non-linear transformations can be parametric or non-parametric. Parametric transformations have the form:

$$\phi(\boldsymbol{x}) = \boldsymbol{x} + \sum_{i=1}^{n} W_{i_1} f_i(\boldsymbol{x})\boldsymbol{e}_1 + W_{i_2} f_i(\boldsymbol{x})\boldsymbol{e}_2 + \ldots \tag{17.29}$$

where

$\{f_i(\boldsymbol{x}) : \Omega \to \mathbb{R}\}$ is a basis, such as the Fourier basis or the B-spline basis;
$\boldsymbol{e}_1, \boldsymbol{e}_2$ are unit vectors in the cardinal coordinate directions;

and W_{i_1}, W_{i_2} are the coefficients of the basis functions.

Usually, a relatively small number of low frequency basis functions are used to represent a parametric transformation. The resulting transformations vary smoothly across Ω. Such transformations are called low dimensional non-linear transformations. Non-parametric transformations do not have such a parametric form. Instead, at every point in Ω, a vector $\boldsymbol{v}(\boldsymbol{x})$ is defined, and the transformation is given simply by:

$$\phi(\boldsymbol{x}) = \boldsymbol{x} + \boldsymbol{v}(\boldsymbol{x}) \tag{17.30}$$

Diffeomorphic transformations are a special class of non-parametric deformable transformation. These transformations are differentiable on Ω and have a differentiable inverse. For example, in one dimension ($n = 1$), diffeomorphic transformations are monotonically increasing (or monotonically decreasing) functions. Diffeomorphic transformations are very useful for medical image registration because they describe realistic transformations of anatomy, without singularities such as tearing or folding. Registration algorithms that restrict deformations to being diffeomorphic exploit the property that the composition of two diffeomorphic transformations is also diffeomorphic. In these algorithms, the deformation between two images is constructed by composing many infinitesimal deformations, each of which is itself diffeomorphic. However, full exposition of the topic of diffeomorphic transformations is beyond the scope of this book.

17.4.2. Registration similarity metrics

Image registration tries to match places in images that are similar. Since true anatomical similarity is not known, surrogate measures based on image intensity are used. Many metrics have been proposed. We will only review three such metrics.

(i) *Mean squared intensity difference*: The similarity is measured as a difference in image intensity. The similarity of images I and J is given by:

$$\mathrm{Sim}(I,J\,|\,\phi) = \frac{1}{|\Omega|}\int_{\Omega}[I(\boldsymbol{x}) - J(\phi(\boldsymbol{x}))]^2\,\mathrm{d}\boldsymbol{x} \tag{17.31}$$

This metric is simple to compute and is appropriate when anatomically similar places can reasonably be expected to have similar image intensity values. It is not appropriate for registration of images with different modalities. It is also not very well suited for MRI registration, because MRI intensity values are not consistent across scans.

(ii) *Mutual information*: This metric is very useful for multimodality image registration. Imagine a pair of images of the body acquired with different modalities. In modality 1, bone may have an intensity range of 100–200 and soft tissue may have a range of 10–20. In modality 2, bone may have an intensity of between 3000 and 5000 and soft tissue may have an intensity of between 10 000 and 20 000. Clearly, the mean square intensity difference metric would return very large values if these two images were aligned properly. Another metric is needed that does not directly compare the intensity values. The mutual information metric is derived from information theory. To compute mutual information between images I and J, we treat the pairs of intensity values (I_k, J_k) as samples from a pair of random variables X, Y. One such sample exists at each pixel. Mutual information is a measure of how dependent the random variables X and Y are on each other. It is given by:

$$\int\int p(x,y)\log\left(\frac{p(x,y)}{p(x)p(y)}\right)\mathrm{d}x\,\mathrm{d}y \tag{17.32}$$

where $p(x, y)$ is the joint density of X and Y, $p(x)$ is the marginal density of X and $p(y)$ is the marginal density of Y. The marginal densities are estimated by the histograms of the images I and J. The joint density is estimated by the 2-D joint histogram of the images I and J (illustrated in Fig. 17.24).

(iii) *Cross-correlation*: The cross-correlation metric is computed as follows. At each pixel index k, we compute the correlation coefficient between the values of image I in a small neighbourhood of pixels surrounding k and the values of image J over the same neighbourhood. The correlation coefficients are summed up over the whole image. The cross-correlation metric is robust to noise because it considers neighbourhoods rather than individual pixels. However, it is expensive in computation time.

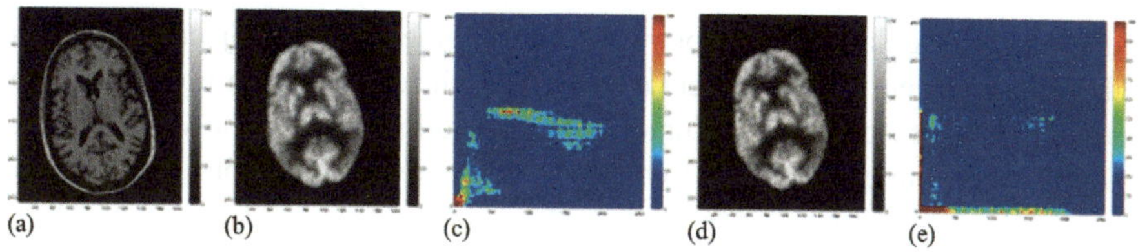

FIG. 17.24. Illustration of the joint histogram used in the computation of the mutual information metric. (a) Axial slice from an MR image. (b) Axial slice from a PET image aligned with the MRI. (c) Joint histogram of the MR and PET images. (d) PET slice rotated out of alignment with the MRI. (e) Joint histogram of the MRI and misaligned PET slice.

17.4.3. The general framework for image registration

The general algorithmic framework for image registration is illustrated in the flowchart in Fig. 17.25.

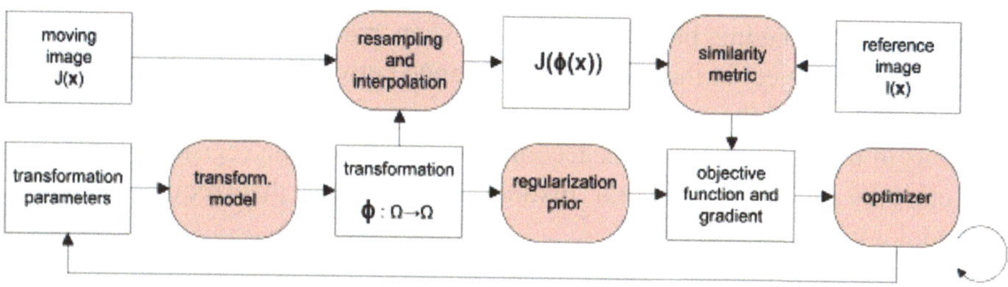

FIG. 17.25. General algorithmic framework for image registration.

Usually, one of the images is designated as a 'reference image' and the other image is the 'moving image'. Transformations are applied to the moving image, while the reference image remains unchanged. The transformation ϕ is defined by some set of parameters, small for linear registration, bigger for parametric non-linear registration, and very large for non-parametric non-linear registration. Some initial parameters are supplied. Usually these initial parameters correspond to the identity transformation. The transformation is applied to the moving image. This involves resampling and interpolation because the values of $\phi(\boldsymbol{x})$ fall between voxel centres. The resampled image $J(\phi(\boldsymbol{x}))$ is compared with the reference image $I(\boldsymbol{x})$ using the similarity metric. This results in a dissimilarity value. Registration seeks to minimize this dissimilarity value. However, in many registration problems, an additional term, the regularization prior, is minimized. This term measures the complexity of the transformation ϕ. It favours smooth, regular transformations over irregular transformations. This term can be thought of as an Occam's razor prior for transformations. Together, the dissimilarity value and the regularization prior value are combined into an objective function value. The gradient of the objective function with respect to the transformation parameters is also computed. Numerical optimization updates the values of the transformation parameters so as to minimize the objective function.

17.4.4. Applications of image registration

There are many image analysis problems that require image registration. Different problems require different transformation models and different similarity metrics. We can group medical image registration problems into two general categories, presented in the next two sections.

17.4.4.1. Registration that accounts for differences in image acquisition

In many biomedical applications, multiple images of the same subject are acquired. These images may have completely different modalities (MRI versus CT, CT versus PET, etc.) or they may be acquired on the same piece of equipment using different imaging parameters. Even when parameters are identical, the position of the subject in the scanner may change between images. To co-analyse multiple images of the same subject, it is necessary to match corresponding locations in these images. This is accomplished using image registration. Within this category, there are several distinct subproblems that require different methodology:

- *Accounting for the subject's motion*: When multiple images of a subject are acquired in a short span of time, the subject may move. For example, in fMRI studies, hundreds of scans are acquired during an imaging session. To analyse the scans, they must first be aligned, so that the differences due to subject motion are factored out. Motion correction typically uses image registration with rigid transformation models. Simple image similarity metrics suffice.
- *Alignment of multimodality 3-D images*: Often, information from different imaging modalities must be combined for purposes of visualization, diagnosis and analysis. For example, CT and PET images are often co-analysed, with CT providing high resolution anatomical detail and PET capturing physiological measures, such as metabolism. Image registration is needed to align the PET image to the CT image. However, the images have very different intensity patterns. This requires specialized image similarity metrics, such as mutual information. Often, rigid transformations suffice. However, some modalities introduce geometric distortions to images. In

this case, low dimensional parametric transformations may be necessary to align images.

- *Alignment of 3-D and 2-D imaging modalities*: Sometimes, registration is needed to align a 2-D image of the subject to a 3-D image. This problem arises in surgical and radiotherapy treatment contexts. A 3-D scan is acquired and used to plan the intervention. During the intervention, X ray or angiographic images are acquired and used to ensure that the intervention is being performed according to the plan. Corrections made to the intervention are based on the imaging. For this to work, image registration must accurately align images of different dimensions and different modality. This is a challenging problem that typically requires the image registration algorithm to simulate 2-D images via data from the 3-D image.

17.4.4.2. Registration that accounts for anatomical variability (image normalization)

The other major application of image registration is to match corresponding anatomical locations in images of different subjects, or in images where the anatomy of a single subject has changed over time. The term commonly used for this is image normalization. Again, there are several different applications:

- *Cross-sectional morphometry*: Often, we are interested in measuring how the anatomy of one group of subjects differs from that of another. In a clinical trial, we may want to compare the anatomy of a cohort receiving a trial drug with the cohort receiving a placebo. We may do so by matching every image to a common 'template' image, using image registration with non-linear transformations. We may then compare the transformations from the template to the images in one cohort with the transformations to the images in the other cohort. Specifically, we may examine the Jacobian of each transformation. The Jacobian of the transformation describes the local change in volume caused by the transformation. If an infinitesimal region in the template has volume δV_0, and the transformation ϕ maps this region into a region of volume δV_1, then the ratio $\delta V_1/\delta V_0$ equals the determinant of the Jacobian of the transformation.
- *Longitudinal morphometry*: When studying the effects of disease, intervention or ageing on human anatomy, we may acquire multiple images of a subject at different time points. To measure the differences over time, we can employ parametric or non-parametric deformable registration. As the overall anatomy does not change extensively between images, the regularization priors and other parameters of registration may need to be different than for cross-sectional morphometry.

17.5. OPEN SOURCE TOOLS FOR IMAGE ANALYSIS

This section briefly reviews several mature image processing and analytical tools that were available freely on the Internet at the time of writing this chapter. The reader can experiment with the techniques described in this chapter by downloading and running these tools. Most tools below run on Apple and PCs (with Linux and Windows operating systems).

- *ImageJ* (http://rsbweb.nih.gov/ij): ImageJ provides a wide array of image processing operations that can be applied to 2-D and 3-D images. In addition to basic image processing (filering, edge detection, resampling), ImageJ provides some higher level image analysis algorithms. ImageJ is written in Java. ImageJ can open many common 2-D image files, as well as DICOM format medical imaging data.
- *ITK-SNAP*[1] (http://itksnap.org): ITK-SNAP is a tool for navigation and segmentation of 3-D medical imaging data. ITK-SNAP implements the active contour automatic segmentation algorithms by Caselles et al. (1997) [17.5] and Zhu and Yuille (1996) [17.6]. It also provides a dynamic interface for navigation in 3-D images. Several tools for manual delineation are also provided. ITK-SNAP can open many 3-D image file formats, including DICOM, NIfTI and Analyze.
- *FSL* (http://www.fmrib.ox.ac.uk/fsl): FSL is a software library that offers many analytical tools for MRI brain imaging data. It includes tools for linear image registration (FLIRT), non-linear image registration (FNIRT), automated tissue classification (FAST) and many others. FSL supports NIfTI and Analyze file formats, among others.
- *OsiriX* (http://www.osirix-viewer.com): OsiriX is a comprehensive PACS workstation and DICOM image viewer. It offers a range of visualization capabilities and a built-in segmentation tool. Surface and volume rendering capabilities are especially well suited for CT data. OsiriX requires an Apple computer with MacOS X.
- *3-D Slicer* (http://slicer.org): Slicer is an extensive software platform for image display and analysis. It offers a wide range of plug-in modules that provide automatic segmentation, registration and statistical analysis functionality. Slicer also includes tools for image guided surgery. Many file formats are supported.

---

[1] Disclaimer: The author is involved in the development of ITK-SNAP.

These are just of few of the many excellent tools available to the reader. The Neuroimaging Informatics Tools and Resources Clearinghouse (http://www.nitrc.org) is an excellent portal for finding free image analysis software.

REFERENCES

[17.1] PERONA, P., MALIK, J., Scale-space and edge detection using anisotropic diffusion, IEEE Trans. Pattern Anal. Mach. Intell. **12** (1990) 629–639.

[17.2] DUDA, R., HART, P., Use of the Hough transformation to detect lines and curves in pictures, Commun. ACM **15** (1972) 11–15.

[17.3] COLLINS, D.L., et al., Design and construction of a realistic digital brain phantom, IEEE Trans. Med. Imaging **17** (1998) 463–468.

[17.4] KASS, M., WITKIN, A., TERZOPOULOS, D., Snakes: Active contour models, Int. J. Comput. Vis. **1** (1988) 321–331.

[17.5] CASELLES, V., KIMMEL, R., SAPIRO, G., Geodesic active contours, Int. J. Comput. Vis. **22** (1997) 61–79.

[17.6] ZHU, S.C., YUILLE, A., Region competition: Unifying snakes, region growing, and Bayes/MDL for multiband image segmentation, IEEE Trans. Pattern Anal. Mach. Intell. **18** (1996) 884–900.

BIBLIOGRAPHY

Image processing

GONZALEZ, R.C., WOODS, R.E., EDDINS, S., Digital Image Processing Using MATLAB, 3rd edn, Pearson-Prentice-Hall, Upper Saddle River, NJ (2004).

LINDEBERG, T., Edge detection and ridge detection with automatic scale selection, Int. J. Comput. Vis. **30** (1998) 117–156.

Image segmentation

COOTES, T.F., EDWARDS, G.J., TAYLOR, C.J., Active appearance models, IEEE Trans. Pattern Anal. Mach. Intell. **23** (2001) 681–685.

COOTES, T.F., TAYLOR, C.J., COOPER, D.H., GRAHAM, J., Active shape models — Their training and application, Comput. Vis. Image Understand. **61** (1995) 38–59.

LI, S.Z., Markov Random Field Modeling in Image Analysis, 3rd edn, Springer, London (2009).

PHAM, D.L., XU, C., PRINCE, J.L., Current methods in medical image segmentation, Annu. Rev. Biomed. Eng. **2** (2000) 315–337.

Image registration

MAINTZ, J.B., VIERGEVER, M.A., A survey of medical image registration, Med. Image Anal. **2** 1 (1998) 1–36.

PLUIM, J.P., MAINTZ, J.B., VIERGEVER, M.A., Mutual-information-based registration of medical images: A survey, IEEE Trans. Med. Imaging **22** 8 (2003) 986–1004.

Chapter 18

IMAGE PERCEPTION AND ASSESSMENT

I. REISER
University of Chicago,
Chicago, United States of America

18.1. INTRODUCTION

The main purpose of a medical image is to provide information to a human reader, such as a radiologist, so that a diagnosis can be reached — rather than to display the beauty of the human internal workings. It is important to understand how the human visual system affects the perception of contrast and spatial resolution of structures that are present in the image. If the image is not properly displayed, or the environment is not appropriate, subtle clinical signs may go unnoticed, which can potentially lead to a misdiagnosis.

This chapter provides an introduction to human visual perception and task based objective assessment of an imaging system. A model for the contrast sensitivity of the human visual system is presented. This model is used to derive the greyscale standard display function for medical displays. Task based assessment measures the quality of an imaging system as the ability of an observer to perform a well defined task, based on a set of images. Metrics for observer performance are introduced, as well as experimental methodologies for the measurement of human performance. The last section of the chapter describes the estimation of task performance based on mathematical observer models.

18.2. THE HUMAN VISUAL SYSTEM

18.2.1. The human eye

The human eye is a complex organ that processes visual images and relays information to the brain [18.1]. The retina is the light sensitive part of the eye. There are two types of photosensitive cell in the retina: rods and cones. In these photoreceptors, light quanta are converted into chemical energy, giving rise to an impulse that is transferred, via neurons, to the optic nerve. Each individual cone is connected to the optic nerve via a single neuron, while several rods are merged into one neuron that connects to the optic nerve. Therefore, the spatial acuity of cones is higher than that of rods, while light sensitivity is greater for rods than for cones.

The highest spatial acuity of the human visual field is in the fovea, a small region of the retina with a visual field of 2°, where the density of cones is largest. In scotopic (night) vision, luminance levels are low and only rods respond to light. Rods are not colour sensitive; therefore, objects appear grey at night. In photopic (day) vision, both rods and cones are activated. Photopic vision occurs at luminance levels of between 1 and 10^6 cd/m^2.

18.2.2. The Barten model

A model for the contrast sensitivity of the human visual system has been developed by Barten. Contrast sensitivity (S) is inversely proportional to the threshold modulation (m_t) (see Fig. 18.1), being the ratio of the minimum luminance amplitude to the mean luminance of a sinusoidal grating such that the grating has a 50% probability of being detected. Contrast sensitivity of the human visual system can be measured by presenting images of a sinusoidal grating to a human observer. From repeated trials, the threshold modulation can be determined.

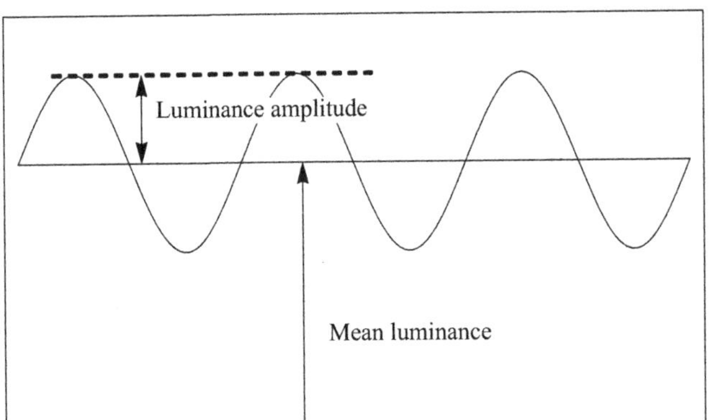

FIG. 18.1. Definition of the modulation of a sinusoidal grating.

Barten's model is based on the observation that, in order to be detected, the threshold modulation of a grating needs to be larger than that of the internal noise, m_n, by a factor of k, where:

$$m_t = km_n \tag{18.1}$$

As the image enters the eye, it is distorted and blurred by the pupil and optical lens. This is described by the optical modulation transfer function (M_{opt}) and is a function of spatial frequency (u) and incident luminance (L):

$$M_{\text{opt}}(u,L) = \mathrm{e}^{-2(\pi\sigma(L)u)^2} \tag{18.2}$$

The width of M_{opt} depends on pupil diameter, which controls the impact of lens aberrations (C_{ab}), where $\sigma(L)^2 = \sigma_0^2 + (C_{ab}d(L))^2$. The dependence of pupil diameter (d (in mm)) on luminance (L (in cd/m^2)) can be approximated by $d(L) = 5–3\tanh(0.4 \lg L)$.

When light quanta are detected by the retinal cells, photon noise (Φ_{ph}) is incurred, with a spectral density given by:

$$\Phi_{\text{ph}}(L) = \frac{1}{\eta p E(L)} \tag{18.3}$$

where

η is the quantum detection efficiency of the eye;
p is the luminous flux to photon conversion factor;

and E is the retinal illumination, where:

$$E(L) = \frac{\pi d(L)^2}{4} L\left[1-(d(L)/9.7)^2 + (d(L)/12.4)^4\right] \tag{18.4}$$

This expression includes the Stiles-Crawford effect that accounts for variations in efficiency as light rays enter the pupil at different locations.

Lateral inhibition occurs in retinal cells surrounding an excited cell, and is described as:

$$M_{\text{lat}}^2(u) = 1 - \mathrm{e}^{-2(u/u_0)^2} \tag{18.5}$$

Neural noise (Φ_0) further degrades the retinal signals. Including these factors in threshold and noise modulation gives:

$$m_{\text{t}} M_{\text{opt}}(u,L) M_{\text{lat}}(u) = 2k\sqrt{\frac{(\Phi_{\text{ph}}(L) + \Phi_{\text{ext}}) M_{\text{lat}}^2(u) + \Phi_0}{XYT}} \tag{18.6}$$

where X, Y and T are the spatial and temporal dimensions of the object. With angular extent of the object X_0 and maximum integration angle of the eye X_{max}, and the maximum number of cycles N_{max} over which the eye can integrate, the complete Barten model for contrast sensitivity of the human eye is given by:

$$S(u,L)=\frac{1}{m_t(u,L)}=\frac{M_{opt}(u,L)/k}{\sqrt{\frac{2}{T}\left(\frac{1}{X_0^2}+\frac{1}{X_{max}^2}+\frac{u^2}{N_{max}^2}\right)\left(\frac{1}{\eta p E(L)}+\frac{\Phi_0}{M_{lat}^2(u)}+\Phi_{ext}\right)}} \tag{18.7}$$

Figure 18.2 shows the contrast sensitivity as a function of both luminance and spatial frequency.

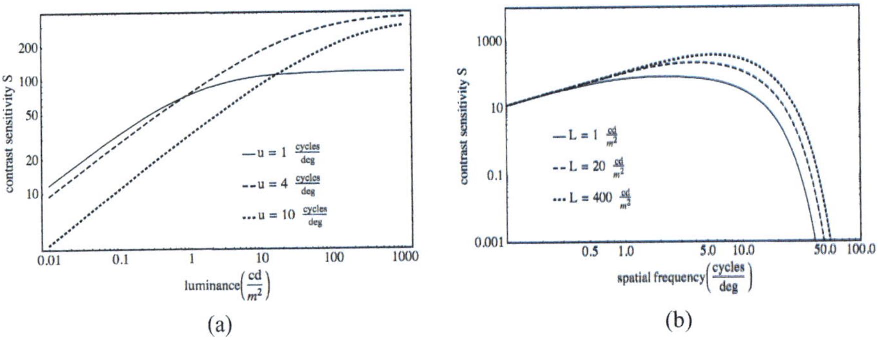

FIG. 18.2. Contrast sensitivity (a) as a function of luminance and (b) as a function of spatial frequency. Parameters used in Eq. (18.7) were k = 3, σ_0 = 8.33 × 10^{-3} deg, C_{ab} = 1.33 × 10^{-3} deg/mm, T = 0.1 s, X_{max} = 12 deg, N_{max} = 15 cycles, η = 0.03, Φ_0 = 3 × 10^{-8} s/deg², u_0 = 7 cycles/deg, X_0 = 2 deg, p = 1 × 10^6 photons/s/deg²/Td. The values for X_0 and p correspond to the standard target used in DICOM14 [18.2].

The Barten model as presented here is valid for foveal vision in photopic conditions and has been verified with a number of experiments covering the luminance range from 0.0001 to 1000 cd/m². Of the parameters in Eq. (18.7), σ_0, η and k were varied to fit measurements. All other parameters were kept fixed; σ_0 varied between 0.45 and 1.1, η from 0.005 to 0.3 and k from 2.7 to 5.0. Extensions of the model to parafoveal and temporal contrast sensitivities have also been described [18.3].

18.2.3. Perceptual linearization

On the basis of the contrast sensitivity, S, a 'just noticeable difference' (JND) in luminance can be defined as twice the luminance amplitude (a_t) of a sine wave grating at threshold modulation:

$$\text{JND}(u,L) = 2*a_t = 2*m_t(u,L)*L = 2*L/S(u,L) \tag{18.8}$$

In a display device such as a liquid crystal display monitor, the relationship between input gray value, n_g, and output luminance is non-linear. The plot of n_g versus luminance is the so-called characteristic curve of the display device.

In a perceptually linearized display, a constant difference of input gray value, Δn_g, results in a luminance difference corresponding to a fixed number, j, of JND indices, $\Delta L = j$*JND, across the entire luminance range of the display. To standardize medical devices, medical displays are required to conform to the Greyscale Standard Display Function (GSDF) (DICOM14 [18.2], also see Chapter 16). The GSDF defines the relationship between JND and display luminance for the standard target, consisting of a sinusoidal pattern of 4 cycles/° over a 2° × 2° area, embedded in a background with mean luminance equal to the mean luminance of the pattern. The GSDF is shown in Fig. 18.3.

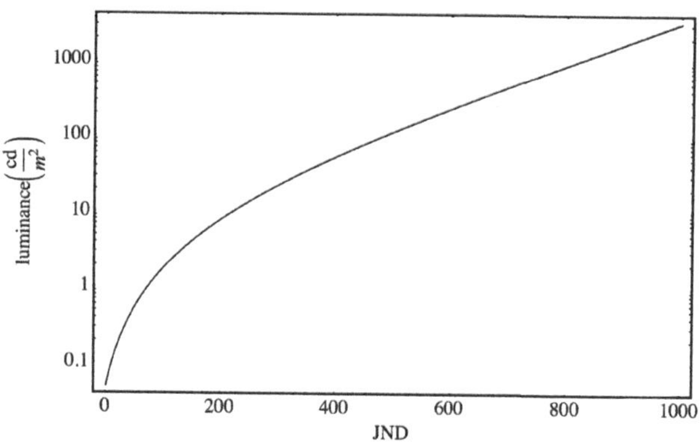

FIG. 18.3. The GSDF.

18.2.4. Viewing conditions

Ambient conditions in a reading room can significantly impact observer performance. Ambient lighting decreases performance and is recommended to be kept below 50 lx in a reading room. It is also recommended that observers allow time for dark adaptation of their eyes (which takes about 5 min) prior to any reading. Further, there are indications that fatigue deteriorates performance.

18.3. SPECIFICATIONS OF OBSERVER PERFORMANCE

18.3.1. Decision outcomes

The following describes outcomes of a binary decision, as exemplified by the task of detecting a signal or abnormality. A detection task is a binary decision because there are two truth states: the 'normal' state or the absence of a signal (H_0), and the presence of an abnormality or a signal (H_1). H_0 is often called a negative or disease free finding.

Presented with an image $\boldsymbol{g}$ of unknown truth state, the observer must decide whether to categorize $\boldsymbol{g}$ as normal (D_0) or as containing a signal (D_1). There are four possible outcomes to this decision (Table 18.1). A 'true positive' (TP) occurs when an image that contains a signal is assigned to H_1. When that same image (an image that actually contains a signal) is assigned to H_0, a 'false negative' (FN) occurs. For an image that does not contain a signal, a 'true negative' (TN) occurs when it is assigned to H_0, and a 'false positive' (FP) occurs when it is assigned to H_1.

TABLE 18.1. POSSIBLE DETECTION DECISION OUTCOMES

Decision	Actual condition	
	H_0: signal absent (negative)	H_1: signal present (positive)
D_0: signal absent (negative)	TN	FN: type II error
D_1: signal present (positive)	FP: type I error	TP

The observer can make two types of decision error. A type I error occurs when the truth state H_0 is rejected when it is actually true (FP). A type II error occurs when the truth state H_0 is not rejected when it is not true (FN). In medical applications, the cost associated with a type I error results in patient anxiety and societal costs because of additional tests and procedures, while the cost of a type II error is a missed cancer or misdiagnosis.

The accuracy of a medical procedure is given by the correct decisions per total number of cases (N):

$$\text{accuracy} = (\text{TP} + \text{TN})/N \tag{18.9}$$

Accuracy does not account for prevalence and can, therefore, be misleading; in screening mammography, the cancer prevalence in some screening populations is about 4 per 1000 patients. The accuracy of a radiologist who classifies all screening cases as normal is 9996/10 000 or 99.96%. Such a radiologist would have missed all cancers! For details, see Ref. [18.4].

More meaningful performance measures are sensitivity, specificity and associated variables. Sensitivity, also known as the true positive fraction (TPF), measures the proportion of actual true (positive) cases that are correctly identified:

$$\mathrm{TPF} = \mathrm{TP}/(\mathrm{TP} + \mathrm{FN}) \tag{18.10}$$

Likewise, specificity measures the proportion of negative cases that are correctly identified. It is conveniently derived through the false positive fraction (FPF), where:

$$\text{specificity} = 1 - \mathrm{FPF} \tag{18.11}$$

and

$$\mathrm{FPF} = \mathrm{FP}/(\mathrm{TN} + \mathrm{FP}) \tag{18.12}$$

Often in the literature, positive predictive values (PPVs) and negative predictive values (NPVs) are also reported:

$$\mathrm{PPV} = \mathrm{TP}/(\mathrm{TP} + \mathrm{FP}) \tag{18.13}$$

$$\mathrm{NPV} = \mathrm{TN}/(\mathrm{TN} + \mathrm{FN}) \tag{18.14}$$

18.3.2. Statistical decision theory and receiver operating characteristic methodology

In a probabilistic approach to perception, which is also referred to as 'decision theory', the observer derives a decision variable, λ, from each image. The conditional probability density functions of the decision variable λ under the truth states H_0 and H_1, $p(\lambda|H_0)$ and $p(\lambda|H_1)$, are shown in Fig. 18.4.

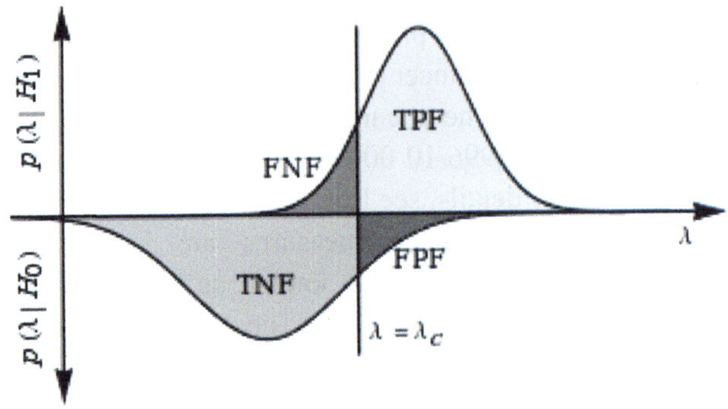

FIG. 18.4. Decision outcomes in the probabilistic decision paradigm. FNF: false negative fraction; TNF: true negative fraction; TPF: true positive fraction; FPF: false positive fraction.

The 'operating point' of the observer, λ_c, is the decision threshold at which an observer will call an image 'normal' (negative) or 'abnormal' (positive). Both the TPF and FPF values depend on λ_c. Often, it is assumed that both probability density functions are normally distributed so that a binormal model can be used. Then, if $p(\lambda|H_0)$ has zero mean and unit variance, and $p(\lambda|H_1)$ has a mean of a/b and variance $1/b$, the FPF and TPF are given by:

$$\mathrm{FPF}(\lambda) = \Phi(-\lambda) \tag{18.15}$$

$$\mathrm{TPF}(\lambda) = \Phi(a - b\lambda) \tag{18.16}$$

where $\Phi(.)$ is the cumulative normal distribution function, $\Phi(y) = \int_{-\infty}^{y} \varphi(x)\,\mathrm{d}x$, and $\varphi(.)$ is the normal probability density function, $\varphi(x) = \frac{1}{\sqrt{\pi}} \mathrm{e}^{-x^2}$.

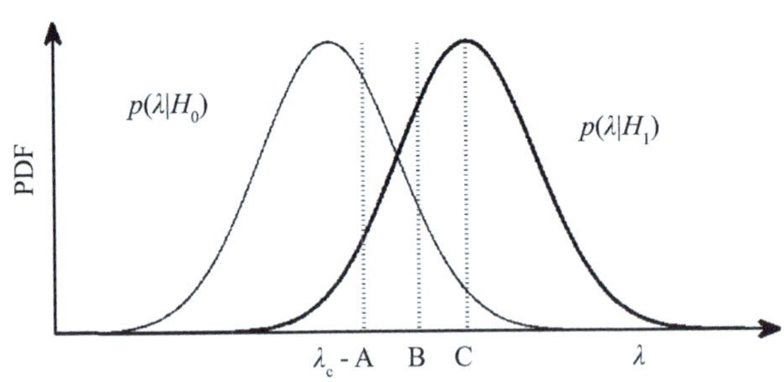

FIG. 18.5. Location of the decision thresholds A, B and C for the probability density functions associated with H_0 and H_1.

Different λ_c values reflect the vigilance of the reader, as illustrated in Fig. 18.5. A low decision threshold ($\lambda_c = A$) corresponds to an 'aggressive' reader, characterized by high sensitivity but low specificity. The choice of operating threshold depends on the task, or on the costs associated with decision errors. When screening for a condition with low prevalence, the reader typically operates at a high specificity (FPF < 0.1) in order to reduce the number of callbacks to an acceptable level (operating points B or C). The associated decrease in sensitivity might be offset by the yearly repeat of the screening procedure. On the other hand, in a diagnostic task, the reader might be more aggressive.

The receiver operating characteristic (ROC) curve reveals the trade-off between sensitivity and specificity. This very useful and much used curve is generated by plotting the TPF versus the FPF for all values of λ. A set of ROC curves is shown in Fig. 18.6. The end points of the ROC curve are (0, 0) and (1, 1). The ROC curve lies on or above the diagonal. If an ROC curve is below the diagonal, it implies that the truth states H_0 and H_1 have been interchanged, which can be caused by an observer misreading instructions. If the probability density functions have equal variance under both truth states, the ROC curve is symmetrical about a line from (0, 1) to (1, 0).

The area under the ROC curve, AUC, quantifies overall decision performance and is the integral of TPF over FPF:

$$\text{AUC} = \int_0^1 \text{TPF(FPF)}\,\text{dFPF} \tag{18.17}$$

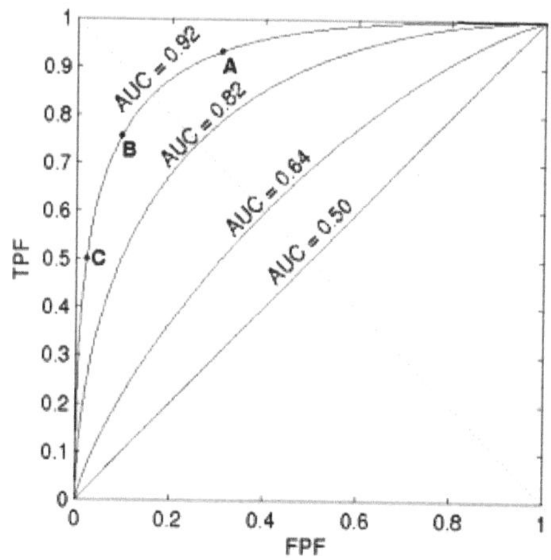

FIG. 18.6. ROC curves for a = 2, 1.3, 0.4 and 0, and b = 1. The underlying probability density functions have equal variance and, therefore, the ROC curves are symmetrical. The operating points A, B and C correspond to the decision thresholds in Fig. 18.5.

The values of AUC vary between 1.0 and 0.5. A perfect decision maker achieves an AUC of 1.0, while random guessing results in an AUC of 0.5. Within the binormal model, AUC is given by:

$$\mathrm{AUC} = \Phi\left(\frac{a}{\sqrt{1+b^2}}\right) \tag{18.18}$$

AUC is an overall measure of decision performance that does not provide any information about specific regions of the ROC curve, which may be important when deciding between two procedures that are to be used in a specific task. As an example, two procedures are to be evaluated for their performance in a screening task. Figure 18.7 shows the ROC curves corresponding to the two procedures. The AUC for both procedures is 0.82. However, in the region of a specificity value of <0.9, which typically would be a limit when screening for a condition with low prevalence such as breast cancer, the performance of procedure A exceeds that of procedure B.

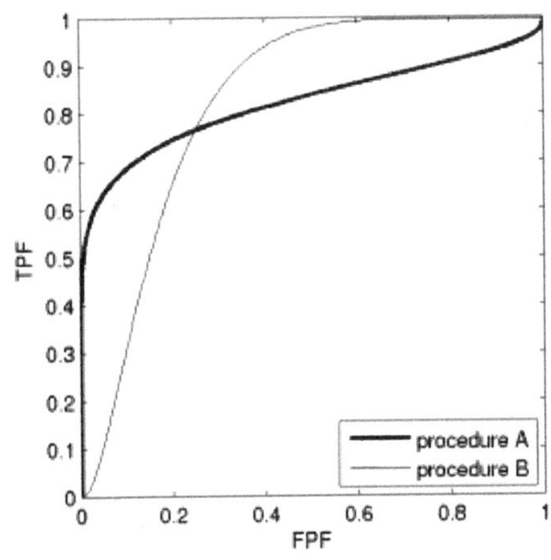

FIG. 18.7. ROC curves for two procedures with equal AUC, but different a and b.

18.3.3. Signal to noise ratio

A frequently used measure of performance is the signal to noise ratio (SNR), estimated from the decision variables λ_0 and λ_1 under the truth states H_0 and H_1, where:

$$\mathrm{SNR} = \frac{<\lambda_1> - <\lambda_0>}{\sqrt{1/2\left(\mathrm{var}(\lambda_0) + \mathrm{var}(\lambda_1)\right)}} \tag{18.19}$$

The computation of SNR by use of this expression is valid only if the decision variables are normally distributed and the distributions of λ are fully characterized by mean and variance. If the distributions depart from normality, SNR values computed by this expression can be misleading.

SNR is related to AUC through:

$$\mathrm{AUC}=\frac{1}{2}(1+\mathrm{Erf}(\mathrm{SNR})/2) \tag{18.20}$$

where Erf is the Gaussian error function:

$$\mathrm{Erf}(z)=\frac{2}{\sqrt{\pi}}\int_0^z \mathrm{e}^{-t^2}\mathrm{d}t \tag{18.21}$$

For a detection task, SNR is often called detectability (d'). In Section 18.5, $d'$is estimated directly from the image statistics.

18.4. EXPERIMENTAL METHODOLOGIES

The preferred metric for quantifying the image quality of a medical imaging system is task performance, which measures the performance of an observer in a well defined task, based on a set of images. Image quality is thus a statistical concept. The following describes methodologies used to measure the performance of a human reader (observer) in the task of detecting a known signal in a noisy background.

18.4.1. Contrast–detail methodology

To provide an objective measure of image quality, imaging systems have been evaluated using contrast–detail phantoms that consist of discs of increasing diameter and contrast arranged in columns and rows, an example of which is shown in Fig. 18.8. The reader's task is to identify the lowest contrast where the signal can be perceived, for each detail size. The reader's threshold contrasts, plotted as a function of disc radius, is called a contrast–detail curve.

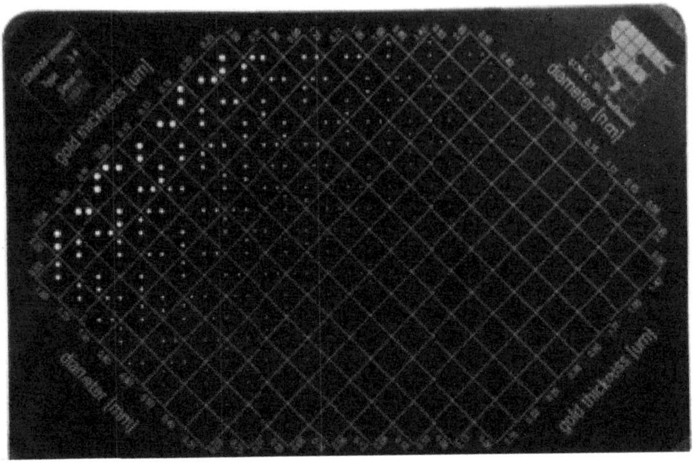

FIG. 18.8. The CDMAM phantom is an example of a contrast–detail phantom consisting of gold discs of exponentially increasing diameter (0.06–2 mm) and thicknesses (0.002–0.03 mm).

The drawback of contrast–detail methodology is that threshold contrast is dependent on the internal operating point of each observer. Therefore, inter- and intra-observer variabilities can be quite high. Furthermore, there may be a memory effect where the observer anticipates, and therefore reports, a signal, while the signal cannot yet be perceived. The CDMAM phantom [18.5] (Fig. 18.8) has been developed to address some of these shortcomings, specifically by placing two signals, one at the centre and a second randomly in one of the four corners of each cell in the array. This arrangement prevents the reader from anticipating the signal. An alternative approach is to determine contrast thresholds using computer reading of the images.

18.4.2. Forced choice experiments

A different way of measuring human observer performance is through alternative forced choice (AFC) experiments. In an AFC experiment, M alternative images are shown to the observer, whose task it is to identify the image that contains the signal. The proportion of correct answers (PC) in an $M-$AFC experiment is computed through:

$$\text{PC} = (\text{number of correctly scored trials})/(\text{total number of trials}) \tag{18.22}$$

The value of PC varies between 1 and the guessing score $1/M$. The number of trials (that is, the number of test images shown) determines the accuracy of PC. Typically, more than 100 trials are required. PC can be predicted based on the assumption that the observer will assign the score to the image that evokes the highest internal response. Then, if the probability density function of negative responses is normal, with zero mean and unit variance, the probability that the response of the $M - 1$ actually negative response exceeds that of the actually positive response (normally distributed with mean d' and unit variance) becomes [18.6]:

$$\mathrm{PC} = \int_{-\infty}^{\infty} \Phi(x)^{M-1} \varphi(x - d')\,\mathrm{d}x \qquad (18.23)$$

where $\varphi(.)$ and $\Phi(.)$ are the normal probability density function and the cumulative normal distribution function, respectively, as defined above. Note that when $M = 2$, PC equals AUC. This can be seen by using the definitions in Eqs (18.15–18.17).

Typically, 2-AFC experiments are carried out so that the observers score 90% of all trials correctly. Depending on the signal, the corresponding amplitude thresholds may need to be very low, which can strain the observer. By increasing the number of alternative locations, the task becomes more difficult and higher threshold amplitudes are required. The alternative signal locations can be centres of individual regions of interest, or all possible signal locations can be contained within one single test image. In this case, one needs to ensure that the distance between locations is larger than the correlation distance of the background, otherwise observer responses might become correlated. If the responses are uncorrelated, d' can be determined from PC by inverting Eq. (18.23). Experiments with values of M between 2 and 1800 have been conducted [18.7].

18.4.3. ROC experiments

In an ROC experiment, a single image is presented to the observer whose task it is to provide a likelihood rating. In a detection experiment, this could be the 'likelihood of signal present'. Researchers have used continuous rating scales, as well as categorical scales. In the latter, typically less than ten categories are provided. For a given number of images available, ROC experiments provide better statistical power because one single image is required per trial, whereas in a 2-AFC experiment, two images are required per trial. Furthermore, an ROC curve can be generated from rating data, which provide more detailed information on task performance in specific regions. On the other hand, ROC experiments are more demanding of the observer, which can increase the reading time and observer fatigue. Curve fitting problems, such as the presence of hooks, can be overcome by using ROC curve fitting software provided by the University of

Chicago [18.8], whose curve fitting model does not allow the ROC curve to fall below the diagonal.

18.4.3.1. Search experiments

A drawback of ROC or 2-AFC experiments is the lack of signal localization. In a clinically realistic task, such as screening for cancer, the radiologist is required to indicate the locations of potential lesions, in addition to an overall rating of the image. To allow for more clinically realistic laboratory experiments, several extensions to ROC methods have been developed.

18.4.3.2. Location ROC

In the location ROC (LROC) methodology, the observer is required to indicate the location of a lesion, as well as to provide a score. In an LROC curve, the FPF is plotted along the *x* axis, but on the *y* axis the TPF with correct signal location is plotted. The upper right end point of the LROC curve is determined by the proportion of correctly located signals.

18.4.3.3. Free response ROC (FROC)/alternative free response ROC (AFROC)/ JAFROC

In free response ROC (FROC) experiments, the observer indicates the location of a lesion, but provides no rating. The FROC curve is the proportion of correctly detected signals plotted as a function of the average number of FP detections per image. FROC analysis is often used in the performance assessment of computer aided detection schemes.

In alternative free response ROC (AFROC) methodology, the proportion of correctly detected signals is plotted as a function of the probability that at least one FP per image is found. In the AFROC methodology, a summary figure of merit exists, A_{1J}, which is the probability that lesions are rated higher than FP marks in normal images. JAFROC is a software package developed to estimate A_{1J} and statistical significance from human observer FROC data [18.9].

18.5. OBSERVER MODELS

Often, it is impractical to determine the task performance of imaging systems from studies involving human observers, simply because reader time is scarce and expensive, especially for a trained reader such as a radiologist. To address this issue, mathematical observer models have been developed, either to predict human performance or to estimate ideal observer performance. The observer models described here assume that images contain exactly known additive signals, i.e. signals of known shape and amplitude that have been added to a noisy background.

18.5.1. The Bayesian ideal observer

In a detection task, the observer is faced with the decision of assigning the unknown data, $\boldsymbol{g}$, to one of the truth states. The ideal Bayesian observer's decision strategy is to minimize the average cost, $<C>$, of the decision. With the observer's decision outcomes D_0 and D_1 (Table 18.1), the average cost, $<C>$, is:

$$\begin{aligned}<C> = {} & C_{00}P(D_0|H_0)P(H_0) + C_{11}P(D_1|H_1)P(H_1) + \\ & C_{01}P(D_0|H_1)P(H_1) + C_{10}P(D_1|H_0)P(H_0)\end{aligned} \tag{18.24}$$

where

C_{00} and C_{11} are costs associated with correct decisions;
C_{01} is the cost of an FN decision;

and C_{10} is the cost of an FP decision.

Minimizing average cost leads to the decision rule, choose H_1 if:

$$\Lambda(\boldsymbol{g}) = \frac{p(\boldsymbol{g}\,|\,H_1)}{p(\boldsymbol{g}\,|\,H_0)} \geq \frac{P(H_0)(C_{10} - C_{00})}{[1 - P(H_0)](C_{01} - C_{11})} \tag{18.25}$$

otherwise choose H_0.

This decision criterion is the so-called Bayes criterion and the resultant minimum cost is the Bayes risk. A test based on the ratio of probabilities is called a likelihood ratio test. It can be shown that the likelihood ratio test maximizes sensitivity for a given FPF. Therefore, the ideal observer maximizes the AUC and provides an upper limit on task performance. However, knowledge of the probability density functions $p(\boldsymbol{g}|H_i)$ is required to form the decision variable (also called the test statistic), $\Lambda(\boldsymbol{g})$, which is generally difficult to determine for realistic imaging tasks. When the underlying data statistics are normal, it is often easier to work with the log-likelihood ratio, where $\lambda(\boldsymbol{g}) = \lg(\Lambda(\boldsymbol{g}))$. Since the logarithmic transformation is monotonic, it does not affect the decision outcome or AUC.

18.5.2. Observer performance in uncorrelated Gaussian noise

If the image noise is zero-mean Gaussian distributed with variance σ^2 and is independent in each image pixel, the ideal observer derives the decision variable λ by convolving each image, $\boldsymbol{g}$, with a template, $\boldsymbol{w}$, where:

$$\lambda = \boldsymbol{w}^{t}\,\boldsymbol{g} \tag{18.26}$$

In this notation, $\boldsymbol{w}$ and $\boldsymbol{g}$ are column vectors created by concatenating all image pixels into a single vector. The ideal observer template for detection of additive exactly known signals is the signal itself, $\boldsymbol{w} = \boldsymbol{s}$. From the test statistic λ, d' can be estimated using Eq. (18.19).

If the image statistics are known, d' can be computed directly from:

$$d'^2 = \frac{\text{SE}}{\sigma^2} \tag{18.27}$$

where $\text{SE} = \sum_{i=1}^{N} s_i^2$ is the signal energy of the signal-only image containing N pixels.

18.5.2.1. Observer model for human performance

Compared with the ideal observer, a human reader performs suboptimally. To account for the signal degradation by the human visual system, the template **w** can be formed by convolving the signal with an eye filter, **e**:

$$\boldsymbol{w} = \boldsymbol{e} \otimes \boldsymbol{s} \tag{18.28}$$

where $\boldsymbol{e}$ is defined in the spatial frequency domain as $E(f) = fe^{-cf}$. Typically, c is chosen so that $E(f)$ peaks at 4 cycles/mm. This model observer model is called a 'non-prewhitening observer with eye filter'. This observer model can predict the performance of a human reader when the image background is uniform [18.10]. However, this model fails when strong background correlations are present, such as in anatomical backgrounds [18.11, 18.12].

18.5.3. Observer performance in correlated Gaussian noise

When the noise in the image is Gaussian correlated, the background statistics can be characterized by the covariance matrix, $\boldsymbol{K}_g$, where:

$$\boldsymbol{K}_{\mathrm{g}} = \langle(\boldsymbol{g} - \langle\boldsymbol{g}\rangle)(\boldsymbol{g} - \langle\boldsymbol{g}\rangle)^{t}\rangle \tag{18.29}$$

where <.> indicates the mean. In this background type, the ideal observer 'prewhitens' the image by using a template that removes the background correlation:

$$\boldsymbol{w} = \boldsymbol{K}_{\mathrm{g}}^{-1}\boldsymbol{s} \tag{18.30}$$

and ideal observer performance becomes:

$$\boldsymbol{d}'^{2} = \boldsymbol{s}^{\mathrm{t}}\boldsymbol{K}_{\mathrm{g}}^{-1}\boldsymbol{s} \tag{18.31}$$

Estimating the background covariance is often difficult. However, when the image data are stationary, the covariance matrix is diagonalized by the discrete Fourier transform (FT). This allows computation of d' using:

$$d'^{2} = \int \frac{|S(\vec{f})|^{2}}{W(\vec{f})} \mathrm{d}\vec{f} \tag{18.32}$$

where $S(\vec{f})$ is the FT of the signal **s** and $W(\vec{f})$ is the background power spectrum. $W(\vec{f}) = \left\langle \left| \mathrm{FT}\{g(\vec{x}) - \langle g(\vec{x})\rangle\} \right|^{2} \right\rangle$, and $\vec{f} = (fx, fy)$ are spatial frequencies along the two image directions. This equation is the basis for estimation of the SNR from noise equivalent quanta in Chapter 4 (image quality).

Another approach to estimating the background covariance is to expand the image into so-called channels or basis functions. The choice of channel function depends on the objective of the observer model. It has been found that Laguerre–Gauss are efficient channel functions when approximating the ideal observer, whereas Gabor channels provide a good estimate of human observer performance. Often, ten channel functions or fewer are sufficient to represent **g**, reducing the dimensionality of the problem considerably, without the assumption of background stationarity. These observer models are often called 'channellized hotelling observers'. The University of Arizona provides software and tutorials for assessing image quality by use of observer models [18.13].

REFERENCES

[18.1] DRAGOI, V., "Chapter 14: Visual processing: Eye and retina", Neuroscience Online, The UT Medical School at Houston, Houston, http://neuroscience.uth.tmc.edu/s2/chapter14.html (accessed on 17 September 2012).

[18.2] NATIONAL ELECTRICAL MANUFACTURERS ASSOCIATION, Digital Imaging and Communications in Medicine (DICOM), Part 14: Greyscale Standard Display Function, NEMA, Rosslyn, VA (2004), http://medical.nema.org/dicom/2004/04_14pu.pdf (accessed on 17 September 2012).

[18.3] BARTEN, P.G., Contrast Sensitivity of the Human Eye and its Effects on Image Quality, SPIE Press, Bellingham, Washington, DC (1999).

[18.4] BARRETT, H.H., MYERS, K., Foundations of Image Science, John Wiley & Sons, Hoboken, NJ (2004)

[18.5] BIJKERK, K.R., THIJSSEN, M.A.O., ARNOLDUSSEN, T.J.M., "Modification of the CDMAM contrast-detail phantom for image quality of full field digital mammography systems", IWDM 2000 (Proc. Conf. Toronto, 2000), Medical Physics Publishing, Madison, WI (2000) 633–640.

[18.6] ECKSTEIN, M.P., WHITING J.S., Visual signal detection in structured backgrounds, I: Effect of number of possible locations and spatial contrast, J. Opt. Soc. Am. A Opt. Image Sci. Vis. **13** (1996) 1777–1787.

[18.7] BURGESS, A.E., GHANDEHARIAN, H., Visual signal detection, II: Signal-location identification, J. Opt. Soc. Am. A Opt. Image Sci. Vis. **1** (1984) 906–910.

[18.8] METZ, ROC Software, http://metz-roc.uchicago.edu

[18.9] DEVCHAKRABORTY.COM, http://www.devchakraborty.com/index.php (accessed on 18 September 2012).

[18.10] SEGUI, J.A., ZHAO, W., Amorphous selenium flat panel detectors for digital mammography: Validation of a NPWE model observer with CDMAM observer performance experiments, Med. Phys. **33** (2006) 3711–3722.

[18.11] BURGESS, A.E., Visual signal detection with two-component noise: Low-pass spectrum effects, J. Opt. Soc. Am. A Opt. Image Sci. Vis. **16** (1999) 694–704.

[18.12] BURGESS, A.E., JACOBSON, F.L, JUDY, P.F., Human observer detection experiments with mammograms and power-law noise, Med. Phys. **28** (2001) 419.

[18.13] THE IMAGE QUALITY WEBSITE, http://www.radiology.arizona.edu/CGRI/IQ/index.html (accessed on 18 September 2012).

BIBLIOGRAPHY

INTERNATIONAL COMMISSION ON RADIATION UNITS AND MEASUREMENTS, Receiver Operating Characteristic Analysis in Medical Imaging, ICRU Rep. 79, ICRU, Bethesda, MD (2008).

Chapter 19

QUALITY MANAGEMENT

P.A. HILES
Glan Clwyd Hospital,
Bodelwyddan, United Kingdom

I.D. McLEAN
International Atomic Energy Agency

S. CHRISTOFIDES
New Nicosia General Hospital,
Nicosia, Cyprus

19.1. INTRODUCTION

This chapter introduces the principles and definitions of quality management systems (QMSs) for radiology facilities, to give a framework to assist in the setting up of such systems and to emphasize the role of the medical physicist in this context. While there is a diversity of terms currently in use to describe quality processes both generally and specifically within radiology, there is broad agreement that the effective management of radiation medicine services demands a quality culture that includes a systematic approach to the elements that govern the delivery of that service.

Therefore, the concept of quality assurance (QA) within the radiological facility covers, in its widest sense, all those factors that affect the intended outcome, that is, a clinical diagnosis. The medical physicist has an important role in the overall QMS, especially, but not exclusively, with respect to the equipment performance.

A worked example of a quality control (QC) programme is included at the end of the chapter, to demonstrate the depth of detail and involvement of the medical physicist.

19.2. DEFINITIONS

Here, only the most relevant definitions are given and explained in the context of a radiology facility. A more comprehensive list of definitions is given in the ISO9000:2000 standard [19.1] and other published literature, such as the ISO 9000 Quality Systems Handbook by D. Hoyle [19.2].

19.2.1. QMS

A QMS is a framework to support the operation of a facility's service, with the objective of continuous quality improvement. A QMS incorporates:

- The organization's objectives and policies;
- Documented procedures consistent with these objectives and policies;
- Written practice instructions for staff;
- Monitoring, recording and auditing of practice.

From the above definition, it may appear that a QMS is simply a collection of procedures, tasks and documents. This is not so. Each QMS must be designed specifically for each individual radiology facility. All the components need to fit together and the inputs and outputs need to be defined and connected; system monitors need to feed information to processes that cause changes in the performance of the facility and all parts need to work together to achieve the common purposes of the facility.

A QMS is therefore a set of interrelated and interacting processes that achieves the quality policy and quality objectives of the radiology facility, which in turn form an integral part of the hospital's management system, to focus on achievement of the many aspects of service provision and quality objectives.

19.2.2. QA

QA includes all planned and systematic actions necessary to provide adequate confidence that an item, process or service will satisfy given requirements with regard to quality. As such, it is wide ranging, covering all relevant procedures, activities and actions, and hence, all groups of staff involved in the process under consideration.

A QA programme is part of a QMS and is focused on providing confidence that the quality needs or expectations, whether stated, generally implied or obligatory, are fulfilled.

A QA programme in diagnostic radiology can be thought of as an organized effort by the staff operating a facility to perform the most appropriate examination, to produce images of sufficiently high quality and consistency, and using the lowest possible dose, to result in the correct diagnosis.

The World Health Organization states that "satisfactory performance in service implies the optimum quality of the entire process, i.e., the consistent production of adequate diagnostic information with minimum exposure of both patient and personnel". A comprehensive QA programme should, therefore, embrace the entire process of radiology.

19.2.3. QC

QC is the process through which the actual quality performance is measured and compared with existing standards, and the actions necessary to keep or regain conformance with the standards. It is one part of overall QA, intended to verify that structures, systems and components correspond to predetermined requirements. It is concerned with operational techniques and activities used:

- To check that quality requirements are met;
- To adjust and correct performance if the requirements are found not to have been met.

19.2.4. Quality standards and good practice

'Quality standards' form a set of accepted criteria against which the quality of particular activities can be assessed.

Recommendations for quality standards for diagnostic radiology have been issued by various national and international organizations such as the IAEA, the World Health Organization and the Pan American Health Organization, and more specifically by other organizations, such as the European Commission, the American Association of Physicists in Medicine and the Institute of Physics and Engineering in Medicine. Where recommended standards are not available, local standards need to be developed, based on a local assessment of requirements.

Good practice is a practice that can be recommended on the basis of the most recent considerations of evidence based data, long term experience and knowledge gained on the necessary structure, process and outcome.

Quality standards and good practice also form a basis for clinical audit and are further discussed in Section 23.4.

19.3. QMS REQUIREMENTS

The requirements to set up an effective QMS are given in detail in ISO 9001 [19.3] and are briefly presented next.

19.3.1. General requirements

The organization is required to establish, document, implement and maintain a QMS, and continually improve its effectiveness in accordance with the following requirements, whereby the organization shall:

(i) Identify the processes needed for the QMS and their application throughout the organization;
(ii) Determine the sequence and interaction of these processes;
(iii) Determine the criteria and methods needed to ensure that both the operation and control of these processes are effective;
(iv) Ensure the availability of resources and the information necessary to support the operation and monitoring of these processes;
(v) Monitor, measure and analyse these processes;
(vi) Implement actions necessary to achieve planned results and continual improvements of these processes.

These general requirements capture key activities that are required to develop an effective system. They are not intended as a sequence, but rather the relationship between the requirements can be represented as a cycle. The cycle, which commences with the organization's purpose from which objectives are developed, is reviewed below. In planning to meet organizational objectives, the processes are identified and their sequence and interaction are determined. Once the relationship between the processes is known, the criteria and methods for effective operation and control can be developed and documented. The processes are described in terms that enable their effective communication, and a suitable way of doing this would be to compile the process descriptions into a quality manual that not only references the associated procedures and records but also shows how the processes interact. Before implementation, the processes need to be resourced and the information necessary to operate and control them deployed and brought under document control. Once operational, the processes need to be monitored to ensure that they are functioning as planned. Measurements taken to verify that the processes are delivering the required output and actions taken to achieve the planned results should be documented. The data obtained from monitoring and measurement that are captured on controlled records need to be analysed, and opportunities for continual improvement identified and the agreed actions implemented.

19.3.2. The role of the medical physicist

As seen in Section 19.2.2, a QA programme has elements for ensuring high and consistent image quality, and for ensuring the use of appropriate low doses in radiological examinations. The complex nature of quantifying image quality in radiological imaging (see Chapter 4) and in determining appropriate dose indicators (see Chapter 22) mandates that a physicist must be involved in the design and oversight of any QA programme and in many cases involving complex equipment, there must also be an involvement in the measurement, recording and analysis of QA data. Further, as described below, many of the processes in the QMS require strong technical knowledge and experience that can be provided by the medical physicist. This has been recognized by the IAEA through the International Basic Safety Standards for Protection against Ionizing Radiation and for the Safety of Radiation Sources (BSS) [9.4] that require registrants and licensees to establish a comprehensive programme of QA for medical exposures by, or under the oversight of, or with the advice of, a medical physicist specialized in the relevant field.

19.4. QA PROGRAMME FOR EQUIPMENT

Much of the complexity of modern radiological imaging comes from the equipment and associated technical processes. A QA programme for equipment is, therefore, essential and, to be effective, demands a strong commitment from the facility and institutional leadership to provide the necessary resources of time, personnel and budget. The programme should cover the entire process from the initial decision to adopt a particular procedure through to the interpretation and recording of results, and should include a systematic control methodology.

This should involve:

- Measurements of the physical parameters of medical radiological equipment at the time of acceptance and commissioning prior to clinical use on patients, and periodically thereafter and after any major maintenance that could affect patient safety;
- Implementation of corrective actions if measured values of the physical parameters are outside control limits;
- Verification of the appropriate physical and clinical factors used in patient diagnosis;
- Records of relevant procedures and results, including a manual that defines clear lines of responsibility, outlines the individual QC tests performed, gives the test frequencies, is useful for staff training, facilitates audit of a service and helps to keep information within the service;
- Verification of the appropriate calibration and conditions of the operation of dosimetry and monitoring equipment;

- Optimization of clinical protocols and equipment operation to achieve the aims of QA as stated above;
- Regular and independent audits of the QA programme for medical exposures (see Section 23.4).

19.4.1. Multidisciplinary team

There are many elements that contribute to the imaging process, and the experience of personnel, whether directly or indirectly involved, is crucial to the final outcome. As responsibilities are shared between different disciplines, they must be clearly defined. Each discipline has an important role in the output of the entire process, and these roles are interdependent and require close cooperation. Each staff member must have appropriate qualifications (education, training and experience) and have access to appropriate opportunities for continuing education and development.

It is important for a multidisciplinary team to lead, develop, maintain, manage and improve such quality systems. Well directed team work is as important as, if not more important than, individual work in delivering systematic improvement for a QA programme.

19.4.2. Structure of an equipment QA programme

There are five stages applicable to QA for imaging equipment:

(i) Equipment specification and tendering process;
(ii) Critical examination;
(iii) Acceptance;
(iv) Commissioning;
(v) Routine performance testing.

All are part of the life cycle of imaging equipment, which is illustrated in Fig. 19.1.

19.4.2.1. Life cycle of an imaging system

The equipment life cycle is sometimes referred to as an equipment QA cycle. The cycle begins with the decision to acquire equipment, whether new or used, to fulfil a need at the facility and is completed with the disposal of the equipment. Maintenance is a vital part of the cycle.

The stages of the life cycle are important for all pieces of radiological equipment. Special attention should be paid to equipment designed for paediatrics or health screening and to equipment that may produce high doses.

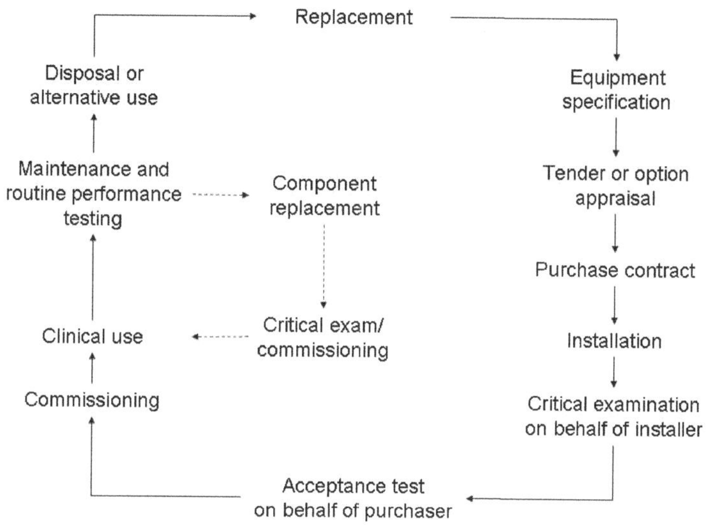

FIG. 19.1. Example of the life cycle of an imaging system.

19.4.2.2. Equipment specification

The cycle begins with an agreed need for equipment, which is translated into an equipment specification, possibly followed by a formal tender or option appraisal process. This is a critical part of the cycle and the facility medical physicist responsible for diagnostic radiology should be a key person on the team that draws up the specifications and would also need to interact with engineering professionals regarding, for example, the construction of the installation site, power supply and air conditioning. Radiation shielding should also be considered at this stage (see Chapter 24) for X ray equipment, while environmental shielding, for example, magnetic and radiofrequency shielding, needs to be considered for other equipment such as magnetic resonance imaging scanners. After a contract is awarded and during construction, the medical physicist should also be actively involved to ensure that architectural plans are being correctly followed and, if possible, that shielding is correctly installed.

19.4.2.3. Critical examination

After installation is complete, the medical physicist, in conjunction with other appropriate personnel (e.g. a representative of the equipment supplier and regulatory inspector), should undertake a critical examination of the installation. At this critical examination, the safety features and warning devices incorporated into the equipment are inspected to ensure correct operation and that there is sufficient protection of the staff, visitors and patients against exposure to ionizing radiation.

A critical examination is appropriate when there could be radiation protection implications associated with an incorrect installation, such as failure of the safety features or warning devices to operate correctly, poor siting or inadequate shielding. This is especially important if it has not been possible to confirm that the correct shielding has been installed during construction, as is often the case. The key word is 'critical'. It requires an examination of the manner in which the equipment is erected and installed. This can be achieved, not necessarily by a long list of prescriptive tests on every part of the system, but rather by an experienced physicist selectively examining those features that are most likely to affect safety, and then probing more deeply, depending on the results obtained.

Since the facility should not allow the equipment to be used for medical exposures unless the results of the critical examination are satisfactory, the medical physicist present at the critical examination should ideally represent the radiology facility. If, however, the medical physicist is representing the installer, this should be agreed to and established, preferably at the contract stage.

19.4.2.4. Acceptance

Acceptance testing involves the verification of equipment specifications and features by representatives of the installer and the facility medical physicist.

Acceptance may be a matter of simply completing a checklist. Any significant discrepancy should be notified formally to the contractor, who should be required to undertake corrective action. During acceptance testing, a qualified person should check the electrical and mechanical safety of any new installation.

19.4.2.5. Commissioning

After acceptance, commissioning is carried out by the facility representative, usually a medical physicist specializing in radiology physics, to ensure that the equipment is ready for clinical use and to establish baseline values against which the results of subsequent routine performance tests can be compared. Testing should include all parameters and conditions of use that are to be expected in clinical use. A check should be made that all relevant testing for the unit has been completed and that no tests have been omitted during either the critical examination or acceptance. In some cases, the equipment commissioning can be most efficient when undertaken jointly by the installer and the medical physicist. Any opportunity for corrective action can then be taken, as necessary, so that the performance of the imaging system is rapidly optimized, which is a major purpose of commissioning.

After major work on the equipment, the relevant commissioning tests may have to be repeated to establish new baseline values, for example, after an X ray tube is replaced or new software is installed.

The same medical physicist may undertake the critical examination, acceptance and commissioning. Although the tests may be combined, their purpose should remain distinct. The commissioning tests undertaken by the medical physicist and the installer should not be considered as exhaustive testing of the system. Following or during training from the product specialist, the radiology facility or users of the X ray equipment should ensure that the system meets the clinical requirements of the users, by checking any clinical protocol settings and image processing and the ergonomics and positioning of the X ray unit.

19.4.2.6. Routine performance testing and QC

Routine performance (or constancy) testing, also known as QC testing, comprises those tests that are undertaken either regularly or after maintenance or repairs, to detect whether any change in the performance of the equipment has occurred that would require corrective action.

Routine performance tests are really a subset of the commissioning tests and will generally involve staff with different levels of expertise, some of whom may be external to the radiology facility.[1] The more frequent tests that are quick to perform are usually undertaken locally with advice from a medical physicist, while the more complex and time consuming tests may require special expertise and instrumentation. A collaborative, multidisciplinary approach to routine performance testing is essential.

---

[1] This would particularly be the case when a regional medical physics unit undertakes the QA programmes for a number of hospitals, with expert staff making periodic visits to a department.

19.4.2.7. Reject analysis

In addition to routine QC testing, reject analysis of images from appropriate systems should be performed. This is important to ensure optimal image quality and to identify faults or long term deterioration. The rejects can be due to many reasons, such as poor film processing, positioning errors or patient movement, incorrect exposure and faulty X ray equipment.

Many of these rejects can be reduced or prevented with adequate training of staff on all appropriate pieces of equipment and careful maintenance of equipment. Reject analysis should be undertaken on a regular basis, with results fed back to staff and action undertaken, as appropriate.

It is important to perform reject analysis in a digital department as well. There should be a simple procedure for rejecting images that do not result in the image disappearing from the system. Ideally, the rejected images should be categorized and stored.

19.4.2.8. Maintenance

In order for a QA programme to be effective, it is important to have a maintenance programme in place that ensures that any malfunction of equipment, revealed by QC testing, is rectified. Tests need to be performed after maintenance or repairs that may affect the equipment's imaging and/or radiation characteristics.

Maintenance and routine performance testing are complementary. Equipment has to be maintained to ensure that it is safe to use and is working properly. Another aim of equipment maintenance is to allow the detailed performance specification, demonstrated at commissioning, to continue to be achieved during the working life of the equipment. Service contractors should be able to demonstrate that they undertake appropriate tests to check the performance against the specification.

Departments can use the results of their own routine performance testing programme to audit the service contract. It is important that service engineers feed back the results of any testing or servicing they undertake, particularly if these could affect the clinical image quality and radiation dose. An additional critical examination or partial commissioning tests may be necessary when the machine has been subject to modification, maintenance, reinstallation or repair.

Mechanical and electrical safety checks are an important part of maintenance and should be described in the maintenance contract. It is recognized that users of X ray equipment have a duty of care to be vigilant as regards obvious signs of mechanical or electrical deterioration.

19.4.2.9. Disposal or alternative use

When imaging equipment no longer meets the required performance specifications, it should be withdrawn from use and may be disposed of and replaced. Alternatively, it may be used for less demanding imaging tasks for which a lower specification of performance may be acceptable.

19.4.3. Outline of QC tests

These tests are intended to verify the stability in the operation of the equipment or elements used to acquire the image. The tests can be described in various ways, with some of their key characteristics described below:

- *Frequency*: The recommended frequency of a routine performance test varies from daily to three yearly. It is often given as a range (e.g. three to six monthly) because the frequency selected should depend on the equipment characteristics (e.g. age, reliability) and the clinical workload to which the equipment is subjected. A lower frequency of tests may be appropriate for simple imaging equipment that is used less frequently, for example, in community hospitals or for equipment where experience shows that parameters are unlikely to change. The frequency of tests may also be designated as essential and desirable, for example, a test may be essential every year but desirable every six months.
- *Priority*: The priorities for indicating whether a routine performance test is recommended may be denoted as:
  - *Essential*: Represents the recommended minimum standard; conformance to this standard of testing would be regarded as good practice.
  - *Desirable*: The inclusion of this level of testing would be regarded as best practice. However, it is recognized that the implementation of these tests may be constrained by test equipment costs, personnel availability, equipment characteristics, clinical workload or other factors.
- *Performance standards*: QC tests help maintain equipment performance through the use of tolerance criteria that are applied to QC test results. These performance standards can be characterized as acceptable and achievable:
  - *Acceptable*: Indicates that performance must be within these tolerances and if it is not, the equipment should not be used.
  - *Achievable*: Indicates the level of performance that should be attained under favourable circumstances; if feasible, this is the level at which a facility should work.

- *Test types*: Many QC tests are designed to measure equipment performance consistency over time. In order to do this effectively, it is essential that the conditions of measurement for test parameters be consistent over time. For this publication, the terms repeatability and consistency are used:
  - *Repeatability*: Indicates that performance must be within given tolerances of self-consistency for a set of measurements taken at one time.
  - *Consistency*: Indicates that performance parameters are not changing over the period between QC tests. Usually, this is achieved through comparison with an established baseline measurement. This test type is also known as reproducibility. In cases where the baseline values are fixed to nominal values, it is known as accuracy testing.

19.4.4. Specification for test equipment

Equipment testers at radiology facilities should have access to the necessary test equipment for the required tests within the QA programme. Test equipment in the form of instruments that measure physical parameters should be calibrated to an appropriate standard prior to use and at suitable intervals (see Chapter 21). Additionally, a range of phantoms may be needed for (i) dosimetry measurements, especially under automatic exposure control (AEC) modes (see Chapter 22), and (ii) image QA of systems. Increasingly, the evaluation of digital images requires access to image datasets and the use of a computer and appropriate software. The evaluation of image display monitors requires specific test patterns and is discussed further in Chapters 9 and 16.

19.5. EXAMPLE OF A QC PROGRAMME

An example of a QC programme is given to illustrate the tests that need to be performed. It is not intended to be a QC manual, but is an example to assist in setting up QC programmes for the individual X ray modalities of a radiology facility.

When considering the composition of such a programme, it is important to remember that many factors influence the selection of QC tests, for example, the age, condition and level of use of the equipment, equipment performance and departmental resources.

While the principles of QC testing are clearly established, it is undesirable to be too prescriptive about the detail of test types, frequencies, priorities and tolerances in a monitoring programme, to allow flexibility in response to the constraints mentioned above. There can be a danger of testing for its own sake, particularly with the increasing reliability of modern X ray systems, which have advanced considerably in the last decade. Finally, the resources for QA should match the need to produce images of sufficiently high quality and consistency while incurring the lowest possible dose needed to provide the required diagnostic information.

Special attention should be given to equipment used for medical exposure of children, health screening programmes or high dose procedures. The QC programme may have to be more extensive and/or frequent for such equipment.

19.5.1. QC programme for X ray tubes and generators

All equipment used for medical radiography requires routine performance testing, and this section suggests a number of physical parameters related to the X ray generation, limitation of the X ray beam and manual control of exposure that need to be assessed on each item of equipment, including fixed installations (both radiographic and fluoroscopic), mobile radiography units and mobile image intensifiers (Table 19.1).

TABLE 19.1. OUTLINE QC PROGRAMME FOR X RAY TUBES AND GENERATORS

Parameter	Frequency	Priority
X ray/light beam alignment	1–2 monthly	Essential
X ray/light beam centring	1–2 monthly	Essential
Light beam/Bucky centring	1–2 monthly	Essential
Distance and scales	1–2 monthly	Desirable
Image receptor alignment and collimation	3–6 monthly	Desirable
Radiation output index: repeatability and consistency	1–2 monthly	Essential
Radiation output: repeatability and consistency	1–2 yearly	Essential
Exposure time repeatability and accuracy	1–2 yearly	Essential
Tube potential repeatability and accuracy	1–2 yearly	Essential

19.5.2. QC programme for screen film radiography

This section addresses screen film radiography for general radiographic procedures, including AEC systems.

19.5.2.1. *Screen film AEC systems*

The use of AEC will reduce the number of repeated films, provided that the AEC system is correctly set up for optimal operation. The optimal setup of an AEC system will be established during the commissioning of the unit. QC tests are continued at a regular frequency to ensure continuing optimal performance (Table 19.2).

TABLE 19.2. OUTLINE QC PROGRAMME FOR SCREEN FILM RADIOGRAPHY SYSTEMS

Parameter	Frequency	Priority
AEC backup timer operation	Annually	Essential
Resultant film density under AEC	1–3 monthly	Essential
Consistency between AEC chambers	Annually	Essential
AEC repeatability and consistency	Annually	Essential
AEC image receptor dose	Annually	Essential

19.5.2.2. *Film processor and darkroom*

Reject analysis for screen film radiography typically reveals that 50% of the reject films will be attributable to poorly controlled automatic film processing systems. It is essential that film processing through automatic processors be monitored through sensitometric control (Table 19.3). This process requires a standardized light source that exposes X ray film with logarithmically increasing intensity[2], known as a sensitometer, and a calibrated densitometer to measure the resulting optical density on the processed control film. One box of film, of the type used clinically, is set aside for daily sensitometric measurement. The resulting optical density for a specified step is recorded as the 'speed index' on a trend chart (for an example, see Fig. 19.2) that has the accepted tolerances identified. Action should be taken to correct the film processing when optical density is outside of tolerance. A similar index for film contrast and 'base + fog' values is also required, and in combination on a trend chart can assist in determining the corrective action needed for the film processors.

---

[2] Each intensity step is typically $\sqrt{2}$, so that the intensity is doubled every two steps.

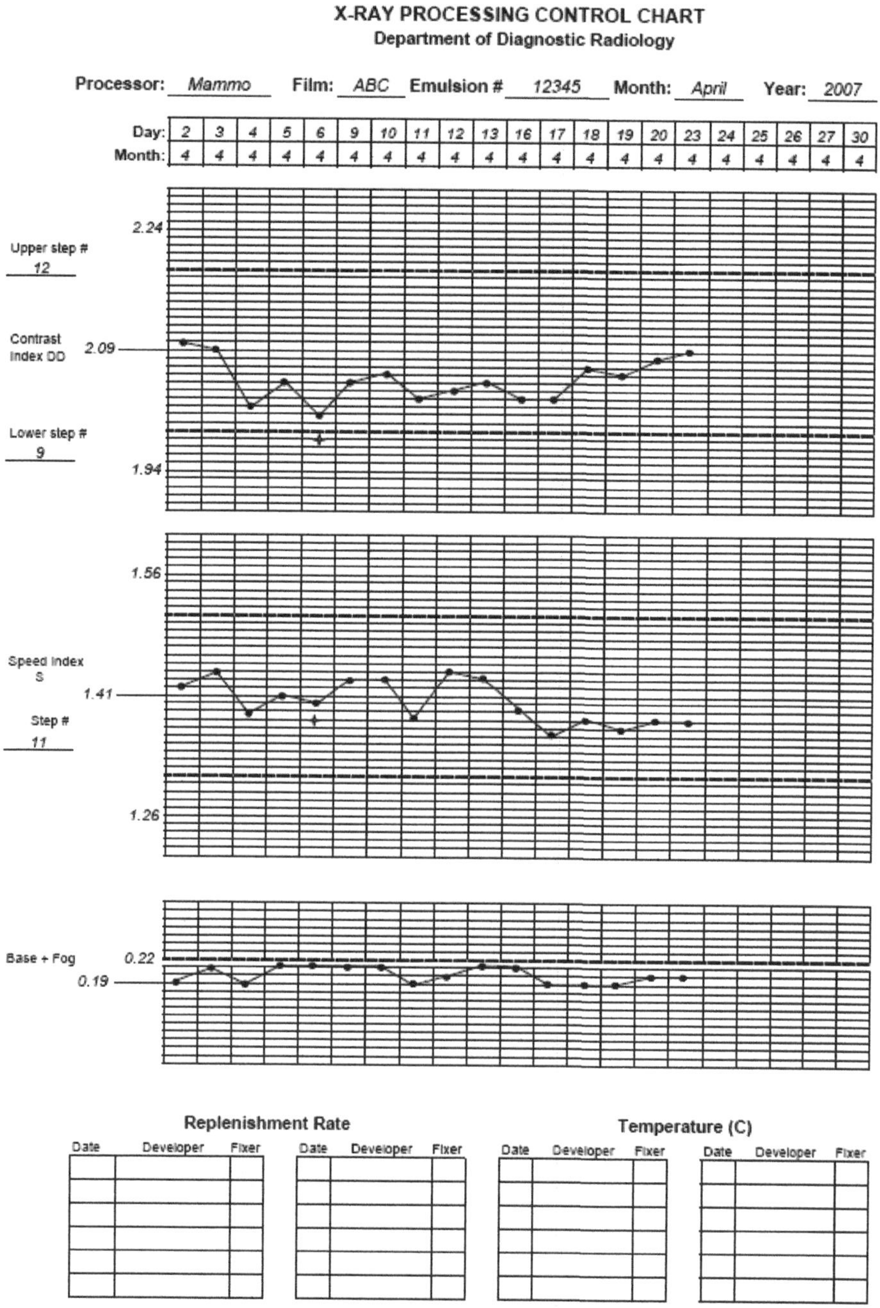

FIG. 19.2. Sample film processor QC chart with upper and lower tolerance limits (dashed lines). Note that the measurement taken on 6 April was repeated because the initial value of the contrast index (DD) was outside the action level.

TABLE 19.3. OUTLINE QC PROGRAMME FOR FILM PROCESSOR AND DARKROOM

Parameter	Frequency	Priority
Developer temperature	Daily to weekly	Essential
Gross fog	Daily to weekly	Essential
Film speed: speed index	Daily to weekly	Essential
Film contrast: contrast index	Daily to weekly	Essential
Replenishment rates	1–3 monthly	Desirable
Fixer pH	1–3 monthly	Desirable
Silver content of fixer	1–3 monthly	Desirable
Condition of cassettes and screens	6–12 monthly	Essential
Relative speed of intensifying screens	Annually	Desirable
Film fogging	Annually	Desirable
Darkroom lightproofing	Annually	Desirable

19.5.2.3. Lightboxes and viewing conditions

Optimal film viewing is critical for the successful reading of radiographic films. This essential part of QA is often overlooked and, once identified, poor viewing conditions can often be quite easily rectified. Care must be taken to ensure that view boxes have uniform luminance and colour temperature, and that the ambient lighting illumination levels are low (Table 19.4).

TABLE 19.4. OUTLINE QC PROGRAMME FOR LIGHTBOXES AND VIEWING CONDITIONS

Parameter	Frequency	Priority
Film viewer condition	6 monthly	Essential
Film viewer luminance	6–12 monthly	Essential
Film viewer uniformity	6–12 monthly	Essential
Variation between adjacent film viewers	6–12 monthly	Desirable
Room illumination	6–12 monthly	Essential

19.5.3. QC programme for digital radiography

This section relates to the image acquisition, processing and display of digital images, primarily through either computed radiography, consisting of a cassette and image processor combination, or digital radiography, which comprises a hard wired digital detector and processor system. Importantly, both include the use of AEC systems.

One fundamental difference between digital and analogue (screen film) systems is the separation of the acquisition, processing and display functions for an image. While this allows a wide dynamic range of the digital receptor, allowing images of varying exposure level to be displayed optimally, it removes the link between image brightness and image receptor or patient exposure. It is, thus, not obvious to users whether the system is operating at an optimal sensitivity or whether exposure levels are changing. This is a major challenge and highlights the need for QC in this area.

Digital systems for radiography have an indication of the detector response to radiation, which is referred to as the exposure index. While digital systems appear to be fully automatic, in fact the functions of acquisition, processing and display can be altered for different examination types to gain optimal image quality for diagnosis at an acceptable dose to the patient. This important area of optimization is discussed in Chapter 23. Once optimization is achieved, QC testing should be undertaken, using consistent image processing, to maintain equipment performance.

Once digital imaging becomes the main form of imaging in an X ray facility, digital communication through a picture archiving and communications system becomes critical to the facility's function. Of first concern is the setting up and performance maintenance of image display devices (for more details, see Chapter 16). Another important consideration is the use of DICOM structures to record equipment generated dose related data. These data are increasingly available to patients and are stored in records. It will be increasingly important for medical physicists to verify the accuracy of such information. A first step is the verification of DICOM dose indicators during the commissioning of new equipment and routine calibration of KAP meters and other dose indicators such as those used in computed tomography, etc., (see Chapters 21 and 22 on instrumentation and dosimetry, respectively).

19.5.3.1. General

TABLE 19.5. OUTLINE QC PROGRAMME FOR DIGITAL RADIOGRAPHY SYSTEMS — GENERAL TESTS

Parameter	Frequency	Priority
EI monitoring	1–3 monthly	Essential
Image uniformity (visual check)	1–3 monthly	Essential
Threshold contrast visibility	4–6 monthly	Desirable
Limiting spatial resolution	4–6 monthly	Desirable
Exposure index repeatability and consistency	Annually	Essential
Image uniformity	Annually	Essential
Threshold contrast detail detectability	Annually	Essential
Limiting spatial resolution	Annually	Desirable
Scaling errors	Annually	Desirable
Dark noise	Annually	Desirable
AEC sensitivity	1–3 monthly	Essential
AEC backup timer operation	Annually	Essential
AEC consistency between chambers	Annually	Essential
AEC repeatability and consistency	Annually	Essential
AEC image receptor dose	Annually	Essential

19.5.3.2. Computed radiography

TABLE 19.6. OUTLINE QC PROGRAMME FOR DIGITAL RADIOGRAPHY SYSTEMS — COMPUTED RADIOGRAPHY SPECIFIC TESTS

Parameter	Frequency	Priority
Condition of cassettes and image plates	1–3 monthly	Essential
Erasure cycle efficiency	Annually	Essential

19.5.3.3. Digital radiography

TABLE 19.7. OUTLINE QC PROGRAMME FOR DIGITAL RADIOGRAPHY SYSTEMS — DIGITAL RADIOGRAPHY SPECIFIC TESTS

Parameter	Frequency	Priority
Sensitivity reproducibility between digital radiography detectors connected to the same generator	Annually	Essential

19.6. DATA MANAGEMENT

Quality management will lead to accumulation of a significant volume of data, which will require a suitable repository for storage and retrieval. However, the data can also be used in an active way to help manage the QC system. This requires the development of a suitable data management system, which could be either paper based or computer based, which is becoming increasingly more common.

The elements of such a system include:

- Policy and control manuals that determine the nature of the QA programmes and the details of the QC testing procedures.
- The results from QC tests, which need to be recorded and compared with the required performance criteria. Some tests involve performance constancy and comparison to baseline data. Graphical representation, such as a trend chart, is very useful to review such data.
- A test report is also often required to document the test results and to initiate action if needed for equipment adjustment (Fig. 19.2).
- Trend analysis, which is an important tool that can be used to assess drifts in performance, highlights tests that consistently fail and enables comparisons of similar types of equipment.

Additional functionality can be achieved with:

- The ability to perform auditing of the data to help determine the suitability of tests and optimum test frequency.
- While automated systems are commonly achieved with spreadsheet software, the use of database software is more likely to accomplish a satisfactory set of outcomes, which can include test scheduling, trend analysis, automated report generation and auditing of performance.

- An efficient data management system that enables QC to function as a dynamic process where the continued analysis of results feeds back into an ongoing QC programme review, including the suitability and relevance of tests performed and the frequency of testing (Fig. 19.3).
- A medical physicist who is involved in the establishment and maintenance of such a data management system.

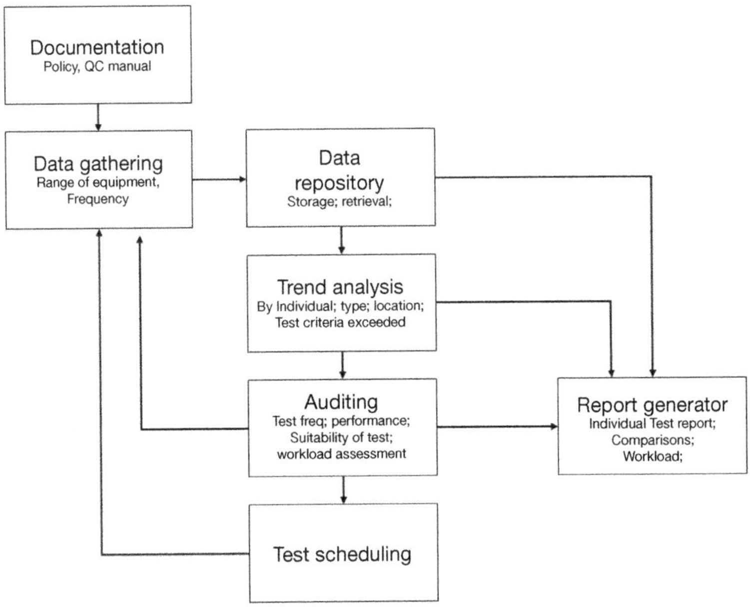

FIG. 19.3. Example of data management structure.

REFERENCES

[19.1] INTERNATIONAL ORGANIZATION FOR STANDARDS, Quality Management Systems – Fundamentals and Vocabulary, Rep. ISO9000:2000, ISO, Geneva (2000).

[19.2] HOYLE, D., ISO 9000 Quality Systems Handbook, 4th edn, Butterworth Heinemann, Oxford (2001).

[19.3] INTERNATIONAL ORGANIZATION FOR STANDARDS, Quality Management Systems – Requirements, Rep. ISO9001:2000, ISO, Geneva (2000).

[19.4] FOOD AND AGRICULTURE ORGANIZATION OF THE UNITED NATIONS, INTERNATIONAL ATOMIC ENERGY AGENCY, INTERNATIONAL LABOUR ORGANISATION, OECD NUCLEAR ENERGY AGENCY, PAN AMERICAN HEALTH ORGANIZATION, WORLD HEALTH ORGANIZATION, International Basic Safety Standards for Protection against Ionizing Radiation and for the Safety of Radiation Sources, Safety Series No. 115, IAEA, Vienna (1996).

BIBLIOGRAPHY

AMERICAN ASSOCIATION OF PHYSICISTS IN MEDICINE, Quality Control in Diagnostic Radiology, AAPM Rep. 74, AAPM, New York (2002),
http://www.aapm.org/pubs/reports/rpt_74.pdf (accessed on 23 August 2012).

AMERICAN ASSOCIATION OF PHYSICISTS IN MEDICINE, An Exposure Indicator for Digital Radiography: Report of AAPM Task Group 116, AAPM Rep. 116, AAPM, New York (2009),
http://www.aapm.org/pubs/reports/RPT_116.pdf (accessed on 23 August 2012).

EUROPEAN COMMISSION, Radiation Protection No. 162: Criteria for Acceptability of Medical Radiological Equipment used in Diagnostic Radiology, Nuclear Medicine and Radiotherapy, EC, Luxembourg (2012).

HAUS, A.G. (Ed.), Advances in Film Processing Systems Technology and Quality Control in Medical Imaging, Medical Physics Publishing, Madison, WI (2001).

HEALTH AND SAFETY EXECUTIVE, Equipment Used in Connection with Medical Exposure, Guidance Note PM77, HSE Books, Sudbury (2006),
http://www.hse.gov.uk/pubns/guidance/pm77.pdf (accessed on 23 August 2012).

INSTITUTE OF PHYSICS AND ENGINEERING IN MEDICINE, Quality Control in Magnetic Resonance Imaging, IPEM Rep. 80, IPEM, York (2002).

INSTITUTE OF PHYSICS AND ENGINEERING IN MEDICINE, Measurement of the Performance Characteristics of Diagnostic X-ray Systems Used in Medicine, IPEM Rep. 32, 2nd edn, Part III: Computed Tomography X-ray Scanners, IPEM, York (2003).

INSTITUTE OF PHYSICS AND ENGINEERING IN MEDICINE, Recommended Standards for the Routine Performance Testing of Diagnostic X-ray Imaging Systems, IPEM Rep. 91, IPEM, York (2005).

INSTITUTE OF PHYSICS AND ENGINEERING IN MEDICINE, The Critical Examination of X-ray Generating Equipment in Diagnostic Radiology, IPEM Rep. 107, IPEM, York (2012).

INTERNATIONAL ATOMIC ENERGY AGENCY, Quality Assurance Programme for Screen Film Mammography, IAEA Human Health Series No. 2, IAEA, Vienna (2009).

INTERNATIONAL ORGANIZATION FOR STANDARDIZATION, Quality Management and Quality Assurance Standards – Part I: Guidelines for Selection and Use, ISO 9000, ISO, Geneva (1994).

PAN AMERICAN HEALTH ORGANIZATION, WORLD HEALTH ORGANIZATION, Organization, Development, Quality Assurance and Radiation Protection in Radiology Services: Imaging and Radiation Therapy, PAHO, Washington, DC (1997).

Chapter 20

RADIATION BIOLOGY

J. WONDERGEM
International Atomic Energy Agency

20.1. INTRODUCTION

Radiation biology (radiobiology) is the study of the action of ionizing radiations on living matter. This chapter gives an overview of the biological effects of ionizing radiation and discusses the physical, chemical and biological variables that affect dose response at the cellular, tissue and whole body levels at doses and dose rates relevant to diagnostic radiology.

20.1.1. Deterministic and stochastic responses

Biological effects of radiation in humans occur either in the irradiated individuals themselves (somatic effects) or in their descendants (hereditary or genetic effects). Somatic effects are divided into deterministic effects (also known as 'tissue reactions') and stochastic effects, where hereditary and genetic effects are all of stochastic origin only.

Deterministic effects result from radiation induced cell loss or damage, e.g. moist desquamation from interventional cardiology. Most organs or tissues of the body are unaffected by the loss of a few cells; however, if the number of cells lost is sufficiently large, there is observable harm and, hence, loss of tissue/organ function. Above a particular level of dose, the so-called threshold dose, the severity of the effect necessarily increases with increasing dose. This threshold varies from one effect to another. Deterministic effects may occur a few hours or days after exposure (i.e. early skin reaction) or may require months or years before expression (i.e. development of cataract of the eye lens).

Stochastic effects, on the other hand, are probabilistic effects. This means that the probability of the occurrence of an effect, but not its severity, is a function of dose — the probability increases with dose. These effects are assumed to exhibit no threshold dose below which they cannot occur. The major stochastic effects of concern at typical diagnostic radiology levels are cancers and genetic effects. These are exclusively late effects because they do not appear until years after radiation exposure.

20.1.2. Diagnostic radiology

The amount of energy deposited in the tissue of patients as a result of diagnostic radiology examinations or interventional procedures is typically a number of orders of magnitude less than that delivered during radiation oncology. Consequently, the detriment caused is largely confined to stochastic effects. The typical effective doses from diagnostic procedures (Fig. 20.1) show that there is a large range in the amount of radiation dose given by various examinations. It should also be considered that, in a small number of procedures, radiation damage to tissue can occur, as seen in skin reactions from long interventional procedures. The occupational dose, although orders of magnitude lower than that of the patient during a single procedure, may become considerable for a worker performing large numbers of procedures, and especially if the shielding precautions needed are not observed. Consequently, there is an increasing incidence of injury to the lens of the eye for some workers, for example, during interventional procedures.

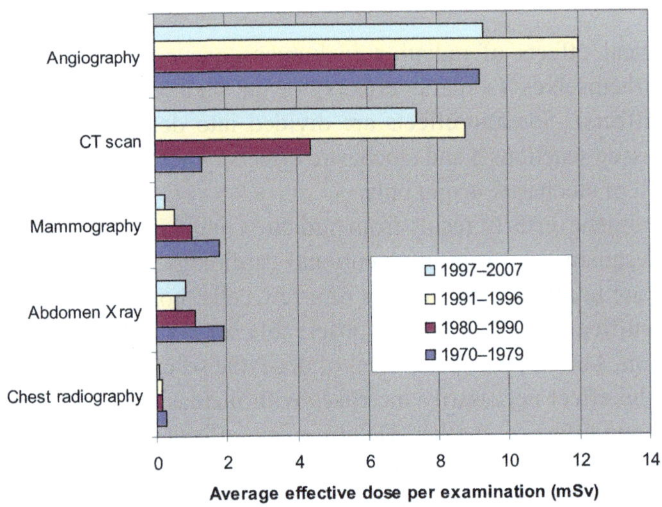

FIG. 20.1. Trends in average effective doses resulting from selected diagnostic medical examinations, health care level I (from Ref. [20.1]).

20.1.3. International organizations on radiation effects

There are two large committees that collect and analyse data from the recent literature regarding biological effects of ionizing radiation. BEIR (Biological

Effects of Ionizing Radiation) and UNSCEAR (United Nations Scientific Committee on the Effects of Atomic Radiation) report periodically on important issues such as risk estimates for radiation induced cancer and hereditary (genetic) effects. In addition, the ICRP (International Commission on Radiological Protection) is involved in recommendation and development of guidelines in the field of radiation protection.

20.2. RADIATION INJURY TO DEOXYRIBONUCLEIC ACID

20.2.1. Structure of deoxyribonucleic acid

Deoxyribonucleic acid (DNA) contains the genetic information of the cell. DNA is a large molecule and has a characteristic double helix structure consisting of two strands, each made up of a sequence of nucleotides. The backbone of the DNA strand is made of alternating sugar and phosphate groups. A nucleotide is a subunit of DNA and is composed of a base linked to a sugar (deoxyribose) and a phosphate group. The four bases are adenine (A), guanine (G), cytosine (C) and thymidine (T). These bases can be classified into two groups: purines (A and G) and pyrimidines (T and C). The unique pairing of the nucleotide bases provides DNA with its identity, which is used in replication. The cell's genetic information is carried in a linear sequence of nucleotides that make up the organism's set of genes.

20.2.2. Radiation chemistry: Direct and indirect effects

The physical interactions of ionizing radiation with matter lead to loss of radiation energy through ionization and the formation of free radicals. These radicals react rapidly (10^{-10} s) with neighbouring molecules and produce secondary DNA or lipid radicals. Free radicals are fragments of molecules with unpaired electrons, which have high reactivity with cellular molecules and, therefore, have a short life. Free radicals are generated in great number by ionizing radiation, owing to the process of energy absorption and breakage of chemical bonds in molecules. Free radicals are known to play a major role in biological tissues and organisms.

When ionizing radiation energy is deposited in a certain macromolecule associated with observable biological effects, such as DNA, it is called a direct effect of radiation. Alternatively, photons may be absorbed in the water of an organism, causing excitation and ionization in the water molecules. The radicals formed, namely the hydrated electron (e_{aq}^{-}), the hydrogen atom (H·) and the hydroxyl radical (OH·), are able to diffuse far enough to reach and damage the critical targets. This is referred to as indirect action of ionizing radiation.

The absorption of energy depends on the abundance of material in the path of the radiation. Water is the most predominant molecule in living organisms (about 80% of the mass of a living cell is water). Therefore, a major proportion of the radiation energy deposited will be absorbed in cellular water. About two thirds of the biological damage caused by low linear energy transfer radiations (sparsely ionizing radiation), such as X rays or electrons, is due to indirect action. A complex series of chemical changes occurs in water after exposure to ionizing radiation; this process is called water radiolysis.

20.2.3. DNA damage

DNA damage is the primary cause of cell death induced by radiation. Radiation exposure produces a wide range of lesions in DNA, such as single strand breaks, double strand breaks (DSBs), base damage, protein–DNA cross-links and protein–protein cross-links (Fig. 20.2). The number of DNA lesions generated by irradiation is large, but there are a number of mechanisms for DNA repair. As a result, the percentage of lesions causing cell death is very small. The numbers of lesions induced in the DNA of a cell by a dose of 1–2 Gy are approximately: base damage >1000; single strand breaks ~1000; DSBs ~40.

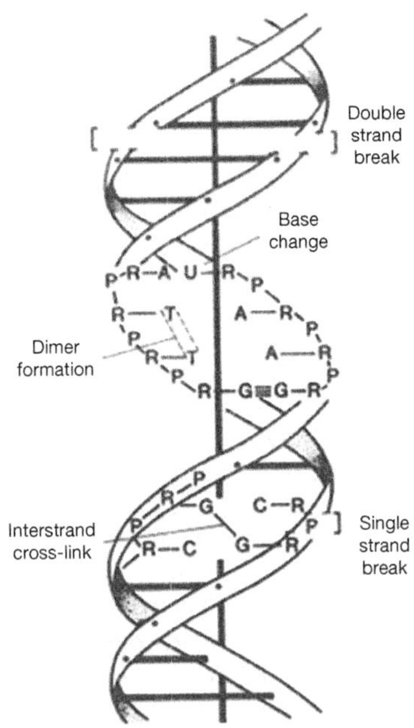

FIG. 20.2. Some possible lesions in DNA (reproduced from Ref. [20.2]; courtesy of R.E. Meyn and R. Humphrey, MD Anderson Cancer Center, Houston).

DSBs play a critical role in cell killing, carcinogenesis and hereditary effects. There are experimental data showing that initially produced DSBs correlate with radiosensitivity and survival at lower dose, and that unrepaired or misrepaired DSBs also correlate with survival after higher doses. Furthermore, there is experimental evidence for a causal link between the generation of DSBs and the induction of chromosomal translocations with carcinogenic potential.

20.3. DNA REPAIR

DNA repair mechanisms are important for the recovery of cells from radiation and other damaging agents. There are multiple enzymatic mechanisms for detecting and repairing radiation induced DNA damage. DNA repair mechanisms, such as base excision repair, mismatch repair and nucleotide excision repair respond to damage such as base oxidation, alkylation and strand intercalation.

Excision repair consists of cleavage of the damaged DNA strand by enzymes that cleave the polynucleotide chain on either side of the damage and enzymes that cleave the end of a polynucleotide chain, allowing removal of a short segment containing the damaged region. DNA polymerase can then fill in the resulting gap using the opposite undamaged strand as a template.

For DSBs, there are two primary repair pathways, non-homologous end joining (NHEJ) and homologous recombination. NHEJ repair operates on blunt ended DNA fragments. This process involves the repair proteins recognizing lesion termini, cleaning up the broken ends of the DNA molecule, and the final ligation of the broken ends. DSB repair by homologous recombination utilizes sequence homology with an undamaged copy of the broken region and, hence, can only operate in late S or G2 phases of the cell cycle (see Section 20.5). Undamaged DNA from both strands is used as a template to repair the damage. In contrast to NHEJ, the repair process of homologous recombination is error free. Repair by NHEJ operates throughout the cell cycle but dominates in G1/S phases (see Section 20.5). The process is error prone because it does not rely on sequence homology.

Unrepaired or misrepaired damage to DNA will lead to mutations and/or chromosome damage in the exposed cell. Mutations might lead to cancer or hereditary effects (when germ cells are exposed), whereas severe chromosome damage often leads to cell death.

20.4. RADIATION INDUCED CHROMOSOME DAMAGE AND BIOLOGICAL DOSIMETRY

In the living cell, chromosomes can be found in the nucleus of the cell. Chromosomes consist of DNA and proteins forming a thread-like structure containing genetic information arranged in a linear sequence. When the repair of DNA DSBs is incomplete, there may be serious implications for a cell, namely, it may lead to chromosomal damage (aberrations). Aberrant (damaged) chromosomes arise when broken ends rejoin with other broken ends to generate rings, dicentrics (chromosomes having two centromeres), translocations and other chromosome aberrations. Symmetrical translocations and small deletions are, in general, non-lethal. Dicentrics and rings are 'unstable' aberrations and are lethal to the cell and, as a consequence, are not passed on to progeny. The incidence of dicentrics and rings declines slowly with time after exposure, since the exposed cells have a finite lifespan and are eliminated from their environment. Translocations are 'stable' aberrations and may persist for many years because they are not lethal to the cell and are passed on to progeny. When these translocations occur in germ cells (testes or ovaries), they may lead to an increase in hereditary effects in the offspring.

Structural chromosome aberrations can be used as an indicator of radiation exposure. Chromosome analysis in mitotic spreads (karyotyping), micronucleus formation and fluorescence in situ hybridization can detect unrepaired damage to DNA in chromatids, caused by radiation and a variety of DNA damaging agents. These cytological techniques are used in biodosimetry (assays to estimate the radiation dose based on the type and/or frequency of chromosomal aberrations in the exposed cells/tissues). Biodosimetry has provided an important tool for assessing doses in known or suspected cases of acute (unwanted) radiation exposure.

20.5. THE CELL CYCLE

The cell cycle has two well defined time periods: mitosis (M), where division takes place, and the period of DNA synthesis (S). The S and M portions of the cell cycle are separated by two periods (gaps), G1 and G2. Cells in a growing population (e.g. skin, gut, bone marrow), but not resting fully differentiated G0 phase cells, participate in the cell cycle and, thus, are more sensitive to radiation. Replication of the genome occurs in the S phase and mitotic propagation to daughter generations occurs in the G2/M phases. Typical cell generation times are 10–40 h, with the G1 phase taking up about 30%, S phase 50%, G2 phase

15% and M phase 5% of the cell cycle time. There are checkpoints at the G1/S and G2/M boundaries that ensure the fidelity of genomic processing.

Radiosensitivity differs throughout the cell cycle with, in general, late S phase being most radioresistant, G2/M being most radiosensitive and the G1 phase taking an intermediate position. The greater proportion of repair by homologous recombination than by NHEJ in late S phase may explain the resistance of late S phase cells. Chromatin compaction and poor repair competence (reduced enzyme access) could explain the high radiosensitivity in G2/M phase. Resting cells, not involved in the cell cycle, are even more resistant to radiation when compared with late S phase cells.

20.6. SURVIVAL CURVE THEORY

20.6.1. Survival curves

The generally accepted standard for measuring the radiosensitivity of a cell population is the 'retention of reproductive integrity', i.e. the ability of a cell to undergo more than 5 or 6 cell divisions and produce a viable colony containing at least 50 cells. This is referred to as 'cell survival' (Fig. 20.3). Survival curves are best shown as a semi-log plot of survival on the ordinate against irradiation dose on a linear scale on the abscissa. When the surviving fraction of irradiated

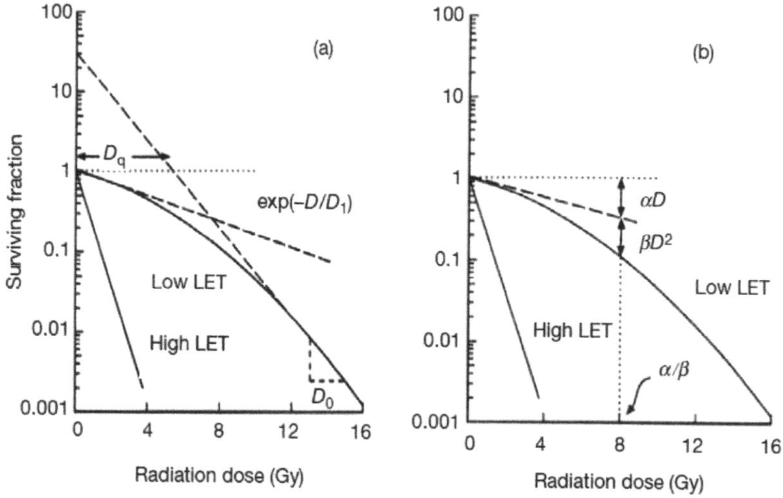

FIG 20.3. Cell survival curves for low linear energy transfer (sparsely ionizing) radiation and high linear energy transfer (densely ionizing) radiation. (a) The earlier multitarget single hit model; (b) the current linear quadratic model. (See text for expalanation of symbols used.)

cells is plotted as a function of dose, the curvature of the survival curve at lower dose is somewhat higher (also called 'shoulder' region), whereas at higher dose the curve is straighter. In order to describe the shape of cell survival curves, several mathematical models of cell killing have been developed, all based on the random nature of energy disposition by radiation.

20.6.2. Linear quadratic model

The most common model used today is the linear quadratic model, where cell death as a function of dose is described by a second order polynomial (Fig. 20.3(b)). This model assumes that there are two components to cell killing by radiation, commonly represented by two constants, α and β. In this model, survival is described as a function of dose by the following equation:

$$S = \mathrm{e}^{-(\alpha D + \beta D^2)} \tag{20.1}$$

A plausible explanation of the linear component is that the majority of DNA interactions are single radiation track events. Under these circumstances, DNA damage can be effectively repaired before possible interaction with another single track, when enough time is available and doses are relatively low. As the dose or dose rate increases, multitrack events, reflecting the quadratic component, will predominate, resulting in an increased probability of misrepair and cell death.

20.6.3. Target theory

An alternative older model is the single-hit single-target model described by:

$$S = \mathrm{e}^{-D/D_0} \tag{20.2}$$

where D_0 is effectively the reciprocal of α (above) and represents the dose that reduces survival to e^{-1} or 37% (Fig. 20.3(a)). The target theory is based upon the idea that there are n targets in a cell, all of which must be 'hit' to kill the cell. The log-linear relationship is consistent with data from some bacteria but it does not apply in eukaryotic cells (except at high linear energy transfer), which show shouldered survival curves that can be accommodated by a single-hit multitarget model described by:

$$S = 1 - [1 - \mathrm{e}^{-(D/D_0)}]^n \tag{20.3}$$

This is reliable at high dose but not at low dose, because it does not accurately describe the shoulder region at low doses, even if another single-hit term is added.

20.7. CONCEPTS OF CELL DEATH

Radiation doses of the order of several sieverts may lead to cell loss. Cells are generally regarded as having been 'killed' by radiation if they have lost reproductive integrity, even if they have physically survived. Loss of reproductive integrity can occur by apoptosis, necrosis, mitotic catastrophe or induced senescence. Although all but the last of these mechanisms ultimately results in physical loss of the cell, this may take a significant time to occur.

Apoptosis or programmed cell death can occur naturally or result from insult to the cell environment. Apoptosis occurs in particular cell types after low doses of irradiation, for example, lymphocytes, serous salivary gland cells and certain cells in the stem cell zone in the testis and intestinal crypts.

Necrosis is a form of cell death associated with loss of cellular membrane activity. Cellular necrosis generally occurs after high radiation doses.

Reproductive cell death is a result of mitotic catastrophe (cells attempt to divide without proper repair of DNA damage), which can occur in the first few cell divisions after irradiation, and occurs with increasing frequency after increasing doses.

Ionizing radiation may also lead to senescence. Senescent cells are metabolically active but have lost the ability to divide.

20.8. CELLULAR RECOVERY PROCESSES

At higher doses and dose rates (i.e. multiple radiation exposures during interventional cardiology), cellular recovery may play an important role in the fixation of the radiation damage.

20.8.1. Sublethal and potentially lethal damage repair

Owing to cellular recovery, an increase in cell survival can be expected when the same dose is given as two fractions separated by two or more hours, compared with the same dose delivered in a single fraction. The greater survival when the dose is split in this way is attributed to sublethal damage repair between dose fractions. The half-time of repair, $T_{½}$, is defined as the time it takes for half the repair to take place and is usually about 30–60 min for cells in culture, but can be longer for tissues. Thus, full repair may take 6–8 h and can be longer in tissues (e.g. in the central nervous system it may be greater than 24 h). The recovery ratio is a measure of sublethal damage repair and is given by the ratio of the survival of cells receiving a split dose to the survival of cells receiving the total dose as a single dose.

Potentially lethal damage repair is another class of repair and is determined by delayed plating experiments. In such experiments, contact inhibited (i.e. confluent cell cultures) cells are irradiated, incubated for various periods and subsequently reseeded. Analysis of cell survival by colony assay then gives a measure of this type of repair.

20.8.2. Fractionation effect

The ‘shoulder’ or the curvature of a survival curve is usually considered to be a reflection of the repair capacity of a cell population. In terms of the target theory, this can be thought of as arising from the concept that sublethal DNA damaging events must be accumulated to allow sublesion interactions for cell killing to occur.

20.8.3. Dose rate effects

The successive increase of cell survival with declining dose rate is consistent with the role of time in repair. The dominance of repair at low dose rate eliminates the shoulder/curvature of the survival curve and results in a straight but shallower line on a semi-logarithmic plot, with good separation of survival between cell lines with different repair capacity. This is due to the cells having different radiosensitivities. Repair during irradiation is negligible at dose rates of 1–5 Gy/min, but is significant at low dose rates (<100 mGy/min).

20.9. RELATIVE BIOLOGICAL EFFECTIVENESS

When the effects of equal doses of different types of radiation are compared, they are shown to produce unequal biological effects. Comparison of the effects of different types of radiation is expressed as relative biological effectiveness and is defined as the ratio of doses of reference radiation and the test radiation required to produce an equal amount of a particular biological effect. Historically, the reference used was 250 kV X rays, but more recently, ^{60}Co radiation has become the standard. Relative biological effectiveness varies with cell system end point and dose and is higher at low doses and at low dose rates. For high doses with a single fraction, the relative biological effectiveness is lower than for multiple small fractions. For radiation protection purposes (at low doses and low dose rates), the ICRP has described the effectiveness of radiations of differing qualities by a series of radiation weighting factors (w_R) (see Chapter 22).

20.10. CARCINOGENESIS (STOCHASTIC)

20.10.1. Mechanism of multistage carcinogenesis

The development of cancer in tissues is assumed to be a multistage process that can be subdivided into four phases: neoplastic initiation, promotion, conversion and progression. This subdivision is an oversimplification, yet it provides a suitable framework for the identification of specific molecular and cellular changes involved.

Neoplastic initiation leads to the irreversible potential of normal cells for neoplastic development by creating unlimited proliferative capacity. There is good evidence that this event results from one or more mutations in a single cell, which is the basis of the clonal evolution of the cancer. Further neoplastic development of initiated cells depends on promotional events, which involves intercellular communication, e.g. by growth factors, hormones or environmental agents. This results in proliferation of the initiated pre-neoplastic cells in a semi-autonomous manner.

During the process of conversion of the pre-neoplastic cells into fully malignant cells, additional mutations in other genes are accumulated, probably facilitated by increasing loss of genomic stability. The subsequent progression into an invasive cancer depends on still more mutations in the unstable genome.

There is strong evidence from animal studies and some human studies that the risk of radiation induced cancer may be influenced by various genes, such as mutations of the RB1 gene (predisposing for retinoblastoma and osteosarcoma) and of the BRCA1 gene (predisposing for early breast cancer and ovarian cancer), or the presence of polymorphisms (single nucleotide polymorphisms) in the gene. However, at the present state of knowledge, the role of genetic susceptibility in individual risk of radiation induced cancer cannot be resolved definitively, although there is general agreement that it is important.

20.10.2. Mechanism of mutation induction

Two classes of cancer associated genes have been identified: proto-oncogenes and tumour suppressor genes. Proto-oncogenes are normal genes involved in growth regulation. Mutations, for example, by the translocation of a promoter, may result in an increased rate of proliferation. Proto-oncogene mutations to oncogenes are, thus, classified as ‘gain of function’ mutations.

Tumour suppressor genes are genes that are involved in growth regulation of normal cells and that prevent excessive cell proliferation. The critical mutations in these genes are ‘loss of function’ mutations, which may be the result of partial or complete loss of the gene structure, e.g. by deletions. Since radiation induced DNA damage preferentially causes deletions, it is generally assumed that the inactivating mutation of tumour suppressor genes is the most probable mechanism for the induction of cancer by radiation.

There is good evidence that many, if not most, cancers are the clonal descendants of a single neoplastic cell and, furthermore, that a single DSB may, although with an extremely low probability, cause a deletion in a specific DNA sequence, e.g. of a tumour suppressor gene. It has, hence, been argued that, in principle, a single mutational event in a critical gene in a single target cell in vivo can create the potential for neoplastic development. Thus, a single radiation track traversing the nucleus of an appropriate target cell has a finite probability, albeit very small, of generating the specific damage to DNA that results in the initiating mutation. This argument would strengthen the hypothesis that the risk of radiation induced cancer increases progressively with increasing dose and there is no lower threshold.

20.10.3. Risk models

In order to evaluate the health effects of radiation on exposed populations or workers (persons), the incidence or frequency of a certain effect is studied in both the exposed and non-exposed control group (same age characteristics, same sex balance, etc.). Risk estimates derived from these studies are generally presented as relative risk (RR), excess relative risk (ERR) or excess absolute risk (EAR) per unit of radiation dose. RR is the ratio of both the frequency of a certain effect (i.e. number of cancer cases) in the exposed group (Rr) and the frequency of the same effect (Ro) in the non-exposed group (RR = Rr/Ro). The ERR is RR minus 1 (ERR = RR – 1). The EAR is the difference between the frequencies observed in the exposed and the non-exposed groups, respectively (EAR = Rr – Ro).

For assessing the risk of radiation induced cancer in humans, two conceptually different models are used: (i) absolute risk models and (ii) RR models. The absolute risk model assumes that radiation induces a ‘crop’ of cancers over and above the natural incidence and unrelated to it. After the latency period has passed, the cancer risk returns to ‘spontaneous’ levels. The RR model assumes that the effect of radiation is to increase the natural incidence at all ages subsequent to exposure by a given factor. As the natural or spontaneous cancer incidence rises significantly in old age, the RR model predicts a larger number of radiation induced cancers in old age. The RR model is favoured by the BEIR committee for estimating risks after radiation exposure.

20.10.4. Time course and latency period

Epidemiological information derived from the lifespan study of the A-bomb survivors in Japan and data from other studies have provided the main source of risk estimates currently used in radiation protection.

The time interval between exposure to irradiation and the appearance of cancer is known as the latency period. Leukaemia has a minimum latency of about 2 years after exposure; the pattern of risk over time peaks after 10 years (most cases occur in the first 15 years) and decreases thereafter. Solid tumours show a longer latency than leukaemia, by anything from 10 to 60 years or even more.

20.10.5. Dose–response relationship for cancer

The linear non-threshold hypothesis was introduced by the ICRP as the best practical approach to managing risk from radiation exposure to low doses/low dose rates. The linear non-threshold model postulates that there is a linear relationship between radiation dose and health risk, without a threshold at low doses. It means that there is no level of radiation exposure that can be assumed to have no associated health risk. The slope of the linear dose–response curve provides the risk coefficient (cancer risk per unit radiation dose received) appropriate for low level exposures.

20.10.6. Dose and dose rate effectiveness factor

Comparison of the Japanese data for A-bomb survivors with those of other epidemiological studies, including studies of multiple medical and chronic exposures, have demonstrated that the risk calculated in proportion to the dose differs. Both the BEIR and UNSCEAR committees considered that there is a dose rate effect for low energy transfer radiation, with fewer malignancies induced if a given dose were to be spread out over a period of time at low dose rate than if it were delivered in an acute exposure.

The dose and dose rate effectiveness factor (DDREF) is defined as the factor by which radiation cancer risks observed from large acute doses should be reduced when the radiation is delivered at low dose rate or in a series of small dose fractions. For general purposes in radiological protection, the ICRP recommends a DDREF of 2 for doses below 200 mSv at any dose rate and for higher doses if the dose rate is less than 100 mSv/h.

20.10.7. Cancer risk

The ICRP recommendations for radiation protection purposes are based on the Japanese study and other epidemiological studies. The risk coefficients for non-radiation workers are given as 5×10^{-2} per sievert for cancer lethality by irradiation and 10×10^{-2} per sievert for high doses and dose rates (higher than 200 mSv and 100 mSv/h). For workers, the risk adopted for radiation protection purposes is 4×10^{-2} per sievert and 8×10^{-2} per sievert for high doses and dose rates. These estimates show mean values for the whole population. However, there is ample evidence that cancer risk is dependent not only on the dose but also on the age at exposure, and to a lesser extent also on sex. In most cases, those exposed at an early age are more susceptible than those exposed at later times (Fig. 20.4) and females are slightly more susceptible than males. Since not all radiation exposures concern the whole body but only a region or part of the body, tissue weighing factors (w_T) should be taken into account (see Chapter 22). Tables are available to calculate the age and sex specific lifetime attributable incidence and mortality risks per organ, after partial body radiation exposure (Tables 20.1 and 20.2).

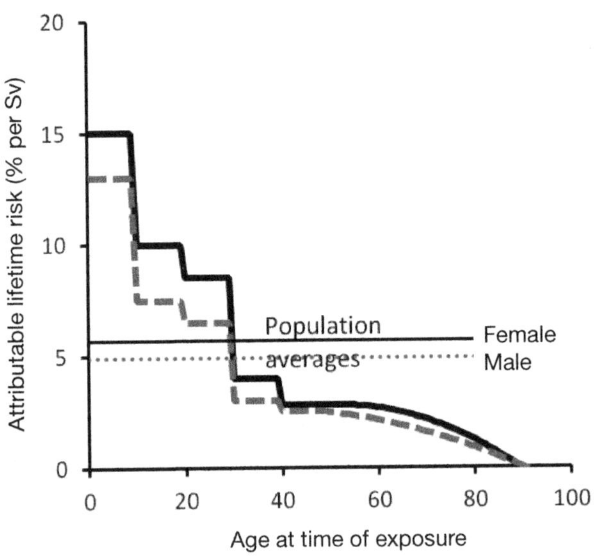

FIG. 20.4. The attributable lifetime risk from a single small dose of irradiation at various ages at the time of exposure. The dramatic decrease in radiosensitivity with age should be noted. The higher risk for the younger age groups is not expressed until late in life. These estimates are based on a relative risk model and on a DDREF of 2 (from Ref. [20.3] and adapted from Ref. [20.4]).

TABLE 20.1. LIFETIME ATTRIBUTABLE RISK OF CANCER INCIDENCE[a] *(from Ref. [20.5])*

Cancer site	Age at exposure (years)										
	0	5	10	15	20	30	40	50	60	70	80
Males											
Stomach	76	65	55	46	40	28	27	25	20	14	7
Colon	336	285	241	204	173	125	122	113	94	65	30
Liver	61	50	43	36	30	22	21	19	14	8	3
Lung	314	261	216	180	149	105	104	101	89	65	34
Prostate	93	80	67	57	48	35	35	33	26	14	5
Bladder	209	177	150	127	108	79	79	76	66	47	23
Other	1123	672	503	394	312	198	172	140	98	57	23
Thyroid	115	76	50	33	21	9	3	1	0.3	0.1	0.0
All solid	2326	1667	1325	1076	881	602	564	507	407	270	126
Leukaemia	237	149	120	105	96	84	84	84	82	73	48
All cancers	2563	1816	1445	1182	977	686	648	591	489	343	174
Females											
Stomach	101	85	72	61	52	36	35	32	27	19	11
Colon	220	187	158	134	114	82	79	73	62	45	23
Liver	28	23	20	16	14	10	10	9	7	5	2
Lung	733	608	504	417	346	242	240	230	201	147	77
Breast	1171	914	712	553	429	253	141	70	31	12	4
Uterus	50	42	36	30	26	18	16	13	9	5	2
Ovary	104	87	73	60	50	34	31	25	18	11	5
Bladder	212	180	152	129	109	79	78	74	64	47	24
Other	1339	719	523	409	323	207	181	148	109	68	30
Thyroid	634	419	275	178	113	41	14	4	1	0.3	0.0
All solid	4592	3265	2525	1988	1575	1002	824	678	529	358	177
Leukaemia	185	112	86	76	71	63	62	62	57	51	37
All cancers	4777	3377	2611	2064	1646	1065	886	740	586	409	214

[a] Number of cases per 100 000 persons exposed to a single dose of 0.1 Gy.

Note: These estimates are obtained as combined estimates based on relative and absolute risk transport and have been adjusted by a DDREF of 1.5, except for leukaemia, which is based on a linear quadratic model.

TABLE 20.2. LIFETIME ATTRIBUTABLE RISK OF CANCER MORTALITY[a] *(from Ref. [20.5])*

Cancer site	Age at exposure (years)										
	0	5	10	15	20	30	40	50	60	70	80
Males											
Stomach	41	34	30	25	21	16	15	13	11	8	4
Colon	163	139	117	99	84	61	60	57	49	36	21
Liver	44	37	31	27	23	16	16	14	12	8	4
Lung	318	264	219	182	151	107	107	104	93	71	42
Prostate	17	15	12	10	9	7	6	7	7	7	5
Bladder	45	38	32	27	23	17	17	17	17	15	10
Other	400	255	200	162	134	94	88	77	58	36	17
All solid	1028	781	641	533	444	317	310	289	246	181	102
Leukaemia	71	71	71	70	67	64	67	71	73	69	51
All cancers	1099	852	712	603	511	381	377	360	319	250	153
Females											
Stomach	57	48	41	34	29	21	20	19	16	13	8
Colon	102	86	73	62	53	38	37	35	31	25	15
Liver	24	20	17	14	12	9	8	8	7	5	3
Lung	643	534	442	367	305	213	212	204	183	140	81
Breast	274	214	167	130	101	61	35	19	9	5	2
Uterus	11	10	8	7	6	4	4	3	3	2	1
Ovary	55	47	39	34	28	20	20	18	15	10	5
Bladder	59	51	43	36	31	23	23	22	22	19	13
Other	491	287	220	179	147	103	97	86	69	47	24
All solid	1717	1295	1051	862	711	491	455	415	354	265	152
Leukaemia	53	52	53	52	51	51	52	54	55	52	38
All cancers	1770	1347	1104	914	762	542	507	469	409	317	190

[a] Number of cases per 100 000 persons exposed to a single dose of 0.1 Gy.

Note: These estimates are obtained as combined estimates based on relative and absolute risk transport and have been adjusted by a DDREF of 1.5, except for leukaemia, which is based on a linear quadratic model.

20.11. RADIATION INJURY TO TISSUES (DETERMINISTIC)

20.11.1. Tissue and organ anatomy

Tissues and organs in the human body are composed of many different cells. The majority of cells in tissues and organs are differentiated and have developed a specific morphology and function. In many tissues and organs, but not all, the rate of death of differentiated cells is balanced by renewal from a 'pool' of tissue stem cells in order to maintain a healthy state and function.

Since radiation may lead to sterilization of dividing cells, in particular tissue stem cells, terminally differentiated (mature) cells can no longer be replaced. Lack of replacement can eventually result in a loss of sufficient numbers of functioning cells and, as a consequence, a loss of tissue and/or organ integrity and function.

Tissue and organ reactions are generally classified under deterministic effects. Above a certain threshold (sufficient number of cells sterilized by radiation), the severity of the effect will increase steeply with increasing radiation dose.

20.11.2. Expression and measurement of damage

Detailed knowledge about radiation induced normal tissue effects comes primarily from experience with patients receiving radiotherapy, radiation accidents and laboratory studies, mainly with rodents. The radiosensitivity of the cells of a number of normal tissues can be determined directly using in situ assays in the laboratory. Considerable variability in sensitivity is apparent within and between the different tissues and organs.

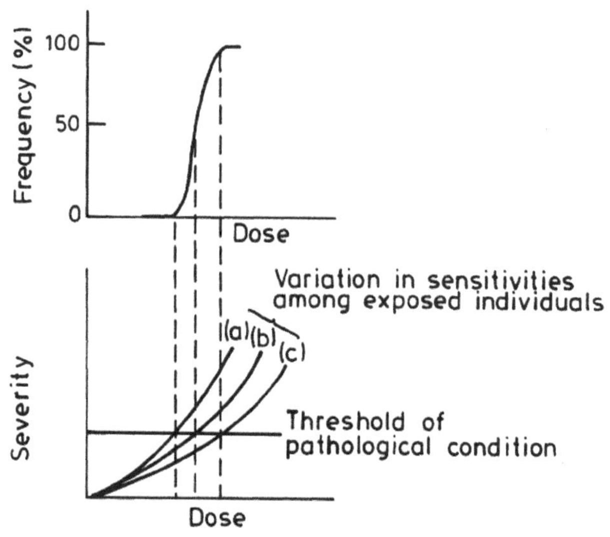

FIG. 20.5. Relationships between dose and the frequency and severity of tissue reactions (deterministic effects). Upper panel: expected sigmoidal increase in frequency in a population of individuals with varying sensitivities. Lower panel: expected dose–severity relationships for three individuals with different sensitivities (from Ref. [20.6]).

For the study of the response of individual organs, one widely used approach is to define a level of functional deficit and to determine the percentage of irradiated individuals (or animals in laboratory studies) that express at least this level of damage following different radiation doses. This approach results in sigmoidal dose–response curves (Fig. 20.5). Dose–response relationships for normal tissues are generally quite steep and well defined. In any exposed population, there is individual variation in radiosensitivity, which is influenced by several factors, including the hormonal status, age and state of health of the individuals.

20.12. RADIATION PATHOLOGY: ACUTE AND LATE EFFECTS

20.12.1. Acute and late responding normal tissues

Radiation induced cell death in normal tissues generally occurs when the cells attempt to divide (mitosis). Tissue tends to respond on a timescale similar to that of the normal rate of loss of functional cells in the tissue.

Traditionally, the effects of radiation on normal tissues have been divided into early (or acute) and late responses, based largely on functional and histopathological end points. Acute responses may manifest clinical symptoms within a few weeks of radiation exposure, while for late responses, clinical symptoms may take many months or years to develop.

20.12.2. Pathogenesis of acute and late effects

Acute responses occur primarily in tissues with rapid cell renewal, where cell division is required to maintain the function of the organ. As many cells express radiation damage during mitosis, there is early death and loss of cells by radiation.

Examples of early responding tissues are bone marrow, gastrointestinal tract and skin. In these tissues, the acute radiation responses have been related to death of critical cell populations, such as stem cells in the crypts of the small intestine, stem cells in the bone marrow, or the cells in the basal layer of the skin.

Radiation induced apoptosis has also been detected in many cells and tissues, such as lymphoid, thymic and hematopoietic cells, spermatogonia, salivary gland and intestinal crypts. In lymphoid and myeloid tissue, a substantial fraction of the functional cells can die by apoptosis and, thus, this mode of death plays an important role in the temporal response of these tissues to irradiation. In the crypts of the small bowel, a fraction of stem cells die by apoptosis, while other cells die via a mitosis linked death.

Late responses tend to occur in organs whose parenchymal cells divide infrequently (e.g. liver or kidney) or rarely (e.g. central nervous system or muscle) under normal conditions. Depletion of the parenchymal cell population owing to entry of cells into mitosis, with the resulting expression of radiation damage and cell death, will, thus, be slow.

Late tissue responses also occur in tissues that manifest early reactions, such as skin/subcutaneous tissue and intestine, but the nature of these reactions (subcutaneous fibrosis, intestinal stenosis) is quite different from the early reactions in these tissues. One common late reaction is the slow development of tissue fibrosis and vascular damage that occurs in many tissues and is often seen in cancer patients a number of years after radiation treatment.

20.12.3. Radiation induced skin reaction

The skin consists of a relatively thin epidermis and the underlying dermis. The epidermis renews rapidly, i.e. within 15–30 d, by terminal differentiation of keratinocytes from the basal layer. The dermis is much thicker, is highly vascularized and contains connective tissue and accessory structures such as hair follicles, sweat glands, sebaceous glands, nerve endings, etc.

A wide variety of expressions of radiation induced skin effects have been described (Table 20.3). Early transient erythema, similar to sunburn, occurs a few hours after irradiation and reaches a peak value within 24 h. The early erythema is believed to be related to the release of 5-hydroxytryptamine by mast cells, increasing vascular permeability. Similar mechanisms may lead to the early nausea and vomiting observed following irradiation of the intestine.

A second and more severe erythema develops after a latent period of 8–10 d. It is bright red in colour, limited to the radiation field, and is accompanied by a sensation of heat and itching. This second wave of erythema is mainly due to an inflammatory reaction of the skin.

Expression of further acute skin reactions (moist desquamation and ulceration) depends on the relative rates of cell loss and cell proliferation of the basal cells, and they occur more rapidly in murine (7–10 d) than in human skin (2–4 weeks). The extent of these reactions and the length of time for recovery depend on the dose received and the volume (area) of skin irradiated, because early recovery depends on the number of surviving basal cells that are needed to repopulate the tissue.

TABLE 20.3. SKIN EFFECTS AFTER A SINGLE EXPOSURE
(data from Ref. [20.7])

Effect	Acute exposure threshold (Gy)	Onset	Peak
Temporary epilation	3	~3 weeks	
Permanent epilation	7	~3 weeks	
Early transient	2	~ hours	~24 h
Erythema:			
Main erythema	6	~10 d	~2 weeks
Dry desquamation	10	~4 weeks	~5 weeks
Moist desquamation	15	~4 weeks	~5 weeks
Secondary ulceration	20	>6 weeks	
Late erythema	15	~6–10 weeks	
Dermal necrosis	18	>10 weeks	
Telangiectasia	12	> 52 weeks	

Transient erythema in human skin occurs after single doses greater than 2 Gy, main erythema occurs at doses greater than about 7 Gy, while moist desquamation (Fig. 20.6) and ulceration occur after single doses of 15–20 Gy (Table 20.3). After the desquamation reaches a peak value, recovery and regeneration of the epidermis will start from islands of surviving cells in the basal layer.

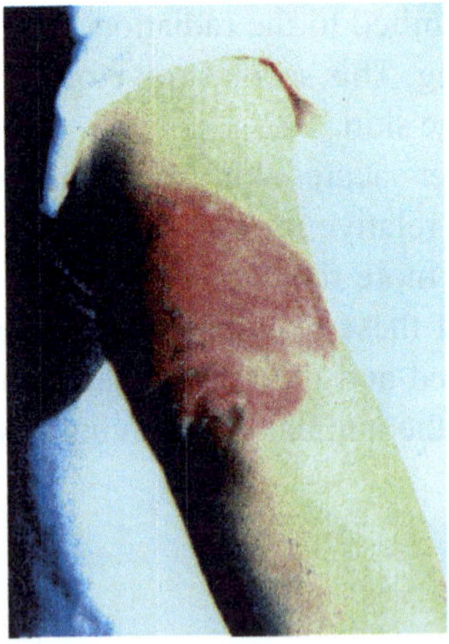

FIG. 20.6. Demarcated erythema above the right elbow at 3 weeks after radiofrequency cardiac catheter ablation (from Ref. [20.8]).

20.12.4. Radiation induced cataract formation

The lens of the eye contains transparent lens fibres and a small number of dividing cells limited to the pre-equatorial region of the epithelium within the lens capsule. During life, the progeny of these mitotic cells differentiate into lens fibres and accrete at the equator.

It has been known for many years that the lens of the eye is very sensitive to radiation. Radiation may even lead to total blindness. If dividing epithelium is injured by radiation, opacity (spots or cloudiness) of the lens (cataract) will develop because there is no mechanism for removal of injured cells and abnormal fibres.

Moderate doses of radiation can produce cataracts or lens opacities in a few individuals, with the incidence increasing to 100% in individuals exposed to a single dose of 2 Gy or higher. The frequency of cataracts varies with exposure to chronic and acute doses, with chronic doses producing a lower frequency of cataracts than acute doses. The time period between exposure and the appearance of lens opacities (cataract) might vary between about 6 months and 30 years. The radiation dose greatly influences the latent period. In general, it can be stated that the higher the dose, the shorter the latent period.

20.13. RADIATION GENETICS: RADIATION EFFECTS ON FERTILITY

20.13.1. Target cells for infertility

The reproductive organs (gonads) of humans are the testes (in males) and the ovaries (in females), in which the gametes are developed (spermatozoa in males and ova in females). Exposure of the gonads to radiation may lead to temporary or permanent sterility or to hereditary effects in the offspring of the exposed individuals, depending on the dose.

20.13.1.1. Effect of irradiation on spermatogenesis

The process by which male spermatogonia develop into mature spermatozoa is called spermatogenesis and starts in puberty. The initial development starts with the spermatogonial stem cells, which first proliferate to spermatogonia (types A and B), and then differentiate into spermatocytes (primary and secondary). The spermatocytes undergo meiosis to become haploid spermatids. Without further cell divisions, the spermatids differentiate into spermatozoa. The whole process takes approximately 74 d in humans.

The primary effect of radiation on the male reproductive system is damage and depopulation of the spermatogonia, eventually resulting in depletion of mature spermatozoa in the testes. The sensitivity of germ cells to a given dose of radiation is strongly related to the stage they are in at the time they are irradiated. Recovery of spermatogenesis will occur from the stem cell compartment when the exposure is below the sterilization dose. Depending on the dose, recovery to pre-irradiation levels of spermatozoa might take from 2–3 months up to several years.

20.13.1.2. Effect of irradiation on oogenesis

The process by which primary oocytes develop into ova (egg cells) is called oogenesis and starts with puberty and ends with menopause. In contrast to spermatogenesis, where new spermatozoa are produced all the time, the female can only produce a limited number of egg cells since, after the fetal stage, oocytes no longer divide. At birth, a fixed number of oocytes is present and their number diminishes steadily with age.

During development from the primary oocyte to ovum, the developing oocytes are very sensitive to radiation, while the primary oocytes and the ova are less sensitive. Maturation from primary oocytes to mature egg cells takes several months. Every month, one mature egg cell (occasionally two or three) is released during the menstrual cycle. In the case of radiation exposure of one or both of the ovaries, it is recommended to delay a wanted pregnancy by at least 6 months, because during this period the developing and more radiosensitive oocytes will have been ovulated.

20.13.2. Doses necessary for temporary and permanent sterility

In males, a dose as low as 1 Gy leads to a temporary reduction in the number of spermatozoa, while 1.5 Gy leads to temporary sterility, whereas a dose of 2 Gy results in temporary sterility lasting several years. Permanent sterility can be produced by an acute radiation dose in the moderate range (5–6 Gy). In females, radiation doses of 0.65–1.5 Gy will lead to reduced fertility. A dose greater than 6 Gy produces sterility. The ‘sterility’ dose is lower for older women who have fewer primary oocytes.

20.13.3. Genetic effects

At low doses, radiation may cause damage to the germinal cells in the gonads, which does not lead to cell death but may lead to DNA damage and, hence, to gene mutations in the exposed cells. These mutations may lead to an increase in hereditary disease in the offspring of exposed parents.

Hereditary diseases are classified into three major categories: Mendelian (mutation in a single gene), chromosomal and multifactorial. Although animal studies clearly show the hereditary effects of radiation, no hereditary effects have been observed in human populations exposed to radiation. For example, no significant increase in heritable diseases was found in a study of 70 000 children of Japanese A-bomb survivors whose parents had received a conjoint radiation dose to their gonads of 0.15 Gy on average.

On the basis of mouse data, the doubling dose (dose necessary to double the spontaneous mutation frequency) for low dose rate exposures is estimated to be 1 Gy. There is no good reason to assume that in humans the doubling dose may differ significantly from that in mice. The mutation doubling dose, however, does not give any useful information about the risk of heritable disease. Therefore, the mouse doubling dose is combined with information derived from human population genetics to estimate the risk of heritable disease in the progeny of irradiated individuals.

For protection purposes, the ICRP recommends a risk factor for hereditary disease of 0.2% per sievert for members of the public and 0.1% per sievert for workers. The lower risk factor for workers than for the whole population is because of the difference in age structure of the two groups.

20.14. FETAL IRRADIATION

20.14.1. Fetal irradiation: Effects versus gestational date

Radiation induced lethality and specific gross abnormalities in the embryo and fetus are dependent on two factors: the radiation dose and the stage of development at the time of exposure.

Between conception and birth, the fetus passes through three basic stages of development: (i) pre-implantation (day 1 to 10), (ii) organogenesis (day 11 to 42) and (iii) growth stage (day 43 to birth). The principal effects of radiation on a fetus are fetal or neonatal death, malformations, growth retardation, congenital defects and cancer induction. Embryos in the pre-implantation stage are very radiosensitive and radiation damage will inevitably lead to death of the conceptus and early spontaneous abortion. However, those embryos that survive this stage develop normally.

In the human early fetus, radiation exposure during the period of major organogenesis will lead to the development of abnormalities, mostly related to the central nervous system (brain defects and/or mental retardation), the skeleton and the organ systems. However, in most cases, the damage to the fetus is too severe for survival, ultimately resulting in neonatal death. During this period, the developing brain is very sensitive to radiation. Irradiation during the fetal period (after week 6) results in a much lower incidence of gross organ malformation abnormalities and mental retardation.

20.14.2. What to do when the fetus has been exposed to radiation

Systematic studies of the effect of radiation on the developing embryo have been conducted on laboratory animals, particularly mice and rats. In experimental studies, no damage to the intrauterine development of animals has been found for doses below 100 mGy. Additionally, in the studies of the Hiroshima children, there is evidence of a threshold dose of >100 mGy. After high doses, the risk of severe mental retardation increases rapidly to a value of 40% at 1 Gy. In the later stages of pregnancy, the threshold dose may be higher.

The findings of a probable threshold of 100 mGy will influence the advice to be given to pregnant women after a diagnostic radiology procedure. After abdominal computed tomography investigations in particular, careful analysis of the radiation dose to the uterus, as well as medical anamnestic exploration, has to be performed.

According to the ICRP [20.9], termination of pregnancy at fetal doses of less than 100 mGy is not justified based upon radiation risk. At fetal doses between 100 and 500 mGy, the decision should be based upon the individual circumstances. The issue of pregnancy termination is undoubtedly managed differently around the world. It is complicated by individual ethical, moral and religious beliefs, as well as perhaps laws or regulations at a local or national level. This complicated issue involves much more than radiation protection considerations and requires the provision of counselling for the patient and her partner. At fetal doses in excess of 500 mGy, there can be significant fetal damage, the magnitude and type of which is a function of the dose and stage of pregnancy. However, one should bear in mind that in a pregnant population *not* exposed to radiation there is always a certain risk of: (i) spontaneous abortion (>15%), (ii) intrauterine growth retardation (~4%), (iii) genetic abnormalities (4–10%) and (iv) major malformation (2–4%).

Regarding the radiation induced risk of cancer, the ICRP [20.6] considers that the lifetime cancer risk following in utero exposure will be similar to that following radiation in early childhood, i.e. at most about three times that of the population as a whole (>15% per sievert). So far, no effect of gestational date on cancer risk has been found.

REFERENCES

[20.1] UNITED NATIONS, Sources and Effects of Ionizing Radiation, Report 2008, Scientific Committee on the Effects of Atomic Radiation (UNSCEAR), UN, New York (2010).

[20.2] TRAVIS, E.L., Primer of Medicine Radiology, 2nd edn, Elsevier Health Sciences Division, Philadelphia, PA (1989).

[20.3] HALL, E., GIACCIA, A.J., Radiobiology for the Radiologist, 6th edn, Lippincott Williams & Wilkins, Philadelphia, PA (2006).

[20.4] NTERNATIONAL COMMISSION ON RADIOLOGICAL PROTECTION, Recommendations of the International Commission on Radiological Protection, Publication 60, Pergamon Press, Oxford and New York (1990).

[20.5] NATIONAL RESEARCH COUNCIL, BEIR VII: Health Risks from Exposure to Low Levels of Ionizing Radiation, The National Academies Press, Washington, DC (2006).

[20.6] INTERNATIONAL COMMISSION ON RADIOLOGICAL PROTECTION, The 2007 Recommendations of the ICRP, Publication 103, Elsevier (2008).

[20.7] WAGNER, L.K., EIFEL, P.J., GEISE, R.A., Potential biological effects following high X-ray dose interventional procedures, J. Vasc. Interv. Radiol. **5** 1 (1994) 71–84.

[20.8] KOENIG, T.R., WOLFF, D., METTLER, F.A., WAGNER, L.K., Skin injuries from fluoroscopically guided procedures: Part 1, Characteristics of radiation injury, Am. J. Roentgenol. **177** 1 (2001) 3–11.

[20.9] INTERNATIONAL COMMISSION ON RADIOLOGICAL PROTECTION, Pregnancy and Medical Radiation, Publication 84, Elsevier (2000).

BIBLIOGRAPHY

BALTER, S., HOPEWELL, J.W., MILLER, D.L., WAGNER, L.K., ZELEFSKY, M.J., Fluoroscopically guided interventional procedures: A review of radiation effects on patients' skin and hair, Radiology **254** 2 (2010) 326–341.

INTERNATIONAL ATOMIC ENERGY AGENCY, Radiation Oncology Physics: A Handbook for Teachers and Students, IAEA, Vienna (2005).

INTERNATIONAL ATOMIC ENERGY AGENCY, Radiation Biology: A Handbook for Teachers and Students, Training Course Series No. 42, IAEA, Vienna (2010).

INTERNATIONAL ATOMIC ENERGY AGENCY, Radiobiology Modules in Applied Sciences of Oncology — Distance Learning Course in Radiation Oncology for Cancer Treatment,
http://www.iaea.org/NewsCenter/News/2010/aso.html (accessed on 23 August 2012).

JOINER, M.C., VAN DER KOGEL, A.J. (Eds), Basic Clinical Radiobiology, 4th edn, Hodder Arnold, London, (2009).

NATIONAL RESEARCH COUNCIL OF THE NATIONAL ACADEMIES, Health Risks from Exposure to Low Levels of Ionizing Radiation: BEIR VII phase 2, Committee to Assess Health Risks from Exposure to Low Levels of Ionizing Radiation, The National Academies Press, Washington, DC (2006), http://www.nap.edu/openbook.php?isbn=030909156X (accessed on 23 August 2012).

STECKER, M.S., et al., Guidelines for patient radiation dose management, J. Vasc. Interv. Radiol. **20** 7 Suppl. (2009) S263–S273.

TANNOCK, I.F., HILL, R.P., BRISTOW, R.G., HARRINGTON, L. (Eds), The Basic Science of Oncology, 4th edn, McGraw Hill, Philadelphia, PA (2005) Ch. 14 and 15.

Chapter 21

INSTRUMENTATION FOR DOSIMETRY

J.C. HOURDAKIS
Greek Atomic Energy Commission,
Athens, Greece

R. NOWOTNY
Medical University of Vienna,
Vienna, Austria

21.1. INTRODUCTION

Measurements of absorbed dose (or air kerma) are required in varying situations in diagnostic radiology. The radiation fields vary from plain, slit and even point projection geometry, and may be stationary or moving, including rotational. Owing to the use of low photon energies for these fields, it is important that dosimeters have a satisfactory energy response. In general, the requirements for dosimeter accuracy are less stringent than those in radiation therapy; however, the dose and dose rate measurements cover a large range.

Patient dosimetry (see Chapter 22) is a primary responsibility of the medical physicist specializing in diagnostic radiology and is required by legislation in many countries. Dose data are also required in the optimization of examinations for image quality and dose. Radiation measurement is also critical for occupational and public exposure control (see Chapter 24). Dose measurements are essential in acceptance testing and quality control (see Chapter 19).

Several types of dosimeter can be used, provided that they have a suitable energy response, but typically, ionization chambers of a few cubic centimetres in volume, or solid state detectors specifically designed for such measurements, are used. If dosimeters are used to make measurements during an examination, they must not interfere with the examination. These devices are also used for determination of the half value layer (HVL). Special types of ionization chamber are employed for computed tomography (CT), mammography and interventional radiology dosimetry.

21.2. RADIATION DETECTORS AND DOSIMETERS

21.2.1. General characteristics of radiation detectors

A dosimeter is an instrument that measures ionizing radiation. It usually comprises a measuring assembly, often referred to as an electrometer, and one or more detector assemblies, which may or may not be an integral part of the measuring assembly. In diagnostic radiology, dosimetric instruments can be classified as either active or passive dosimeters. Active dosimeters display the dose value directly and include ionization chambers and/or semiconductor detectors[1] (sometimes loosely referred to as solid state detectors) used to measure air kerma (K_i), air kerma rate ($\dot{K}_i$), air kerma–length product (P_{KL}) and air kerma–area product (P_{KA}) in the primary beam conditions.

Measurements involving scatter radiation, such as patient exit dose measurements and CT phantom measurements, are also performed with ionization chambers; however, care must be taken if attempting to use semiconductor detectors for this purpose. Passive dosimeters cannot display the dose value directly, but record a dose signal when exposed to radiation, which must be subsequently retrieved and converted to dose (or air kerma) by a reading device. These include solid state devices such as thermoluminescent dosimeters (TLDs), optically stimulated luminescent (OSL) dosimeters and film dosimeters (including radiochromic film) that may be placed on a patient's skin or inside cavities to measure the skin or organ doses. Similar measurements can be performed in phantoms.

Other instruments are needed to measure the X ray tube voltage and exposure time that can be used without direct connection into the electrical circuits of the X ray units. These non-invasive instruments are often called kV meters and timers. There are also a variety of devices used for occupational and public dose assessment, including ionization chambers for direct measurements and TLD, OSL and film for indirect use, as either personal dosimeters or area monitors (see Chapter 24).

21.2.2. Properties of diagnostic radiology dosimeters

Many types of diagnostic radiology dosimeter are commercially available for the measurement of air kerma (and its derivatives). They incorporate either ionization chambers or semiconductor detectors. Although ionization chambers have been the standard instruments for diagnostic radiology dosimetry for many years, semiconductor detectors have found widespread use recently in the area

[1] Detectors need to conform to the IEC-61674 standard [21.1].

of quality control measurements, mainly because of their small size, ruggedness and convenience of use. The measurement assembly analyses and processes the electrical signals from the detector, in order to display the value of the radiological quantity being measured (K, $\dot{K}_{\mathrm{i}}$, P_{KL}, P_{KA}) and its units, i.e. Gy, Gy/s, Gy·m or Gy·m^2, with SI subunit prefixes, e.g. m, μ or n. When an ionization chamber is used, the electrometer provides the appropriate polarizing voltage.

Most commercial dosimeters can be used for both radiographic and fluoroscopic applications, using either the accumulated air kerma over time (integrate mode) or air kerma rate mode. Some commercial dosimeters automatically perform conversion and/or corrections to their reading, in order to display the actual air kerma value. In most cases, the calibration coefficient is applied through the system's software, to convert the measured charge (current) to air kerma at a given beam quality. Some dosimeter models have internal sensors for measurement of the environmental temperature and pressure, in order to perform corrections for the air density automatically.

The air kerma, K (or any other associate dosimetric quantity), is obtained from:

$$K = M_Q k_{\mathrm{TP}} N_{K,Q_0} k_Q \prod k_j \tag{21.1}$$

where the dosimeter's reading M_Q is corrected for air density by k_{TP}, converted to air kerma at an appropriate reference radiation quality by the N_{K,Q_0} calibration coefficient and corrected for the applied X ray spectrum by the k_Q factor. Further corrections for other influencing quantities may be applied by multiplication factors k_j, for example, corrections for ion recombination, polarizing voltage, radiation incident angle or humidity (see Section 21.6).

Since the dosimeters are used for various types of X ray unit and exposure conditions, the choice of the appropriate instrument is important, in order for the radiation measurement to be sufficiently accurate. Irrespective of the application, radiation dosimeters must exhibit several desirable properties, as discussed below.

21.2.2.1. Sensitivity

Sensitivity is the minimum air kerma required to produce a signal output (charge or current produced by the detector and collected by the measuring assembly). The better the sensitivity of the dosimeter, the higher the charge (or current) produced for the same air kerma (rate) and consequently the better the air kerma (rate) resolution and detectability. Ionization chambers with larger active (effective) volumes exhibit higher sensitivity than those with smaller volumes.

For this reason, large ionization chambers are preferred for low air kerma rate measurements, such as in fluoroscopy or for scattered radiation. In radiography, where the air kerma rates are higher, smaller chambers can be used, allowing better measurement of spatial resolution.

In general, semiconductor detectors have a sensitivity that can be orders of magnitude higher than that of ionization chambers. This property, among others, makes the use of these detectors advantageous for a wide range of applications. However, their intrinsic energy dependence makes their use problematic in non-calibrated beams and for scatter radiation measurements.

21.2.2.2. Linearity

The dosimeter reading M should be linearly proportional to the air kerma (rate). All dosimeters exhibit a linear response for a certain range of air kerma (rate). The linearity range and the non-linear behaviour depend on the type of dosimeter and its physical characteristics. Among other factors, the scale/reading resolution of the measuring assembly, the sensitivity and the leakage/dark current of the dosimeter restrict the rated range to a lower value, while saturation (over ranging) effects determine the upper value. The air kerma (rate) range in which the dosimeter performance is linear (rated range) should be stated by the manufacturer; the linearity of the dosimeter over this range should be tested by the user.

According to IEC-61674 standard [21.1], the non-linearity of a dosimeter is expressed by the ratio $(R_{max} - R_{min})/(R_{max} + R_{min})$, which should be less than 0.02 over the whole rated range of air kerma (rate). Values R_{max} and R_{min} are the maximum and minimum dosimeter response, respectively, over the rated range of air kerma (rate). The response is the quotient of the indicated value (dosimeter reading) to the true value of air kerma (rate).

21.2.2.3. Energy dependence

For diagnostic dosimeters, the X ray spectrum (often referred to as the radiation or beam quality) is specified by the beam HVL and is one of the important quantities affecting the response of a dosimeter. Within the range of the clinical X ray radiation qualities (25–150 kV), the variation in the dosimeter response with energy may be significant. This depends on the detector type and its physical and structural properties. The variation in response to different radiation qualities is taken into account by the use of a beam quality correction factor k_Q (see Eq. (21.1)). For a radiation quality Q, k_Q is the ratio of the calibration factors for quality Q to the reference radiation quality (RQR 5, for example, see Section 21.6). By definition, k_Q is unity at the reference beam quality. Figure 21.1

shows the variation in k_Q with HVL for six commercial dosimeters, including both ionization chambers and semiconductor detectors. Simple semiconductor detectors generally have more pronounced variation of k_Q with energy; however, modern semiconductor detectors incorporate multiple semiconductor elements covered by filters (typically copper) that allow the necessary compensation to reduce the effect of radiation quality. The IEC-61674 standard [21.1] imposes a ±5% upper limit on variation of energy response in the 50–150 kV range, while the IAEA [21.2] proposes the stricter limit of ±2.6% for dosimeters used as reference instruments at calibration laboratories.

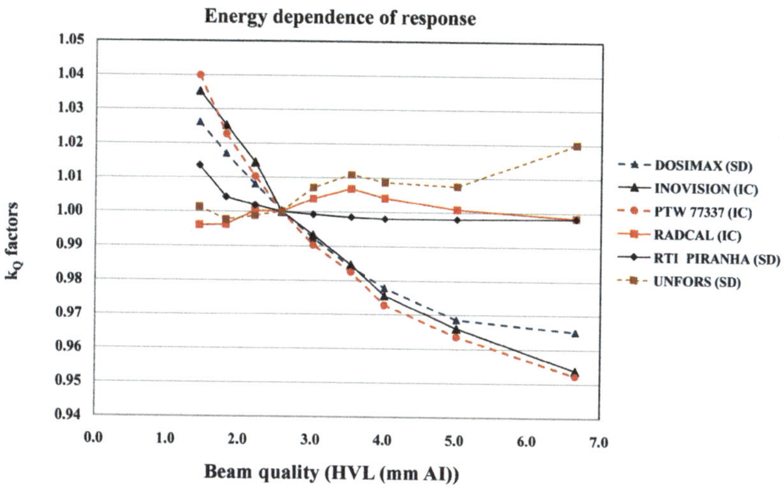

FIG. 21.1. Energy dependence of response of six commercial dosimeters incorporating ionization chambers (IC) or solid state detectors (SD): Dosimax plus (IBA, Schwarzenbruck, Sweden), Inovision Triad 35050A (FLUKE Biomedical Corp., Everett, WA, USA), PTW Unidos with 77337 (PTW GmbH, Freiburg, Germany), Radcal 2025 with 20x5-6 (Radcal Corporation Monrovia, CA, USA), RTI Piranha (RTI Electronics AB, Molndal, Sweden) and Unfors Xi (Unfors Instruments AB, Billdal, Sweden). The beam qualities (x axis) correspond to the RQR series described in the IEC-61674 standard [21.1].

21.2.2.4. Directional dependence

The response of a dosimeter may vary when the radiation is incident on the detector from different angles. The directional or angular dependence primarily depends on detector construction and physical size but will also depend on the energy of the incident radiation. The directional dependence of cylindrical or spherical ionization chambers is negligible, while parallel plate chambers might exhibit significant dependence at large incident angles. Most commercial semiconductor detectors are mounted on lead backing plates, to attenuate radiation incident from the rear, while some models incorporate several semiconductor elements covered with filters to attenuate the radiation. In such cases, the directional dependence is important and care should always be taken to ensure that the radiation is incident on the elements through the filters at right angles. The IEC-61674 standard [21.1] imposes a ±3% upper limit of variation of response at incident angles of ±5° from the normal direction.

21.2.2.5. Leakage current

Leakage current refers to any signal change recorded by the measuring assembly that is not generated by radiation. This could be electronic noise, current from resistor–capacitor circuits, damaged cables or bad cable connections, lack of electronic or environmental equilibrium or humidity, etc. According to the IEC-61674 standard [21.1], the leakage current shall not exceed 5% of the minimum effective air kerma rate for the range in use. When a dosimeter is left in measurement mode after being exposed to the maximum effective air kerma value, the indicated value shall not change by more than 1% per minute.

21.3. IONIZATION CHAMBERS

The ionization detector is an air filled chamber in which an electric field is formed by the application of a polarizing voltage across two electrodes to collect all charges liberated by the ionization of the air contained within the chamber. The electric field is sufficient to collect almost all of the liberated charges that reach the electrodes (i.e. there is very little recombination) but insufficient to induce gas/charge multiplication and collision ionization of other molecules (in contrast with Geiger Müller and proportional counters). The number of ions collected, or the rate of their collection, is the recorded signal, which is multiplied by the mean energy required to produce an ion pair in dry air, $\bar{W}_{\text{air}} = 33.97\,\text{eV/ion pair} = 33.97\,\text{J/C}$ (see Eq. (3.12) and Sections 3.2.2 and 3.2.4 to deduce the energy transferred (ε_{tr}) from the radiation to the mass of air in the chamber). The ratio of ε_{tr} and the mass of air corresponds to the air kerma (rate) (Eq. (3.3)).

Figure 21.2 shows the internal structure of typical ionization chambers. In parallel plate chambers, the electrode separation is of the order of 1 cm and the electrodes are parallel to each other and to the entrance window. In cylindrical and spherical shaped chambers, the central electrode stands at the geometrical centre of the cavity, while the wall (outer shell) of the chamber is coated by a conductive material, which is often at ground potential (ground electrode). The wall (ground) and the collecting electrode are separated by a high quality insulator to reduce the leakage current. A third electrode, the guard, reduces chamber leakage current by allowing any leakage to flow to ground, bypassing the collecting electrode and ensuring high uniformity of the electrical field in the chamber volume.

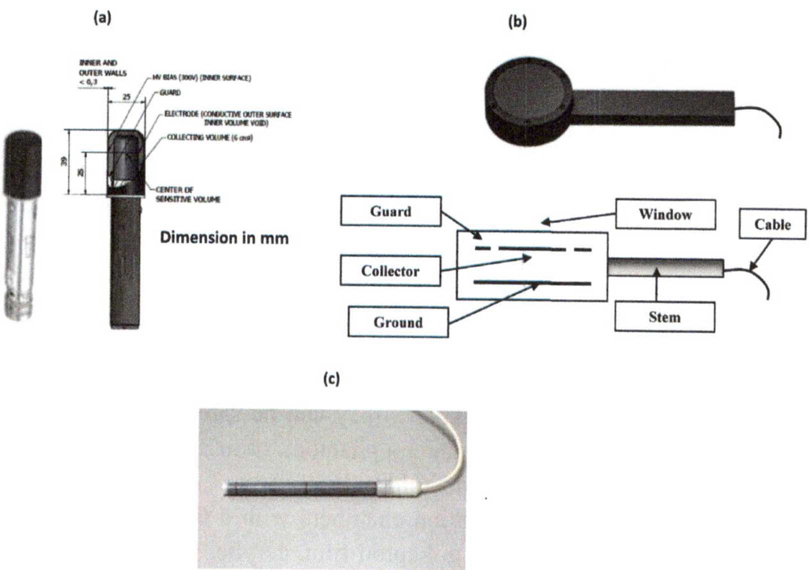

FIG. 21.2. Schematic diagram of (a) specialized chamber design (from http://www.radcal.com); (b) parallel plate (from http://www.standardimaging.com and Ref. [21.2]); and (c) cylindrical pencil type ionization chambers (from Ref. [21.2]).

Current high performance ionization chambers used in diagnostic radiology can have a more complex design, with the principle of keeping the gap between the ground and collecting electrodes small to prevent ion recombination at high dose being a primary consideration.

Ionization chambers used in diagnostic radiology should be vented, i.e. the air inside the volume communicates with the environment, rendering the mass of air dependent on temperature, pressure and humidity conditions. Humidity has an insignificant effect on air mass changes, but temperature and pressure affect the air mass within the chamber significantly. Therefore, the air density correction factor, k_{TP}, should always be applied to the dosimeter's readings. This factor is calculated from the formula $k_{TP} = (P_0 \cdot T)/(P \cdot T_0)$; where P_0 and T_0 are the values of the calibration reference conditions for pressure and temperature (usually 101.3 kPa (1 atm) and 293.2 K (273.2 + 20°C) or 295.2 K (273.2 + 22°C)), and P and T are the ambient pressure (kPa) and temperature (K) during the air kerma measurement. According to the IEC-61674 standard [21.1], sealed chambers, in which the air volume does not change, are not suitable for diagnostic radiology dosimetry; their necessary wall thickness may cause unacceptable energy dependence, while the long term stability of the chambers is not guaranteed.

21.3.1. Clinical application of ionization chambers

21.3.1.1. Chambers for air kerma (dose) measurements

Determination of the air kerma (dose) in common diagnostic radiology applications (radiography, fluoroscopy and mammography) is performed by ionization chambers, either cylindrical or parallel plate. There is a large variety of chamber types and vendors.

Commercial parallel plate (p-p) chambers are disc shaped, with a diameter of several centimetres and a thickness of a few centimetres. The most common chambers with effective volumes (air cavity) from about 1 cm^3 to several hundreds of cubic centimetres are then suitable for application over a wide range of exposure rates. Owing to their shape, they can be safely inserted in hollow spaces, such as on the X ray table under a phantom, or in contact with the image intensifier, or inside the film cassette holder (Bucky), etc.

In mammography, p-p ionization chambers with a thin entrance window, made of a low density material (e.g. kapton film, acrylic, mylar) of micrometre thickness (20–50 μm, 3–10 mg/cm^2), are used. The major disadvantage of p-p chambers is the directional dependence of their response. The p-p chamber should always be placed perpendicular to the radiation beam.

Cylindrical chambers are uniformly sensitive around their central geometrical axis. The chambers used for measurement in the X ray beam have an effective volume of a few cubic centimetres (3–6 cm^3).

21.3.1.2. Cylindrical pencil type chambers

Cylindrical pencil type ionization chambers are used in several diagnostic radiology applications for the measurement of the air kerma length product, P_{KL}. For the last few decades, these chambers have mainly been used in CT dosimetry (see Section 22.4.7), but they are also used in dental applications (see Section 22.4.8). This chamber type is a long cylinder with a typical effective active length of 100 mm. The physical dimensions are about 15 cm in length and 1 cm in diameter (Fig. 21.2(c)). In contrast to other detectors used in diagnostic radiology, the chamber is partially irradiated. It is positioned with its axis at right angles to the central beam axis. The response of the active volume should be uniform along its entire axial length. Special procedures and radiation qualities are used for their calibration.

21.3.1.3. KAP chambers

Air kerma area product (KAP) chambers have a large surface area and are transparent to both radiation and light. They are usually mounted on the tube housing after the beam collimation (see Sections 22.4.4 and 22.4.5) and encompass the entire radiation field. KAP chambers measure the integral of the air kerma over the area of the chamber and should have a uniform response throughout their entire area. Theoretically, P_{KA} is the same along the central X ray beam; however, in practice, scatter radiation, extra focal radiation and other factors affect the measurements. The requirement for the electrodes of the chamber to be transparent to light results in the use of materials that have a significant energy dependence over the diagnostic energy range.

Depending on their use and calibration, the KAP chambers measure the incident radiation, i.e. the radiation that falls on the chamber, or the transmitted radiation, i.e. the radiation that emerges from the chamber. The latter includes the attenuation of the radiation by the KAP chamber. Special procedures and radiation qualities are applied for the calibration of KAP meters.

KAP chambers are usually used for patient dosimetry in interventional radiology, fluoroscopy and general radiography, and are beginning to be used in pantomographic dental radiography (see Section 10.2.2.2). This is reflected in the use of KAP for diagnostic reference levels. Owing to the presence of extrafocal and scatter radiation, they should be calibrated in situ.

21.3.2. Application hints for ionization chambers

The practical points listed below should be considered:

- Appropriate ionization chambers should be selected for the application and the measuring procedure required (Table 21.1).
- Corrections for air density should always be applied to the dosimeter reading. Great care should be taken with dosimeters that incorporate internal sensors for automatic temperature and/or pressure corrections, in order to interpret the reading correctly.
- In general, ionization chambers detect radiations from all directions; thus, they measure all scatter, extrafocal and leakage radiation. When the incident air kerma is being measured, the chamber should be at a distance from the X ray couch or other supporting devices, in order to avoid backscatter

TABLE 21.1. BASIC CHARACTERISTICS OF DIAGNOSTIC RADIOLOGY DOSIMETERS

Application	Type of detector	Range of X ray tube voltage (kV)	Range of air kerma or air kerma rate	Intrinsic error (%)	Variation of energy response (%)	K rate dependence (%)	Angular dependence (%)
General radiography	Cylindrical, spherical or plane parallel IC[a] Solid state detectors	60–150	10 μGy to 1 Gy 1 mGy/s to 500 mGy/s[b] 10 mGy/s to 5 mGy/s[b]	5	±5	±2	±3 @ ±5°
Fluoroscopy, interventional radiology[c]	Plane parallel IC[a] Solid state detectors	50–120	10 μGy/s to 10 mGy/s[b,d] 0.1 μGy/s to 100 μGy/s[b,d]	5	±5	±2	±3 @ ±5°
Fluoroscopy, interventional radiology[e]	KAP meters	50–150	10^{-1} to 10^{6} μGy m^2 10^{-1} to 10^{3} μGy m^2/s	10	±8	±5	—
Fluoroscopy	Plane parallel IC[a] Solid state detectors KAP meters	50–150	10 mGy/s to 10 mGy/s[a, b] 0.1 mGy/s to 100 mGy/s[b]	5	±5	±2	±3 @ ±5°
Mammography	Plane parallel IC Solid state detectors	22–40	10 μGy to 1 Gy 10 μGy/s to 10 mGy/s[b]	5	±5	±2	±3 @ ±5°
CT	Cylindrical pencil type IC[a] of 100 mm active length[f]	100–150	0.1–50 mGy/s	5	±5	±2	±3 @ ±180°
Dental radiography	Cylindrical, spherical or plane parallel IC[a] Solid state detectors KAP meters Cylindrical pencil type IC[a]	50–100	10 μGy to 100 mGy 1–10 mGy/s	5	±5	±2	±3 @ ±5°

[a] IC: ionization chamber.
[b] Unattenuated beam.
[c] For air kerma rate measurements.
[d] In the light of new CT technologies and the revision of CT dosimetry methodology, new types of detector may be proposed that will be suitable for measuring pulsed radiation as well.
[e] For air kerma area product (rate) measurements.
[f] Attenuated beam.

radiation; other objects should not be allowed to interfere with the X ray beam.

- The ionization chamber should be totally covered by the radiation field, except for pencil type and KAP chambers. It is good practice to use field sizes at least twice the detector cross-section, and to check complete coverage of the detector through imaging methods if practicable. All chambers, especially p-p chambers, should be placed perpendicular to the radiation beam axis.
- Ionization chambers should be calibrated at several qualities. This is especially important for chambers with a large energy dependence. At least the qualities RQR 3 (50 kV), RQR 5 (70 kV) and RQR 9 (120 kV) should be used for radiography and fluoroscopy, and RQR-M1 (25 kV), RQR-M2 (28 kV) and RQR-M4 (35 kV) for mammography. For CT chambers, calibration should be performed at least at RQT 9 (120 kV) (see Section 21.6.2).
- The user should know the limitations and the rated ranges of all the quantities affecting the measurements. It is important to check that the leakage (dark) current is negligible and does not affect the measurements.
- Prior to use, users should check whether the battery voltage of the measuring assembly is within the manufacturer's rated range.

21.4. SEMICONDUCTOR DOSIMETERS

Diagnostic radiology dosimeters based on semiconductor technology have found widespread use. Two types are used: silicon diodes or metal oxide semiconductor field effect transistors (MOSFETs). Owing to their small size and rigidity, they are convenient for use in many applications.

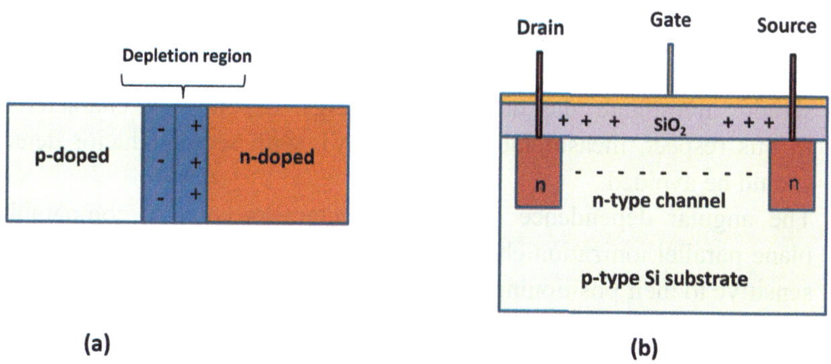

FIG. 21.3. Cross-sectional diagram of (a) p–n junction and (b) MOSFET.

21.4.1. Theory of operation

A silicon diode dosimeter is a p–n junction diode. In most cases, p type (rather than n type) diodes are used for diagnostic radiology dosimeters, since they are less affected by radiation damage and have a much smaller dark current (noise). When radiation falls on the diode, it produces electron hole pairs in the body of the diode and a current is generated in the reverse direction in the diode. The number of such pairs is proportional to the incident radiation dose. Owing to the diode structure and the intrinsically formed potential difference, there is no need to apply a bias voltage across the p and n type diode regions to collect the charge liberated by the radiation.

A MOSFET is a miniature silicon transistor. Its structure is equivalent to a planar capacitor with one of the electrodes replaced by a semiconductor. MOSFET dosimeters are based on the production of electron hole pairs in the SiO_2 of the MOSFET gate region (Fig. 21.3) resulting from incident radiation. The positive charge carriers move in the direction of the Si–SiO_2 interface, where they are trapped, building up a positive charge, which causes changes to the current in the n type channel and leads to a change of the gate bias voltage (shift in the threshold voltage). The threshold voltage shift is a linear function of absorbed dose. The integrated dose may be measured during (in real time) or after irradiation. MOSFETs often require a connection to a bias voltage during irradiation. They are mainly used in patient dosimetry.

21.4.2. Application hints for semiconductors

The practical points listed below should be considered:

- The response of semiconductors (diodes and MOSFETs) generally has a more pronounced energy dependence than that of ionization chambers. Although modern dosimeters typically use compensation methods to correct the energy dependence at specified beam qualities, the energy dependence for non-specified beam characteristics may be unpredictable. The user should investigate the dosimeter's energy dependence characteristics. In this respect, measurements of the HVL with semiconductor detectors should be avoided.
- The angular dependence of semiconductor detectors is comparable to plane parallel ionization chambers. However, semiconductor detectors are sensitive to their positioning in the X ray field, especially to the direction of the heel effect.
- When a semiconductor detector is used for dose measurements on a surface of a phantom (or patient), backscatter and sidescatter radiation may not

contribute significantly to the dosimeter reading, owing to the presence of backing plates.

- The semiconductor detector response does not depend on temperature or pressure. For the sake of a standard dosimetric formalism, k_{TP} in Eq. (21.1) is set to unity.
- Semiconductors have a limited useful life, owing to accumulated radiation damage. Although the doses measured in diagnostic radiology dosimetry are low, it is good practice to recalibrate the detectors at regular intervals.
- Research with MOSFET devices is currently in the experimental stages for dose measurements in some aspects of diagnostic radiology. These may be potentially beneficial in some high dose applications, such as interventional radiology, where high skin doses need to be avoided. However, they exhibit a high energy dependence and, therefore, frequent calibration is essential in order to achieve adequate measurement accuracy.

21.5. OTHER DOSIMETERS

21.5.1. Film dosimetry: Radiographic film and radiochromic film

21.5.1.1. Radiographic film

Radiographic film still finds application as a dosimeter in personal radiation monitoring using film badges (see Section 24.5.3). The structure of radiographic film and the basic principles of densitometry are described in Section 7.3.3. The emulsion in a film dosimeter directly absorbs ionizing radiation and can be correlated to the optical density of the developed film. However, the sensitometric curve is very different to that for screen film systems. A radiographic emulsion is far from tissue equivalent and the energy response of a film badge is, therefore, modified by addition of several filters. The provision, processing and analysis of such dosimeters are the tasks of specialized departments and companies and are not commonly within the duties of a medical physicist.

21.5.1.2. Radiochromic film

Radiochromic films (e.g. Gafchromic®) contain colourless dyes (diacetylene) that become blue after exposure because of radiation induced polymerization. This process is self-developing and requires no chemical process but needs some time for full development. Depending on the material, a density increase of about 10% from 1 to 24 h after exposure is typical.

The film comprises an active dye layer (15–20 μm thick) sandwiched between two transparent polyester sheets, each containing a yellow dye. The yellow dye enhances visual contrast and reduces the effects of exposure to blue and ultraviolet light. Some films use an opaque white backing sheet. Film optical density is measured with densitometers or film scanners. For films with an opaque backing, a reflective densitometer is needed. The blue coloured polymer exhibits a maximum in optical absorption at around 635 nm. Accordingly, a densitometer with a red light source should be used.

The composition of the film is near tissue equivalence. Some types of film incorporate barium compounds in the white backing to increase radiation absorption and sensitivity. Several types of radiochromic film are optimized for applications in diagnostic radiology. Their energy response and other properties can differ and the specifications should be collected from the supplier or from the literature. Sensitivity ranges from ~1 mGy to ~50 Gy, depending on film type. The sensitometric response is not linear and suitable calibration curves need to be applied. For film calibration and dose measurements, it is essential to use the same protocol and densitometer. The handling of radiochromic films is simple. Darkrooms are not required and ambient conditions are of little concern except for exposure to intensive light sources or humidity. The film can be obtained in large format (35 cm × 43 cm, maximum) and can be bent and cut to size as required.

Radiochromic films can be used for relative dosimetry in diagnostic radiology. The measurement and mapping of patient skin dose in interventional procedures is one such application (see Chapter 8).

21.5.2. Thermoluminescent dosimetry

A large and growing number of solid state materials exhibit the phenomenon of thermoluminescence (TL), which can be harvested for dosimetric purposes. This process consists of two stages: the first stage is the transference of an equilibrium TLD material to a metastable state through irradiation, and the second stage is application of energy (through heat) to reduce the metastable state back to equilibrium. Figure 21.4 demonstrates these two stages, using a semiconduction model for the solid state material. In this model, electron energies are not localized and the narrow energy gap between the valency and conduction bands is populated with midgap or trap sites that are caused by defects within the material.[2] Irradiation creates free electrons with enough energy to cross the gap into the conduction band, with the possibility of some of the electrons being

[2] Trap sites are usually formed by the addition of dopant material. For example, TLD100 is made of LiF with additional dopant materials of Mg and Ti.

trapped in a midgap site. By subsequently adding energy to the material, trapped electrons with sufficient energy can escape the trap site into the conduction band and might return towards the valency band to recombine with a trapped hole, accompanied by a radiative TL emission (see Chapter 7 for a discussion on photostimulable phosphors).

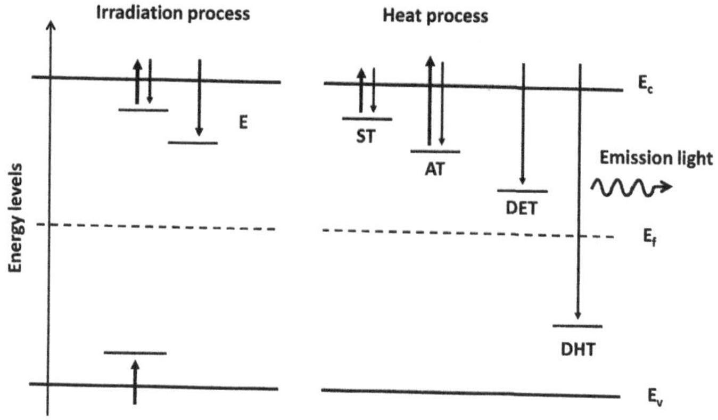

FIG. 21.4. Electron energy levels in a TLD material showing on the left the process of free electron and hole creation, followed by non-radiative charge trapping. On the right, the release of thermally stimulated electrons is shown for energy level $E_c - E$. The released electron may be retrapped or recombine with trapped holes. If this process is radiative, TL emission occurs. E_c and E_v are the conduction and valency band edges, E_f is the Fermi level, ST and AT are shallow and active traps, respectively, while DET and DHT are deep electron traps and deep hole traps, respectively.

The stability of the trapped electrons depends largely on the energy level (depth) of the traps. Shallow traps require little energy for electron release and are thermally unstable, leading to signal fading at ambient temperatures. Trap energy levels corresponding to higher excitation temperatures are desirable to obtain stable signals.

In a typical TLD reader (Fig. 21.5), the dosimeters are placed on a planchet heated directly by an electric current. The temperature is measured with a thermocouple welded to the planchet. Other methods of heating the TLD are also used, such as hot nitrogen jets, laser heating or infrared lamps. The TL signal is detected with a photomultiplier tube.

If a linear temperature ramp is applied, the TL signal (glow curve) shows various peaks at characteristic temperatures attributable to the traps present. Figure 21.6 shows a typical glow curve for LiF:Mg,Cu,P. Besides peaks at lower temperatures, the main peak useful for dosimetric measurements appears at ~210°C. Each type of TLD requires a specific optimized reading cycle.

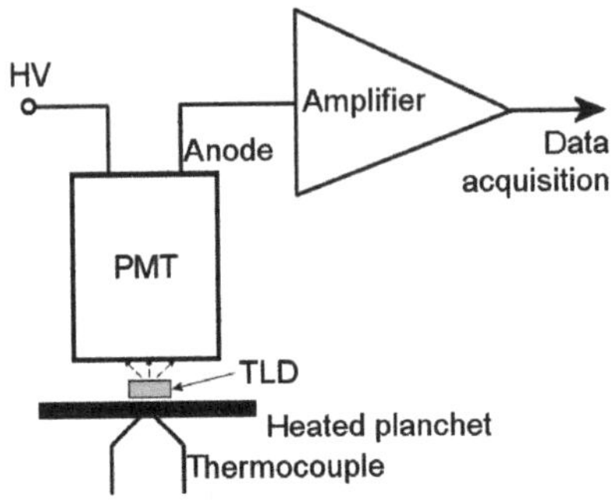

FIG. 21.5. Principal elements of a TLD reader (PMT: photomultiplier tube).

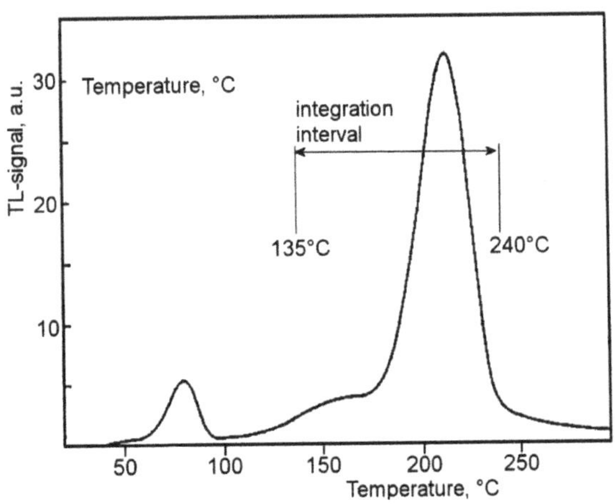

FIG. 21.6. Typical glow curve for LiF:Mg,Cu,P (a.u.: arbitrary units).

The reading cycle of a TLD is divided into preheat, signal integration and annealing. During preheat, the dosimeter is maintained for some seconds at a constant temperature sufficient to remove all low temperature signals. The temperature is then raised up to the maximum value. In that period, the TL signal from the dosimeter is integrated to give a dose relevant signal. A typical integration temperature interval is shown in Fig. 21.6. Finally, the dosimeter is annealed in a dedicated oven to remove all remaining signals, thus resetting the dosimeter to zero. The reading cycle parameters depend on the TLD material, and good reproducibility is essential to achieve accurate results using the same reader for calibration and measurements.

The commonly used LiF:Mg,Ti (e.g. TLD100) is a well standardized dosimeter but less sensitive than LiF:Mg,Cu,P (GR200, TLD100H, MCP-N), which has a detection threshold of about 0.1 μGy. TLDs are available in many forms and shapes (chips, rods, cubes, ribbons and powder). The relationship of dose to TL signal is linear up to doses of <1 Gy. For higher doses, correction factors for a non-linear response can be applied.

21.5.3. OSL

OSL is the luminescence emitted from an irradiated solid state material (OSL dosimeter) after being illuminated by stimulating light. The absorbed radiation causes ionization of the valence electrons and creation of electron pairs and leads the material to a metastable state. Pre-existing defects within the material localize the free electrons and holes through non-radiative trapping transitions. The illumination of the irradiated material with stimulating light (visible or infrared) leads to electron transition from the localized trap into the conduction band (equilibrium, ground energy level) and subsequent radiative emission and luminescence (OSL). In terms of energy levels, the two processes (excitation and transition) are quite similar to those described in Fig. 21.4. In fact, OSL is closely related to TL, with the basic difference being the use of light instead of heat as the added energy for the trapped electron. OSL should not be confused with photoluminescence, where the electron excitation is caused by light absorption rather than radiation dose.

Usually, the stimulating light used for OSL has a lower photon energy than the emitted light. The intensity of the emitted light is related to the rate at which the system returns to equilibrium, resulting in a characteristic luminescence–time curve. The integral of this curve corresponds to the trapped charged concentration, which is proportional to the radiation absorbed dose.

In a typical measurement using an OSL dosimeter, the sample material is illuminated with an appropriate light source. The emitted light is passed through an optical filter to suppress unwanted (PL) light and then detected with a photomultiplier tube. The arrangement of such a reader is similar to a TLD reader. An improvement in signal to noise ratio can be achieved by pulsing the stimulating light.

One OSL dosimeter that is commercially available uses aluminium oxide doped with carbon (Al_2O_3:C) and its dominant OSL trapping levels require thermal energies above 200°C to create thermoluminescence. Consequently, the OSL signal is thermally stable and signal fading is negligible. Some transient signals due to shallow traps will disappear after a few minutes. Dominant emission occurs in a band centred at around 420 nm. Stimulation of OSL is carried out by green light, from either green light emitting diodes or a laser. Since a single reading with useful signal intensities requires only 0.05% of the signal, stored re-reading or intermittent reading of the dosimeter is feasible and the dosimeter can be kept as a permanent dose record. Care must be taken to avoid exposure of the dosimeter to light (particularly ultraviolet), as electrons from deep traps could be transferred to dosimeter traps (phototransfer), changing the response of the dosimeter. Doses in the range 10 μGy to 15 Gy can be measured using commercial systems. The OSL principle is also utilized for imaging using computed radiography[3] systems (see Chapter 7).

21.5.4. Dosimetric applications of TLD and OSL

Solid state dosimeters can be used for patient dosimetry external to the body or phantom in the same way as an ionization chamber, for internal measurements, typically in a phantom and also for occupational and public exposure monitoring (Section 24.5.3).

For internal dosimetry, a near tissue equivalent composition may have advantages in determining energy deposition within the body (see Section 2.4.2). The effective atomic numbers of tissue and water are 7.22 and 7.42, respectively. TLD materials such as LiF:Mg,Ti and $Li_2B_4O_7$:Mn have effective atomic numbers of 8.31 and 7.4, respectively, while the main OSL material, Al_2O_3, and the promising OSL material, BeO, have effective atomic numbers of 11.3 and 7.21, respectively.

It must be remembered, however, that the primary dosimetry system in diagnostic radiology is based on air kerma and not absorbed dose to water or tissue. Further, solid state dosimetry is a relative methodology that requires standardized calibration procedures. Care must be exercised if TLD or OSL dosimeters are used in radiation fields that differ from the calibration conditions. Consequently, careful consideration must be given before LiF and Al_2O_3 dosimeters are used in applications such as CT phantom dosimetry.

21.6. DOSIMETER CALIBRATION

All instruments used for dosimetric measurement in the clinical environment should have a calibration traceable to a recognized dosimetry standard. A prerequisite for measurement of a dosimetric quantity, such as air kerma, is that there be an international measurement system that determines the quantity and its unit. Primary standards dosimetry laboratories (PSDLs) employ free air ionization chambers for the measurement of absorbed dose traceable to the fundamental SI absorbed dose unit (Gy). The secondary standards dosimetry laboratories (SSDLs) calibrate their reference class instruments at PSDLs and

---

[3] In the case of computed radiography, the active imaging material is known as a photon stimulable phosphor.

use these as their local dosimetry standards. Therefore, the traceability of the measurements to the specific PSDL is maintained. The main role of the SSDL is to bridge the gap between a PSDL and the dosimeter user.

21.6.1. Standard free air ionization chamber

Free air ionization chambers are often used by PSDLs as the primary standard for the determination of air kerma against which the secondary standard chambers from SSDLs are calibrated. Such a chamber is shown in Fig. 21.7. The charge liberated by X rays in the mass of the air inside the chamber volume is measured. The air kerma is deduced according to its definition, $K = (\mathrm{d}E_{\mathrm{tr}})/\mathrm{d}m = (\mathrm{d}Q \cdot W_{\mathrm{air}})/\mathrm{d}m$, from measurements of basic physical quantities (charge and mass) and applying physical constants and relative correction factors (see Chapter 3).

21.6.2. SSDL calibration

Most SSDLs apply the substitution method for the dosimeter calibration. At a given beam quality, Q, the true value of air kerma, $K_Q^{\text{'true'}}$ is measured using the reference dosimeter. The reference point of the user's dosimeter is placed at the same point and the dosimeter's reading is used to derive the calibration coefficient from the ratio $N_{K,Q}^{\text{user}} = K_Q^{\text{'true'}} / M_Q^{\text{user}}$, where M_Q^{user} is the reading of the user's instruments corrected for air density.

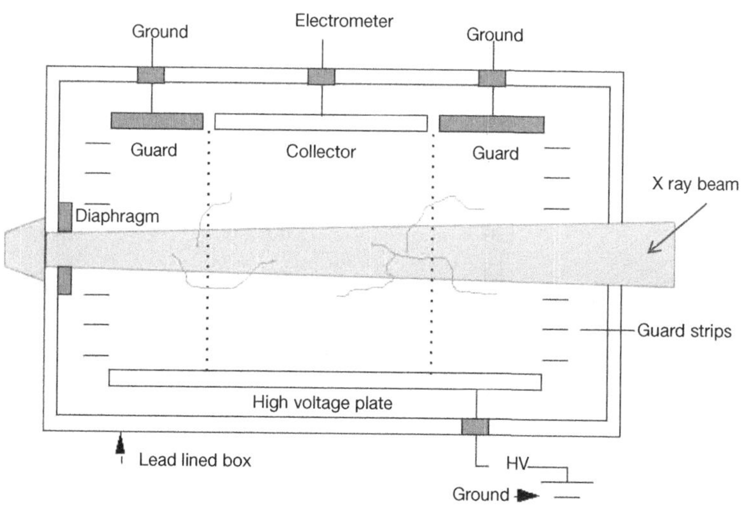

FIG. 21.7. The standard free in air ionization chamber used for calibration of an ionization chamber dosimeter.

The calibration of diagnostic radiology dosimeters is performed according to the radiation qualities that are described in the IEC-61267 standard [21.3] and is achieved using appropriate tube filtration at the specified tube voltage. Depending on the application of the dosimeter, a series of different beam qualities is used. For example, the RQR series simulates the primary beams incident on the patient, the RQT series simulates the beam qualities used in CT, the RQA series simulates the transmitted radiation qualities through the patient, and the RQR-M series simulates mammography beams. Each series consists of several beams with different combinations of tube voltage and filtration (see Tables 21.2–21.4). For the RQR, RQA and RQT quality series, an X ray tube with a tungsten (W) target and aluminium (Al) and/or copper (Cu) filters is used, while for the RQR-M quality series, an X ray tube with a molybdenum (Mo) target and molybdenum (Mo) filter is used.

TABLE 21.2. CHARACTERIZATION OF RADIATION QUALITY SERIES RQR USED FOR UNATTENUATED BEAMS FOR GENERAL RADIOGRAPHY APPLICATIONS
(according to IEC-61267[21.3])

Radiation quality	X ray tube voltage (kV)	First HVL (mm Al)	Homogeneity coefficient (*h*)
RQR 2	40	1.42	0.81
RQR 3	50	1.78	0.76
RQR 4	60	2.19	0.74
RQR 5[a]	70	2.58	0.71
RQR 6	80	3.01	0.69
RQR 7	90	3.48	0.68
RQR 8	100	3.97	0.68
RQR 9	120	5.00	0.68
RQR 10	150	6.57	0.72

[a] This quality is generally selected as the reference of the RQR series.

TABLE 21.3. CHARACTERIZATION OF RADIATION QUALITY SERIES RQR-M USED FOR UNATTENUATED BEAMS FOR MAMMOGRAPHY APPLICATIONS
(according to IEC-61267[21.3])

Radiation quality	X ray tube voltage (kV)	First HVL (mm Al)
RQR-M1	25	0.28
RQR-M2[a]	28	0.31
RQR-M3	30	0.33
RQR-M4	35	0.36

[a] This quality is generally selected as the reference of the RQR-M series.

TABLE 21.4. CHARACTERIZATION OF RADIATION QUALITY SERIES RQT USED FOR CT APPLICATIONS
(according to IEC-61267[21.3])

Radiation quality	X ray tube voltage (kV)	First HVL (mm Al)
RQT 8	100	6.90
RQT 9[a]	120	8.40
RQT 10	150	10.1

[a] This quality is generally selected as the reference of the RQT series.

A general purpose dosimeter should be calibrated in terms of air kerma at the RQR (RQR 2 to RQR 10) radiation qualities. According to common practice, the calibration coefficient, N_K, of a dosimeter is obtained at the RQR 5 (70 kV). For the other radiation qualities of the RQR series, further correction factors (k_Q) are provided to take into account the energy dependence of the dosimeter response. For a given radiation quality Q, k_Q is defined as the ratio of the calibration coefficients at radiation quality Q to that at radiation quality RQR 5. By definition, $k_Q = 1$ at RQR 5. For mammography, the standard beam quality is RQR-M2 (28 kV) and for CT it is RQT 9 (120 kV).

21.6.3. Field calibration

In some cases, for practical, economic and other reasons, users may calibrate their field instruments themselves. For example, when many dosimeters are being used in a large hospital, the user may prefer to calibrate them against a reference dosimeter, rather than send all of them to an SSDL. Some dosimetry equipment, such as KAP meters, is permanently installed on X ray systems and must be calibrated on-site. Generally, cross-calibration of a field instrument refers to its direct comparison in a suitable user's beam quality, Q, against a reference instrument that has been calibrated at an SSDL. The calibration coefficient is obtained from Eq. (21.2):

$$N_{K,Q}^{\text{field}} = \frac{K_{i,Q}^{\text{ref}}}{M_Q^{\text{field}}} = \frac{M_Q^{\text{ref}} N_{K,Q_0}^{\text{ref}} k_Q^{\text{ref}}}{M_Q^{\text{field}}} \tag{21.2}$$

where 'field' and 'ref' refer to the field and the reference instruments, respectively. The M values are readings of the reference and the field instruments and have been corrected for the influence of all quantities except beam quality. Since the calibration coefficient refers to a specific beam quality, the cross-calibration should be performed at the whole range of beam qualities that are used in the hospital. It is important to note that other essential elements of traceability of measurement, such as uncertainty evaluation, evidence of competence, documentation, etc., should be taken into account and be declared for cross-calibrations.

21.7. INSTRUMENTS FOR MEASURING TUBE VOLTAGE AND TIME

Measurement of the X ray tube voltage and exposure duration (often referred to as 'exposure time') is usually performed with non-invasive, portable electronic devices, often called kV meters and timers.

Figure 21.8 shows a typical X ray tube voltage waveform from a three-phase, six-pulse generator operating at an 80 kV tube voltage and a 165 ms exposure time. Depending on the model, the kV meter measures the absolute peak voltage (the maximum value of the voltage during the exposure — circled point in Fig. 21.8), the average peak voltage (average of all peak values), the average voltage (average of all voltage values), the effective peak voltage (the voltage that would give the same image contrast as a constant potential X ray system) and the practical peak voltage (defined as the equivalent value of a voltage of any waveform related to an ideal X ray generator that provides a constant voltage and that produces the same image contrast). Practical peak voltage has been proposed as the standard quantity for the X ray tube voltage.

The kV meter is positioned in the primary X ray beam and measures the X ray tube voltage with methods based on attenuation measurements. Such instruments usually incorporate two (or more) detectors covered with filters (usually made of copper) of different thickness. When exposed to radiation, the detectors produce different signals, owing to the different attenuation of the X ray beam by the filters. The signal ratio (or any other relationship of the signals) is a function of the incident X ray energy and consequently of the tube voltage. During the initial calibration of the kV meter at the factory, the signal output and/or the reading is appropriately adjusted to the 'correct' tube voltage value. Many kV meters digitize, process and store their detector signals and can supply voltage and/or exposure waveforms. The kV meter detectors' long geometrical axis should be positioned perpendicular to the tube anode–cathode direction, to eliminate the influence of the 'heel' effect to the kV measurement.

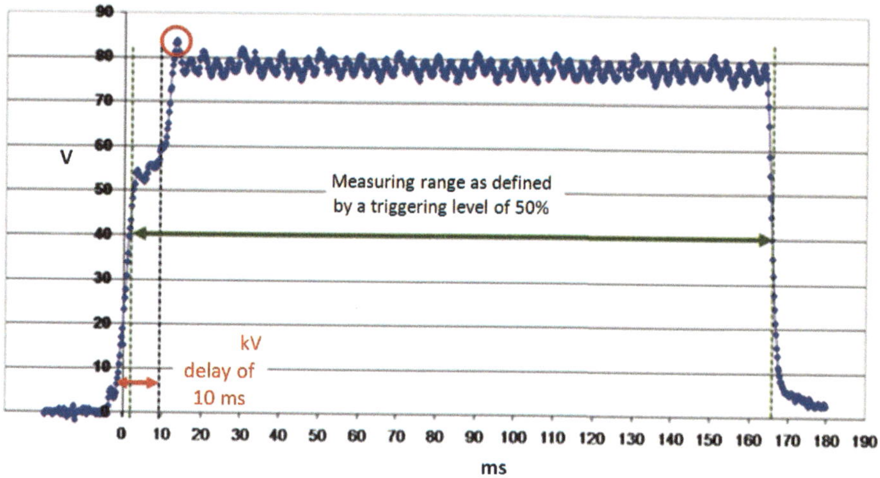

FIG. 21.8. Typical X ray tube voltage waveform from a three-phase six pulse generator operating at 80 kV tube voltage and 165 ms exposure time.

The IEC-61676 standard [21.4] specifies the performance requirements of instruments used for the non-invasive measurement of the X ray tube voltage up to 150 kV. It recommends that the relative intrinsic error of the practical peak voltage measurement should not be greater than ±2% over the effective voltage ranges, i.e. 60–120 kV for diagnostic and 24–35 kV for mammography. In addition, it recommends that a 1.5% limit of variation in response is acceptable at tube filtration ranges of 2.5–3.5 mm Al (for diagnostic radiology applications).

Exposure time is the time during which the radiation X ray beam is produced. It is measured as the radiation pulse width (time difference) between an 'initial' and 'final' point of the exposure, which are defined by a preset triggering level. The proposed convention for timing measurements of X ray systems is to measure the pulse width at a height of 50% of the waveform peak (full width at half maximum). Some manufacturers use different values of the triggering level (e.g. 10% or 75%). The exposure time may be measured using either invasive or non-invasive equipment.

21.8. INSTRUMENTS FOR OCCUPATIONAL AND PUBLIC EXPOSURE MEASUREMENTS

Radiation monitoring is performed in diagnostic radiology facilities to determine the radiation levels in and around work areas and radiology equipment, and to assess the radiation protection of the workplace and individuals (see Chapter 24). Such monitoring devices should typically measure in integrate mode and for direct (or real time) measurements include ionization chambers and some specifically developed semiconductor detectors suitable for scattered radiation. Longer term monitoring is typically achieved with film, or increasingly with solid state devices (for personal dosimeters (such as TLDs or film badges), see Section 21.5).

The use of survey meters (such as Geiger Müller counters or proportional counters) is not recommended for diagnostic radiology. Such devices are typically designed to detect isotope emissions; they are used extensively for radiation detection in nuclear medicine and have some application in radiation therapy, particularly for ^{60}Co units, brachytherapy usage and radioactive iodine treatment of patients. The two main difficulties in diagnostic radiology for the use of Geiger Müller counters, for example, is the response time of several seconds, when diagnostic X ray exposures have a duration of only small fractions of a second, and they also exhibit a strong energy dependence at low photon energies. Detailed descriptions of these instruments can be found elsewhere [21.5].

Detectors used for occupational and public exposure measurement should be traceable to appropriate calibration standards at suitable X ray energies (e.g. ISO Narrow series N40 to N80 [21.6]). While the user should know or estimate the mean energy of the X ray beam for calibration factor application, in some situations, such as the mean energy of radiation transmitted through protective barriers, this can be difficult. In these cases, it is acceptable to use measurements directly from a detector with a small energy response variation. The uncertainty of measurement should be assessed with reference to the variation in the calibration coefficients within the energy range used.

REFERENCES

[21.1] INTERNATIONAL ELECTROTECHNICAL COMMISSION, Medical Electrical Equipment – Dosimeters with Ionization Chambers and/or Semi-conductor Detectors as used in X-Ray Diagnostic Imaging, IEC-61674, IEC, Geneva (1997).

[21.2] INTERNATIONAL ATOMIC ENERGY AGENCY, Dosimetry in Diagnostic Radiology: An International Code of Practice, Technical Reports Series No. 457, IAEA, Vienna (2007).

[21.3] INTERNATIONAL ELECTROTECHNICAL COMMISSION, Medical Diagnostic X-Ray Equipment – Radiation Conditions for Use in the Determination of Characteristics, IEC-61267, IEC, Geneva (2005).

[21.4] INTERNATIONAL ELECTROTECHNICAL COMMISSION, Medical Electrical Equipment – Dosimetric Instruments Used for Non-invasive Measurement of X-ray Tube Voltage in Diagnostic Radiology, IEC-61676, IEC, Geneva (2002).

[21.5] INTERNATIONAL ATOMIC ENERGY AGENCY, Radiation Oncology Physics: A Handbook for Teachers and Students, IAEA, Vienna (2005).

[21.6] INTERNATIONAL ORGANIZATION FOR STANDARDIZATION, X and Gamma Reference Radiation for Calibrating Dosemeters and Doserate Meters and for Determining their Response as a Function of Photon Energy – Part 1: Radiation Characteristics and Production Methods, ISO 4037-1:1996(E), ISO, Geneva (1996).

BIBLIOGRAPHY

BØTTER-JENSEN, L., McKEEVER, S.W.S., WINTLE, A.G., Optically Stimulated Luminescence Dosimetry, Elsevier Science, Amsterdam (2003).

CHEN, R., McKEEVER, S.W.S., Theory of Thermoluminescence and Related Phenomena, World Scientific Publishing, Singapore (1997).

McKEEVER, S.W.S., MOSCOVITCH, M., TOWNSEND, P., Thermoluminescence Dosimetry Materials: Properties and Uses, Nuclear Technology Publishing, Ashford, UK (1995).

YUKIHARA, E.G., McKEEVER, S.W.S., Optically Stimulated Luminescence: Fundamentals and Applications, Wiley, Singapore (2011).

Chapter 22

PATIENT DOSIMETRY

D.R. DANCE
Royal Surrey County Hospital,
Guildford, United Kingdom

I. CASTELLANO
The Royal Marsden Hospital,
London, United Kingdom

22.1. INTRODUCTION

Patient exposures arising from radiological procedures form the largest part of the population exposure from artificial sources of radiation. According to United Nations Scientific Committee on the Effects of Atomic Radiation, the annual frequency of X ray examinations is 360 per 1000 individuals worldwide. Owing to the associated risk of radiation detriment to the patient, there is a clear need to monitor and control these exposures, and to optimize the design and use of the X ray imaging equipment so that the patient dose is reduced as far as possible, consistent with achieving the required clinical image quality.

The dosimetric quantities used in diagnostic radiology can be divided into two broad groups:

(i) *Application specific quantities*: These are practical dosimetric quantities, which may be directly measured and which may be tailored to specific situations or modalities. Examples include incident air kerma (IAK), air kerma–area product (KAP) and computed tomography (CT) air kerma indices.
(ii) *Risk related quantities*: These are dosimetric quantities, which can be used to estimate radiation detriment or risk and are thus measures of absorbed dose. Examples include organ dose and mean glandular dose (MGD) (for mammography).

In some situations, it is desirable to make direct measurements of the application specific quantities, but for others it is preferable to make measurements using a standard phantom to simulate the patient. Examples of the latter include quality control (QC), the comparison of different systems and optimization studies. The measurement methodology used depends upon the type of examination. In this chapter, we consider in outline methods used for general projection radiology, fluoroscopy, mammography, CT and dental radiography. A detailed description of the measurement methodology can be found in Ref. [22.1].

In contrast to application specific quantities, risk related quantities are usually difficult to measure directly and are generally estimated from application specific quantities using tables of dose conversion coefficients, determined either by Monte Carlo calculation or by measurements using phantoms.

Dose limits are used to control the exposure of workers and members of the public to ionizing radiation. However, dose limits for medical exposures could have a detrimental effect on patient health through failure to obtain essential clinical information. Therefore, patient doses are managed rather than controlled and the primary tool for this is the diagnostic reference level. In this chapter, we provide an overview of patient dose data and diagnostic reference levels available and of how the latter can be used in dose audit.

22.2. APPLICATION SPECIFIC QUANTITIES

Application specific quantities are practical dosimetric quantities for particular X ray modalities that are used for measurements in diagnostic radiology. Various specific quantities have been found useful in the past but there has been ambiguity in the names of the quantities and their (sometimes incorrect) use. In this handbook, we follow the recommendations in Ref. [22.2]; also adopted in Ref. [22.1]. Air kerma[1] is used as the basis for all application specific quantities. The SI unit for air kerma is the gray (Gy).

22.2.1. IAK

The IAK, K_i, is the simplest application specific quantity to measure and is particularly useful in situations such as plain film radiography, where the X ray field parameters remain unchanged throughout the exposure. It is defined as the kerma to air from an incident X ray beam measured on the central beam axis at the position of the patient or phantom surface (Fig. 22.1). Only the radiation incident on the patient or phantom and not the backscattered radiation is included.

---

[1] In the past, the quantity exposure (old unit: roentgen (R)) was used instead of air kerma. Values of exposure in roentgen can be converted to air kerma in gray using the conversion 0.876×10^{-2} Gy/R [22.3].

22.2.2. Entrance surface air kerma

In some situations (for example measurements at the patient surface), backscattered radiation is included in the measurement. The measured quantity is then known as the entrance surface air kerma, K_e. This is defined as the kerma to air measured on the central beam axis at the position of the patient or phantom surface (Fig. 22.1). The radiation incident on the patient or phantom and the backscattered radiation are included in the measurement. When the entrance surface air kerma is not measured directly, it can be estimated using the relationship:

$$K_e = K_i B \tag{22.1}$$

where B is the backscatter factor, which depends on the field size, radiation quality and backscatter material. It can be obtained from published tables or it can be measured using phantoms.

For procedures such as fluoroscopy and fluorography, where the exposure time can vary considerably from patient to patient, it can be important to

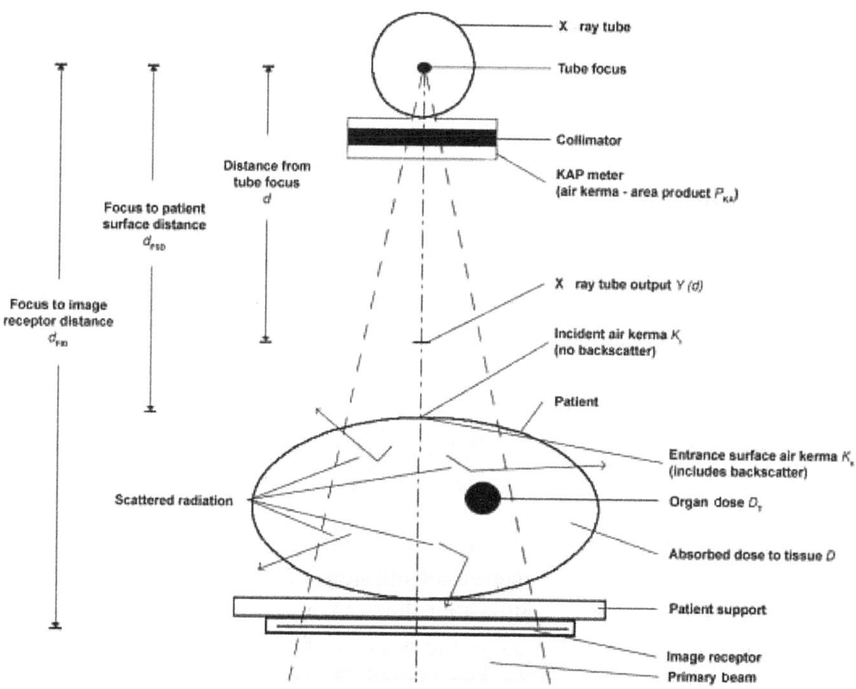

FIG. 22.1. Application specific quantities.

determine the entrance surface air kerma rate because of the potential for giving very high skin doses. In most countries, there will be a limit to the maximum air kerma rate that can be used for fluoroscopy.

22.2.3. X ray tube output

It is sometimes not possible to measure the incident or entrance surface air kerma directly and when this is the case, these quantities can then be estimated from measurements of the tube output using knowledge of the exposure parameters for the examination. The X ray tube output, $Y(d)$, is defined as the quotient of the air kerma, $K(d)$, at a specified distance, d, from the X ray tube focus, and the tube current–exposure time product, P_{It} (Fig. 22.1). It is given by:

$$Y(d) = \frac{K(d)}{P_{\mathrm{It}}} \tag{22.2}$$

Tube output is usually expressed in units of mGy·mA·s^{-1}. The tube current–exposure time product is sometimes referred to as the 'tube loading' or the 'mAs'.

The IAK for a particular exposure, X, is readily estimated from the tube output and the tube loading for the exposure $P_{\mathrm{It}}(X)$ by applying the inverse square law (see Eq. (22.16)).

22.2.4. KAP

In examinations such as fluoroscopy, where the beam direction, tube voltage, field size and tube current vary throughout the exposure, the IAK is not a good measure of radiation detriment. The KAP P_{KA} may be used instead. It is defined as the integral of the air kerma over the area of the X ray beam in a plane perpendicular to the beam axis (Fig. 22.1):

$$P_{\mathrm{KA}} = \int_A K(x,y)\,\mathrm{d}x\,\mathrm{d}y \tag{22.3}$$

KAP is usually expressed in units of cGy·cm^2, μGy·cm^2 or mGy·cm^2 and is generally measured using a plane transmission ionization chamber known as a KAP meter. In the approximation that the air kerma does not vary across the radiation field, P_{KA} is indeed equal to the product of the air kerma and field area.

The KAP has the useful property that it is approximately independent of the distance from the X ray tube focus (when interactions in air and extrafocal radiation can be neglected), as long as the planes of measurement and calculation are not so close to the patient or phantom that there is a significant contribution from backscattered radiation.

22.2.5. Air kerma–length product

In some situations, a useful alternative to P_{KA} is the air kerma–length product, P_{KL}, which is the integral of the air kerma along a line, L:

$$P_{KL} = \int_L K(x)\,dx \tag{22.4}$$

where P_{KL} is usually expressed in units of mGy·cm. It is used for dosimetry in CT and in panoramic dentistry, where it is also referred to as the KLP.

22.2.6. Quantities for CT dosimetry

The irradiation conditions in CT are quite different from those in planar imaging and it is necessary to use special dosimetric quantities and techniques. Measurements may be made free-in-air or in-phantom.[2] The dosimetric quantities for both are referred to as CT kerma indices and are based on measurements of P_{KL}. A pencil ionization chamber is generally used. The CT kerma index, $C_{a,100}$, measured free-in-air for a single rotation of a CT scanner is the quotient of the integral of the air kerma along a line parallel to the axis of rotation of a CT scanner over a length of 100 mm and the product of the number of simultaneously acquired tomographic sections, N, and the nominal section thickness, T. The integration range is positioned symmetrically about the volume scanned:

$$C_{a,100} = \frac{1}{NT}\int_{-50}^{+50} K(z)\,dz \tag{22.5}$$

---

[2] Both types of measurement have, in the past, been expressed in terms of a 'computed tomography dose index' (CTDI). However, for measurements 'in-phantom' using an air kerma calibrated ionization chamber, the measured quantity is air kerma. The absorbed dose to an air cavity within a phantom arises from a situation without secondary electron equilibrium and is difficult to measure. For these reasons, the terminology 'computed tomography kerma index' is used in this handbook for both free-in-air and in-phantom measurements. This is in accordance with the International Commission on Radiation Units and Measurements (ICRU) 74 [22.2]. All of the CT kerma indices used correspond directly with those previously referred to as CTDI related quantities.

where $C_{a,100}$ is usually expressed in units of mGy. For in-phantom measurements, the notation $C_{PMMA,100}$ is used.

It can be seen from Eq. (22.5) that the CT air kerma index is the height of a rectangular air kerma profile of width equal to the product of the number of sections, N, and the nominal section thickness, T, which has the same value as the line integral. This is illustrated in Fig. 22.2 for a single slice scanner.

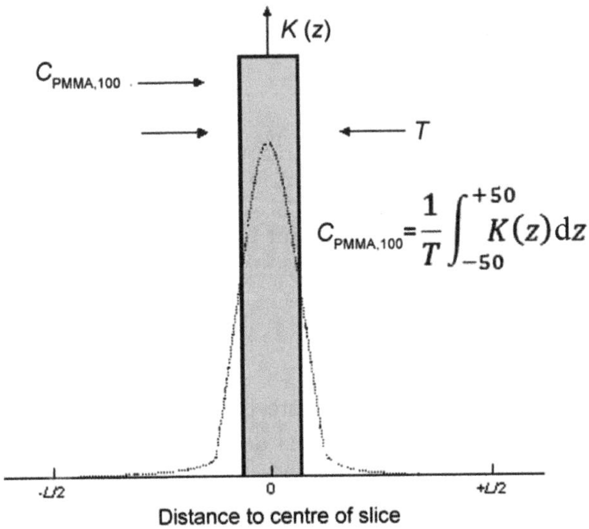

FIG. 22.2. Profile of the air kerma K(z) in a CT dosimetry phantom along an axis (z) for a single CT slice of nominal thickness T mm. The CT kerma index, $C_{PMMA,100}$, is obtained by integrating the air kerma over a length of 100 mm.

Unlike some areas of dosimetry, only two phantoms have found common application. We refer to these as the standard head and body phantoms. The weighted CT kerma index, C_W, combines values of $C_{PMMA,100}$ measured at the centre and periphery of these phantoms. It is given by:

$$C_W = \frac{1}{3}\left(C_{PMMA,100,c} + 2C_{PMMA,100,p}\right) \tag{22.6}$$

The quantity $C_{PMMA,100,c}$ is measured at the centre of the standard CT dosimetry phantom and $C_{PMMA,100,p}$ is the average of values measured at four positions around the periphery of the phantom. The weighted CT kerma index is an approximation to the average air kerma in the volume of the phantom interrogated by a single rotation of the scanner.

A further quantity, C_{VOL}, provides a volume average, which takes account of the helical pitch or axial scan spacing, where:

$$C_{\mathrm{VOL}}=\frac{C_{\mathrm{W}}NT}{l}=\frac{C_{\mathrm{W}}}{p};\quad {}_{\mathrm{n}}C_{\mathrm{VOL}}=\frac{C_{\mathrm{VOL}}}{P_{\mathrm{It}}} \tag{22.7}$$

where

N is the number of simultaneously acquired tomographic sections;
T is the nominal slice thickness;
l is the distance moved by the patient couch per helical rotation or between consecutive scans for a series of axial scans;

and P_{It} is the tube loading for a single axial scan.

The quantity:

$$p=\frac{l}{NT} \tag{22.8}$$

is known as the CT pitch factor (or pitch) for helical scanning. The quantity ${}_{\mathrm{n}}C_{\mathrm{VOL}}$ is normalized to unit tube current–exposure time product.

The CT air kerma index, C_{VOL} (or equivalently the index C_{W}), can be combined with patient exposure parameters to provide a dose measure for a complete patient examination. This is the CT air kerma–length product $P_{\mathrm{KL,CT}}$, which is given by:

$$P_{\mathrm{KL,CT}}=\sum_{j}{}_{\mathrm{n}}C_{\mathrm{VOL}}\,l_{j}P_{\mathrm{It}_{j}} \tag{22.9}$$

where

index j represents each serial or helical scan sequence forming part of the examination;
l_j is the distance moved by the patient couch between or during consecutive scanner rotations;

and P_{It_j} is the total tube loading for scan sequence j.

It has been found that the above CT kerma quantities will lead to underestimates of patient dose when the width of the rotating X ray field approaches or exceeds 40 mm. In this case, C_{W} can be determined using the following formulation:

$$C_{\mathrm{W},NT} = C_{\mathrm{W,Ref}} \times \left(\frac{C_{\mathrm{a},100,NT}}{C_{\mathrm{a,100,Ref}}} \right) \tag{22.10}$$

where

$C_{\mathrm{W},NT}$ is the weighted CT air kerma index for a beam width of NT mm (if NT is >40 mm);
$C_{\mathrm{W,Ref}}$ is the weighted CT air kerma index for a reference beam width of 20 mm (or closest possible below 20 mm);
$C_{\mathrm{a},100,NT}$ is the CT air kerma index measured free in air for a beam width of NT mm;

and $C_{\mathrm{a,100,Ref}}$ is a similar quantity at the reference beam width.

The methodology used to measure $C_{\mathrm{a},100,NT}$ can be found in recent publications [22.4].

22.3. RISK RELATED QUANTITIES

The detriment arising from medical X ray examinations can be stochastic or non-stochastic (deterministic) and depends upon the dose to individual organs. For stochastic effects, the total risk is the sum of the organ and tissue doses multiplied by appropriate risk coefficients. For deterministic effects, the nature and magnitude of the effect are determined by the dose to the organs or tissues concerned. Thus, the dose to individual organs and tissues has to be quantified in order to assess detriment. With the exception of localized skin dose, it is not possible, or at best very difficult, to measure such doses directly, and use is made instead of the application specific quantities described in Section 22.2, in combination with absorbed dose conversion coefficients derived from Monte Carlo calculations or phantom measurements. However, in practice, phantom measurements of coefficients are little used because of the general availability of Monte Carlo calculated factors and the practical difficulties associated with such measurements. The calculation and measurement of conversion coefficients is described in Section 22.5.1 and the use of the coefficients is described in

Section 22.5.3. First, we present the dosimetric quantities we will need and then consider the measurement of the application specific quantities (Section 22.4).

22.3.1. Organ and tissue dose

The mean absorbed organ dose, D_T, in a specified organ or tissue is equal to the ratio of the energy imparted, $\bar{\varepsilon}_\mathrm{T}$, to the tissue or organ and the mass, m_T, of the tissue or organ:

$$D_\mathrm{T} = \frac{\bar{\varepsilon}_\mathrm{T}}{m_\mathrm{T}} \tag{22.11}$$

The mean absorbed dose to a specified organ or tissue is sometimes simply referred to as the organ dose. Organs that commonly require individual dose determination include the uterus and the lens of the eye.

It is important to remember that organs may only be partly exposed to the incident radiation field and that the dose distribution within the body is far from homogeneous. In some situations, the local absorbed dose in an organ or tissue may considerably exceed the mean absorbed dose. In such a situation (e.g. coronary angiography), it can be desirable to estimate local dose values as well as the mean organ dose.

Assessment of the absorbed dose to the most exposed area of the skin is essential in interventional radiology because of the possibility, for complicated procedures, of exceeding the threshold for deterministic effects. Knowledge of the skin dose during such procedures is necessary to avoid deterministic effects and reduce their severity. Knowledge after the procedure is necessary for appropriate management of the patient.

22.3.2. MGD

The International Commission on Radiological Protection (ICRP) and the ICRU recommend the use of the mean (or average) dose to the glandular tissues[3] within the breast for breast dosimetry in diagnostic radiology. These are the tissues that are at the highest risk of radiation induced carcinogenesis. This recommendation has been generally adopted.

The acronym MGD for the mean glandular dose is used in this handbook.

---

[3] The term glandular tissues includes the acinar and ductal epithelium and associated stroma.

22.3.3. Equivalent dose

Different types of ionizing radiation can cause stochastic effects of different magnitudes for the same value of the absorbed dose. To allow for this, the equivalent dose, H_T, to an organ or tissue, T, is used. For a single type of radiation, R, it is the product of a radiation weighting factor, w_R, for radiation R and the organ dose, D_T:

$$H_T = w_R D_T \tag{22.12}$$

The radiation weighting factor, w_R, represents the relative biological effectiveness of the incident radiation in producing stochastic effects at low doses in tissue or organ T. In diagnostic radiology, w_R is usually taken to be unity. The SI unit for equivalent dose is the sievert (Sv).

22.3.4. Effective dose

The radiation exposure of the organs and tissues of the human body results in different probabilities of detriment for the different organs and for different individuals. For radiation protection purposes, the ICRP has introduced the effective dose, E, as a measure of the combined detriment from stochastic effects for all organs and tissues for an average adult. It is the sum over all the organs and tissues of the body of the product of the equivalent dose, H_T, to the organ or tissue and a tissue weighting factor, w_T, for that organ or tissue:

$$E = \sum_{T} w_T H_T \tag{22.13}$$

The tissue weighting factor, w_T, for organ or tissue T represents the relative contribution of that organ or tissue to the total 'detriment' arising from stochastic effects for uniform irradiation of the whole body. The sum over all the organs and tissues of the body of the tissue weighting factors, w_T, is unity.

The SI unit for effective dose is the sievert (Sv). This is the same unit as for equivalent dose and care should be taken to indicate which quantity is being used.

TABLE 22.1. TISSUE WEIGHTING FACTORS ACCORDING TO ICRP 103 [22.5]

Tissue or organ	Tissue weighting factor (w_T)	$\sum w_T$
Bone marrow, colon, lung, stomach, breast, remainder tissues[a]	0.12	0.72
Gonads	0.08	0.08
Bladder, oesophagus, liver, thyroid	0.04	0.16
Bone surface, brain, salivary glands, skin	0.01	0.04

[a] The tissue weighting factor for remainder tissues is applied to the arithmetic mean of the doses to the following 14 organs/tissues: adrenals, extrathoracic region, gall bladder, heart, kidneys, lymphatic nodes, muscle, oral mucosa, pancreas, prostate, small intestine, spleen, thymus and uterus/cervix.

The current values of the weighting factors, w_T, are given in Table 22.1. They have been estimated by the ICRP on the basis of population studies of cancer induction and hereditary effects and are averaged over age and sex for a particular population. Owing to this averaging process, the risk factors used can be quite different from the values appropriate to a particular individual undergoing an X ray examination. It is, therefore, strongly emphasized that the effective dose should not be used directly to estimate detriment for individual medical exposures. Risk values for the individual tissues and organs at risk and for the age distribution and sex of the individual or population being exposed (such as those tabulated in BEIR VII [22.6]) should be used instead.

Notwithstanding this caveat, the effective dose can be very useful for comparative purposes[4], for example, between procedures carried out with different exposure parameters or carried out on a given population.

22.4. MEASURING APPLICATION SPECIFIC QUANTITIES

22.4.1. General considerations

In Section 22.2, we defined the application specific quantities used in diagnostic radiology. In this section we consider the methods used for their

[4] Care should be taken when comparing values of effective dose to ensure that the same values of the tissue weighting factors, w_T, have been used. Prior to the publication of ICRP 103 [22.5], effective dose was calculated using tissue weighting factors taken from ICRP 60 [22.7], which are different from those in ICRP 103.

measurement. The treatment is intended to be reasonably comprehensive, but the reader wishing to know the 'fine details' of the measurements should consult Ref. [22.1], which may also be consulted for details of calibration procedures.

There are two general approaches for the measurements:

(i) Direct measurement on patients or phantoms.
(ii) Indirect measurements on patients and phantoms. These use free-in-air measurements to characterize X ray output, which are then scaled for exposure and geometry using actual patient or phantom exposure factors.

Application specific quantities can be measured using ionization chambers (including KAP meters) or, in some cases, semiconductor detectors (see Sections 21.3 and 21.4, respectively). For direct patient measurement, KAP meters or TLDs are often the dosimeters of choice because they are radiolucent in the diagnostic radiology energy range (except mammography). It is important that TLDs are of high sensitivity and capable of detecting an air kerma of 0.1 mGy. It is good practice to construct a TLD dosimeter comprising at least three TLD chips.

In each case, the following equation is used to calculate the relationship between the air kerma related quantity K and the measurement M:

$$K = N_{K,Q_0} k_Q M k_{\mathrm{TP}} \tag{22.14}$$

where N_{K,Q_0} is the dosimeter calibration at the calibration radiation quality Q_0 and the factor k_Q corrects this to the radiation quality Q of the actual measurement. The factor k_{TP} corrects for temperature and pressure. Its value is unity for semiconductor dosimeters and for ionization chambers is given by (see Section 21.3):

$$k_{\mathrm{TP}} = \left(\frac{273.2 + T}{273.2 + T_0}\right)\left(\frac{P_0}{P}\right) \tag{22.15}$$

The quantities T and P are the temperature and pressure at the time of measurement, and T_0 and P_0 are the corresponding values for the calibration. Depending on the measurement uncertainty required, this equation may be applied using either the normal pressure for the altitude of the measurement and the average temperature in the room of measurement, or the actual values at the time of measurement.

The measurement uncertainty desirable in the application specific quantities depends on the use to be made of the measurement.[5] Advice is given in Ref. [22.1] as follows:

- For estimation of absolute stochastic risk: 10%.
- For estimation of relative risks (comparative dose measurements): 7%.
- For estimation of the dose to the embryo/fetus: 7%.
- For quality assurance: 7%.

These uncertainties all correspond to an expanded uncertainty[6], $k = 2$, and are in addition to any uncertainties in conversion coefficients used for the calculation of risk related quantities. It is important to estimate uncertainties for each measurement. It is doubtful whether the above uncertainties can be achieved in all cases [22.9].

22.4.2. Measurements using phantoms and patients

Measurements using phantoms are useful for:

- The control of technical parameters, including equipment under automatic exposure control;
- The comparison of the same system at different times;
- The comparison of different systems;
- Optimization of individual components of the imaging system or of the whole system.

However, they cannot provide a direct estimate of the average dose for a given patient population or the variation that occurs in practice because of variations in patient size and composition. They also provide no information on how technique factors may vary with the operator. It is important, therefore, that measurements made using phantoms are complemented with measurements made on patients, though the measurement frequency will be different for the two types of measurement.

Phantoms vary in degree of complexity and anatomical accuracy. Generally, the more realistic they are, the more expensive they will be. However,

---

[5] The uncertainty of measurement for secondary standard dosimetry laboratories is discussed in Ref. [22.8].

[6] Appendix 1 of Ref. [22.1] provides a detailed discussion of measurement uncertainties. An expanded uncertainty, $k = 2$, corresponds to a 95% confidence limit for the quantity in question.

if it is just total attenuation that is to be matched, simple plastic phantoms can often be used. A good example is the use of PMMA phantoms for dosimetry in mammography. Simple plastic phantoms are also used for dosimetry in CT, although in this case they cannot be considered to be representative of typical patients. Other examples include the simple phantoms designed by the Center for Devices and Radiological Health (CDRH) in the USA. These are available for the dosimetry of chest and lumbar spine/abdomen examinations. For example, the CDRH abdomen phantom is designed to correspond to an average US citizen in the anterior–posterior projection (average thickness 230 mm).

Dosimetric quantities obtained from patient exposures will include variations in equipment performance and operator technique, as well as patient related differences. Therefore, a single measurement will not be representative of clinical practice. Instead, dosimetric data will need to be collected from a patient cohort so that a median and/or average value can be calculated. Such values can be used for comparative studies at local, regional, national and international levels, always provided that the median values of the patient size are similar. It is evident that the patient cohort selected must be representative and sufficiently large to reduce statistical fluctuations in the median or the average dose for the sample to an acceptable level. Sample sizes of between 10 and 50 have been used. The use of the sample median rather than the sample mean has the advantage that it is little influenced by outlying values arising from very large or very small patients. If the sample average is to be used, patient selection based on mass can be helpful if the sample size is small. In any case, the recording of patient mass and height is recommended, to aid the interpretation of the results.

Finally, it is noted that the relationship between risk related quantities and measured application specific quantities will, in general, depend upon field size and beam quality. Information regarding these parameters should be recorded as appropriate.

22.4.3. Free-in-air measurements

In situations where the exposure parameters for a radiographic examination are known, the IAK can be calculated directly from knowledge of these parameters and measurements of the tube output, $Y(d)$ (Eq. (22.2)), using:

$$K_{\mathrm{i}} = Y(d) P_{\mathrm{It}}(X) \left[\frac{d}{d_{\mathrm{FSD}}} \right]^2 \qquad (22.16)$$

where

d_{FSD} is the focus skin (or phantom surface) distance;
d is the distance from the focus to the point of measurement of the tube output;

and P_{It} is the tube loading (mAs) for the exposure.

For the tube output measurement, the dosimeter is placed free-in-air on the central axis of the X ray beam and sufficiently high above the table to reduce the effects of backscatter to a low level. A solid state dosimeter with shielding from backscatter may instead be placed on the patient table or floor. A typical set-up is shown in Fig. 22.3.

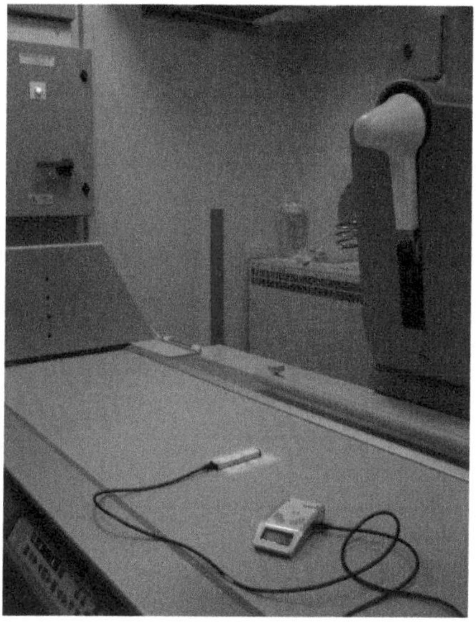

FIG. 22.3. Measurement of tube output using a calibrated solid state dosimeter which has lead backing to reduce backscatter. Note that the dosimeter and control unit are both shown in the photograph for demonstration purposes. In an actual situation, they would be further apart and the cable would not be coiled.

Tube output is measured at a range of tube voltages and for the filters in clinical use. For the purposes of interpolation, the output for each filter can be fitted to:

$$Y(d) = a(\text{kV})^n \tag{22.17}$$

where

$Y(d)$ is the X ray tube output;
kV is the tube voltage;

and a and n are constants.

As regards the constants, a is specific to the filter in use and n has a value of approximately 2 for tungsten targets and 3 for molybdenum targets. An example of such a fit is shown in Fig. 22.4 for a tungsten target.

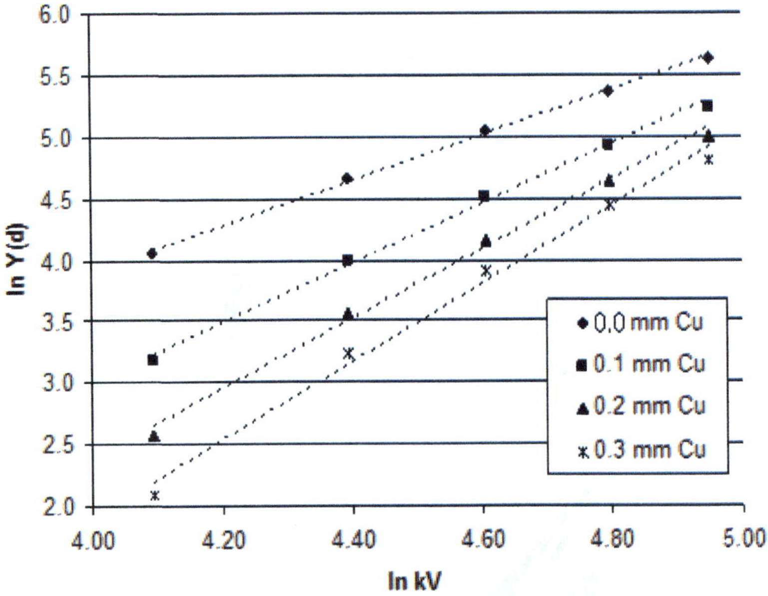

FIG. 22.4. Variation of tube output with tube voltage for an X ray tube filtered with various thicknesses of copper.

22.4.4. Radiography

The application specific quantities used for dosimetry in radiography are IAK, entrance surface air kerma and KAP. In practice, for patient dosimetry, the KAP is the simplest to obtain as long as a calibrated KAP meter is fitted to the X ray tube. Where this dosimeter is provided by the X ray manufacturer, the reading will usually be displayed on the X ray console. Occasionally, the displayed KAP is calculated by the X ray generator microprocessor from the exposure factors, the settings of the collimator blades and a generic value for the tube output. It is, therefore, important to check the calibration of the KAP meter before using it for patient dosimetry.

In the absence of a KAP meter, IAK or entrance surface air kerma are reasonable alternatives to the KAP. Both can be most easily obtained using indirect calculation from recorded exposure parameters (see Section 22.2.4), although direct measurement is also possible. If the entrance surface air kerma is required, direct measurements may sometimes be preferred, as these will include backscatter. In this case, the dosimeter is placed on the entrance surface of the patient or phantom at the centre of the X ray field and the exposure is taken in accordance with normal clinical practice.

22.4.5. Fluoroscopy

Fluoroscopic examinations are, by their nature, very variable. Changes in mode (i.e. fluoroscopy, image acquisition), exposure factors, filtration, projection, collimation and body part irradiated may all take place during such examinations. The patient dose will depend on the size of the patient, the operator selections and the complexity of the case. Therefore, dosimetric quantities based on patient exposures are essential. Phantom exposures are of use for simple procedures and for QC to ensure suitable set-up and optimization of the equipment. Owing to this variability, the KAP is the dosimetric quantity of choice for the estimation of radiological risk. The use of IAK and entrance surface air kerma is, however, needed for examinations where there is a risk of skin injury, or exposure of the eye to unattenuated beam.

For fluoroscopy systems, the total KAP for the examination and the total fluoroscopy time are displayed on the X ray console. The former is usually measured with a KAP meter but can also be calculated (in which case the caveat noted in the previous section should be observed). In the case of undercouch units, the measured KAP overestimates the KAP to the patient, owing to attenuation of the X ray beam by the patient couch. Accurate correction for couch attenuation is often not practical as it is a function of beam quality and X ray projection.

Modern interventional fluoroscopy units will report the IAK at a reference point calculated from the KAP, the collimator settings and the exposure geometry. The reported IAK can be used to estimate the maximum value of the entrance surface air kerma. This is the maximum value because changes in projection angle during the examination have been ignored. Nevertheless, this may be a useful quantity for monitoring skin dose, but must be fully understood and calibrated before being used in patient dose surveys.

Measurements of entrance surface air kerma rates on phantoms for selected clinical protocols and typical projections can be combined with fluoroscopy times, image acquisition parameters and selected field sizes to yield an estimate of the total entrance surface air kerma for simple examinations. Such a measurement is illustrated in Fig. 22.5.

22.4.6. Mammography

The application specific quantities appropriate for dosimetry in mammography are IAK and entrance surface air kerma. However, the latter is little measured or used, as it is the IAK that is required for the calculation of MGD. This is further discussed in Section 22.5.3.2

The standard method of determining the IAK for both patient and phantom exposures is to calculate it using Eq. (22.16), measurements of tube output and knowledge of the exposure parameters used for the examination (tube charge (mAs), tube voltage and filtration, and breast or phantom thickness). Direct measurements are little used for dosimetry with phantoms and are not possible for patient exposures because of the visibility of even small dosimeters such as a TLD on the image.

Measurements made using PMMA phantoms are included in national QC programmes for mammography, but patient measurements are needed to determine the actual distributions of MGD. Unlike many situations in radiography and fluoroscopy, the standard phantoms are well defined for mammography, so

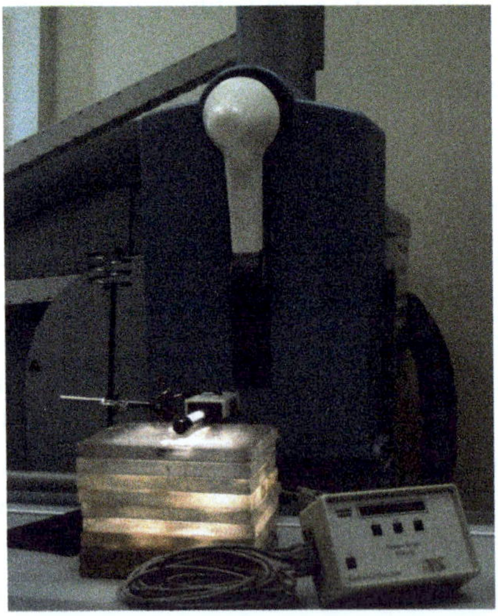

FIG. 22.5. Measurement of entrance surface air kerma using a PMMA slab. Note that the dosimeter and control unit are both shown in the photograph for demonstration purposes. In an actual situation, they would be further apart and the cable would not be coiled.

that comparisons between different sites at national and international levels are feasible.

22.4.7. CT

The application specific dosimetric quantities that can be used for patient dosimetry in CT are introduced in Section 22.2.6 and comprise:

- The free-in-air CT kerma index, $C_{a,100}$;
- The in-phantom CT kerma indices, $C_{PMMA,100,p}$ and $C_{PMMA,100,c}$;
- The weighted CT kerma index, C_W;
- The volumetric CT kerma index, C_{VOL};
- The kerma–length product, $P_{KL,CT}$.

The first two quantitities can be measured using a pencil ionization chamber or a stack of TLDs, though for practical reasons the former is generally preferred. The standard active length of the chamber is 100 mm, to match the integration limits for the CT kerma indices measured free-in-air (Eq. (22.5)) or in-phantom. In general, the measured indices are normalized by the exposure time–tube current product (mAs) and can be scaled where necessary to match the exposure parameters for a given procedure.

The free-in-air CT kerma index is useful for characterizing the tube output of the CT scanner and for QC. It is easy to measure (by aligning the pencil chamber with the scanner axis of rotation as shown in Fig. 22.6), but is not influenced by the shaped beam filtration that is present in the scanner (see Section 11.3.4). It is also required as a scaling factor when using some tabulated conversion factors to calculate absorbed organ dose or effective dose.

In-phantom measurements give a better measure of patient dose. The weighted CT kerma index provides an estimate of the average dose within a slice for a single scanner rotation without translation. It is obtained by combining measurements of $C_{PMMA,100}$ taken in the centre and peripheral positions of a standard CT dosimetry phantom (Eq. (22.6)). Two phantoms are used: standard head and body phantoms (Fig. 22.7 shows the body phantom; the head phantom forms the inner portion of this phantom). These phantoms comprise circular cylinders constructed from PMMA. They have bores at the centre and at the cardinal points 1 cm below the surface to facilitate measurement. Their diameters are 16 cm and 32 cm, respectively.

The weighted CT kerma index is useful for characterizing the dosimetric performance of the CT scanner, but not for patient dosimetry, as it applies to a single scan rotation, not the whole scan. However, once measured, it can be used to calculate the volumetric CT kerma index and, hence, the air kerma–length product (via Eqs (22.7, 22.9) and using the pitch and tube loading), which are widely used to describe patient doses in CT, sometimes without the care they deserve (see Section 22.5.3.3). Nevertheless, they can be considered to be the preferred quantities for patient dosimetry in CT.

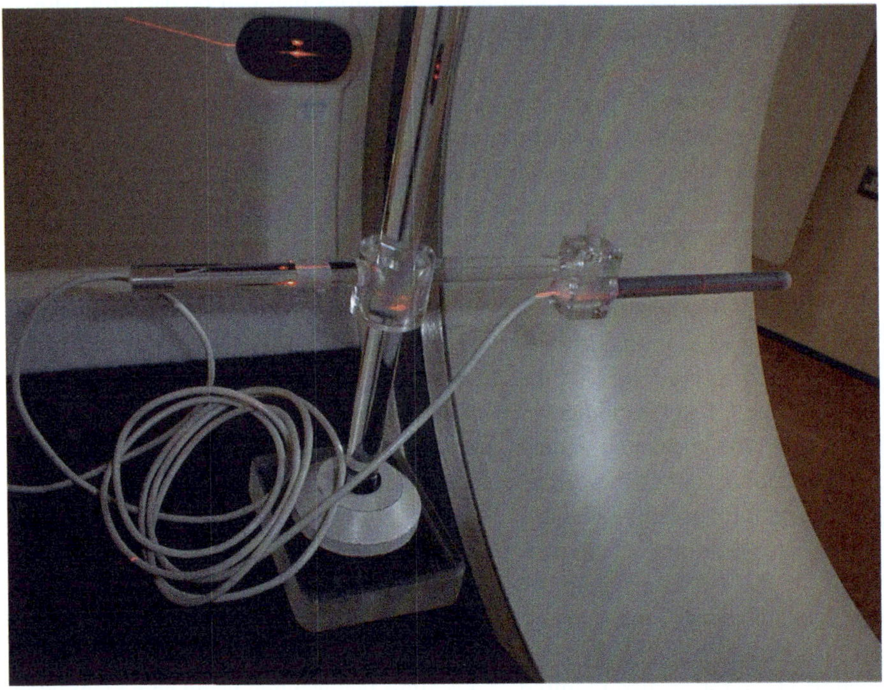

FIG. 22.6. Arrangement for the measurement of the CT air kerma index free-in-air, $C_{a,100}$. The chamber is clamped in a specially designed support and aligned so that it is coaxial with the scanner rotation axis and that its sensitive volume is bisected by the scan plane. In this particular example, alignment has been achieved with the aid of laser lights. The cable is shown coiled in the demonstration photograph, but in an actual situation this would not be the case.

Patient dosimetry in CT is unique in diagnostic radiology in that the CT specific dosimetric quantities are defined in terms of standard phantoms, yet are applied to patient exposures. The size of the patient may be different to the size of the phantom, so that the dosimetric quantities may overestimate or underestimate the air kerma in the patient. In other words, the volumetric CT kerma index and the air kerma–length product cannot be measured on a patient in the way that IAK, KAP, etc., can be. It is, therefore, vital to remember that in CT, dosimetric quantities refer to phantoms.

Modern CT scanners report the volumetric CT kerma index on the scanner console; it is shown as the ‘CTDI_{vol}’. It is more convenient to record this displayed value than to calculate it from measurements of weighted CT kerma index and the scan parameters. In the case of CT scanners with tube current modulation, the average volumetric CT kerma index for the scan can realistically be obtained only from the display. This approach is acceptable if the volumetric CT kerma index calculation has been validated against measurement.

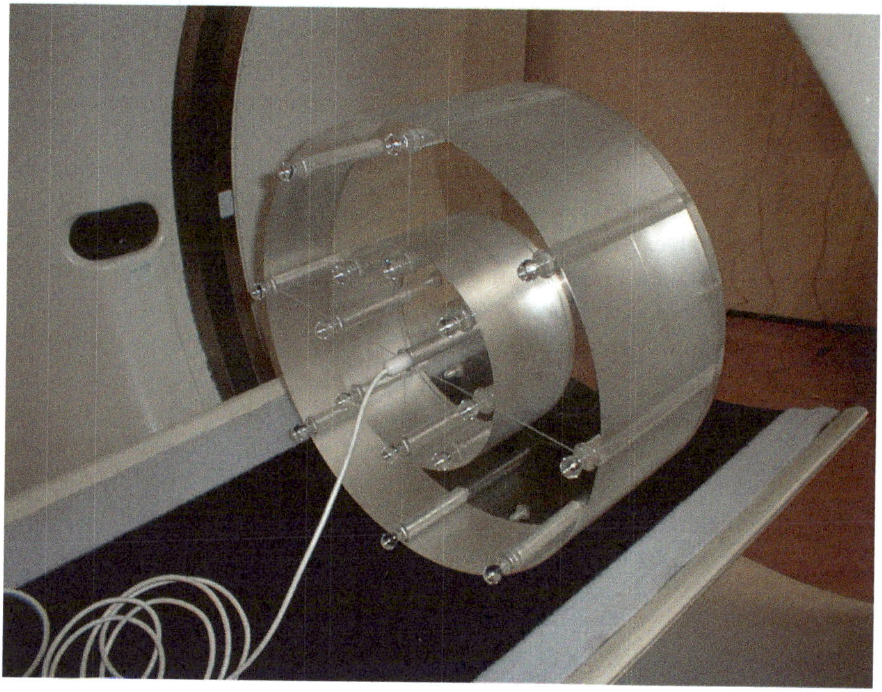

FIG. 22.7. Arrangement for the measurement of the CT air kerma index, $C_{PMMA,100,c}$, in the standard body phantom. The phantom is positioned on the couch top and the chamber is positioned in the central hole of the phantom. A plastic sleeve is placed over the chamber to ensure a good fit within the phantom. The central plane of the phantom has still to be aligned with the position of the scan slice. The cable is shown coiled in the demonstration photograph, but in an actual situation this would not be the case.

Modern CT scanners also report the air kerma–length product on the scanner console; it is shown as the dose–length product. This approach is also acceptable if the dose–length product calculation has been validated against measurement.

22.4.8. Dental radiography

In dental radiography, dosimetric measurements are normally taken based on patient exposures rather than using phantoms. The dosimetric quantities generally used are the IAK, which is readily measured for intraoral examinations, and the KLP and the KAP, which are used for panoramic examinations.

On dental radiography equipment, exposures are generally set manually by the operator, or selected from default protocols. The exposure factors are, therefore, not dependent on the subject. However, in the case of panoramic units that use AEC, typical exposure parameters must be recorded so that the exposure can be duplicated under manual control for dosimetry purposes.

Direct measurements are preferred in dental radiography as they are easy to implement. The number of protocols used clinically is generally small, so measurements for each protocol are more time efficient than characterizing the tube output as described in Section 22.4.3.

For intraoral examinations, the IAK may be measured by placing the dosimeter free-in-air at the end of the spacer/alignment cone in the centre of the X ray beam. The exposure is taken using a standard clinical protocol.

For panoramic dentistry, the KLP can be measured using a cylindrical ionization chamber or a stack of TLDs that are longer than the width of the X ray beam. The ionization chamber is most easily affixed to the detector housing across the centre of the secondary X ray beam slit, with the exposure taken using a standard clinical protocol. The KAP can be estimated from the KLP by multiplying by the height of the X ray beam at the position of the dosimeter. This can be measured using an X ray film or a computed radiography plate.

22.5. ESTIMATING RISK RELATED QUANTITIES

In Section 22.3, it was noted that the absorbed dose to individual organs and tissues has to be quantified in order to assess radiation detriment. Owing to the difficulty of direct measurement, an organ or tissue dose is generally estimated from the measurement (or calculation) of an application specific quantity (such as IAK or KAP) in combination with an absorbed dose conversion coefficient, c, defined as:

$$c = \frac{\text{organ or tissue dose}}{\text{measured or calculated quantity}} \tag{22.18}$$

Suffixes are added to c to denote the particular quantities used, so that, for example, to relate IAK K_i to organ dose D_T, we use:

$$c_{D_T,K_i} = \frac{D_T}{K_i} \tag{22.19}$$

Section 22.5.1 considers the use of Monte Carlo calculations and measurements to determine organ dose conversion coefficients. Section 22.5.3 briefly covers their practical use and Section 22.5.2 discusses backscatter factors, which are used to convert between IAK and entrance surface air kerma (see Eq. (22.1)).

22.5.1. Determination of organ dose conversion coefficients

22.5.1.1. Monte Carlo methods

The key features of a Monte Carlo model for the calculation of absorbed dose conversion coefficients are the simulation of the radiation field incident on the patient (including field size, direction and X ray spectrum), the simulation of photon transport through the patient and the simulation of the patient themself. Once such a program has been developed, it is usually possible to simulate a wide range of examinations and X ray spectra. As a consequence, Monte Carlo methods are generally a much more powerful tool for the production of tables of conversion coefficients than measurements taken using anthropomorphic phantoms.

The methodology for the simulation of photon histories is well established. For the diagnostic energy range, it is sufficient in most cases to assume that energy deposited after a photon interaction is locally absorbed so that organ doses may be estimated by recording the energy depositions that take place when many individual photon histories are followed. An important exception to this is the energy deposition in the red bone marrow, where the range of secondary electrons may be comparable to the size of the marrow cavities and electron transport must then be considered. A correction may be applied for this effect.

Two approaches have been adopted for the simulation of the human body. One is to use a mathematical phantom (also known as a geometrical phantom) in which the body and the organs it contains are constructed as combinations of various geometrical solids. The first such phantom was based on the ICRP Reference Man of 1975 and a series of other phantoms have subsequently been developed which represent, for example, children (neonate and 1, 5, 10 and

15 years old) and adult males and females. Mathematical phantoms can be criticized as being unrealistic in terms of organ position and shape.

An alternative and more realistic approach is to use one or more voxel phantoms based on the anatomy of individuals. Such phantoms may be obtained, for example, from whole body CT or magnetic resonance images, which have been segmented voxel by voxel into different organs and tissue types.

As a result of the statistical nature of Monte Carlo simulations, the organ dose conversion coefficients have associated statistical errors. In general, the statistical uncertainties in the doses to organs lying within the radiation field will be less than those for organs lying outside the radiation field. For the latter case, the relative uncertainty will increase with distance from the edge of the field.

Organ dose conversion coefficients calculated using Monte Carlo techniques have been published by various authors. The most extensive tabulations are those of the CDRH in the USA, the National Research Centre for Evironment and Health (GSF) in Germany and the National Radiological Protection Board (NRPB) in the UK.

The choice of tabulation for a particular situation will depend upon data availability and how well the situation modelled (including the radiation field parameters and the patient or patient population) matches the situation for which the organ doses are required. All conversion coefficients are dependent on beam quality. In most situations, it is adequate to interpolate linearly between values of the conversion coefficients at different beam qualities. Figure 22.8 illustrates

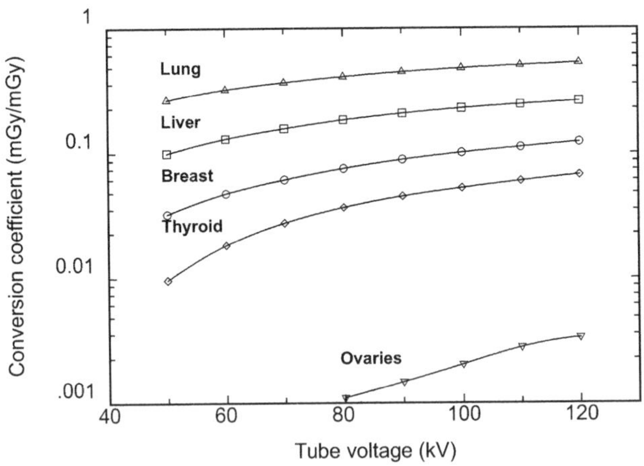

FIG. 22.8. Variation with tube voltage of organ dose conversion coefficients, c_{D_T,K_e}, *for lung, liver, breast, thyroid and ovaries. Chest posterior–anterior examination; X ray spectra have total filtration of 3 mm Al (data from Ref. [22.10]).*

this by showing the variation of selected organ dose conversion coefficients with tube voltage for a chest posterior–anterior examination.

For CT, it is important to match the data to the particular scanner used. Figure 22.9 shows how the conversion coefficient for absorbed dose to the lung, thyroid and ovaries varies with CT slice position for single CT slices, 5 mm thick.

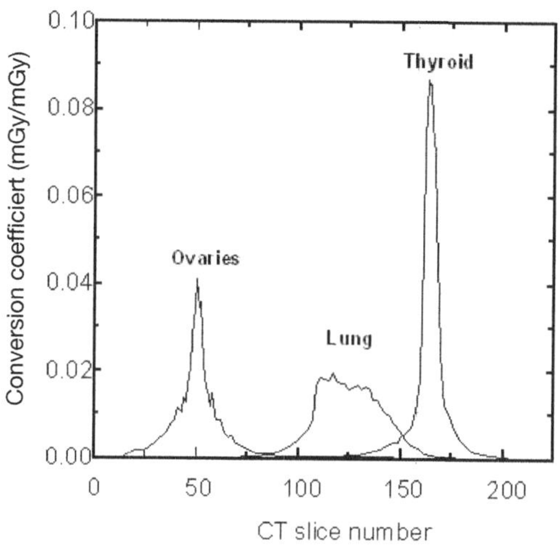

FIG. 22.9. Variation along the length of the patient of organ dose conversion coefficients, $c_{D_T,C_{100,a}}$, *per 5 mm CT slice for lung, thyroid and ovaries for a particular CT scanner (data based on Ref. [22.11]).*

22.5.1.2. Phantom measurements

For situations where no appropriate Monte Carlo calculated conversion coefficients are available, it may be necessary to take custom measurements of organ dose using a suitable anthropomorphic phantom. The measurement of local skin dose for a fixed radiation field is quite straightforward, provided that the IAK varies slowly across the field. For the measurement of organ dose for internal organs, TLDs are often used. This presents more difficulty, particularly for large organs (such as lungs) or widely distributed tissues (such as red bone marrow) because of the difficulty of obtaining adequate spatial sampling This difficulty may arise from either or both of two effects: the rapid decrease of dose with depth in tissue and the partial irradiation of some organs by the primary beam.

22.5.2. Backscatter factors

The backscatter factor relates the IAK K_i and entrance surface air kerma K_e, in accordance with Eq. (22.1). There are several circumstances where it is necessary to convert from IAK to entrance surface air kerma or vice versa. One is when organ dose conversion coefficients are available normalized to IAK, but only measurements of entrance surface air kerma are available. A second circumstance is when measurements of the two air kerma quantities need to be compared. A third (and sometimes very important) example is when the IAK is known and the local skin dose has to be estimated.

As with organ dose conversion coefficients, backscatter factors can be calculated using Monte Carlo methods, or measured using a suitable phantom (to provide backscatter).

Table 22.2 illustrates the dependence of the backscatter factor on field size and beam quality. The effect of backscatter material is also significant. For a 150 kV spectrum filtered by 3 mm Al and a 250 mm × 250 mm field, the values of B for water, ICRU tissue and PMMA backscatter materials are: 1.52, 1.53 and 1.63, respectively, showing that PMMA is not a good tissue substitute material in this case.

22.5.3. Use of data

As noted above, to estimate risk related quantities such as organ dose and effective dose for a given examination and patient size, appropriate conversion coefficients are selected from tabulated data by matching the projection, radiation field and beam quality of the examination. The selected conversion coefficient is then multiplied by the value of the application specific quantity (say Q_i) measured for the examination:

$$D_T = Q_i \, c_{D_T,Q_i} \tag{22.20}$$

TABLE 22.2. BACKSCATTER FACTORS FOR WATER
(data from Ref. [22.12])

Tube voltage (kV)	Filtration (mm Al)	Backscatter factor (B)	
		100 mm × 100 mm field	250 mm × 250 mm field
50	2.5	1.24	1.26
100	3.0	1.36	1.45
150	3.0	1.39	1.52

TABLE 22.3. ORGAN DOSE CONVERSION COEFFICIENTS PER UNIT IAK, CALCULATED FOR TWO VOXEL PHANTOMS AND ONE MATHEMATICAL PHANTOM
(chest posterior–anterior examination; tube voltage: 141 kV; total filtration: 5.7 mm Al; focus image distance: 1500 mm; field size at the image plane: 350 mm × 400 mm) (data from Ref. [22.12])

Organ	Organ dose per unit IAK (mGy/mGy)		
	Voxel GOLEM	Voxel VISIBLE HUMAN	Mathematical ADAM
Colon	0.09	0.04	0.008
Testes	—	—	—
Liver	0.38	0.30	0.27
Lung	0.57	0.51	0.79
Pancreas	0.27	0.19	0.32
Red bone marrow	0.26	0.21	0.21
Skeleton	0.40	0.33	0.39
Spleen	0.77	0.52	0.39
Small intestine	0.09	0.04	0.01
Stomach wall	0.30	0.24	0.14
Thyroid	0.28	0.18	0.14
Surface (entrance)	1.27	1.40	1.39
Surface (exit)	0.10	0.07	0.09

It is important to note that it may not be possible to get a good match between the size of the modelled patient, the position and size of the modelled organs and the position and size of the radiation field, and those of the real situation. Significant errors can arise as a consequence. Whole organs may lie wholly within or partly within the field for one case and wholly outside the field for another, and their depth within the body can be quite different. Table 22.3 demonstrates the differences in organ dose conversion coefficients for a posterior–anterior chest examination at 141 kV when three different phantoms that simulate an adult male are used. The ADAM mathematical phantom and GOLEM voxel phantom have similar external dimensions, but the coefficients for several organs including lung, liver and thyroid are significantly different, owing to differences in the size, shape and position of the internal structures of the two phantoms. The VISIBLE HUMAN voxel phantom is much larger than the GOLEM phantom and comparisons of the results for these two phantoms show that the conversion coefficients in general decrease with increasing patient size, owing to the increased shielding offered to most organs as the body size increases.

22.5.3.1. Radiography and fluoroscopy

Conversion coefficients that are suitable for radiography and fluoroscopy are available, which have been normalized to KAP, IAK and entrance surface air kerma.

Software is available for some of the data tabulations, which can greatly facilitate the calculation of organ or effective dose. In addition, a PC based Monte Carlo computer program (Personal Computer X ray Monte Carlo Program) is available from the Radiation and Nuclear Safety Authority (STUK) in Finland [22.13], which can directly compute organ doses for user specified radiation fields, with the added feature of adjusting the size of the patient, including sizes appropriate for paediatric dosimetry.

A potential source of further error is the use of the KAP in situations where the X ray field extends beyond the patient. A useful check on the accuracy of the calculation is to estimate the IAK from the KAP with knowledge of the X ray beam area and repeat the calculation of organ or effective dose using the former.

In the case of paediatric dosimetry, it is unlikely that the subjects will match the paediatric phantoms used to calculate existing tables of conversion coefficients. This problem can be avoided by using the Personal Computer X ray Monte Carlo Program. Alternatively, tabulated conversion coefficients can be plotted against a measure of phantom size — not age — and the conversion coefficient appropriate for the size of the subject deduced by interpolation.

22.5.3.2. Mammography

Different approaches have been adopted for patient dosimetry in mammography in Europe and the USA, and the methodology is still developing. The methodology discussed in Ref. [22.1] followed European practice at that time and is outlined here. The same general approach is also used in the more recent IAEA report (see Ref. [22.14]).

The MGD for a patient examination is calculated for full field contact mammography using the following equation, which is based on Ref. [22.15]:

$$D_{\mathrm{G}} = g c s K_{\mathrm{i}} \tag{22.21}$$

where

K_i is the IAK for the patient exposure;
g is the conversion coefficient from IAK to MGD for a standard breast of 50% glandularity;
c corrects for differences in glandularity between the patient breast and the standard breast;

and s corrects for differences in the spectrum used.

The factors g and c depend on the beam quality used to image the breast and are tabulated as a function of the half value layer. The standard breast model used for the Monte Carlo simulations was semicircular in cross-section and of radius 80 mm and had a central region that was a uniform mixture of adipose and glandular tissues. Such a model is clearly not representative of all breasts, but provides a reasonable indication of a typical dose for a breast of given glandularity. The same tables of factors are used for craniocaudal and oblique projections.

In order to apply Eq. (22.21), it is necessary to know the glandularity of the breast. This will, in general, not be known, and typical values can be used instead, where these are available. Such data are available from a number of countries. For example, data for women aged 50–64 years attending for breast screening in the UK are given in Table 22.4.

Dosimetry using phantoms is much used in mammography. PMMA is a suitable tissue substitute and Ref. [22.1] uses a standard phantom 45 mm thick to simulate a breast 50 mm thick and of 50% glandularity. As the equivalence is not exact, a small correction term is included in the conversion coefficient used. In the IAEA report on quality assurance for screen film mammography [22.14], this phantom is used to simulate a standard breast 53 mm thick and of 29% glandularity. In this case, the equivalence between the phantom and the standard breast is exact.

An alternative approach that avoids the use of the small correction term involves finding the typical breast that gives the same IAK as a given thickness of PMMA. The resulting thicknesses and compositions are given in Table 22.4 for the above mentioned UK population. Equation (22.21) can then be used directly.

For magnification mammography, the MGD can be approximated by calculating in accordance with Eq. (22.21) and then scaling the result by the ratio of the breast area directly irradiated to that of the compressed breast.

Very occasionally, the effective dose is required for mammography examinations. In this instance, it is reasonable to assume that the absorbed radiation dose in other organs is negligible. Note that the organ weighting factor for the breast must be doubled when calculating the effective dose for a woman, as it is based on the average risk for men and women.

TABLE 22.4. EQUIVALENCE BETWEEN TYPICAL BREASTS AND PMMA PHANTOMS
(data from Ref. [22.15])

PMMA thickness (mm)	Equivalent breast thickness (mm)	Equivalent breast glandularity (%)
20	21	97
30	32	67
40	45	40
45	53	29
50	60	20
60	75	9
70	90	4
80	103	3

22.5.3.3. CT

Monte Carlo conversion factors for CT are fundamentally different from those available for projection radiology because they are tabulated for a sequence of contiguous transverse slices through the phantom, rather than per CT examination. The most widely used Monte Carlo data for CT are the conversion coefficients available from the NRPB and the GSF, which are summarized in Table 22.5.

TABLE 22.5. SUMMARY OF CONVERSION COEFFICIENTS AVAILABLE FOR CT

Source of tabulated conversion factors	Phantom	Datasets	Application specific quantity	Risk related quantity
NRPB	ADULT CRISTY	23	CT kerma index for ICRU muscle	Organ dose
GSF	ADAM, EVA, CHILD, BABY	3	$C_{a,100}$	Organ dose

The practical use of these tabulations is greatly facilitated by software that can integrate the conversion coefficients for individual slices to obtain organ doses for a complete scan. The CT Patient Dosimetry Calculator (ImPACT) is commonly used to manipulate the NRPB data sets. CT-Expo is based on the GSF datasets.

Note that the NRPB and GSF calculations were reported in 1993 and 1991, respectively, using scanners that are no longer in use. Owing to the large diversity of scanner types and their continual change, it is necessary to utilize scanner correction factors as well as conversion coefficients to estimate accurately the organ dose for CT. Extensive work to establish a basis for these factors has been carried out by ImPACT. Third party providers of CT dose calculators have incorporated these scanner correction factors into their calculation algorithms, but their use is not always readily apparent.

Care is needed when using third party CT dose calculators. The potential pitfalls are:

- *Setting the scan range*: Radiosensitive organs should be covered to the same extent as in the patient examination, and the range should include overscanning in spiral mode.
- *Scan mAs*: In spiral mode, the scan mAs can be the actual mAs or the equivalent mAs for a pitch of 1, depending on the manufacturer.
- *Tube current modulation*: The average mAs for the CT scan can be used without incurring significant errors.

Organ and effective doses for paediatric patients can be estimated by calculating the dose for the ADAM, EVA, CHILD and BABY phantoms, plotting the results against patient weight, establishing the best fit function and calculating the organ or effective dose at the appropriate weight.

Finally, it is noted that an approximate relationship between KLP and effective dose calculated has been established using the NRPB CT conversion factors. This empirical relationship facilitates a rough estimate of effective dose directly from $P_{\mathrm{KL,CT}}$ as follows:

$$E = c_{\mathrm{E,KLP}} P_{\mathrm{KL,CT}} \tag{22.22}$$

where E is the effective dose and $c_{\mathrm{E,KLP}}$ is a conversion factor that is specific to the phantom size and anatomical site and is broadly independent of CT scanner model.[7]

[7] Values of $c_{\mathrm{E,KLP}}$ can be found in Refs [22.16, 22.17].

22.5.3.4. *Dental radiography*

The radiation risk associated with dental radiography is very small, so the calculation of organ dose or effective dose is not carried out routinely. Therefore, no extensive tabulations of conversion coefficients exist for dental radiography.

22.5.3.5. *Fetal dose calculations*

From time to time, it is necessary to estimate the fetal dose for a given examination, for example, when the fetus is in the primary beam. For a gestational age of between 0 and 12 weeks, the dose to the uterus can be used as a surrogate for fetal dose. For gestational ages greater than 12 weeks, appropriate conversion coefficients should be used, but only limited data are available.

22.6. DOSE MANAGEMENT

The ICRP states in publication 105 [22.18] that it is inappropriate to set dose limits or dose constraints for patient exposures because the medical condition is invariably more significant than the potential for radiation harm arising from any justified exposure. Instead, the ICRP recommends that justification and dose optimization be used as the primary tools for radiological protection of the patient. Dose management is implicit in the optimization task. Patient doses can only be successfully managed if information is available on the magnitude and range of doses encountered in clinical practice, and DRLs are set using these data. Local practice can then be improved by comparison with appropriate DRLs.

22.6.1. Population based dose surveys

A number of countries have rolling programmes of patient dose surveys for common X ray and CT examinations, such as the Nationwide Evaluation of X ray Trends programme in the USA, and the five-yearly reviews of the UK national patient dose database. Their findings are published on their web sites and as scientific papers. Several other countries conduct ad hoc patient dose surveys, the results of which can be found in the scientific literature. A variety of methodologies (e.g. patient measurements, phantom measurements) and dose quantities (e.g. entrance surface air kerma, IAK) are reported, so care must be exercised when undertaking comparisons.

Figure 21.10 shows the UK distribution of X ray room mean entrance surface doses for the anterior–posterior abdomen examination for 1995. The shape of the distribution is typical of patient dose surveys: a broad, skewed distribution with a high dose tail. The mean entrance surface dose for this examination is 5.6 mGy, but the doses range between 0.75 and 16.6 mGy, and the ratio of the third to the first quartile is 2.0. The range in doses encountered can be explained in part by the differences in screen film systems in clinical use, which ranged in ISO speed from less than 200 to more than 600.

Figure 21.10 also shows a comparison of the distributions of the X ray room mean entrance surface doses in the UK for anterior–posterior abdomen in 2000 and 2005. The dotted lines show the national reference dose set at the third quartile of the distribution. A downward trend in the mean entrance surface dose and the national reference dose is evident over time. This was achieved by improvements in screen film speed: in the 1995 survey, 40% of the rooms used ISO speeds lower than 400; in 2005 this figure was 13%. The high dose tail was less prolonged in 2005 than 1995, providing evidence that national reference doses work as a dose management tool, by encouraging outliers to review their practices. Nevertheless, some X ray rooms still exceeded the 1995 national reference dose in 2005. The ratio of the third quartile to the first quartile did not change with time, suggesting that dose optimization (which would result in a narrowing of the dose distribution) is not taking place, or is less influential than the range in detector technology.

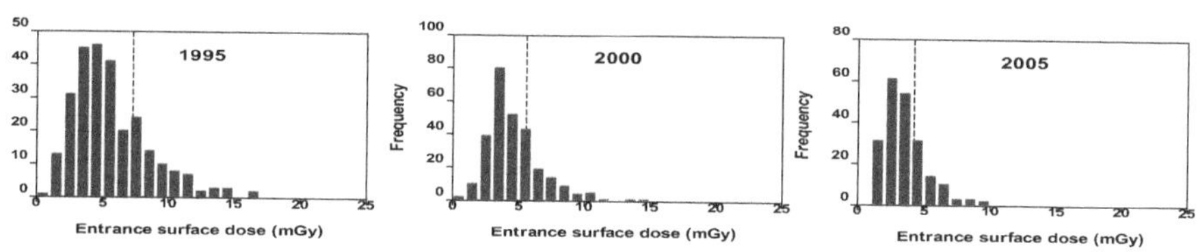

FIG. 22.10. Distributions of X ray room mean entrance surface doses for anterior–posterior abdomen in the UK between 1995 and 2005 (data from Ref. [22.19]).

22.6.2. DRLs

The ICRP identifies DRLs as an essential tool in the management of patient dose. DRLs provide the means of deciding whether the typical patient dose for a particular medical imaging procedure is too high or too low. Note that DRLs are not intended for the management of individual patient doses. In some countries, for example, in the EU, DRLs are required by law.

DRLs are obtained from patient dose surveys or from exposures of standard phantoms, and those obtained from patient dose surveys apply to standard patients only, e.g. 70 kg for an adult in some countries. DRLs are most useful for dose management if they are set in terms of the application specific quantities defined in Section 22.2, because they will then match the data available from dose surveys. For simple X ray examinations, for example, where the tube voltage does not vary, a single exposure factor such as the tube current–exposure time product may be sufficient as the DRL. Computed radiography and digital radiography systems display an exposure index; the exact quantity being manufacturer dependent. These exposure indices refer to irradiation of the detector, not the patient, and correlate poorly with patient dose because of susceptibility to other variables, such as anatomical region and collimation. They are not, therefore, useful for patient dose management.

DRLs can be set at international, national, regional and local levels. In many countries, national DRLs are set for common X ray and CT examinations at the 75% centile of the national patient dose distributions. For example, Table 22.6 summarizes Sweden's DRLs for common adult X ray examinations, Table 22.7 presents Austria's DRLs for a selection of common paediatric examinations, and Tables 22.8 and 22.9 give the DRLs recommended in 2003 for a selection of CT examinations in the UK on adult and paediatric patients, respectively.

TABLE 22.6. SWEDISH DRLs FOR COMMON ADULT X RAY EXAMINATIONS
(data from Ref. [22.20])

Examination	DRL	Quantity
Chest	0.6 Gy·cm^2	KAP
Coronary angiography	80 Gy·cm^2	KAP
Barium enema	50 Gy·cm^2	KAP
Urography	20 Gy·cm^2	KAP
Lumbar spine	10 Gy·cm^2	KAP
Pelvis, hip joints (anterior–posterior or posterior–anterior view)	4 Gy·cm^2	KAP
Mammography (complete examination)	4 mGy	MGD

TABLE 22.7. AUSTRIAN DRLs FOR A SELECTION OF COMMON PAEDIATRIC X RAY EXAMINATIONS
(data from Ref. [22.21])

Examination	Age (years)	IAK (μGy)	KAP (μGy·m^2)
Chest anterior–posterior/ posterior–anterior	0	50	1.7
	1	60	2.3
	5	70	2.6
	10	90	3.7
	15	110	7.3
Skull anterior–posterior/ posterior–anterior	0	350	15
	1	600	25
	5	750	35
	10	900	45
	15	1000	50

TABLE 22.8. REFERENCE DOSES RECOMMENDED FOR USE AS DRLs IN THE UK FOR COMMON ADULT CT EXAMINATIONS
(data from Ref. [22.16])

Examination	DRL (mGy·cm)		Quantity
	Single slice CT	Multiple detector CT	
CT head	760	930	KLP
CT chest	430	580	KLP
CT abdomen and pelvis	510	560	KLP

TABLE 22.9. REFERENCE DOSES RECOMMENDED FOR USE AS DRLs IN THE UK FOR COMMON PAEDIATRIC CT EXAMINATIONS
(data from Ref. [22.16])

Examination	Age (years)	DRL (mGy·cm)	Quantity
CT chest	0–1	200	KLP
	5	230	KLP
	10	370	KLP
CT head	0–1	270	KLP
	5	470	KLP
	10	620	KLP

Radiology departments should set local DRLs with regard to appropriate international or national DRLs. Local dose audit is used to check compliance with the local DRL; each time a dose audit is carried out, the mean value is compared with the local and national DRLs. If the local DRL is exceeded, an investigation should be triggered.

It is important to bear in mind that the national DRL may, from time to time, be derived from technology no longer in use in the radiology department. For example, the national DRL may have been derived from the audit of screen film radiography whereas the radiology department uses computed radiography.

22.6.3. Local dose audit

The dosimetric techniques described in Section 22.5 form the basis of dose audit. Patient data can be collected every 3–5 years for each common X ray and CT examination, and a few months after a new X ray installation. In many situations, a sample can be selected to best represent the population being studied and large enough to reduce the statistical error to an acceptable value. If patient throughput is not sufficient to provide such a sample, constraints may be placed on the range of the appropriate anatomical parameter that is accepted for the survey (e.g. patient weight or breast thickness). The dose for a typical patient may then be found from the median of this distribution or by interpolation of the sampled data to a standard patient size. For paediatric patients, it is necessary to use several size groupings.

REFERENCES

[22.1] INTERNATIONAL ATOMIC ENERGY AGENCY, Dosimetry in Diagnostic Radiology: An International Code of Practice, Technical Reports Series No. 457, IAEA, Vienna (2007).

[22.2] INTERNATIONAL COMMISSION ON RADIATION UNITS AND MEASUREMENTS, Patient Dosimetry for X rays Used in Medical Imaging, Publication 74, ICRU, Bethesda, MD (2006).

[22.3] INTERNATIONAL COMMISSION ON RADIATION UNITS AND MEASUREMENTS, Measurement of Dose Equivalents from External Photon and Electron Radiations, Publication 47, ICRU, Bethesda, MD (1992).

[22.4] INTERNATIONAL ATOMIC ENERGY AGENCY, Status of Computed Tomography Dosimetry for Wide Cone Beam Scanners, IAEA Human Health Reports No. 5, IAEA, Vienna (2011).

[22.5] INTERNATIONAL COMMISSION ON RADIOLOGICAL PROTECTION, The 2007 Recommendations of the International Commission on Radiological Protection, Publication 103, Elsevier (2008) 13–32.

[22.6] NATIONAL RESEARCH COUNCIL, BEIR VII: Health Risks from Exposure to Low Levels of Ionizing Radiation, The National Academies Press, Washington, DC (2006).

[22.7] INTERNATIONAL COMMISSION ON RADIOLOGICAL PROTECTION, The 1990 Recommendations of the International Commission on Radiological Protection, Publication 60, Pergamon Press, Oxford and New York (1991).

[22.8] INTERNATIONAL ATOMIC ENERGY AGENCY, Measurement Uncertainty: A Practical Guide for Secondary Standards Dosimetry Laboratories, IAEA-TECDOC-1585, IAEA, Vienna (2008).

[22.9] INTERNATIONAL ATOMIC ENERGY AGENCY, Implementation of the International Code of Practice on Dosimetry in Diagnostic Radiology (TRS 457): Review of Test Results, IAEA Human Health Reports No. 4, IAEA, Vienna (2011).

[22.10] HART, D., JONES, D.G., WALL, B.F., Normalised Organ Doses for Medical X-ray Examinations Calculated Using Monte Carlo Techniques, Rep. NRPB-SR262, National Radiological Protection Board, Chilton, UK (1994).

[22.11] JONES, D.G., SHRIMPTON, P.C., Normalised Organ Doses for X-ray Computed Tomography Calculated Using Monte Carlo Techniques, Rep. NRPB-SR250, National Radiological Protection Board, Chilton, UK (1996).

[22.12] PETOUSSI-HENNS, N., ZANKL, M., DREXLER, G., PANZER, W., REGULLA, D., Calculation of backscatter factors for diagnostic radiology using Monte Carlo methods, Phys. Med. Biol. **43** (1998) 2237–2250.

[22.13] STUK — RADIATION AND NUCLEAR SAFETY AUTHORITY, PCXMC – A PC-based Monte Carlo Program for Calculating Patient Doses in Medical X-ray Examinations, STUK, Helsinki, http://www.stuk.fi/sateilyn_kaytto/ohjelmat/PCXMC/en_GB/pcxmc/ (accessed on 23 August 2012).

[22.14] INTERNATIONAL ATOMIC ENERGY AGENCY, Quality Assurance Programme for Screen Film Mammography, IAEA Human Health Reports No. 2, IAEA, Vienna (2009).

[22.15] DANCE, D.R., SKINNER, C.L., YOUNG, K.C., BECKETT, J.R., KOTRE, C.J., Additional factors for the estimation of mean glandular breast dose using the UK mammography dosimetry protocol, Phys. Med. Biol. **45** (2000) 3225–3240.

[22.16] SHRIMPTON, P.C., HILLIER, M.C., LEWIS, M.A., DUNN, M., Doses from Computed Tomography (CT) Examinations in the UK – 2003 Review, Rep. NRPB-W67, National Radiological Protection Board, Chilton, UK (2005).

[22.17] AMERICAN ASSOCIATION OF PHYSICISTS IN MEDICINE, The Measurement, Reporting, and Management of Radiation Dose in CT, AAPM Rep. 96, AAPM, College Park, MD (2008).

[22.18] INTERNATIONAL COMMISSION ON RADIOLOGICAL PROTECTION, Radiological Protection in Medicine, Publication 105, Elsevier (2008) 1–108.

[22.19] HART, D., HILLIER, M.C., WALL, B.F., Doses to Patients from Radiographic and Fluoroscopic X ray Imaging Procedures in the UK – 2005 Review, Rep. HPA-RPD-029, Health Protection Agency, Chilton, UK (2007).

[22.20] SWEDISH RADIATION SAFETY AUTHORITY, The Swedish Radiation Protection Authority's Regulations and General Advice on Diagnostic Standard Doses and Reference Levels within Medical X-ray Diagnostics, SSMFS 2008:20, Swedish Radiation Safety Authority, Stockholm (2008).

[22.21] BILLINGER, J., NOWOTNY, R., HOMOLKA, P., Diagnostic reference levels in pediatric radiology in Austria, Eur. Radiol. **20** 7 (2010) 1572–1579.

BIBLIOGRAPHY

CONFERENCE OF RADIATION CONTROL PROGRAM DIRECTORS, Nationwide Evaluation of X-ray Trends (NEXT) Program, 2nd edn, Conference of Radiation Control Program Directors, Frankfort, KY (2007), http://www.crcpd.org/pubs/next_docs/next99dental.pdf (accessed on 23 August 2012).

ZANKL, M., PANZER, W., DREXLER, G., The Calculation of Dose from External Photon Exposures Using Reference Human Phantoms and Monte Carlo Methods, Part VI: Organ Doses from Computed Tomographic Examinations, GSF-Bericht 30/91, National Research Centre for Environment and Health, Neuherberg, Germany (1991).

Chapter 23

JUSTIFICATION AND OPTIMIZATION IN CLINICAL PRACTICE

M. SANDBORG
Linköping University,
Linköping, Sweden

M. BÅTH
Sahlgrenska University Hospital,
Göteborg, Sweden

H. JÄRVINEN
Radiation and Nuclear Safety Authority (STUK),
Helsinki, Finland

K. FAULKNER
North East Strategic Health Authority,
Newcastle upon Tyne, United Kingdom

23.1. INTRODUCTION

All medical exposures must be subject to the principles of justification and optimization of radiological protection, which are common to all practices dealing with potential exposures of humans to ionizing radiation. Justification of medical exposures requires that all medical imaging exposures must show a sufficient net benefit when balanced against possible detriment that the examination might cause.

For patients undergoing medical diagnosis or treatment, there are different levels of justification (see Section 23.2). The practice involving exposure to radiation must be justified in principle through the endorsement of relevant professional societies, as matters of effective medical practice will be central to this judgement. Also, each procedure should be subject to a further, case by case, justification by both the referring clinician who is responsible for the management of the patient and the radiologist who selects the most appropriate imaging examination to answer the referrer's question.

In addition to the requirements of optimization of radiological protection, the concept of optimization of clinical practice in diagnostic radiology must also be considered (see Section 23.3). This is the process requiring a diagnostic outcome for a patient from an imaging procedure while minimizing the factors that cause patient detriment. Along with radiation related considerations, these factors include adverse patient contrast media reactions in computed tomography (CT) and interventional radiology.

Optimization is a multidisciplinary task involving the medical physicist, radiologist, radiographer, hospital or vendor engineer and department management. It is a cyclical process comprising:

- Evaluation of clinical image quality and patient dose to identify the need for action;
- Identification of the possible alternatives to maintain necessary image quality and minimize patient absorbed doses;
- Selection of the best imaging option under the given circumstances;
- Implementation of the selected option;
- Regular review of image quality and patient dose to evaluate if either requires further action.

One key element in managing quality in health care is clinical audit. Clinical audit is a systematic review of the medical procedures against agreed standards for good procedures, seeking to improve the quality and outcome of patient care. It is applicable to justification and optimization and is reviewed in Section 23.4.

23.2. JUSTIFICATION

Justification of medical exposures is the responsibility of both the radiological medical practitioner and the referring medical practitioner. A medical exposure is justified if it provides a benefit to the patient in terms of relevant diagnostic information and a potential therapeutic result that exceeds the detriment caused by the examination. Imaging methods with lower patient effective dose should be considered if the same diagnostic information can be obtained. This is true for all patients, but is especially important for younger patients. No new imaging modality should be established unless the exposed individual or society receives a net benefit to offset the detriment. Justification of medical exposures should be made on three levels (Table 23.1).

TABLE 23.1. LEVELS OF JUSTIFICATION OF MEDICAL EXPOSURES

Level	Description
1	Use of radiation for diagnosis in medicine is generally accepted.
2	Use of radiation in a specific procedure for a specific objective is justified, for example, mammography to follow up after breast cancer. It is important to evaluate if the radiological examination will improve the accuracy of the diagnosis and the treatment of patients. Justification may need to be re-evaluated if new information or new imaging techniques are made available. For example, plain radiography of the lumbar spine for acute back pain or disc hernia except for osteoporotic collapse may not be justified, but magnetic resonance imaging or CT could be considered instead.
3	Use of radiation for an individual patient should be justified prior to the examination. Here, the specific reasons for the exposure and the explicit conditions of the patient should be considered. Referral guidelines (see Section 23.2.1) are an important tool in this evaluation. The request for a radiological examination should convey all relevant information in order for the radiologist to decide on the best radiological procedure. Communications between the referring clinician and the radiologist are very important. Pregnancy and allergy to contrast media should also be considered, as should any relevant previous examination or information in the patient's medical record.

23.2.1. Referral guidelines for imaging

Referral guidelines for imaging are precise statements to help the clinician in making correct decisions on which type of radiological examination is most appropriate, given the clinical conditions. While such guidelines are not absolute rules, there must be good reasons for ignoring them, as they are examples of good practice. The objectives of the referral guidelines are to improve clinical practice, to reduce the number of unnecessary examinations and, hence, to reduce unnecessary medical exposure. The main target group of the guidelines is referring clinicians. Medical physicists can, however, also benefit from studying the general scheme of the guidelines, in order to cooperate better with medical staff in using the guidelines.

In Europe, referral guidelines for imaging have evolved from the Royal College of Radiologists publication Making the Best Use of Clinical Radiology Services [23.1]. Radiological societies in EU States have contributed to an evidence based booklet adopted by the expert groups entitled Referral Guidelines for Imaging [23.2]. The American College of Radiology has published Appropriateness Criteria [23.3], which are evidence based guidelines to assist referring clinicians in making the most appropriate imaging decision.

Guidelines are important, since not all medical imaging examinations give results that alter management of the patient or add confidence to the clinician's diagnosis, and, hence, may add unnecessary radiation dose. There are several causes of unnecessary examinations, including:

- A repeated examination when relevant information was available but not obtained;
- Performing an irrelevant examination;
- Too frequent use of a particular examination;
- Inadequate clinical information preventing important clinical questions from being answered.

The recommendations in the Referral Guidelines for Imaging [23.2] are classified as indicated when the examination is likely to contribute to clinical diagnosis and management of the patient. Other recommendations are specialized examinations that are complex, expensive and require individual discussion with an expert radiologist. Finally, the recommendations can be not indicated initially, routinely or not recommended at all. The guidelines further classify the typical effective doses into five groups from 0 to IV, where group 0 are examinations without ionizing radiation (e.g. ultrasound and magnetic resonance imaging) and group I are examinations where the effective dose is less than 1 mSv (e.g. limb and plain chest radiography). In groups II–IV, the effective doses are 1–5 mSv (e.g. intravenous urogram), 5–10 mSv (e.g. CT chest) and higher than 10 mSv (e.g. positron emission tomography/abdominal CT), respectively.

23.2.2. Sensitive populations

It is recognized that the cancer excess mortality by age of exposure is approximately two to three times higher for children than for the average population. It is, therefore, particularly important to optimize the imaging conditions for children. Typically, however, lower patient doses are used in paediatric radiology simply because the body or body part of the child is smaller than that of the adult. European guidelines with image criteria and criteria for radiation dose are available for common paediatric examinations, but surveys show that the dose to the child can, in some cases, be reduced further.

Contrast media are sometimes necessary to visualize different soft tissues and vessels, since the object contrast is inherently too low. The ideal contrast media will attenuate the X ray beam more than the surrounding tissue but will otherwise leave body organs unaffected. However, in practice, this is not always possible. Some patients react adversely to injected iodine contrast media, with acute (i.e. within 2 h) or late (i.e. within 2 weeks) side effects, which may be severe. Special caution needs to be taken with patients suffering from kidney problems or diabetes. The use of contrast media must be evaluated prior to imaging such patients.

23.2.3. High skin dose examinations

Some interventional radiological procedures may, in addition to high equivalent doses to internal organs, also result in such high local skin or eye lens doses that there is deterministic (acute) radiation damage (see Chapter 20). Examples of deterministic radiation damage include skin erythema and temporary epilation, or lens cataract with visual impairment. The International Commission on Radiological Protection gives guidance on how to identify and manage patients with potential high doses to their skin (see Chapter 24). In these situations, it is important that staff document the measures of absorbed dose that the imaging equipment provides after the procedure, so that any subsequent radiation injury can be managed properly.

23.2.4. Population screening

Diagnostic procedures are examinations of individuals who exhibit some signs or symptoms of disease. Population screening, on the other hand, is a systematic testing of asymptomatic individuals for a disease between its actual onset and manifestation of its symptoms. The objective of screening is to detect the disease while treatment will still have the greatest effect. Therefore, specific guidelines and criteria for screening procedures and for selecting individuals for screening are particularly important. The problem of selecting the proper screening procedure lies in the imaging procedure's ability to separate an early manifested disease in a healthy population. The adverse effects of, for example, cancer screening are the radiation dose and the potential cancer it may induce later in life, the risk of false positive cases with possible anxiety and unnecessary and potentially harmful subsequent examinations and, of course, potentially harmful treatment.

23.2.5. Informed consent

Before the examination, patients undergoing medical imaging procedures should be informed of the potential risk associated with the examination. This includes the risk of allergic reactions to intravenous injected contrast media and potentially high skin doses following sometimes lengthy imaging sessions, for example, percutaneous coronary intervention. Healthy volunteers or patients undergoing alternative or experimental imaging procedures must also be properly informed of the risks. The scientist managing such research must seek and obtain approval from the ethics committee in advance, in accordance with national legislation.

23.3. OPTIMIZATION

Working as a medical physicist with responsibility for optimization of radiographic procedures, it is necessary to use a strategy to perform the optimization work in an efficient way. Different approaches for such strategies exist. For example, it could be argued that it is most important that the examinations that result in the highest patient doses — on an individual level or a population level — are optimized first. An alternative strategy is to focus on examinations that have questionable image quality, as such examinations run the risk of not providing the necessary diagnostic information. No matter what strategy is chosen, it is obvious that examinations that have questionable image quality and those that are of high importance for the patient, as well as those that result in high radiation doses, should be optimized first. It is then important to consider carefully what methods to use for the actual optimization. As optimization involves both radiation dose and image quality, it is necessary to decide what relevant measures to use. Since, for most radiographic procedures, it is the stochastic risk of radiation that is of interest, a dose measure that can be used to estimate this risk should be used. Effective dose is, therefore, often the natural choice. Although the use of effective dose for individual patients is not appropriate, it is suitable for groups of patients and for the purpose of comparing the relative risk between different radiological examinations, or comparing doses before and after a change in imaging conditions (see Chapter 24). The age and sex of the patients need to be considered for a proper risk evaluation. For mammography, the mean glandular dose to the breast tissues is generally used. It could be argued that for procedures for which there is a risk of sustaining deterministic injury, such as interventional radiological procedures, other dose measures, such as skin dose, are also relevant. However, such injuries are rare events and can, in most situations, be avoided if the personnel are adequately trained and the imaging system is not malfunctioning.

Regarding image quality, as described in Chapter 4, there is a large variety of methods intended for evaluation of this somewhat diffuse measure. No matter what method is chosen, it is important to bear in mind that the validity of the results is limited by the validity of the method. Thus, the method used should preferably incorporate the entire imaging chain. As the current gold standard for determining image quality is receiver operating characteristic (ROC) based methods (see Chapter 18), the use of such methods may be advocated for optimization. However, conducting ROC studies may be a cumbersome task and they may, therefore, not be best suited for daily optimization work.

Visual grading is a common and very practical methodology used for the determination of image quality in optimization as an alternative to the ROC approach (see Chapter 4). It uses observers' ratings of the visibility of structures in the image. The ratings are then used to establish a measure of image quality. Visual grading has the strength that the entire imaging chain can be included in the evaluation. The task of the observer resembles that of the radiologist in everyday work, i.e. deciding whether a given image can be used for the required task of detecting abnormality or not. A successful visual grading study is based on letting the observers judge the visibility of those structures that are important in being well visualized in the examination. Commonly reported weaknesses with visual grading are that it is somewhat subjective and that it is prone to bias. This is definitely true. However, radiologists rely on their subjective impression in their daily diagnostic work and it is difficult to remove this limitation without excluding the radiologist from the image quality assessment.

There are many differences between analogue screen film systems and digital systems for optimization. The most important is the fact that while the film constitutes detector, processing and display medium with almost fixed properties, the digital system not only consists of independent detector, processing and display media, but also that many relevant properties of these components are adjustable. For a given screen film system, optimization is a limited task, owing to the fixed sensitivity and latitude of the system. Therefore, the most important task is to choose exposure settings for obtaining a correct exposure. The optimization process consists of choosing the optimal beam quality (tube voltage and filtration) and tube charge (mAs) to match the input signal to the latitude and sensitivity of the screen film system. The sensitivity and latitude of the screen film system can be altered by changing the screen and film, respectively. In this way, a noise level or spatial resolution suitable for a given examination can be obtained.

For digital systems, the displayed image contrast can be adjusted without clinically relevant restrictions, which can be interpreted as if the system has adjustable sensitivity and latitude. The two most important tasks for optimization of a screen film system (using the correct detector dose to obtain optimal optical density (OD) and the correct beam quality to adapt the attenuation differences in the object to the latitude of the system) are, therefore, of little relevance for digital systems. Instead, optimization of digital equipment can be more focused on actually finding the parameter combination (exposure parameters, image processing parameters, etc.) that results in the best image quality for a given effective dose (or other relevant dose measure). Finally, it is necessary to decide on the appropriate tube charge (mAs) that provides sufficiently low noise given the clinical requirements (see Section 23.3.3). In this way, the necessary image quality is obtained at the lowest possible exposure of the patient.

23.3.1. Equipment, guidelines and image criteria

For some common radiographic examinations, the EU has published guidelines that give diagnostic requirements, criteria for radiation dose and examples of good radiographic technique. The requirements include both image criteria and important image details and apply to 'standard sized' patients with the usual symptoms for that type of examination. The image criteria relate to important anatomical structures that should be visible in the images. Typically, the criteria are expressed in several degrees of visibility. For example 'visually sharp reproduction' means that the details are clearly defined, whereas 'visualization' reflects a situation where the details are detected but not fully reproduced. The list of important image details gives the minimum dimensions in the image at which normal or abnormal anatomical details should be recognizable. The criteria have been further developed over the years to be more specific to changes in the imaging condition for use in visual grading evaluation of clinical images (see Section 23.3.3).

The criteria given in the above guidelines for radiation doses to the patient are expressed in terms of entrance surface dose. However, the IAEA code of practice recommends the use of kerma area product, P_{KA}, as the dosimetric quantity in fluoroscopy. The advantage of P_{KA} over entrance surface dose is that the radiation beam size is directly included in the measurement and that P_{KA} values for different projections can be added together with reasonable validity. Adding entrance surface dose from different projections is not meaningful. Internationally, the concept of diagnostic reference levels (see Chapters 22 and 24) has been implemented in some countries and diagnostic standard doses are periodically measured locally in the hospitals and compared with the reference levels. If the reference level is exceeded in a particular X ray room, the hospital needs to review their imaging conditions and to consider, and possibly implement, corrective action to reduce the dose if the clinical image quality requirements can still be met. The implementation of diagnostic reference levels has led to a reduction in patient absorbed doses and must be considered as a successful radiological protection action and a first step towards achieving optimal imaging conditions.

As an example, Table 23.2 lists the European guidelines for an examination of the urinary tract in the anterior–posterior view and with a screen film system. The important details in the urinary tract examination relate to 1 mm calcifications. The image criteria require reproduction of the area of the whole urinary tract, from the upper pole of the kidney to the base of the bladder, as well

TABLE 23.2. AN EXAMPLE OF GOOD RADIOGRAPHIC TECHNIQUE FOR A URINARY TRACT EXAMINATION USING A SCREEN FILM IMAGING SYSTEM

Technique variable	Technique value
Nominal focal spot size	≤1.3 mm
Total filtration	≥3.0 mm Al
Tube voltage	75–90 kV
Exposure time	<200 ms
Imaging system sensitivity	400 (or air kerma 2.5 μGy at detector)
Automatic exposure control	Central or lateral chamber
Antiscatter grid	Grid ratio 10 and strip frequency 40/cm
Focus image detector distance	115 cm (100–150 cm)
Protective shielding	Gonad shields for male patients where appropriate

Note: Optimal radiographic techniques for a digital image detector may differ from those listed here.

as reproduction of the kidney outlines. The psoas outlines should be visualized and visually sharp reproduction of the bones is required. The criterion for entrance surface dose for a standard sized patient is 10 mGy.

23.3.2. Good practice

Listed below are some aspects associated with good radiological practice. Some are related to the management of procedures for examining pregnant patients and of handling patients receiving high absorbed doses, others are related to performing the examination, such as positioning of the patient and radiation field and selecting the most appropriate examination technique and circumstances for reading the images.

23.3.2.1. Pregnant patient and fetus protection

Prior to an examination in the region of the lower abdomen, women should be asked if they are pregnant. If the woman is pregnant or pregnancy cannot be ruled out, and if the primary beam is located close to the fetus, the examination should be postponed until the baby is born, provided this is acceptable from a clinical point of view. If postponing the examination is not possible, an examination without ionizing radiation should be considered if sufficient diagnostic information could be expected to be obtained. If this is also not possible, the examination should be performed, with special measures taken to minimize the dose to the fetus. The decision should be noted in the patient's medical records. This applies especially to an examination with relatively high dose (e.g. CT of the lower abdomen, urography, colon and interventional procedures in that region). Methods to minimize the dose to the fetus should be listed in the procedure documentation and may include limiting the number of projections, use of low dose irradiation protocols and careful collimation of the primary radiation beam. If the fetus is exposed during either a planned or accidental medical exposure, for example, trauma CT of an unconscious pregnant woman, the medical physicist should be contacted to estimate the fetus dose, in order for the clinician to inform the woman in due course of the risks involved.

23.3.2.2. Adapting the exposure setting to patient size

As the relationship between the exposure setting used and the resultant image quality and patient dose is dependent on the size of the patient, it is important to adjust the exposure setting to the size of the patient. This is of particular importance for paediatric CT. Prior to the introduction of tube current modulation in CT, the radiation dose levels used in paediatric CT were often too high. If adult settings were employed, small children would obtain radiation doses several times higher than adults. Tube current modulation has partially solved this problem, as the tube current used is automatically adjusted according to patient size and density (see Chapter 11). However, it is still necessary to find the optimal dose level, as different tube current modulation techniques behave in different ways. Also, the relationship between image quality and noise level is dependent on patient size. This is mainly because the internal structures of children are smaller, but also because children typically have less intra-abdominal fat, which requires that the image noise be lower (and dose higher) to permit delineation of the organs.

23.3.2.3. Managing high local skin doses

The patient should be placed close to the image detector, with the tube as far from the patient as possible in order to minimize local entrance skin dose and to reduce the effect of geometrical unsharpness. This is particularly important in interventional radiological procedures, as long fluoroscopy times and multiple exposures can be anticipated (see Chapter 8). The local skin dose can be high if the same projection is maintained throughout the whole procedure or for a large fraction of it. Changing the projection slightly may reduce the local skin dose below that for deterministic skin injuries, but will not necessarily reduce the dose to internal organs or the stochastic radiation risk. To further reduce local skin dose, additional copper filtration can be dynamically inserted into the X ray beam, provided the generator power is sufficient. Additional copper filtration increases the mean energy of the primary X ray beam and increases the relative transmission through the part being imaged, and hence, for fixed image detector dose, decreases the dose to the skin of the patient. The documentation of high skin dose is facilitated by use of the cumulative dose at the so-called ‘interventional reference point’. For fluoroscopy units, this point is located 15 cm from the isocentre, towards the X ray tube.

23.3.2.4. Positioning of the patient

The patient should be accurately positioned by the radiographer to allow the area of interest to be properly imaged. To minimize patient movement, immobilization equipment should be readily available when needed. In paediatric radiology, correct positioning of the child may be more difficult than for an adult patient. An accompanying person, for example, a parent, should preferably assist in immobilizing the child to ensure that the radiographic projection is properly centred and collimated. The parent should be given appropriate protective clothing (i.e. protective apron, thyroid shield) and their hands should not be directly exposed to the primary beam. In CT, it is particularly important to place the patient in the centre of the gantry to match the shape of the CT beam shaping bowtie filters, otherwise the patient will be overexposed and image artefacts may appear.

23.3.2.5. Limiting the radiation field

Limiting the radiation field to the area of interest will both reduce the radiation risk and improve image quality (as, for a smaller irradiated volume, less scattered radiation will reach the image detector). For example, in fluoroscopy, reducing the radius of the primary beam from 12 cm to 9 cm will almost halve the kerma area product, P_{KA}.

The primary radiation field should not extend beyond the active area of the image detector. This may not always be properly considered in dental radiology (with rectangular image detectors and circular primary beam collimation (see Chapter 10)) and CT, where the dose profile, in some cases, is much wider than the sensitivity profile (see Chapter 11).

23.3.2.6. Protective shielding

Protective shielding should not typically be used on patients, with a few exceptions, for example, a thyroid shield in intraoral radiography and male gonad shields whenever the testicles are in, or a few centimetres outside, the primary radiation beam. In such situations, their use is recommended when the protective shield does not obscure any important radiological structure or result in image artefacts.

23.3.2.7. Compression

Examination of a small body or body part typically results in lower absorbed doses, owing to the shorter path length through tissue and decreased attenuation of the primary beam. Methods to reduce this path length by 'compressing' the body or body part can, therefore, result in significant dose reduction. For example, if the patient's abdomen can be made 3 cm thinner in the central beam direction, the tube charge (mAs) can be reduced by approximately 50%, while maintaining the dose at the image detector. Positioning a patient scheduled for a lumbar spine frontal view in a posterior–anterior position will allow the patient to 'compress' themself. By doing so, the irradiated volume may be reduced and the degrading effect of scattered radiation on image quality will also be reduced. Furthermore, some tissue may be displaced out of the primary X ray beam and hence receive a reduced dose. Compression is generally used in mammography, where, in addition to reducing the mean glandular dose, it has many other benefits (see Chapter 9).

23.3.2.8. Photon energy

The energy of the X ray beam should be adapted to the thickness of the part of the patient being imaged and the diagnostic tasks. Traditionally, lower tube voltages (25–60 kV) are used for thin body sections, such as the extremities and female breast, intermediate tube voltages (60–120 kV) for imaging of the abdomen and when iodine contrast media are used, and high tube voltages (>120 kV) for chest radiography and CT. However, the selection of tube voltage is, in many cases, based on empirical data from screen film radiography, where image contrast is not adjustable after exposure and total exposure (i.e. tube charge) is determined by properly exposing the film to achieve an appropriate OD. In digital radiography, these restrictions do not apply and the tube voltage and tube charge should be selected on the basis of other principles, e.g. detection of pathology. When a fixed energy imparted per unit area to the image detector is required for properly exposing a screen film system, the combination of higher tube voltages and lower tube charges typically results in lower effective dose to the patient. In digital radiography, the opposite combination may be optimal. There are some indications (see Section 23.3.3) that tube voltages lower than those typically used in skeletal examinations and in examinations with iodine contrast media are more appropriate.

23.3.2.9. Low attenuating materials

Any absorbant material between the patient and the image detector will reduce the radiation fluence rate at the image detector and lead to a loss of image information. If an automatic exposure control (AEC) system is used, the exposure time will automatically increase with increasing amounts of absorbing material between the patient and image detector, to compensate, leading to an increase in patient dose. Consequently, efforts should be made to reduce this absorption. Such materials are the image detector protective coating, AEC chambers, couch, cushion and antiscatter grid. Nowadays, most of these are made from low atomic number, low density materials such as plastic or carbon fibre, with the exception, of course, of the lead strips in the antiscatter grid. Also, without AEC, the exposure setting may need to be altered, but this will need to be done manually by the radiographer. It should be noted that if the X ray tube is situated below the patient (as is common in fluoroscopy and interventional radiology), the couch and cushion add extra beam filtration, but do not necessarily increase patient exposure.

23.3.2.10. Scatter rejection methods

The majority of the photons exiting the patient have been scattered in the patient and have changed direction before reaching the image detector plane. These photons will not convey information about the patient and will reduce the contrast and add noise to the image if they are not removed before being absorbed in the image detector. Three main methods are used to minimize the contribution of scattered photons to image formation (see Chapter 6). The most dose efficient method is a scanning fan beam assembly. Here, only a small fraction of the patient is irradiated at a time, with one or several narrow moving collimators in front of and behind the patient, allowing all primary photons, but only a small fraction of the scattered photons, to reach the image detector.

The second method is to increase the separation between the patient and image detector to 20–40 cm, to allow the scattered photons to miss the image detector to some extent. This method is often used when small volumes are irradiated, such as in limb radiography and in the case of small children. In these situations, this air gap technique is also more dose efficient than the third and most common method, the antiscatter grid technique. The grid consists of thin lead strips separated by a low density material, to allow a large fraction of the primary photons to pass through but selectively absorb the scatter. With increasing grid ratio, the solid angle that allows scattered photons to pass decreases and the efficiency of the grid increases, provided the interspace material between the lead strips is made of low atomic number and low density material such as fibre material, but not aluminium. The optimal grid ratio and lead strip width increase with increased scattering volume. The optimal grid ratio also increases with increased lead strip frequency (lead strips per centimetre), although proper alignment of the grid becomes more critical. For this reason, in bedside chest radiography, grids with low strip frequency and grid ratio and large focusing distance are used.

23.3.2.11. AEC setting

The setting of the AEC is important for both patient dose and image quality and should be evaluated for each type of examination. The AEC system usually consists of ionization chambers located behind the grid but before the image detector. During the exposure, the signal is read from the chamber and when the required air kerma is reached, a signal is sent to the X ray generator to terminate the exposure. The AEC system was initially designed for screen film radiography, to assist the radiographer in obtaining the correct exposure of the film by matching the patient structures of interest to the linear part of the film characteristic curve (see Chapters 6 and 7). Digital image detectors have a wider useful dynamic range and can, to some extent, manage over- or underexposure. Figure 23.1 shows that a variation in exposure by a factor of 25 still results in a digital image with appropriate greyscale image contrast. However, quantum noise is very visible in the image to the left, with a fifth of the exposure compared with the one in the middle, which has the clinically used exposure level.

Similar exposure correction systems exist in fluoroscopy units and are known as automatic brightness control. The area used to monitor the signal level from the image intensifier is outlined in the live view monitor and can, to some extent, be changed in size and location to adapt to different projection requirements and field of view.

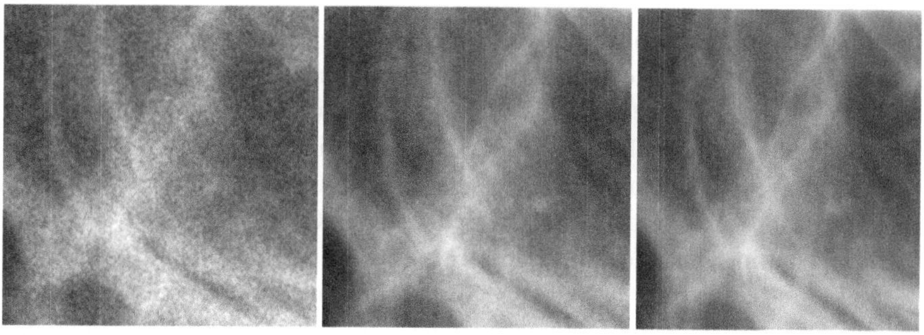

FIG. 23.1. Digital radiographs with different quantum noise levels, showing anatomical structures of the temporal bones in an anthropomorphic head phantom. The dose and quantum noise level in the middle image are used clinically and the consequences of increasing and reducing the dose by a factor of five are shown to the right and to the left, respectively.

23.3.2.12. Appropriate film OD or background quantum noise level

In screen film radiography, the OD of the developed film influences image quality, since the radiographic contrast depends on the OD, i.e. the film characteristic curve (see Chapter 7). Regular control of the film processing is important for maintaining a consistent image quality and dose. However, consistent film processing is not a sufficient requirement for good radiographic practice, as the processing temperature may be set too low, resulting in too low an OD and contrast. This may result in having to increase the required tube charge to maintain sufficient image quality.

The sensitivity of a screen film system depends on the sensitivity of both the fluorescent screen and the film and will influence the amount of quantum noise for a given OD. The sensitivity of the screen can be altered by either increasing the thickness of the fluorescent screen material (absorbing a larger fraction of the photons) or increasing the light yield (emitting more light photons per X ray photon) (see Chapter 7). The latter, however, increases quantum noise.

23.3.2.13. Viewing conditions

Appropriate viewing conditions will aid in reading the diagnostic images. The maximum luminance of monitors ranges between 250 and 450 cd/m^2. With lightboxes, the luminance ranges between 1500 and 6000 cd/m^2 — with the higher values used for mammography. The ambient light in the reading room should be kept low and reflections of other light sources in the monitor minimized. The reader must be able to magnify the image two to four times in order to resolve submillimetre details, as the resolution of the image display is typically less than that of the image itself. Viewing stations for digital images should also be properly calibrated to match the sensitivity of the human eye. Nowadays, common practice is to calibrate diagnostic monitors according to the Grayscale Standard Display Function (GSDF) described in DICOM part 14 (see Chapter 16). The GSDF aims at allowing the rendition of an image with similar appearance on all display systems that are both GSDF calibrated and have the same luminance ratio. Furthermore, based on the assumption of variable adaptation, a calibration using the GSDF results in a perceptually linearized system. This means that a luminance change corresponding to a given number of pixel values has the same probability of being detected over the entire image.

23.3.3. Optimization — two practical examples

23.3.3.1. Example 1: Optimal tube charge in lumbar spine radiography

The European Commission image criteria can be used for simple optimization studies, together with anthropomorphic phantoms or with patients. In the example that follows, an anthropomorphic pelvis phantom and seven image criteria in the lumbar spine anterior–posterior projection (Table 23.3) were used to assess clinical image quality and to identify the required tube charge. Eight images of the pelvis phantom were obtained with different tube charge but the same tube voltage, filtration, field of view and post-processing, etc. The images were assessed by a group of four radiologists and the seven criteria were scored as either fulfilled or not fulfilled. The average fraction of fulfilled criteria

TABLE 23.3. IMAGE CRITERIA FOR LUMBAR SPINE RADIOGRAPHY

Criterion	Description
1	Visually sharp reproduction[a] of the upper and lower plate surfaces, represented as lines in the central beam area
2	Visually sharp reproduction[a] of the pedicles
3	Reproduction[b] of the intervertebral joints
4	Reproduction[b] of the spinous and transverse processes
5	Visually sharp reproduction[a] of the cortex and trabecular structures
6	Reproduction[a] of the adjacent soft tissues, particularly the psoas shadow
7	Reproduction[a] of the sacro-iliac joints

[a] Visually sharp reproduction: Anatomical details are clearly defined; details clear.
[b] Reproduction: Details of anatomical structures are visible but not necessarily clearly defined; details emerging.

was then plotted as a function of the tube charge, which in this case is directly proportional to the effective dose.

Figure 23.2 shows that the average fraction of fulfilled criteria is independent of the tube charge down to approximately 100 mAs, but that this fraction then rapidly decreases to 0.5 with decreasing tube charge. It was primarily the fifth image criterion and secondly the first and second image criteria listed in Table 23.3 that were rated as unfulfilled when the dose was reduced and the quantum noise increased. These three criteria are evaluated on a higher level of image quality 'visually sharp reproduction' than the others. Limitations of the phantom did not allow the sixth criterion to be properly evaluated by the radiologists. In this simple example, a tube charge of approximately 100 mAs minimizes the absorbed dose but maintains clinical image quality in terms of fulfilment of the criteria.

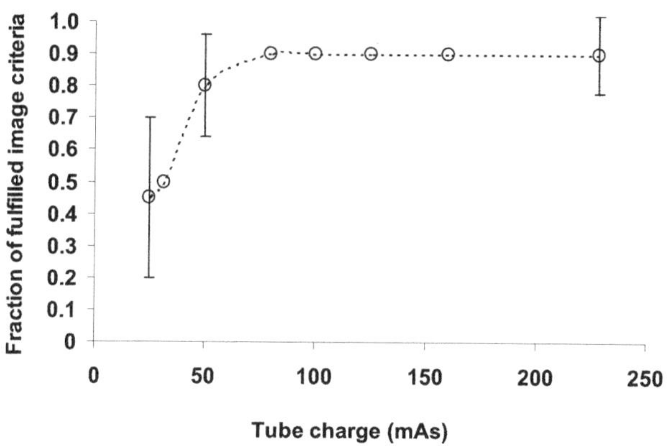

FIG. 23.2. The figure shows the average fraction of fulfilled image criteria assessed by radiologists for images of an anthropomorphic pelvis phantom as function of the tube charge, lumbar spine anterior–posterior, 70 kV. The error bars represent ±1 standard error of the mean.

23.3.3.2. Example 2: Optimal tube voltage for conventional urography

In the second example, it was identified that with the increasing use of CT for urography examination, the indications for conventional urography, when still performed, had changed and were more focused on high contrast details. Therefore, it could not be assumed that the existing tube voltage setting (73 kV) remained optimal for the Gd_2O_2S based flat panel image detector used, although the image quality was acceptable. The purpose of the work was, therefore, to optimize the tube voltage for the urography examination for the new conditions of the examination, so that the necessary image quality could possibly be obtained at a lower effective dose. As a first step, a phantom study was performed to investigate a wide range of tube voltages. Images of an anthropomorphic pelvis phantom, containing simulated contrast filled kidneys and ureters, were collected with the system at tube voltages ranging from 50 to 90 kV at constant effective dose (Fig. 23.3). The images were analysed by radiologists in a visual grading study (see Chapter 4), where the reproduction of the simulated renal pelvises, calyces and ureters was rated. The tube voltage resulting in the best image quality was 55 kV, which, therefore, was selected as the clinical setting.

After using the new setting for some time, images from a number of patients collected with the new setting were selected for comparison with images previously collected with the old setting of 73 kV. The 55 kV images underwent simulated dose reduction to represent images collected at 80, 64, 50, 40 and 32% of the original dose level. All images were included in a visual grading study, where radiologists once again rated the visibility of the renal pelvises, calyces and ureters. The analysis of the given ratings is presented in Fig. 23.4 and shows that for images collected at 55 kV, an effective dose of approximately 85% resulted in the same image quality as images collected at 73 kV at 100% dose. It was, therefore, concluded that the low tube voltage should be used for conventional urography focused on high contrast details and that by using a tube voltage of 55 kV instead of 73 kV, the effective dose could be reduced by approximately 10–20% without adversely affecting image quality.

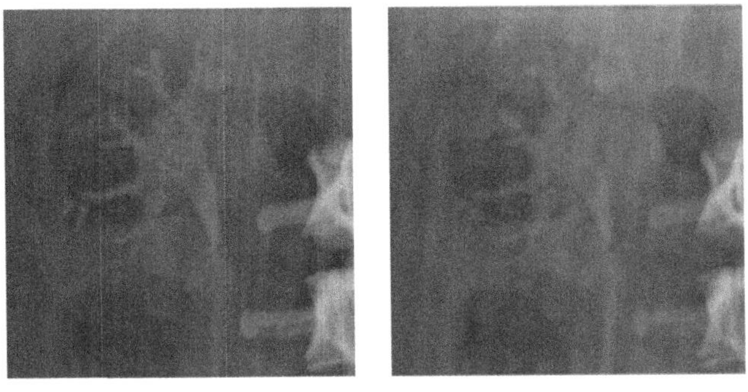

FIG. 23.3. Two X ray images of a pelvis phantom with an added contrast filled kidney collected at 50 kV (left) and 90 kV (right) that were post-processed to achieve similar image contrast. As the image display stage is separated from the image collection stage for a digital radiography system (contrary to a screen film system), the dependence of the displayed image contrast on tube voltage can be much less. Hence, the selection of optimal tube voltage in digital radiography can be different from screen film radiography.

Interestingly, the European Commission guidelines suggest a tube voltage of between 75 and 90 kV for urography (cf. Table 23.2). This shows that the recommended technique settings for screen film systems are not automatically valid for digital radiography and that the exposure parameters need revision after the diagnostic requirements have changed.

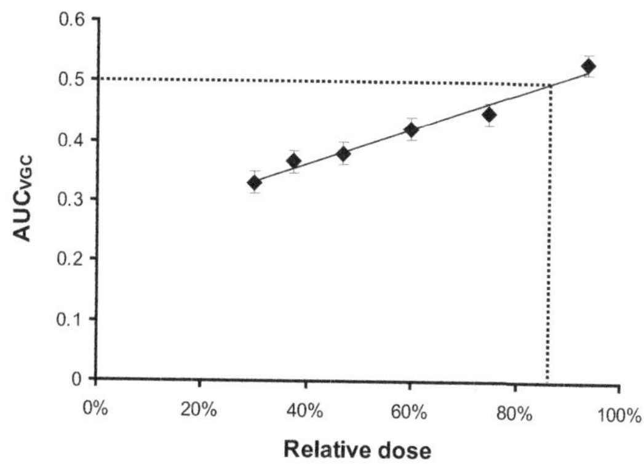

FIG. 23.4. The image quality measure, AUC_{VGC}, for each simulated dose level at 55 kV in the patient study with 73 kV and 100% dose as reference. The error bars represent the standard error of the mean. The AUC_{VGC} can be interpreted as the proportion of comparisons for which the image quality for the evaluated system (in this case the 55 kV images at different dose levels) is rated higher than the reference (the only alternatives for each comparison are higher or lower image quality). An AUC_{VGC} of 0.5 thus corresponds to equal image quality between the evaluated system and the reference. The figure indicates that with 55 kV, only 85% of the dose is needed to obtain the same image quality as with 73 kV.

23.4. CLINICAL AUDIT

23.4.1. Objectives

In the European Commission Medical Exposures Directive [23.4], clinical audit is defined as:

> "a systematic examination or review of medical radiological procedures which seeks to improve the quality and outcome of patient care through structural review whereby radiological practices, procedures and results are examined against agreed standards for good medical radiological procedures, with modification of practices where indicated and the application of new standards if necessary."

In general, the objectives of clinical audit can be distinguished as follows:

(i) Improvement in the quality of patient care;
(ii) Promotion of the effective use of resources;
(iii) Enhancement of the provision and organization of clinical services;
(iv) Further professional education and training.

With these objectives, clinical audit is an integral part of the overall quality improvement process and should be considered as an integral part of quality management and clinical governance.

Clinical audit is a truly multidisciplinary, multiprofessional activity. It must be carried out by auditors with extensive knowledge and experience of the radiological practices to be audited, i.e. they must generally be professionals involved in clinical work within these practices. Further, the general understanding of the concept 'audit' implies that the review or assessment is carried out by auditors who are independent of the organizational unit or practice to be audited.

Clinical audit aims at continuous improvement of the medical practices. Therefore, it should be carried out regularly and it should be ensured that the audit cycle is completed. The general audit cycle consists of selecting a standard of good practice, assessing and comparing local practice with accepted standards, implementing change when necessary, and re-auditing after a certain time. Regular re-audits will improve the quality or give reassurance that good quality is maintained.

Clinical audit should comprise both internal and external assessments and these should supplement each other. Internal audits are undertaken within a given healthcare setting by staff from the same institution, whereas the audit findings can be externally reviewed. In small health care units, internal audits would, rather, be self-assessments. External audits involve the use of auditors who are independent of the radiology department/institution. External audits bring added perspectives to the audit process, because internal auditors might not be able to identify all the weaknesses in their own institution. External auditors should also possess better benchmarking skills in relation to the assessment.

Clinical audit should yield multiple benefits to the health care system, such as:

- Provision of a tool for quality improvement;
- Recognition of quality, good practices and outdated practices;
- Motivation of staff to improve quality;
- Improvement of practice and local standards;
- Adherence to national standards;

- Avoidance of litigation;
- Improvement of communication within the institution;
- Identification of weak points;
- Promotion of development of quality systems.

Clinical audit should, therefore, be able to identify the strengths of a radiology department, as well as areas requiring improvement, and the main beneficiary of this will eventually be the patient.

Comprehensive guidance for clinical audits has been published by the European Commission [23.5] and the IAEA [23.6]. The former provides a general framework for establishing sustainable national systems of audit, while the latter supplements this framework for diagnostic radiology by introducing very practical advice for implementing external clinical audits.

23.4.2. Coverage of radiological practices

Clinical audit should cover the whole clinical pathway and address the three main elements of the radiological practices:

(i) *Structure*: The attributes of the setting in which care occurs, including material resources (e.g. facilities, equipment), human resources (e.g. number, grade and qualification of staff) and organizational structure.
(ii) *Process*: The delivery of patient care.
(iii) *Outcome*: The impact of the department on the health status of patients.

A single clinical audit can assess either the whole clinical pathway of the radiological process, from referral to follow-up (comprehensive audit), or can be limited to specific critical parts of it (partial audit). It can assess the parts of the practices that are generic either to all radiological procedures or to a given specialty (e.g. for CT), or it can go deeper to a selected individual examination.

Clinical audits should address the critical issues for the radiological protection of the patient and the key components of the overall quality system. These include justification and optimization (see Sections 23.2. and 23.3) as essential parts of the process.

Auditing the examination specific practices can usually mean only a few selected examination types per audit. Full details of the procedures should be assessed, at least for the parts where a reasonable consensus on a good practice can be achieved, such as:

- Indications;
- Image criteria, reproduction of anatomical structures;

- Patient position and imaging parameters;
- Protective shielding.

Before starting the clinical audit, the critical areas should be identified and the objectives agreed. For internal audits, the objectives are set by the management of the health care unit to be audited. For external audits, the detailed objectives should be agreed between the auditing organization and the unit to be audited and should be based on any legal requirements on audit programmes, as well as on any recommendations by national coordinating organizations or by health professional and/or scientific societies, when available.

In practice, the process may be subdivided into four sections:

(i) Quality management procedures and infrastructure;
(ii) Patient related procedures;
(iii) Technical procedures;
(iv) Teaching, training and research.

The audit of quality management procedures and infrastructure includes the mission and vision of the radiology unit, its business plan and long term objectives and the departmental workload/patient demographics, the department's organizational structure, staff management processes such as programmes for continuing professional development, working practice instructions and protocols/procedures, departmental premises and equipment.

The audit of patient related procedures includes processes to ensure the appropriateness of examination (referral guidelines used, risk–benefit considerations, contraindications, etc.), processes to ensure relevant clinical conditions are taken into account prior to undertaking an examination (asking about allergies, anticoagulant therapy, pregnancy, etc.), patient identification procedures and failsafes, policies to respect patient confidentiality, and the protocols and procedures for imaging techniques, clinical care, image quality reporting, accidents/incidents, image and record retention, etc.

The audit of technical procedures includes the quality assurance infrastructure and equipment quality assurance procedures. Particular attention is paid to personnel, instrumentation, management support and documentation.

If the centre undertakes research and/or teaching, the programmes for these activities should also be assessed.

23.4.3. Standards of good practice

Good practice is the practice that can be recommended based on the most recent considerations of evidence based data, long term experience and knowledge gained on the necessary structure, process and outcome. These can be based on:

- Legal requirements;
- Ethical principles;
- Results of research;
- Consensus statements;
- Recommendations by learned societies;
- Local agreement (if there is no more universal reference).

The definition of clinical audit presumes that suitable written criteria for good practice are available for the assessments. The guidelines published by the IAEA [23.6] include basic criteria and also reference other publications that can be used as a basis for the establishment of extended criteria. International medical/scientific/professional societies could play an important role in developing such standards.

23.4.4. Relationship with other quality assessment and regulatory control

For external clinical audit, it is important to recognize that this is a different concept to other activities of external quality assessment, such as quality audits for certification of a quality system or audits for accreditation or regulatory inspections. Therefore, when defining the aims and objectives of external clinical audits, it is important to ensure that these will supplement rather than duplicate those of other activities. The relationship of clinical audit with other quality assessments and regulatory control is discussed in detail in the European Commission guidelines [23.5].

23.4.5. Methods and practical organization

Partial audits can be carried out externally by the collection of recordable or measurable data via mail or Internet, with central assessment of the data.

For comprehensive audits, a site visit is needed and should comprise a series of interviews, observations, document and data reviews, measurements, collection of data samples and analysis. Owing to the multidisciplinary nature of the audit, a team of auditors is usually needed, comprising different professionals (radiologist, medical physicist, radiographer, etc.) depending on the scope of the audit. Besides the basic clinical competence, the auditors should receive specific training on the general audit procedure and techniques, as well as the agreed audit programme and the criteria of good practices to be applied.

Once the clinical audit has been completed and the auditor's report with recommendations is available to all staff, the unit should respond to the recommendations with an agreed timeline for improvement. This is important not only to achieve maximum benefit from the audit but also to retain the respect and motivation of the staff for subsequent re-audits.

23.4.6. Role of the medical physicist

In collaboration with the other professionals, the medical physicist has an important role in the planning, preparation and conduct of clinical audits of radiological practices. Medical physics expertise is inevitably required for judging the adequacy and quality of equipment, and assessing patient dose and physical image quality, as well as establishing and running the quality assurance and quality control programmes for equipment. Medical physicists often play a key role in the arrangements and provisions for radiation safety of patients and staff, which are among the major areas for clinical audits of radiological practices.

When the audit involves specific measurements or tests, the physicist member usually takes care of these tests. Further, physicists are usually well practised in making use of different mathematical and statistical tools, which can be of great value in organizing and analysing the audit data. For all these reasons, the audit team should include a medical physicist.

REFERENCES

[23.1] THE ROYAL COLLEGE OF RADIOLOGISTS, Referral Guidelines: Making the Best Use of Clinical Radiology Services, 6th edn, MBUR6, Royal College of Radiologists, London (2007), http://www.rcr.ac.uk/content.aspx?PageID=995 (accessed on 23 August 2012).

[23.2] EUROPEAN COMMISSION DIRECTORATE-GENERAL FOR THE ENVIRONMENT, Referral Guidelines for Imaging, Radiation Protection No. 118, EC, Luxembourg (2000), http://ec.europa.eu/energy/nuclear/radioprotection/publication/doc/118_en.pdf (accessed on 23 August 2012).

[23.3] AMERICAN COLLEGE OF RADIOLOGY, Appropriateness Criteria (2013), http://www.acr.org/Quality-Safety/Appropriateness-Criteria

[23.4] EUROPEAN COMMISSION, Council Directive 97/43/Euratom of 30th June 1997 on Health Protection of Individuals against the Dangers of Ionizing Radiation in Relation to Medical Exposure, and repealing Directive 84/466 Euratom, Off. J. Eur. Comm. Rep. L. **180** (1997) 22–27.

[23.5] EUROPEAN COMMISSION, European Commission Guideline on Clinical Audit for Medical Radiological Practices (Diagnostic Radiology, Nuclear Medicine and Radiotherapy), Radiation Protection No. 159, EC, Luxembourg (2009), http://ec.europa.eu/energy/nuclear/radiation_protection/publications_en.htm (accessed on 23 August 2012).

[23.6] INTERNATIONAL ATOMIC ENERGY AGENCY, Comprehensive Clinical Audits of Diagnostic Radiology Practices: A Tool for Quality Improvement — Quality Assurance Audit for Diagnostic Radiology Improvement and Learning (QUAADRIL), IAEA Human Health Series No. 4, IAEA, Vienna (2010).

BIBLIOGRAPHY

EUROPEAN COMMISSION, European Guidelines on Quality Criteria for Diagnostic Radiographic Images, EUR 16260 EN, EC, Luxembourg (1996).

EUROPEAN COMMISSION, European Guidelines on Quality Criteria for Diagnostic Radiographic Images in Paediatrics, EUR 16261 EN, EC, Luxembourg (1996).

EUROPEAN COMMISSION, European Guidelines on Quality Criteria for Computed Tomography, EUR 16262 EN, EC, Luxembourg (1996).

INTERNATIONAL ATOMIC ENERGY AGENCY, Radiological Protection for Medical Exposure to Ionizing Radiation, IAEA Safety Standards Series No. RS-G-1.5, IAEA, Vienna (2002).

INTERNATIONAL ATOMIC ENERGY AGENCY, Dosimetry in Diagnostic Radiology: An International Code of Practice, Technical Reports Series No. 457, IAEA, Vienna (2007).

INTERNATIONAL COMMISSION ON RADIOLOGICAL PROTECTION, Managing Patient Dose in Digital Radiography, ICRP Publication 93, Elsevier (2004) 1–75.

INTERNATIONAL COMMISSION ON RADIOLOGICAL PROTECTION, Assessing Dose of the Representative Person for the Purpose of Radiation Protection of the Public and the Optimisation of Radiological Protection: Broadening the Process, ICRP Publication 101, Elsevier (2006) 1–114.

INTERNATIONAL COMMISSION ON RADIOLOGICAL PROTECTION, The 2007 Recommendations of the International Commission on Radiological Protection, ICRP Publication 103, Elsevier (2008) 1–332.

Chapter 24

RADIATION PROTECTION

D. SUTTON
Ninewells Hospital,
Dundee, United Kingdom

L.T. COLLINS
Westmead Hospital,
Sydney, Australia

J. LE HERON
International Atomic Energy Agency

24.1. INTRODUCTION

Chapter 21, in describing basic radiation biology and radiation effects, demonstrates the need to have a system of radiation protection that allows the many beneficial uses of radiation to be realized while ensuring detrimental radiation effects are either prevented or minimized. This can be achieved with the twin objectives of preventing the occurrence of deterministic effects and of limiting the probability of stochastic effects to a level that is considered acceptable.

In a radiology facility, consideration needs to be given to the patient, the staff involved in performing the radiological procedures, members of the public and other staff that may be in the radiology facility, carers and comforters of patients undergoing procedures, and persons who may be undergoing a radiological procedure as part of a biomedical research project.

This chapter discusses how the objectives given above are fulfilled through a system of radiation protection and how such a system should be applied practically in a radiology facility.

24.2. THE ICRP SYSTEM OF RADIOLOGICAL PROTECTION

The means for achieving the above objectives of radiation protection have evolved to the point where there is consensus on a system of radiological protection under the auspices of the International Commission on Radiological

Protection (ICRP). The detailed formulation of the system and its principles can be found in ICRP publications and they cannot easily be paraphrased without losing their essence. However, a brief, although simplified, summary is given in this section, especially as it applies to diagnostic radiology and image-guided interventional procedures.

24.2.1. Situations, types and categories of exposure

There are many terms associated with the ICRP system and some are introduced below. In publication 103 [24.1], the ICRP divides all possible situations where radiological exposure can occur into three types: (i) planned exposure situations, (ii) emergency exposure situations and (iii) existing exposure situations. Fortunately, we need only worry about the first of these. The use of radiation in radiology is a planned exposure situation and must be under regulatory control, with an appropriate authorization in place from the regulatory body before operation can commence. It should be noted that the ICRP has previously used the term 'practice' to describe a planned exposure situation, such as the operation of a radiology facility.

In the daily operation of a radiology facility, there will be some radiation exposures with reasonably predictable magnitudes and these are referred to as normal exposures. In addition, unintended exposures or accidents can give rise to what is called potential exposure. These potential exposures remain part of the planned exposure situation as their possible occurrence is considered in the granting of an authorization.

The ICRP [24.1] places the exposure (both normal and potential) of individuals into three categories: occupational exposure, public exposure and medical exposure. All three exposure categories need to be considered in the radiology facility. Medical exposure itself is divided into three components: (i) patient exposure, (ii) biomedical research exposure and (iii) carers and comforters exposure, all of which are relevant to this chapter. An individual person may be subject to one or more of these categories of exposure, but for radiation protection purposes, each is dealt with separately.

24.2.1.1. Occupational exposure

Occupational exposure is defined by the ICRP as including all radiation exposures incurred by workers as a result of their work, in situations which can reasonably be regarded as within the responsibility of the employing or operating management.

24.2.1.2. *Public exposure*

Public exposure includes all exposures other than occupational or medical exposures, and covers a wide range of sources, of which natural sources are by far the largest. Exposures of the embryo and the fetus of pregnant workers are considered public exposures.

Public exposure in a radiology facility would include exposure to persons who may happen to be close to, or within, the facility and potentially subject to radiation penetrating the walls of an X ray room.

24.2.1.3. *Medical exposure*

Medical exposures are intentional exposures for the diagnostic or therapeutic benefit of the patient. As already stated, medical exposure is divided into three components: (i) patient exposure, (ii) biomedical research exposure and (iii) carers and comforters exposure. All three are considered below.

Medical exposures are a very significant source of exposure, and increasingly so. Developed countries have shown an increase of 58% between the 2000 [24.2] and 2008 [24.3] reports of the United Nations Scientific Committee on the Effects of Atomic Radiation (UNSCEAR). Of the diagnostic exposures, computed tomography (CT) was by far the greatest contributor, accounting for 7.9% of examinations, but 47% of the dose. For the whole world population, the annual effective dose per person from medical sources is 0.62 mSv compared with 2.4 mSv for natural sources.

This rapid growth emphasizes the need for effective implementation of the radiation protection principles of justification and optimization.

24.2.2. Basic framework for radiation protection

The ICRP system has three fundamental principles of radiation protection, namely:

- (i) *The principle of justification*: Any decision that alters the radiation exposure situation should do more good than harm.
- (ii) *The principle of optimization of protection*: The likelihood of incurring exposures, the number of people exposed, and the magnitude of their individual doses should all be kept as low as reasonably achievable, taking economic and societal factors into account.
- (iii) *The principle of limitation of doses*: The total dose to any individual from regulated sources in planned exposure situations other than the

medical exposure of patients should not exceed the appropriate limits recommended by the ICRP (see Table 24.1).

In a radiology facility, occupational and public exposures are subject to all three principles, whereas medical exposure is subject to the first two only. More details on the application of the ICRP system for radiological protection as it applies to a radiology facility is given in the remainder of this chapter.

TABLE 24.1. RECOMMENDED DOSE LIMITS IN PLANNED EXPOSURE SITUATIONS[a] [24.1]

Type of limit	Occupational (mSv)	Public (mSv)
Effective dose	20 mSv per year, averaged over defined periods of 5 years[b]	1 mSv in a year[c]
Annual equivalent dose in:		
Lens of the eye	20	15
Skin[d,e]	500	50
Hands and feet	500	n.a.

[a] Limits on effective dose are for the sum of the relevant effective doses from external exposure in the specified time period and the committed effective dose from intakes of radionuclides in the same period. For adults, the committed effective dose is computed for a 50-year period after intake, whereas for children it is computed for the period up to age 70 years.

[b] With the further provision that the effective dose should not exceed 50 mSv in any single year. Additional restrictions apply to the occupational exposure of pregnant women.

[c] In special circumstances, a higher value of effective dose could be allowed in a single year, provided that the average over 5 years does not exceed 1 mSv per year.

[d] The limitation on effective dose provides sufficient protection for the skin against stochastic effects.

[e] Averaged over 1 cm^2 area of skin, regardless of the area exposed.

24.3. IMPLEMENTATION OF RADIATION PROTECTION IN THE RADIOLOGY FACILITY

24.3.1. Introduction

IAEA Safety Standards Series No. GSR Part 3, Radiation Protection and Safety of Radiation Sources: International Basic Safety Standards (the BSS), was published in 2011 [24.4]. The purpose of the standard is to establish basic requirements for protection against exposure to ionizing radiation and for the safety of radiation sources that may deliver such exposure. The requirements of the BSS underpin the implementation of radiation protection in a radiology facility, supplemented by the relevant IAEA safety guides and reports. In particular, specific guidance on applying radiation safety standards in diagnostic radiology and interventional procedures using X rays can be found in Ref. [24.5]. All IAEA publications are downloadable from the IAEA web site.

The ICRP has addressed recommendations for radiological protection and safety in medicine, specifically in publication 73 [24.6], and reaffirmed them in publications 103 [24.1] and 105 [24.7]. Additional ICRP publications on specific aspects of radiation protection in radiology are given in the bibliography.

24.3.2. Responsibilities

Implementation of radiation protection in the hospital or medical facility must fit in with, and be complementary to, the systems for implementing medical practice in the facility. Radiation protection must not be seen as something imposed from 'outside' and separate from the real business of providing medical services and patient care.

To achieve a high standard of radiation protection, it is very important to establish a safety based attitude in every individual such that protection and accident prevention are regarded as natural parts of everyday duties. This objective is primarily achieved by education and training and by encouraging a questioning and enquiring attitude, and also by encouraging a positive and cooperative attitude from the national authorities and the employer in supporting radiation protection with sufficient resources, in terms of both personnel and financial reward. A feeling of responsibility can only be achieved if the people involved regard the rules and regulations as being necessary, and that these are regarded as a support and not a hindrance to their daily work. Every individual should also know their responsibilities through formal assignment of duties.

For an effective radiation protection outcome, the efforts of various categories of personnel engaged in the medical use of ionizing radiation must be coordinated and integrated, preferably by promoting teamwork, where every individual is well aware of their responsibilities and duties.

24.3.3. Responsibilities of the licensee and employer

The licensee of the radiology facility, through the authorization issued by the radiation protection regulatory body, has the prime responsibility for applying the relevant national regulations and meeting the conditions of the licence. The licensee bears the responsibility for setting up and implementing the technical and organizational measures that are needed for ensuring radiation protection and safety. The licensee may appoint other people to carry out actions and tasks related to these responsibilities, but they retain overall responsibility. In particular, the radiological medical practitioner[1], the medical physicist, the medical radiation technologist[2] and the radiation protection officer all have key roles and responsibilities in implementing radiation protection in the radiology facility and these will be discussed in more detail below.

The BSS need to be consulted for details on all the requirements for radiation protection that are assigned to licensees. The employer, who often may not be the licensee, has joint responsibilities, in cooperation with the licensee, with respect to occupational radiation protection.

With respect to medical exposure, the licensee's key responsibilities include ensuring that:

(a) The necessary personnel (radiological medical practitioners, medical physicists and medical radiation technologists) are employed and that the individuals have the necessary education, training and competence to assume their assigned roles and to perform their respective duties.
(b) No person receives a medical exposure unless there has been appropriate referral that it is justified and that the radiation protection has been optimized.

---

[1] Radiological medical practitioner is the generic term used in the revised BSS and is defined as a health professional with education and specialist training in the medical uses of radiation and who is competent to perform independently or oversee procedures involving medical exposure in a given specialty. In the radiology facility, a radiologist is the most common radiological medical practitioner, but many other medical specialists may also serve in this role, including, for example, interventional cardiologists, urologists, gastroenterologists, orthopaedic surgeons and dentists.

[2] Medical radiation technologist is the generic term used in the revised BSS to cover the various terms used throughout the world, such as radiographer and radiological technologist.

(c) All practicable measures are taken to minimize the likelihood of unintended or accidental medical exposures, and to investigate promptly any such exposure, with the implementation of appropriate corrective actions.

With respect to occupational exposure, key responsibilities of the employer and licensee include ensuring that:

(a) Occupational radiation protection and safety are optimized and that the dose limits for occupational exposure are not exceeded.
(b) A radiation protection programme is established and maintained, including local rules and provision of personal protective equipment.
(c) Arrangements are in place for the assessment of occupational exposure through a personnel monitoring programme.
(d) Adequate information, instruction and training on radiation protection and safety are provided.

The licensee also has responsibility for radiation protection of the public, which includes ensuring that:

(a) There are restrictions in place to prevent unauthorized access to functioning X ray rooms.
(b) Area monitoring is carried out to ensure consistency with public exposure standards and that appropriate records are kept.

24.3.4. Responsibilities of other parties

Radiological medical practitioner: The general medical and health care of the patient is the responsibility of the individual physician treating the patient. However, when the patient is in the radiology facility, the radiological medical practitioner has the particular responsibility for the overall radiological protection of the patient. This means assuming responsibility for the justification of the given radiological procedure for the patient, in conjunction with the referring medical practitioner, and also responsibility for ensuring the optimization of protection in the performance of the examination.

Medical physicist: The medical physicist provides specialist expertise with respect to radiation protection of the patient. The medical physicist in diagnostic radiology has responsibilities in the implementation of optimization of radiation protection in medical exposures, including calibration of imaging equipment, and responsibilities with regard to image quality and patient dose assessment, and physical aspects of the quality assurance programme, including medical radiological equipment acceptance and commissioning. The medical physicist is also likely to have responsibilities in providing radiation protection training for medical and health personnel. In addition, the medical physicist may also perform the role of the radiation protection officer, with responsibilities primarily in occupational and public radiation protection (see below).

Medical radiation technologist: The medical radiation technologist has a key role, and their skill and care in the choice of techniques and parameters determine to a large extent the practical realization of the optimization of a given patient's exposure in many modalities.

Radiation protection officer: The radiation protection officer for a radiology facility has responsibilities to oversee and implement radiation protection matters in the facility, but noting (as above) that specialist responsibilities for patient radiation protection lie with the medical physicist. Of course, the radiation protection officer might also be a medical physicist. Duties of the radiation protection officer include: ensuring that all relevant regulations and licence conditions are followed; assisting in the preparation and maintenance of radiation safety procedures (local rules); assisting in shielding design for the facility; arranging appropriate monitoring procedures (individual and workplace); and overseeing education and training of personnel in radiation protection.

All personnel: Notwithstanding the responsibilities outlined above, all persons working with radiation have responsibilities for radiation protection and safety; they must follow applicable rules and procedures, use available protective equipment and clothing, cooperate with personnel monitoring, abstain from wilful actions that could result in unsafe practice, and undertake training as provided.

24.3.5. Radiation protection programme

The BSS require a licensee (and employer where appropriate) to develop, implement and document a protection and safety programme commensurate with the nature and extent of the risks of the practice to ensure compliance with radiation protection standards. Such a programme is often called a radiation protection programme and each radiology facility should have one.

The radiation protection programme for a radiology facility is quite complex as it needs to cover all relevant aspects of protection of the worker, the patient and the general public. Reference [24.5] provides more detailed information on radiation protection programmes.

For a radiation protection programme to be effective, the licensee needs to provide for its implementation, including the resources necessary to comply with the programme and arrangements to facilitate cooperation between all relevant parties. Often, radiology facilities will have a radiation protection committee, or similar, to help supervise compliance with the radiation protection programme.

24.3.6. Education and training

Education and training in radiation protection underpins much of the practice of radiation protection. Such education and training needs to occur before persons assume their roles in the radiology facility, with refresher training occurring subsequently at regular intervals. The radiologists, medical radiation technologists and medical physicists would normally receive this education and training in radiation protection as part of their professional training. However, there are other medical specialists who assume the role of radiological medical practitioner, such as interventional cardiologists, orthopaedic surgeons, etc. These persons must also have the appropriate education and training in radiation protection, and this typically needs to be arranged outside their professional training. Often, this will fall to the medical physicist associated with the radiology facility. The training in all cases needs to include practical training. Nurses may also be involved in radiological procedures and appropriate education and training in radiation protection needs to be given to them. Details on appropriate levels of training are given in Ref. [24.5].

24.4. MEDICAL EXPOSURES

24.4.1. Introduction

The detailed requirements given in the BSS are applicable to medical exposure in the radiology facility. Furthermore, the IAEA Safety Guide on Radiological Protection for Medical Exposure to Ionizing Radiation [24.8] describes strategies to involve organizations outside the regulatory framework, such as professional bodies (e.g. radiologists, cardiologists, medical physicists, radiographers), whose cooperation is essential to ensure compliance with the BSS requirements for medical exposures. Examples that may illustrate this point include acceptance testing processes for radiation equipment and protocols for quality assurance and for reporting accidental medical exposure. Reference [24.5] provides further specific advice. A summary of the most relevant issues for diagnostic radiology and image-guided interventional procedures is given in this section.

As discussed above, dose limits are not applied to patients undergoing medical exposures. The reason for the differences between the treatment afforded to medical and occupational or public exposures is that there is both a benefit and a detriment associated with medical exposures whereas for the others there is only a detriment. However, as outlined in Section 24.2, there is a class of medical exposure that is concerned with exposures to volunteers in biomedical research programmes and another to carers and comforters. For these groups, some type of constraint needs to be applied since they receive no direct medical benefit from their exposure. (The concept of a source related dose constraint was first introduced in ICRP publication 60 [24.9] and is taken to mean a dose that should not be exceeded from a single specific source, and below which optimization of protection should take place.)

Notwithstanding this exception, the philosophical basis for the management of medical exposures differs from that for occupational or public exposure and, in diagnostic radiology, is concerned with the avoidance of unnecessary exposure through the application of the principles of justification and optimization (see Chapter 23 for more details).

Calibration and clinical dosimetry are two activities that support the implementation of optimization. The licensee of the radiology facility needs to ensure that a medical physicist calibrates all sources used for medical exposures, using dosimeters that have a calibration traceable to a standards dosimetry laboratory. Further, the medical physicist needs to perform and document an assessment of typical patient doses for the procedures performed in the facility.

As mentioned earlier, dose limits do not apply to medical exposure. However, a very important tool in the optimization process is the use of diagnostic reference levels (DRLs), which are discussed in the next section.

24.4.2. DRLs

DRLs are dose levels for typical examinations of groups of standard-sized patients or standard phantoms and for broadly defined types of equipment (see Section 22.6). They do not represent a constraint on individual patient doses but give an idea of where the indistinct boundary between good or normal practice and bad or abnormal practice lies. DRLs are usually set using a threshold in a distribution of patient doses or related quantities. When implemented at national or international level, this is frequently the 75th percentile of the observed distribution of doses (or an indicator of dose, such as fluoroscopic screening time) to patients or phantoms for a particular examination. The 75th percentile is by no means ‘set in stone’, for example, some authors suggest that reference levels set at a local level may be defined as being the mean of a locally measured distribution of doses. Reference levels set using a distribution of doses implicitly accept that all elements in the distribution arise from exposures that produce an image quality that results in the correct diagnosis being given.

In the radiology facility, the DRL is used as a tool to aid dose audit and to serve as a trigger for investigation. Periodic assessments of typical patient doses (or the appropriate surrogate) for common procedures are performed in the facility and comparisons made with the DRLs. A review is conducted to determine whether the optimization of protection of patients is adequate or whether corrective action is required if the typical average dose for a given radiological procedure:

(a) Consistently exceeds the relevant DRL; or
(b) Falls substantially below the relevant DRL and the exposures either do not provide useful diagnostic information or do not yield the expected medical benefit to patients.

If a local dose review demonstrates that doses do not, on average, exceed a DRL established nationally or internationally, it does not mean that that particular radiological procedure has been optimized; it just means that the practice falls on one side of a divide. There may well be scope for improvement and by establishing and setting their own DRLs based on local or regional data, radiology facilities may well be able to adapt local practice and optimize exposures more effectively. Details on the operational aspects of the use of DRLs are given in Chapter 22.

24.4.3. Quality assurance for medical exposures

The BSS require the licensee of the radiology facility to have a comprehensive programme of quality assurance for medical exposures. The programme needs to have the active participation of the medical physicists, radiologists and radiographers, and needs to take into account principles established by international organizations, such as the World Health Organization and the Pan American Health Organization, and relevant professional bodies. Chapter 19 provides more details on quality management.

24.4.4. Examination of pregnant women

As discussed in Chapter 20, different types of biological effect are associated with irradiation of the unborn child. Therefore, special consideration should be given to pregnant women.

As a basic rule, it is recommended that radiological procedures for women who are likely to be pregnant should be avoided unless there are strong clinical indications to the contrary. There should be signs in the waiting area, cubicles and other appropriate places requesting a woman to notify the staff if she is pregnant or thinks she is. Further, for radiological procedures that could lead to a significant dose to an embryo or fetus, there should be systems in place to ascertain pregnancy status. The justification for the radiological procedure would include consideration of the patient being pregnant. If, after consultation between the referring medical practitioner and the radiologist, it is neither possible to substitute a lower dose or non-radiation examination nor postpone the examination, then the examination should be performed. Even then, the process of optimization of protection also needs to consider protection of the embryo/fetus.

Fetal doses from radiological procedures vary enormously, but clearly are higher when the examination includes the pelvic region. At the higher end, for example, routine diagnostic CT examinations of the pelvic region with and without contrast injection can lead to a fetal absorbed dose of about 50 mGy. The use of a low dose CT protocol and a reduction in the scanning area to a minimum will lower the fetal dose.

If a fetal dose is suspected of being high (e.g. >10 mGy), it should be carefully determined by a medical physicist and the pregnant woman should be informed about the possible risks. The same procedure should be applied in the case of an inadvertent exposure, which could be incurred by a woman who was later found to have been pregnant at the time of the exposure, and/or in emergency situations.

Irradiation of a pregnant patient at a time when the pregnancy was not known often leads to her apprehension because of concern about the possible effects on the fetus. Even though the absorbed doses to the conceptus are generally small, such concern may lead to a discussion regarding termination of pregnancy because of the radiation risk. It is, however, generally considered that for a fetal dose of less than 100 mGy, as in most diagnostic procedures, termination of pregnancy is not justified from the point of view of radiation risk (see Section 20.14 and Refs [24.10, 24.11]).

24.4.5. Examination of children

Special consideration needs to be given to the optimization process for medical exposures of children, especially in the case of CT. The CT protocol should be optimized by reducing the mAs and kV without compromising the diagnostic quality of the images. Careful selection of the slice width and pitch as well as the scanning area should also be made. It is important that individual protocols based on the size of the child are used, derived by a medical physicist and the responsible specialist.

24.4.6. Helping in the care, support or comfort of patients

Certain patients, such as children, the elderly or the infirm, may have difficulty during a radiological procedure. Occasionally, people knowingly and voluntarily (other than in their employment or occupation) may offer to help in the care, support or comfort of such patients. In such circumstances, the dose to these persons (excluding children and infants) should be constrained so that it is unlikely that the dose would exceed 5 mSv during the period of a patient's diagnostic examination.

24.4.7. Biomedical research

Diagnostic radiological procedures may form part of a biomedical research project, typically as a means of quantifying changes in a given parameter under investigation or in assessing the efficacy of a treatment under investigation. An exposure as part of biomedical research is treated as medical exposure and therefore is not subject to dose limits. The BSS require the use of dose constraints, on a case-by-case basis, in the process of applying optimization to exposures arising from biomedical research. Typically, the ethics committee would specify such dose constraints in granting its approval.

24.4.8. Unintended and accidental medical exposures

In any radiology facility, there is always the potential for unintended or accidental medical exposures. These include any diagnostic or image guided interventional procedure that irradiates the wrong individual or the wrong tissue of the patient, any exposure for diagnostic purposes or arising from an image guided interventional procedure substantially greater than intended, any inadvertent exposure of the embryo or fetus in the course of performing a radiological procedure, or any equipment, software or other system failure, accident, error or mishap with the potential for causing a patient exposure substantially different from that intended.

If an unintended or accidental medical exposure occurs, then the licensee is required to determine the patient doses involved, identify any corrective actions needed to prevent a recurrence and implement the corrective measures. There may be a requirement to report the event to the regulatory body.

24.5. OCCUPATIONAL EXPOSURE

Detailed requirements for protection against occupational exposure are given in the BSS, and recommendations on how to meet these requirements are given in Refs [24.12, 24.13]. Both of these IAEA safety standards are applicable to the radiology facility and, in addition, Ref. [24.5] provides further specific advice. A summary of the most relevant issues for a radiology facility is given in this section.

24.5.1. Control of occupational exposure

Control of occupational exposure should be established using both engineering and procedural methods. Examples of engineering controls include room shielding specified prior to the installation, whilst procedural controls include the establishment of controlled areas and use of local rules.

It is the joint responsibility of the employer and licensee to ensure that occupational exposures for all workers are limited and optimized and that suitable and adequate facilities, equipment and services for protection are provided. This means that appropriate protective devices and monitoring equipment must be provided and used properly and consequently that appropriate training be made available to staff. In turn, staff themselves have a responsibility to make best use of the equipment and procedural controls instigated by the employer or licensee.

In general, controlled areas should be established in any area in which a hazard assessment identifies that measures are required to control exposures during normal working conditions, or to limit the impact of potential exposures. Designation of controlled areas will depend on the magnitude of the actual and potential exposures to radiation.

In practice, all X ray rooms should be designated as being controlled, whereas the extent of a controlled area established for the purposes of mobile radiography will be the subject of a hazard assessment. Warning signs should be displayed at the entrance to controlled areas and wherever possible entrance to the area should be controlled via a physical barrier such as a door, although this may well not be possible in the case of mobile radiography. There should be local rules available for all controlled areas. The rules should identify access arrangements and also provide essential work instructions to ensure that work is carried out safely, including instruction on the use of individual dosimeters. The local rules should also provide instruction on what to do in the case of unintended and accidental exposures. In this context, the local rules should also identify an occupational dose above which an investigation will be triggered (investigation level).

24.5.2. Operational quantities used in area and personal monitoring

For a monitoring programme to be simple and effective, individual dosimeters and survey meters must be calibrated using a quantity that approximates effective or equivalent dose (see Section 22.3). Effective dose represents the uniform whole body dose that would result in the same radiation risk as the non-uniform equivalent dose, which for X rays is numerically equivalent to absorbed dose. In concept at least, it is directly related to stochastic radiation risk and provides an easy to understand link between radiation dose and the detriment associated with that dose. However, it is an abstract quantity that is difficult to assess and impossible to measure directly. The need for readily measurable quantities that can be related to effective dose and equivalent dose has led to the development of operational quantities for the assessment of external exposure. Defined by the International Commission on Radiation Units and Measurements, the operational quantities provide an estimate of effective or equivalent dose that avoids both underestimation and excessive overestimation in most radiation fields encountered in practice.

The operational quantities are defined for practical measurements in both area and individual monitoring. In radiation protection, radiation is often characterized as being either weakly or strongly penetrating, depending on which dose equivalent is closer to its limiting value. In practice, the term 'weakly penetrating' radiation usually applies to photons below 15 keV and to β radiation.

There are two operational quantities used for monitoring external radiation: ambient dose equivalent and directional dose equivalent. The unit of both is the sievert (Sv). For the purpose of area monitoring, the ambient dose equivalent, $H^*(\mathrm{d})$, and the directional dose equivalent, $H'(\mathrm{d},\Omega)$, are defined. They relate the external radiation field to the effective dose equivalent in the International Commission on Radiation Units and Measurements sphere phantom at depth d, on a radius in a specified direction Ω. For strongly penetrating radiation, a depth, d, of 10 mm is used; the ambient dose equivalent being $H^*(10)$ and the directional dose equivalent being $H'(10,\Omega)$. For weakly penetrating radiation, the ambient and directional dose equivalents in the skin at $d = 0.07$ mm can be used but are not likely to be encountered in the radiological environment.

The operational quantity used for individual monitoring is the personal dose equivalent $H_\mathrm{p}(d)$, measured at a depth, d, in millimetres of soft tissue. The unit of personal dose equivalent is the sievert. Use of the operational quantity $H_\mathrm{p}(10)$ results in an approximation of effective dose. $H_\mathrm{p}(0.07)$ provides an approximate value for the equivalent dose to the skin whilst $H_\mathrm{p}(3)$ is used for the equivalent dose to the lens of the eye. Since $H_\mathrm{p}(d)$ is defined in the body, it cannot be measured directly and will vary from person to person and also according to the location on the body where it is measured. However, practically speaking, personal dose equivalent can be determined using a detector covered with an appropriate thickness of tissue equivalent material and worn on the body.

24.5.3. Monitoring occupational dose

The main purposes of a monitoring programme are to assess whether or not staff doses exceed the dose limits and, through regular review, to asses the effectiveness of strategies being used for optimization. It must always be stressed that the programme does not serve to reduce doses; it is the results of those actions taken as a result of the programme that reduce occupational exposures.

Individual monitoring should be undertaken for workers who are normally exposed to radiation in controlled areas. In the X ray department, this would include radiologists, medical physicists, radiographers and nurses. Other staff groups such as cardiologists and other specialists who perform image guided interventional procedures are also candidates for individual monitoring.

Individual monitors (dosimeters) will be designed to estimate either the effective dose or an equivalent dose to an organ such as the fingers. There are many types of individual dosimeter; technologies include thermoluminescent dosimeters, optically stimulated luminescent dosimeters, film and a variety of electronic devices (see Chapter 21).

Whole body dosimeters measure $H_p(10)$ (and usually $H_p(0.07)$) and should be worn between the shoulders and the waist, and worn *under* any protective clothing, such as an apron, whenever one is used. When it is thought that doses might be high as, for example, in interventional radiology, two dosimeters might be required: one under the apron at waist level and one over the apron at collar level.

There are many published algorithms for utilizing dosimeter values from one or more dosimeters to estimate the effective dose, E. One commonly used algorithm is $E = 0.5H_W + 0.025H_N$, where H_W is the dose at waist level under the protective apron, and H_N is the dose at neck level outside the apron. In all cases, it is important that the wearing position, the presence or not of protective clothing, and the reported dosimeter dose quantities be known. Dosimeters worn at the collar can also give an indication of the dose to the thyroid and to the lens of the eye, but in the latter case it should be noted that this really is indicative only and should not be recorded as an accurate dose to that particular organ.

Individual dosimeters intended for assessing extremity doses usually come in the form of ring badges or finger stalls which slip over the end of the finger (Fig. 24.1). The usual reporting quantity for these devices is $H_p(0.07)$. Both types will measure the dose at different places on the hand and care must be taken when deciding which type to use. It is very important to choose the digit and hand that are going to be monitored; the dominant hand may not be the one which receives the greatest exposure. For example, a right handed radiologist may place his left hand nearer the patient when performing an interventional procedure.

In all cases, whether whole body or extremity monitoring is to be used, the monitoring period should ideally be one month, and should not exceed three months. The exact period should be decided by a hazard assessment.

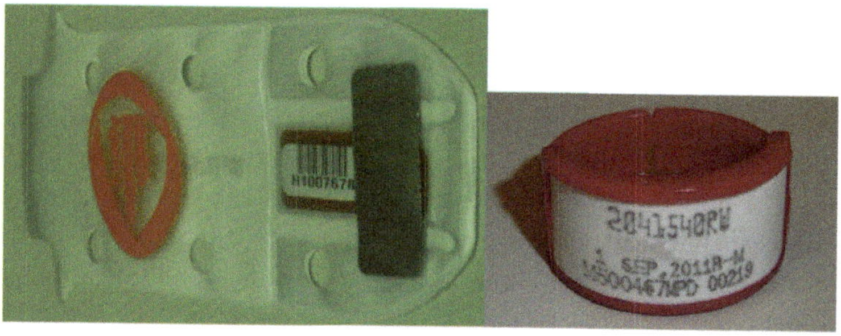

FIG. 24.1. Finger stall and ring badge, both used for extremity monitoring.

To ensure that the monitoring programme is carried out in the most efficient manner, the delay between the last day on which an individual dosimeter is worn and the date of receipt of the dose report from the approved dosimetry service should be kept to a minimum. For the same reason, it is imperative that workers issued with dosimeters return them on time. Results of the monitoring programme should be shared with staff and used as the basis for implementing and reviewing dose reduction strategies.

If, on receipt of a dose report, an employee is found to have either a cumulative or single dose that exceeds the investigation level specified in the local rules, an investigation should be initiated to determine the reason for the anomalous exposure and to ensure that there is no repeat of the occurrence. The investigation level should have been set at a level considerably lower than that of the regulatory dose limit and the opportunity should be taken to alter practice to ensure that doses are kept as low as possible. In the unlikely event that a regulatory dose limit is breached, the regulatory authorities should be informed in the manner prescribed locally.

24.5.4. Occupational dose limits

The recommended occupational dose limits for planned exposure situations, as given by the ICRP, are presented in Table 24.1. The IAEA incorporates the ICRP recommended dose limits into its safety standards. The BSS also add stronger restrictions on occupational doses for 'apprentices' and 'students' aged 16–18 years, i.e. dose limits of an effective dose of 6 mSv in a year, an equivalent dose to the lens of the eye of 20 mSv in a year, and an equivalent dose to the extremities or to the skin of 150 mSv in a year. These stronger dose limits would apply, for example, to any 16–18 year old student radiographer.

24.5.5. Pregnant workers

The BSS require a female worker, on becoming aware that she is pregnant, to notify her employer in order that her working conditions may be modified, if necessary. The notification of pregnancy is not considered a reason to exclude a female worker from work; however, the employer of a female worker who has been notified of the pregnancy must adapt the working conditions in respect of occupational exposure so as to ensure that the embryo or fetus is afforded the same broad level of protection as that required for members of the public. In other words, the dose to the embryo or fetus should not normally exceed 1 mSv.

The possibility of a dose to the embryo or fetus approaching 1 mSv should be assessed once pregnancy has been declared. In general, in diagnostic radiology, it will be safe to assume that provided the dose to the employee's abdomen is less than 2 mSv, then the doses to the fetus will be lower than 1 mSv. The departmental manager, in conjunction with the radiation protection officer, should also decide on whether it is appropriate to reallocate staff duties, or to apply additional protective measures.

Depending on the result of a hazard assessment, which considers the type of work being performed by the pregnant employee, it may prove valuable to issue the member of staff with an electronic personal dosimeter so that patterns of exposure can be identified in real time.

24.5.6. Accidental and unintended exposure

In the case of equipment failure, severe accident or error occurring that causes, or has the potential to cause, a dose in excess of the annual dose limit, an investigation must be instigated as soon as possible. The purpose of the investigation will be to:

(a) Identify how and why the occurrence took place;
(b) Assess what doses were received;
(c) Identify corrective actions;
(d) Make recommendations on actions required to minimize the possibility of future unintended or accidental exposures occurring.

24.5.7. Records

The BSS require that employers and licensees retain exposure records for each worker. The exposure records should include:

(a) Information on the general nature of the work involving occupational exposure;
(b) Information on doses at or above the relevant recording levels and the data on which the dose assessments have been based;
(c) Information on the dates of employment with each employer and the doses received in each employment;
(d) Details of any doses due to emergency exposure situations or accidents, which should be distinguished from doses received during normal work;
(e) Details of any investigations carried out.

Employers and licensees need to provide workers with access to their own exposure records.

24.5.8. Methods of reducing occupational exposure

Reduction of staff and public doses follows the basic principles of time, distance and shielding, which are:

(a) Restrict the time for which a person is exposed to radiation as much as possible. The longer the exposure, the greater the cumulative dose.
(b) Ensure that the distance between a person and the X ray source is kept as large as practicable. Radiation from a point source follows the inverse square law, i.e. the fluence is inversely proportional to the square of the distance from the source. Double the distance means a quarter of the dose, but half the distance means four times the dose. For larger sources, such as scatter from a patient, the inverse square law will not be accurate over short distances and a smaller power than two will be needed. However, as an approximation, and at distances normally used for protection purposes, the inverse square law can be used.
(c) Employ appropriate measures to ensure that the person is shielded from the source of radiation. Materials of high atomic number and density such as lead or steel are commonly used for facility shielding.

It is not always necessary to adopt all three principles. There will be occasions when only one or two should be considered, but equally, there will also be instances when application of the 'as low as reasonably achievable' principle requires the use of all three.

The level of occupational exposure associated with radiological procedures is highly variable and ranges from potentially negligible in the case of simple chest X rays to significant for complex interventional procedures.

From the occupational perspective, there are two 'sources' of radiation exposure. Clearly, the X ray tube is the true source of radiation, but in practice, with proper shielding of the X ray head, there should be very few situations where personnel have the potential to be directly exposed to the primary beam.

This leaves the other source, which is the patient. Interaction of the primary X ray beam with the part of the patient's body being imaged produces scattered radiation, i.e. radiation that emanates from the patient in all directions. Thus, the main source of occupational exposure in most cases is proximity of personnel to the patient when exposures are being made. Further, the level of scatter is determined largely by the dose to the patient, meaning that a reduction in patient dose to the minimum necessary to achieve the required medical outcome also results in a lowering of the potential occupational exposure. A common and useful guide is that by looking after the patient, personnel will also be looking after their own occupational exposure.

24.5.8.1. Working at some distance from the patient

For many situations, such as radiography, mammography and general CT, there is usually no need for personnel to be physically close to the patient. This enables good occupational radiation protection to be achieved through maximizing the distance between the patient and personnel and the use of structural shielding.

Appropriate room design with shielding specification by a radiation protection officer (see Section 24.3.4) should ensure that for these X ray imaging situations, occupational exposure will essentially be zero.

24.5.8.2. Working close to the patient

There are some situations, typically in fluoroscopic examinations and in image guided interventional procedures, where it is necessary to maintain close physical contact with the patient when radiation is being used. Distance and structural shielding are no longer options.

Scattered radiation can be attenuated by protective clothing worn by personnel, such as aprons, glasses and thyroid shields, and by protective tools, such as ceiling suspended protective screens, table mounted protective curtains or wheeled screens, placed between the patient and the personnel. Depending on its lead equivalence (typically 0.3–0.5 mm lead) and the energy of the X rays, an apron will attenuate 90% or more of the incident scattered radiation. Protective aprons come in different thicknesses and shapes, ranging from the simple front-only apron to a full coat, the former being effective only if the wearer is always facing the source of the scattered radiation. Protective clothing should be checked for shielding integrity (not lead equivalence) annually. Depending on the use to which the protective clothing is put, this can be done visually or by using X ray (fluoroscopic) screening.

The lens of the eye is highly sensitive to radiation. For persons working close to the patient, doses to the eyes can become unacceptably high. Wearing protective eye wear, especially that incorporating side protection, can give a reduction of up to 80 or 90% for the dose to the eyes from scatter, but to achieve maximum effectiveness, careful consideration needs to be given to issues such as viewing monitor placement to ensure that the glasses intercept the scatter from the patient. Backscatter from the patient's head is the limiting factor for the dose reduction potential of lead eyewear [24.14]. Measures to protect the eyes will receive increasing attention as a result of the reduction in the annual dose limit for the lens of the eye from 150 mSv to 20 mSv [24.15].

Ceiling suspended protective screens can provide significant protection, but their effectiveness depends on their being positioned correctly. They provide protection to only part of the body — typically the upper body, head and eyes — and their use is in addition to wearing protective clothing. However, their use can remove the need for separate eye shields. Sometimes a protective screen cannot be deployed for clinical reasons. Table mounted protective curtains also provide additional shielding, typically to the lower body and legs.

There are some situations, usually associated with image guided interventional procedures, when the hands of the operator may inadvertently be placed in the primary X ray beam. Protective gloves may appear to be indicated, but wearing such gloves can prove to be counterproductive, as their presence in the primary beam leads to an automatic increase in the radiation dose rate, offsetting any protective value, and they can inhibit the operator's 'feel', which can be dangerous. Gloves may slow the procedure down and also create a false sense of safety; it is better to be trained to keep hands out of the primary beam. Ensuring that the X ray tube is under the table provides the best protection when the hands have to be near the X ray field, as the primary beam will have been been attenuated by the patient's body.

Since radiological workloads can be very different for the different specialties, the necessary protective tools need to be specified by a radiation protection officer. For example, a person with a high workload in a cardiac laboratory should use all the described protective tools; on the other hand, a person in an orthopaedic suite may need only a simple front-only protective apron.

A further factor of direct importance for occupational exposure is the orientation of the X ray tube and image receptor. For near vertical orientations, having the X ray tube under the couch leads to lower levels of occupational exposure because operators are being exposed to scatter, primarily from the exit volume of the patient, where scatter is lowest. Similarly, for near lateral projections, standing to the side of the patient opposite the X ray tube again leads to lower occupational exposure for the same reason. It is essential that personnel performing such procedures have had effective training in radiation protection so that they understand the implications of all the factors involved. Lastly, because of the wide variability in potential occupational exposures from these procedures, it is also essential that individual monitoring be performed continuously and correctly.

More information on fluoroscopic procedures can be found in Chapter 9.

24.6. PUBLIC EXPOSURE IN RADIOLOGY PRACTICES

24.6.1. Access control

Unauthorized access by the public to functioning X ray rooms must be prohibited. Visitors must be accompanied in any controlled area by a person knowledgeable about the protection and safety measures for that area (i.e. a member of the radiology staff), and visitors must be provided with adequate information and instruction before they enter a controlled area so as to ensure their appropriate protection and that of other persons who could be affected by their actions.

24.6.2. Monitoring of public exposure

The programme for monitoring public exposure due to radiology should include dose assessment in the areas surrounding radiology facilities that are accessible to the public. Dose constraints may be applied, if appropriate, in the design stage (see below). A dose constraint applied during operation of a radiology facility can be used as a trigger to examine reasons for the constraint being exceeded and whether there is a need for remedial measures.

Monitoring can easily be achieved by the use of passive devices, such as thermoluminescent dosimeters, placed at critical points for a short period (e.g. two weeks) annually or as indicated. Alternatively, active monitoring of dose rate or integrated dose external to an X ray room for a typical exposure in the room can be used to check shielding design and integrity (see Section 24.7.7). Monitoring is especially indicated and useful when new equipment is installed in an existing X ray room, or where the X ray procedure is altered significantly.

24.6.3. Dose limits

Some regulatory authorities or individual licensees/registrants may wish to apply source related dose constraints. This would take the form of a factor applied to the public dose limit (see Table 24.1). Typically, a value of 0.3 is commonly used. The purpose of the constraint is to ensure, within reason, that the public can be exposed to multiple sources without the dose limit being exceeded.

For shielding calculations, the relevant annual limit is often expressed as a weekly limit, being the annual limit divided by 50 for simplicity.

24.7. SHIELDING

The design of radiation shielding for diagnostic installations can be approached in a number of different ways. However, there are two common approaches used internationally, one based on National Council Radiation Protection and Measurements (NCRP) report 147 [24.16] and one based on the British Institute of Radiology (BIR) report Radiation Shielding for Diagnostic Radiology [24.14]. These are each briefly discussed to give the reader an idea of the different methodologies, and examples of using each approach are provided. The reader is, however, advised to refer to the original sources if either method is to be used, as the necessary tabulated data are not provided here.

24.7.1. Dose and shielding

Dose limits and associated constraints are expressed in terms of effective or equivalent dose. Most X ray output and transmission data are measured in terms of air kerma using ionization chambers. As a result, it is neither practical nor realistic to use effective dose (or its associated operational quantities) when calculating shielding requirements. The relationship between effective dose and air kerma is complex, depending on the X ray spectrum, and, in the case of effective dose, on the distribution of photon fluence and the posture of the exposed individual. Nevertheless, in the energy range used for diagnostic radiology, air kerma can be shown to represent an overestimate of the effective dose. Thus, the assumption of equivalence between air kerma and effective dose will result in conservative shielding models.

It should be noted that since $H_p(10)$ and $H^*(10)$ overestimate effective dose at diagnostic energies [24.17], caution should be used if instruments calibrated in either of these quantities are used to determine, for example, levels of scattered radiation around a room as part of a shielding assessment exercise.

24.7.2. Primary and secondary radiations

The primary beam consists of the spectrum of radiation emitted by the X ray tube prior to any interaction with the patient, grid, table, image intensifier, etc. The fluence of the primary beam will be several orders of magnitude greater than that of secondary radiation. In most radiographic exposures, the primary beam will be collimated so that the entire beam interacts with the patient. Exceptions include extremity radiography, some chest films and skull radiography.

There are two components to secondary radiation, scatter and tube leakage:

(i) Scattered radiation in diagnostic radiology is a direct result of the coherent and incoherent scattering processes (see Chapter 3). The amount of scatter produced depends on the volume of the patient irradiated, the spectrum of the primary beam and the field size employed. Both the fluence and quality of the scattered radiation have an angular dependence.

(ii) Tube leakage radiation arises because X rays are emitted in all directions by the target, not just in the direction of the primary beam. The tube housing is lined with lead but some leakage radiation is transmitted. This component will be considerably harder (i.e. higher half value layer) than the primary beam, but should have a very low intensity relative to the primary beam.

Barriers are often considered as being either primary or secondary in nature, depending on the radiation incident on them. It is possible for a barrier to be both.

24.7.3. Distance to barriers

It is always prudent to take the shortest likely distance from the source to the calculation point. NCRP report 147 [24.16] recommends that distances be measured to a point no less than 0.3 m from the far side of a barrier. For sources above occupied spaces, the sensitive organs of the person below can be assumed to be not more than 1.7 m above the lower floor. For occupied areas above a source, the distance can be measured to a point 0.5 m above the floor.

24.7.4. Shielding terminology

The BIR and NCRP methodologies use the following factors in the calculations, all of which affect the radiation dose to an individual to be shielded:

(a) The design or target dose, P, to a particular calculation point, expressed as a weekly or annual value;
(b) The workload, W (see Section 24.7.6);
(c) The occupancy factor, T (see Section 24.7.8);
(d) The distance, d, from the primary or secondary source to the calculation point.

In addition, the NCRP method employs the factor U, which is termed the use factor. This is the fraction of time that the primary beam is directed towards a particular primary barrier. It ranges from 0 for fluoroscopy and mammography (where the image receptor is the primary barrier) to 1 for some radiographic situations.

24.7.5. Basic shielding equation

With the above information, the required shielding transmission, B, can be calculated for primary and secondary barriers. This value can later be used to determine the barrier thickness. The basic transmission calculation is:

$$B = (P/T) \cdot (1/K) \tag{24.1}$$

where B is the primary or secondary barrier transmission required to reduce air kerma in an occupied area to P/T, which is the occupancy modified design dose. K is the average air kerma per patient at the calculation point in the occupied area. K is determined from the workload, W. The main difference between the two methods described here is the manner in which K is determined.

24.7.6. Workload

In order to determine the amount of shielding required, it is necessary to determine the amount of radiation (primary and secondary) that is incident on the barrier to be shielded. The two methods use different, although fundamentally related, ways of deriving these data. Both utilize measures of tube output, but with different metrics to characterize it.

For all but CT shielding, the NCRP report advocates the use of the total exposure expressed as the sum of the product of exposure time and tube current measured in milliampere-minute as a measure of workload. Workload varies linearly with milliampere-minute. The way the workload is distributed as a function of kV is referred to as the workload distribution. The NCRP report tabulates some workload distributions that are representative of practice in the United States of America.

The BIR report uses patient entrance surface air kerma (K_e) and kerma area product (KAP, P_{KA}) as indicators of workload, where K_e is used as an indicator of primary radiation and KAP to derive the amount of scattered radiation. If a local dose audit is not performed, values of K_e and KAP are readily available in the literature for a large number of examinations. The BIR report provides a conversion factor for KAP to K_e for over-table examinations. Many countries have DRLs which can be used as a basis for calculation should other data not be available and which should result in conservative shielding models. A potential disadvantage of this method is that many facilities do not have access to KAP meters. The BIR method does not use the concept of predetermined workload distribution.

In the case of shielding for CT, the NCRP report advocates the use of either dose length product (DLP, P_{KL}) or computed tomography dose index (CTDI) as a measure of workload, whilst the BIR report recommends the use of DLP only.

24.7.7. Design criteria and dose constraints

Both occupationally exposed employees and members of the public, including employees not directly concerned with the work of the X ray rooms, need to be considered when shielding is being designed. The design methodology must satisfy the radiation protection requirements for both groups.

For members of the public, the BIR approach applies the concept of dose constraints, with the rationale that the public should not receive more than 30% of their maximum permissible dose from any one source. Thus, 0.3 mSv per year is the upper limit on radiation dose in any shielding calculation involving the public. It may be possible to employ a different constraint for employees, depending on local regulatory circumstances, but it would be conservative to use the same dose constraint as a design limit for both groups.

The NCRP report does not advocate the use of dose constraints when determining shielding to members of the public. The design limit is therefore 1 mSv per year to these ‘uncontrolled areas’ (NCRP term). The NCRP approach uses a design limit of 5 mSv per year when considering protection of employees. Areas where this design limit is used are termed ‘controlled areas’ in the NCRP approach and are considered to be subject to access control. Persons in controlled areas should have some training in radiation safety and should be monitored for radiation exposure. This nomenclature is specific to the legislative framework of the USA and does not reflect the BSS.

24.7.8. Occupancy

It is important that the occupancy of surrounding areas be taken into consideration. The occupancy factor is the fraction of an 8 hour day, (2000 hour year or other relevant period, whichever is most appropriate) for which a particular area may be occupied by the single individual who is there the longest. The best way to determine occupancy is to use data derived from the site for which the shielding is being designed, taking into consideration the possibility of future changes in use of surrounding rooms. This is not always possible and therefore suggested figures for occupancy levels are provided in both the BIR and NCRP reports. Suggested values for occupancy factors from the BIR report are reproduced in Table 24.2. The NCRP report gives suggested values ranging from 1 for offices and X ray control areas, to 1/40 for outdoor areas such as car parks or internal areas such as stairwells and cleaner's cupboards. One particular situation deserves mention: the suggested occupancy for a corridor adjacent to an X ray room is 1/5, while for the door from the room to the corridor the value is 1/8, on the basis of the door's small dimensions compared with the length of a wall.

TABLE 24.2. SUGGESTED OCCUPANCY FACTORS (BIR)
(reproduced from Ref. [24.17] with permission of the BIR)

Occupancy and location	Suggested range
Full occupancy:	100%
Control rooms	
Reception areas, nurses stations	
Offices, shops, living quarters, children's indoor play areas, occupied space in nearby buildings	
Partial occupancy:	20–50%
Staff rooms	
Adjacent wards, clinic rooms	
Reporting areas	
Occasional occupancy:	5–12.5%
Corridors	
Store rooms, stairways	
Changing rooms, unattended car parks	
Unattended waiting rooms	
Toilets, bathrooms	

The product of the design constraint and the reciprocal of the occupancy factor should not exceed any dose limit used to define a controlled area. An example is the situation where an occupancy factor of 2.5% is used for an uncontrolled area. Corresponding regulation required that areas with annual doses greater than 6 mSv be controlled. The actual dose outside the barrier, neglecting the occupancy factor, is 12 mSv per year (0.3 mSv (constraint for the public) multiplied by 40 (one divided by the occupancy factor)) and consequently the area would need to be designated as controlled. Presumably this would not have been the designer's intention.

24.7.9. Methodologies for shielding calculations

24.7.9.1. BIR method: Conventional radiography and fluoroscopy

The BIR approach is perhaps more empirical than that advocated in the NCRP report in that the shielding designer is required to evaluate the kerma incident on the barrier using methods derived from the actual workload, and then determine the required transmission to reduce it to the design limit required. However, the underlying principles are the same for both methodologies.

Primary radiation

In fluoroscopy, mammography and CT, the primary beam is intercepted entirely by an attenuator and is not incident directly on any barrier and so need not be taken into account in shielding calculations. However, in the case of plain radiography, this is not the case.

The recommended method assumes that the primary beam is incident on the barrier without any attenuating structure lying in the pathway. In these circumstances, the primary air kerma (K_b) at the barrier can be calculated from the sum of the entrance surface air kerma (K_e) for all exposures (n_i). Inverse square law correction (using the focus to skin distance (FSD) and focus to barrier distance (FBD)) can then be applied to determine the kerma at the barrier or calculation point using:

$$K_b = \Sigma_i(n_i \times K_e) \times (\text{FSD}/\text{FBD})^2 \tag{24.2}$$

The K_e values should be divided by a backscatter factor to convert to incident air kerma (K_i). A backscatter factor of 1.4 is appropriate for larger field sizes and for tube potentials of 80 kV or greater. If it can be assumed that the grid assembly will always intercept the beam, then allowance can be made for attenuation in the assembly. This can be done by subtracting the lead equivalence of the assembly from the total lead equivalence that was calculated for the unattenuated beam. A further possibility is that the beam may also be fully intercepted by the detector. The transmission through the detector is dependent on beam energy, image phosphor and manufacturer and is generally of the order of 10–20%. In these circumstances, the lead equivalence of the imaging device may be added to that of the grid assembly in the shielding assessment. The above will always be the situation in the case of table radiography, but may well not be so when chest radiography is considered.

Secondary radiation

(1) *Scatter*: The BIR treatment of scattered radiation relies on the fact that scatter kerma is proportional to the KAP (P_{KA}) and can be described by:

$$K_{scat} = S \times P_{KA}/d^2 \tag{24.3}$$

where K_{scat} is the scatter kerma at distance d and S is the angle and energy dependent scatter fraction used to derive the scatter air kerma at 1 m. Experimental measurements and Monte Carlo simulation have demonstrated that S follows the shape shown in Fig. 24.2.

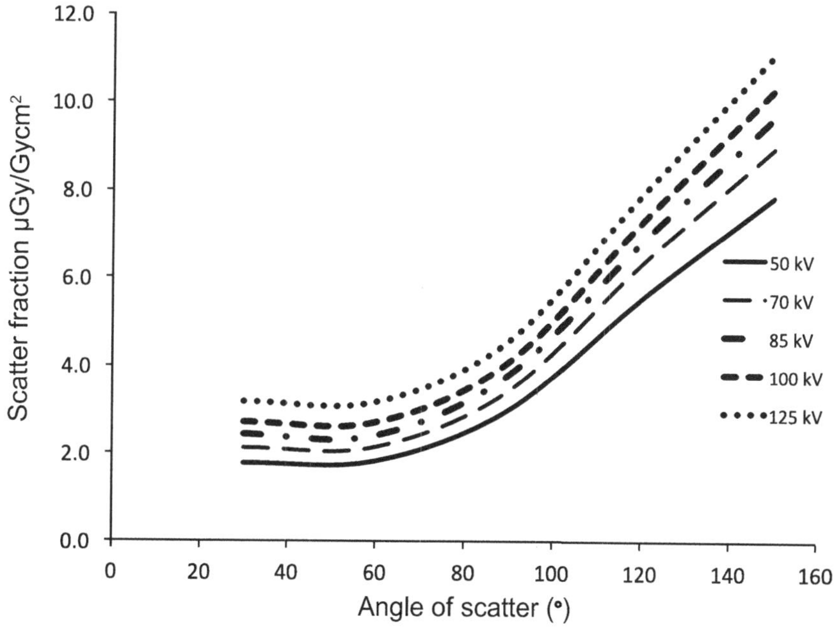

FIG. 24.2. Scatter factor as a function of scatter angle. Reproduced from Ref. [24.17] with permission of the BIR.

The maximum scatter kerma from a patient occurs at scattering angles of between 115° and 120°. This scatter fraction, which can be used in all shielding calculations, can be determined using:

$$S_{max} = (0.031\ \text{kV} + 2.5)\ \mu\text{Gy}\ (\text{Gy}\cdot\text{cm}^2)^{-1} \tag{24.4}$$

to establish K_{scat} in Eq. (24.3).

The use of KAP to predict scatter kerma has several advantages over the method of using a measure of workload, such as milliampere-minute product, as:

(i) No assumptions are required for field size.
(ii) KAP meters are increasingly prevalent on modern fluoroscopic and radiographic equipment, with a significant amount of published data.
(iii) The KAP value is measured after filtration.

When X ray beams filtered by additional copper are used, as for example in an interventional or cardiac catheterization laboratory, S_{max} will exceed the predictions of Eq. (24.4). However, if it is assumed that the accelerating potential never exceeds 85 kV, the scatter kerma at 1 m can be taken as being 8 μGy $(\text{Gy}\cdot\text{cm}^2)^{-1}$.

(2) *Leakage component of radiation*: Leakage is usually defined at the maximum operating voltage of an X ray tube and continuously rated tube current (typically 150 kV and 3.3 mA). At accelerating voltages of less than 100 kV, the leakage component of secondary radiation is at least one order of magnitude less than that of scattered radiation. As the kV decreases, this ratio rises to a factor of 10^8. However, the leakage component of the radiation is considerably harder than that in the primary beam since it has passed through at least 2 mm of lead. The majority of radiological examinations are performed at less than 100 kV, and consequently, it is safe to assume that the amount of leakage radiation will be substantially less than that of scattered radiation and can be neglected. However, at 100 kV or above, transmission curves generated by taking leakage radiation into account should be used.

(3) *Workload*: The most appropriate source of information for estimating workload is local dose audit. If this information is not available, then national survey or published data can be used. Alternatively, entrance surface air kerma or KAP can be calculated using output data obtained for the type of equipment to be used in the room to be shielded.

It is not always easy to identify the potential workload of a facility, but some simplified assumptions can often be made. For example, in the UK at least, the majority of the workload (in dose terms) arises from abdominal, chest and spine examinations. Also, since most of the examinations performed on a wall Bucky are of the chest, only these examinations need be considered when determining protection of the wall against primary radiation. In addition, for table radiography, examinations of the abdomen, pelvis, hips and spine contribute over 80% of the total KAP (in the UK) and the average entrance surface air kerma to KAP ratio for these examinations, weighted for examination frequency, is 2.6 mGy $(Gy \cdot cm^2)^{-1}$. This ratio can be applied to the total KAP workload to give an approximate entrance surface air kerma workload to use in shielding calculations. Thus, the data required for a shielding calculation can be estimated from the projected number of chest examinations and the total KAP workload anticipated from table examinations.

24.7.9.2. NCRP method: Conventional radiography and fluoroscopy

The easiest way to use the NCRP method is to make use of the tabulated data on workload distributions found in the report. The installations for which data are provided range from mammography through general radiography/fluoroscopy to interventional angiography. The tables in the report provide values of unshielded air kerma, K, at a nominal focus to image receptor distance, d_{FID}, for a nominal field area, F, and a nominal value of W. These can then be used, in conjunction with the transmission equations described below (Section 24.7.10) to determine the required shielding. There are also tables and transmission curves which extend the scope to describe the transmission requirements for particular types of X ray room, for example, radiographic and fluoroscopic installations and dedicated chest rooms.

The tables of unshielded kerma (K) and the extended data are based on surveys carried out in the USA and may not be representative of practice in other countries or reflect changes resulting from subsequent advances in technology or practice. The user can, however, modify K for their own particular values of W, F and d_{FID} either manually or by using software that can be obtained from the authors of the NCRP report to produce a user specific workload distribution.

It should be noted that the use of additional beam filtration, such as copper, while reducing both patient entrance dose and scatter, will also result in an increase in mA. In this case, the use of milliampere-minute as a measure of workload may be misleading.

24.7.9.3. CT

A simple relationship between DLP and scattered kerma in mGy is proposed in both the BIR and NCRP reports (the NCRP report also provides data relating

CTDI and scattered kerma). This makes the determination of scattered radiation incident on a barrier straightforward. The person designing the shielding must identify the total DLP from all of the body and head scan procedures carried out in a year and then determine the scattered kerma using the different constants of proportionality assigned to each. There are small differences between the constants recommended by the two reports. The NCRP recommendation is 0.09 μGy $(mGy \cdot cm)^{-1}$ for head scans and 0.36 μGy$(mGy \cdot cm)^{-1}$ for body scans; the BIR recommendation is 0.14 μGy$(mGy \cdot cm)^{-1}$ for head scans and between 0.3 and 0.36 μGy $(mGy \cdot cm)^{-1}$ for body scans.

If there are no DLP data available for the facility, then national DRLs or other appropriate published data can be used. The authors of the NCRP report point out that a considerable number of examinations are repeated with contrast, but using the same procedure identifier. If the number of scans performed with contrast cannot be identified, they suggest using a multiplier of 1.4 for all DLP data. The calculation of barrier requirements can then be made using Eq. (24.1) and the transmission equation described below.

24.7.9.4. Radiation scattered over barriers

The BIR report presents methods for estimation of the amount of radiation scattered over a barrier (so-called tertiary scatter). This issue is not discussed here but further details can be found in Ref. [24.18].

24.7.9.5. Intraoral radiography

The BIR approach makes the simple assumption that the sum of scattered and attenuated radiation at 1 m from the patient is 0.5 μGy. It is further assumed that the beam is fully intercepted by the patient. This makes calculation of barrier thickness a simple matter [24.19].

24.7.9.6. Mammography

Both reports use the same approach and assume that it is conservative to assume a constant scatter fraction for all angles of scatter and all target filter combinations. The NCRP report recommends 36 μGy per patient (four images) at 1 m from the isocentre, while the BIR recommendation is 7.6 μGy per image at 1 m from the isocentre.

24.7.10. Transmission equations and barrier calculations

The determination of the transmission of X rays through a material is not a simple task, given that it takes place under broad beam conditions and that the X ray spectrum is polyenergetic. The so-called Archer equation describes the transmission of broad beam X rays through a material [24.20]:

$$B=\left[\left(1+\frac{\beta}{\alpha}\right)\exp(\alpha\gamma x)-\frac{\beta}{\alpha}\right]^{-\frac{1}{\gamma}} \tag{24.5}$$

where

B is the broad beam transmission factor;
x is the thickness of shielding material required in mm;

and α, β and γ are empirically determined fitting parameters. The parameters α and β have dimensions mm^{-1} while γ is dimensionless.

The equation can be inverted to enable the calculation of the amount of material required to provide the desired transmission:

$$x=\frac{1}{\alpha\gamma}\ln\left[\frac{B^{-\gamma}+\frac{\beta}{\alpha}}{1+\frac{\beta}{\alpha}}\right] \tag{24.6}$$

Provided that the parameters α, β and γ are known, it is a simple matter to incorporate either equation into a spreadsheet and derive either the transmission through a specified material or the amount of material required to provide the desired transmission. Values of α, β and γ are tabulated in the BIR and NCRP reports for a variety of common materials. Note that the tabulated values are for concrete with a density of 2350 kg/m^3. The required thickness for a different density of concrete (± approximately 20%) can be determined using a density ratio correction.

For primary barriers, the total calculated shielding will include any ‘pre-shielding’ provided by the image receptor and the table (if the beam intersects the table). The NCRP and BIR reports give suggested values for pre-shielding which can be subtracted from the result of Eq. (24.6) to obtain the required barrier thickness.

24.7.11. Worked examples

The following examples show how the NCRP and BIR methods may be used in various situations. These are illustrative only; space does not allow for a detailed treatment of each case. All internal walls are assumed to be newly built, with no existing shielding.

Although the examples show how to apply the process to different walls, the final specification must be pragmatic. It is accepted practice to shield all walls to the same specification to avoid errors in the construction process and to allow for future changes in room layout. The specification chosen will be that of the wall that requires most shielding.

24.7.11.1. Radiographic room

Figure 24.3 shows the layout of a simple radiographic room and is used to demonstrate shielding calculations for both the BIR and NCRP methodologies. In the example, the shielding requirements for walls A and B are determined. For the sake of simplicity, it is assumed that there is no cross-table radiography performed in the direction of wall A. Readers are advised to refer to the original sources for details on how to carry out calculations involving primary and secondary beams.

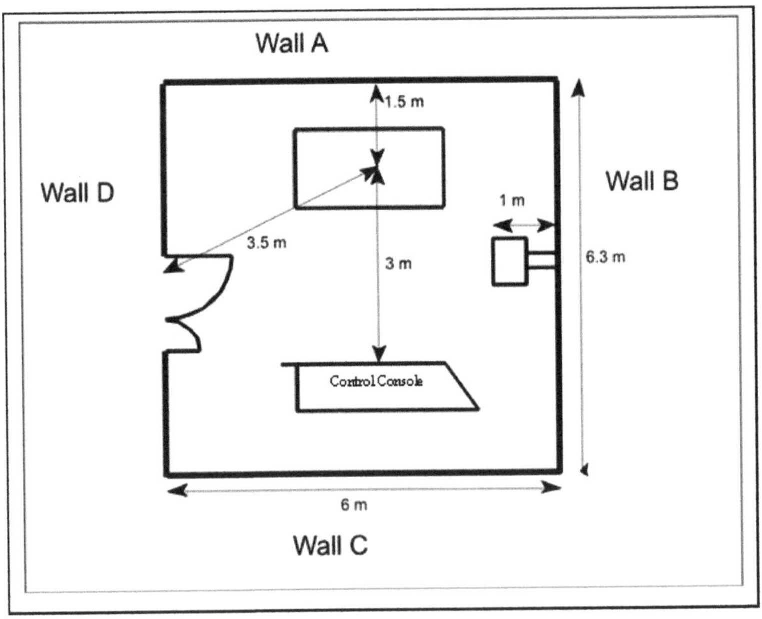

FIG. 24.3. Simple radiographic room.

For the workload calculation it is assumed that:

(a) 200 patients are examined in this room per week.
(b) An average of 1.5 images (or X ray exposures) per patient are taken.
(c) There are 150 chest films and 150 over-table exposures.
(d) The chest films are routinely carried out at 125 kV.

For the purposes of shielding calculations, the workload excludes any extremity examinations that take place. Wall A is adjacent to an office that is assumed to have 100% occupancy. The annual dose limit for occupants will be 1 mSv. Wall B is next to a patient treatment room and occupancy of 50% is assigned. Again, the annual dose limit for occupants will be 1 mSv. Assumptions made for the BIR method are:

(a) The KAP for abdomen and spine/pelvis examinations can be taken as 2 Gy·cm^2 per patient. The accelerating potential can be taken as 85 kV (UK data).
(b) The average KAP per chest exposure is 0.11 Gy·cm^2 (UK data).
(c) K_e for a chest radiograph is 0.15 mGy (UK data).

The NCRP calculations use the assumptions made in NCRP report 147; these are explained where appropriate.

Example calculations for wall A

This wall is exposed to secondary radiation only. The steps in the calculations are:

BIR method

The total weekly KAP from the table exposures is 2 (Gy·cm^2 per exam) × 150 (exams) = 300 Gy·cm^2 and the total weekly KAP from the chest exposures is 16.5 Gy·cm^2. The annual scatter kerma at 1 m from the table (with S calculated from Eq. (24.4)) is:

$$K_{scat} = 50 \text{ weeks} \times 300 \text{ Gy·cm}^2\text{/week} \times (0.031 \times 85 + 2.5) = 77\,000 \text{ µGy}$$

Similarly, the annual scatter kerma at 1 m from the wall Bucky is:

$$K_{scat} = 50 \text{ weeks} \times 16.5 \text{ Gy·cm}^2\text{/week} \times (0.031 \times 125 + 2.5) \approx 5250 \text{ µGy}$$

Since the wall Bucky is 5.5 m from the calculation point and the table only 1.8 m, the scatter contribution from the wall Bucky can be ignored and, therefore, for shielding purposes, the annual scatter kerma at the calculation point is given by:

$$K_{scat} = (77\ 000/1.8^2)\ \mu\text{Gy} = 24\ 000\ \mu\text{Gy} = 24\ \text{mGy}$$

The required transmission will depend on the dose constraint used in the design:

(a) If a constraint of 1 mSv is used, the required transmission, $B = 1/24 = 4.2 \times 10^{-2}$.
(b) If a constraint of 0.3 mSv is used, then $B = 0.3/24 = 1.25 \times 10^{-2}$.

The BIR report advocates using parameters for 90 kV in Eq. (24.6). For lead, these are $\alpha = 3.504$ mm^{-1}, $\beta = 20.37$ mm^{-1} and $\gamma = 0.755$. The resulting solutions of Eq. (24.6), corresponding to the required shielding, are:

(a) For a dose constraint of 1 mSv per year, 0.34 mm lead shielding;
(b) For a dose constraint of 0.3 mSv per year, 0.6 mm lead shielding.

NCRP method

The NCRP method uses the number of patients examined in the room, i.e. 200, as the basis for calculation. Wall A is a secondary barrier, so the use factor (U) is zero. Table 4.7 of the NCRP report indicates that the secondary air kerma factor (leakage plus side scatter) to use in this case is 3.4×10^{-2} mGy per patient at 1 m. A workload of 200 patients per week results in a total annual secondary kerma at the calculation point of:

$$K_{sec} = 3.4 \times 10^{-2}\ \text{mGy/patient} \times 50\ \text{weeks} \times 200\ \text{patients/week}/1.8^2 = 104.9\ \text{mGy}.$$

Again, the required transmission will depend on the dose constraint used in the design. If a constraint of 1 mSv is used, B will be 9.53×10^{-3}, and if a constraint of 0.3 mSv is used, B will be 2.86×10^{-3}. The NCRP report recommends using workload spectrum specific parameters to solve the transmission equation. For a radiographic room, these are (for lead) $\alpha = 2.298$ mm^{-1}, $\beta = 17.3$ mm^{-1} and $\gamma = 0.619$. The resulting solutions are:

(a) For a dose constraint of 1 mSv per year, 0.77 mm lead shielding;
(b) For a dose constraint of 0.3 mSv per year, 1.17 mm lead shielding.

Example calculations for wall B

BIR method

Protection is required for primary transmission through the wall behind the chest stand. An air gap is used and the focus to film distance is 3 m, so the focus to calculation point distance is 4.3 m, as the Bucky extends 1 m from the wall, and the calculation point is defined as being 0.3 m behind wall B (Section 24.7.3). The patient entrance surface to film distance is estimated at 0.5 m; thus, the focus to skin distance is 2.5 m.

As one cannot always be certain that the patient will always intercept the X ray beam, K_e is used to determine the air kerma at the calculation point. Incorporating a backscatter factor of 1.4, the inverse square law indicates a primary air kerma of:

$150\ \mu Gy \times [(2.5/4.3)^2]/1.4 = 36\ \mu Gy$ per chest X ray.

The annual primary kerma at the calculation point, in the absence of the barrier will therefore be:

$36\ \mu Gy \times 150$ X rays/week $\times$ 50 weeks $= 27 \times 10^4\ \mu Gy = 270$ mGy.

The required transmission will depend on the dose constraint used in the design:

(a) If a constraint of 1 mSv is used, the required transmission, $B = 1/270 = 3.7 \times 10^{-3}$.
(b) If a constraint of 0.3 mSv is used, then $B = 0.3/270 = 1.1 \times 10^{-3}$.

At 125 kV, the transmission coefficients for lead are $\alpha = 3.504$ mm^{-1}, $\beta = 20.37$ mm^{-1} and $\gamma = 0.755$. The resulting solutions are:

(a) For a dose constraint of 1 mSv per year, 1.4 mm lead shielding;
(b) For a dose constraint of 0.3 mSv per year, 1.8 mm lead shielding.

NCRP method

Again, the NCRP method uses the total number of patients examined in the room as the basis for calculation. In this case, the number is 200 *not* 100, the number of patients who undergo chest examinations alone. This may appear counter intuitive but should be used since the fraction of patients who receive examinations on the chest stand is accounted for in the workload spectra provided in the report. The interested reader should consult the original report for more details.

Table 4.5 of the NCRP report indicates that for a chest stand in a radiographic room, the unshielded primary air kerma is 2.3 mGy per patient at 1 m. The annual unshielded primary kerma at the calculation point is:

$$2.3 \text{ mGy} \times 50 \text{ weeks} \times 200 \text{ patients/week}/4.3^2 = 1244 \text{ mGy}.$$

The required transmission, B, for a constraint of 1 mSv is $2/1244 = 1.6 \times 10^{-3}$, and for a constraint of 0.3 mSv is $0.6/1244 = 4.82 \times 10^{-4}$. The workload specific fitting parameters for a chest stand in a radiographic room are given in NCRP report 147 as $\alpha = 2.264$ mm^{-1}, $\beta = 13.08$ mm^{-1} and $\gamma = 0.56$. The resulting solutions are:

(a) For a dose constraint of 1 mSv per year, 1.45 mm lead shielding;
(b) For a dose constraint of 0.3 mSv per year, 1.93 mm lead shielding.

Notes on the final specification for wall B

The prefiltration provided by a wall mounted imaging receptor (attenuation by grid, cassette and image receptor supporting structures) is 0.85 mm of lead. If there is certainty that the beam will always intercept the detector, then this can be taken into account and the specification for the entire wall should be the specification for the primary beam.

If it cannot be taken into account, then there is no need to shield all of wall B to the extent required for the primary beam. In cases such as this, the BIR report recommends the entire wall be shielded against secondary radiation and that additional shielding be provided for the primary beam. The scatter kerma resulting from table radiography at the calculation point behind wall B, at approximately 5.2 m from the patient on the table, is 2.8 mGy and from chest radiography it is 1.6 mGy. This will be more than adequately attenuated by the thinnest commercially available lead in the UK, 1.32 mm (code 3), so this would be the design specification for the wall. An additional sheet of code 3 lead should be attached to the wall behind the chest stand and should extend 50 cm either side of the centre of the stand and not exceed 2 m in height. Different countries will have different lead specifications and the designer must consider local conditions (see Section 24.7.12).

24.7.11.2. Mammography

Mammography installations are much simpler and are treated in a similar manner in both reports. Consider the following room design for a screening facility:

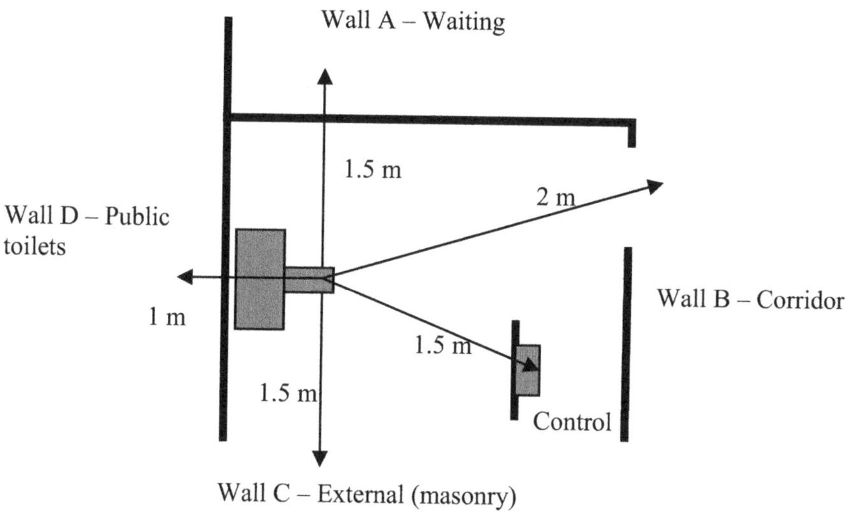

FIG. 24.4. Example plan of a mammography installation.

The following is an outline of how the shielding for the room would be carried out using the methodology in the BIR report:

Assumptions:

(a) The unit, used for mammographic screening, operates at 30 kV.
(b) There are two views taken of each breast and 80 patients are imaged per day.
(c) All primary radiation is intercepted by the detector.

Calculation:

(a) The daily scatter kerma at the wall is
μGy/view × 4 views/patient × 80 patients/1.52 = 1080 μGy ≈ 1.1 mGy
(b) The annual scatter kerma at the wall is
5 days/week × 50 weeks × 1.1 mSv/week = 275 mGy

(c) For a dose constraint of 0.3 mSv, and assuming an occupancy of 50%, the required transmission will be 0.3/(0.5 × 275) = 2.2 × 10^{-3}.

(d) Transmission factors at 30 kV for gypsum wallboard are $\alpha = 0.1198$ mm^{-1}, $\beta = 0.7137$ mm^{-1} and $\gamma = 0.379$. Transmission of 2.2 × 10^{-3} can be achieved with 18 mm wallboard, which is less than the amount used in a standard stud partition wall.

A similar calculation shows that wall D (distance 1 m, occupancy 10%) will not require any greater shielding than wall A, and neither will wall B. The standard 0.25 mm lead equivalence protective screen will provide adequate protection for the operator. Shielding of doors can be an issue in mammographic installations. In this example, an 8 cm thick wooden door would apparently be required. Solid core wooden doors are not necessarily easy to source nor are they cheap. Lead is very effective, but lead doors are heavy and the throughput of patients in a mammographic installation can be high. One solution is to position the equipment as in Fig. 24.4, where it is mounted on a wall opposite the door, which is therefore shielded by the patient. If possible, the door should be constructed of solid core wood, but it is strongly recommended that the adequacy of any door, and in general of any shielding barrier, is verified after installation.

The NCRP approach is very similar to that outlined above, but all curves presented are predicated on the use of a constraint of 1 mSv.

24.7.11.3. Cardiac catheterization laboratory

Both the NCRP and BIR reports include indicative calculations showing how the respective methods can be utilized in a catheterization laboratory. In the example below, the calculation is repeated to demonstrate that each method is applied using (i) a dose constraint of 0.3 (design to 0.3 mSv) and (ii) no dose constraint (design to 1.0 mSv). The geometry chosen is that of a public area with 100% occupancy at a distance of 4 m from the isocentre, as used in NCRP report 147.

The workload used is that detailed in the NCRP report, with 25 patients per week undergoing cardiac angiography. The NCRP method predicts a total secondary air kerma of 3.8 mGy per patient at 1 m. In the BIR report, the highest average KAP identified per patient is 40 Gy·cm^2, which results in a maximum scatter kerma of 224 μGy per patient at 1 m. There is an obvious discrepancy between the two values of scatter kerma. If it is assumed that copper filtration is used in all cases, the scatter kerma predicted by the BIR report rises to 320 μGy per patient at 1 m.

Barrier requirements are calculated using the secondary transmission parameters at 100 kV ($\alpha = 2.507$ mm^{-1}, $\beta = 15.33$ mm^{-1}, $\gamma = 0.912$) for the BIR example with no copper filtration, 85 kV (α = 3.504 mm^{-1}, β = 20.37 mm^{-1}, γ = 0.755) for the example with additional copper filtration and using the coronary angiography specific parameters (α = 2.354 mm^{-1}, β = 14.94 mm^{-1}, γ = 0.748) for the NCRP example. The results of the calculations are given in Table 24.3.

TABLE 24.3. BARRIER THICKNESS (mm LEAD) NEEDED TO PROVIDE SAME DEGREE OF PROTECTION USING CALCULATIONS BASED ON DATA IN BIR AND NCRP REPORTS

Design limit	Method		
	NCRP	BIR	BIR (copper filtration)
0.3 mSv	1.80 mm	0.8 mm	0.6 mm
1.0 mSv	1.3 mm	0.45 mm	0.35 mm

It can be seen that the BIR method calculates that far less shielding is needed than the NCRP approach. The discrepancy is mostly due to the estimates for scatter at 1 m from the patient; 3.8 mGy for the NCRP method and 0.224 mGy or 0.32 mGy (without and with copper filtration) for the BIR approach. The KAP value of 40 Gy·cm^2 per patient used in the BIR report is consistent with several dose surveys published by European centres. The NCRP workload data, measured in milliampere-minute, are not consistent in this case with workloads in Europe and care should be taken if the NCRP method is utilized in this type of calculation.

24.7.11.4. Intraoral radiography

The BIR report makes the assumption that the patient always intercepts the primary beam. Provided that this is the case, the weighted average primary plus scatter dose at a distance of 1 m is of the order of 0.5 μGy per film. Using a dose constraint of 0.3 mSv per annum, no shielding is required if the X ray unit is situated 2 m or more from a barrier. Even when this is not the case, partition walls with 10 mm gypsum plasterboard on each side will provide adequate protection in the majority of situations.

24.7.11.5. CT

The design of CT scanner shielding should take the following factors into account:

(a) The X ray beam is always intercepted by the patient and detector, thus only scattered radiation needs to be considered.
(b) The X ray tube operating voltage is high, ranging from 80 to 140 kV.
(c) The X ray beam is heavily filtered (high half value layer).
(d) The total workload is very high, measured in thousands of mAs/week.
(e) The scattered radiation is not isotropic (and has more of an 'hourglass' distribution).

Workload measures

DLP is employed by both reports as a measure of workload. All the user needs are the DLP values and the average number of each procedure per week. This information should ideally be obtained from an audit of local practice. However, if local DLP data are not available a DRL or another value obtained from the literature may be used. The NCRP report provides typical US data for DLP. The BIR report provides similar information for UK installations.

Calculation

There are only slight differences in the calculation methods advocated by the reports. For brevity, the calculation outlined here follows the methodology used in the NCRP report.

Once the scatter kerma incident on the barrier has been determined, barrier requirements can be determined using the secondary CT transmission parameters for lead at 120 kV ($\alpha = 2.246$ mm^{-1}, $\beta = 5.73$ mm^{-1}, $\gamma = 0.547$) or 140 kV ($\alpha = 2.009$ mm^{-1}, $\beta = 3.99$ mm^{-1}, $\gamma = 0.342$). Parameters for concrete are for 120 kV ($\alpha = 0.0383$ mm^{-1}, $\beta = 0.0142$ mm^{-1}, $\gamma = 0.658$) or at 140 kV ($\alpha = 0.0336$ mm^{-1}, $\beta = 0.0122$ mm^{-1}, $\gamma = 0.519$). In the (common) case where both 120 and 140 kV are used clinically, it would be prudent to use transmission data for 140 kV. This approach assumes isotropy of scattered radiation, but errs on the side of conservatism.

In order to reduce the scatter kerma appropriately, it is important that all barriers extend as close as possible to the roof (underside of the soffit), not just to the standard 2100 mm above the floor.

Scatter estimation

The NCRP report estimates the scatter fraction per centimetre at 1 m from a body or head phantom as:

$k_{head} = 9 \times 10^{-5}\ \mathrm{cm}^{-1}$

$k_{body} = 3 \times 10^{-4}\ \mathrm{cm}^{-1}$

The total kerma from scatter and leakage at 1 m can then be estimated as:

$$K_{sec}\ (\text{head}) = k_{head} \times \text{DLP} \times 1.4 \qquad (24.7(\text{a}))$$

$$K_{sec}\ (\text{body}) = 1.2 \times k_{body} \times \text{DLP} \times 1.4 \qquad (24.7(\text{b}))$$

where the factor 1.4 corrects for repeated examination with contrast agents (see Section 24.7.9). The factor 1.2 in Eq. (24.7(b)) arises from the assumptions made by the authors of the NCRP report.

Example CT shielding calculation

Consider the CT room design illustrated in Fig. 24.5.

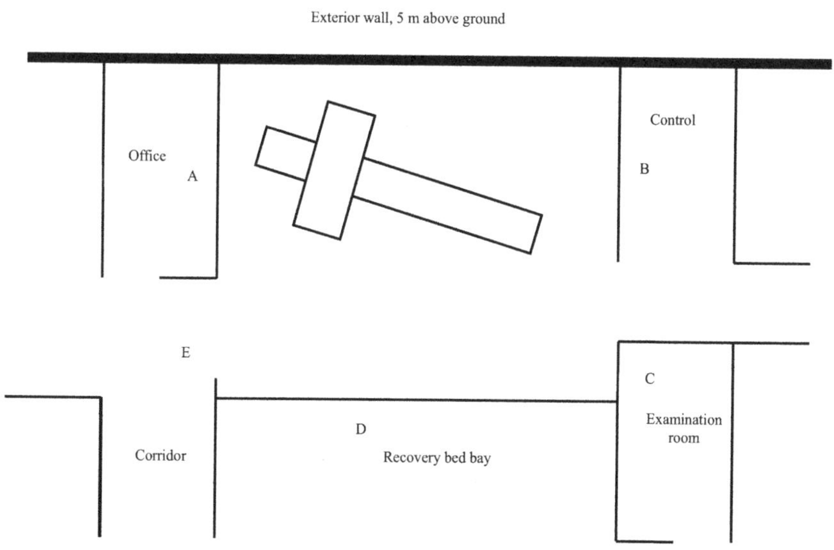

FIG. 24.5. CT room layout.

Assume that:

(a) 30 head and 45 body examinations are performed per week (actual average).
(b) The mean DLP for head examinations is 1300 mGy·cm.
(c) The mean DLP for body examinations is 1250 mGy·cm.
(d) The distances from scan plane to calculation points are (i) A = 2.5 m, (ii) B = 4.5 m, (iii) C = 6.5 m, (iv) D = 4 m and (v) E = 3.5 m.

The total kerma from scatter and leakage at each point can be calculated from Eqs (24.7(a), 24.7(b)), corrected for the corresponding distance. For example, take point B (control room). The total weekly scatter (occupancy factor of 1) is:

K_{sec} (head) = 9×10^{-5} cm^{-1} $\times$ 1300 mGy·cm $\times$ 30 scans/week $\times$ 1.4 $\times$ $(1^2/4.5^2)$ = 0.24 mGy/week

K_{sec} (body) =$1.2 \times 3 \times 10^{-4}$ cm^{-1} $\times$ 1250 mGy·cm $\times$ 45 scans/week $\times$ 1.4 $\times$ $(1^2/4.5^2)$ = 1.4 mGy/week

The total scatter is thus 1.64 mGy/week.

An annual dose constraint of 1 mSv would require 1 mm of lead and an annual dose constraint of 0.3 mSv, i.e. 1.5 mm lead. In all cases, the viewing window must have at least the same lead equivalence as the wall.

For other rooms, the target dose will be dependent on the dose constraint used for members of the public in the shielding design. In this example, an occupancy factor of 1 will be assumed for the office, recovery bay and examination room, while an occupancy factor of 1/8 is assumed for the corridor, as suggested in the NCRP report. A dose constraint of 1 mSv per year will be used. The required shielding can then be calculated:

- Office – 1.5 mm lead;
- Control room – 0.6 mm lead;
- Examination room – 0.8 mm lead;
- Recovery bay – 1.2 mm lead;
- Entry door – 0.6 mm lead.

In practice, it would not be unusual to specify all walls at 1.5 mm lead, in order to avoid errors during construction and to allow for future layout changes. The principal cost of shielding is in the construction and erection, rather than the cost of the lead itself.

24.7.12. Construction principles

Irrespective of the calculation methodology, the construction of shielding barriers is essentially the same.

24.7.12.1. Shielding materials

While lead is an obvious choice for shielding, there are other materials such as concrete, steel and gypsum wallboard (both standard and high density). Masonry bricks may also be used, but the user must be aware of the pitfalls. The most obvious problem is voids in the brick or block material. These must be filled with grout, sand or mortar. Even then, the actual attenuation will depend on the formulation of the masonry and filling.

Lead will come in the form of sheet bonded to a substrate such as gypsum wallboard or cement sheet. Sheet lead alone must never be used as it is plastic in nature, and will deform and droop over time.

Milled or rolled lead is manufactured to defined standards and is often specified by the manufacturer in terms of mass density (kg/m^2 or lb/in^2). This is the product that should be used for shielding and is available in densities such as 10, 15, 20, 25 and 30 kg/m^2. The equivalent lead thickness in millimetres is determined by dividing by 11.34 (density of lead = 11 340 kg/m^3). Some standards assign available thicknesses of milled lead codes; so in the UK, code 3 lead to BSEN 12588 has a density of 15 kg/m^2 and is 1.32 mm thick.

24.7.12.2. Interior walls

Interior walls are easily constructed using a ‘sheet on frame’ process. Lead sheet is supplied commercially.

Gypsum wallboard is of minimal use for shielding except for mammography and dental radiography, as it provides little attenuation at typical X ray energies. Gypsum may also contain small voids and can have non-uniform attenuation. In some countries, high density wallboard (usually provided by barium in the plaster) is available. Each sheet may be equivalent to about 1 mm of lead at typical tube voltages.

Joints between sheets must have an overlap in the shielding of at least 10 mm. Sheets of shielding may be applied using normal fasteners. Gaps in the barrier, however, such as those for power outlets, should only be made in secondary barriers, and even then must have a shielded backing of larger area than the penetration (to allow for angled beams). In general, penetrations should ideally be located either close to the floor, or >2100 mm above the floor, which is often above the shielding material.

24.7.12.3. Doors

Doors are available with lead lining. The builder must be aware that there can be discontinuities in the shielding at the door jamb, and in the door frame in particular. This can be addressed by packing the frame with lead sheet of the appropriate thickness glued to the frame, as in Fig. 24.6.

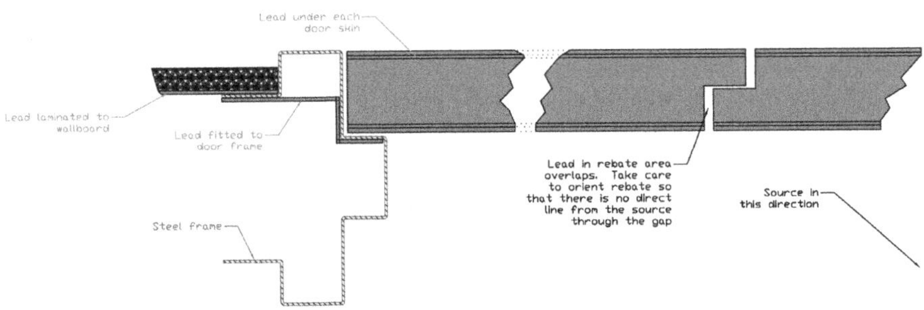

FIG. 24.6. Example of continuity of shielding from wall to door. Reproduced from Ref. [24.21] with permission.

24.7.12.4. Floors and ceilings

Concrete is a common building material for floors. It is cast either in constant thickness slabs (except for load-bearing beams), or with the assistance of a steel deck former with a 'W' shape. Slabs are of varying thickness and the slab thickness must be taken into account if it is to act as a shielding barrier. Formers can have a small minimum thickness and knowledge of this is essential. The minimum thickness is all that can be used in shielding calculations. For diagnostic X ray shielding, most slabs provide sufficient attenuation, but the barrier attenuation must still be calculated.

The designer of shielding must also be aware that, unless poured correctly, voids can form within a concrete slab. In some cases, the floor may be of timber construction, which will sometimes require installation of additional shielding. Another factor that must be determined is the floor-to-floor distance, or pitch, as this will have an influence on doses both above and below.

24.7.12.5. Windows

Observation windows must provide at least the same radiation attenuation as the adjacent wall or door. Normal window glass is not sufficient (except where the required attenuation is very low, such as in mammography), and materials such as lead glass or lead acrylic must be used. Lead acrylic is softer than glass and may scratch easily.

Where lead windows are inserted into a shielded wall or door, the builder must provide at least 10 mm overlap between the wall/door shielding and the window. This may, in some cases, need to be greater, for example, when there is a horizontal gap between the shielding materials.

24.7.12.6. Height of shielding

As a general rule, shielding need only extend to 2100 mm above finished floor level, but as already stated, this will not be the case in all installations, the most notable exception being for CT, where a degree of shielding should extend to the roof slab.

24.7.13. Room surveys

After construction of shielding, the room must be surveyed to ensure that the shielding has been installed as specified.

24.7.13.1. Visual verification

The simplest way to verify construction of shielding according to the design is to perform a visual inspection during construction. For example, if the barrier is to be constructed from lead wallboard on one side of a timber or steel frame, as is commonly the case, the shielding can be inspected before the second side is covered. This is quick and allows problems to be dealt with during construction. Additional shielding over penetrations can also be seen, and the lead sheet thickness can be measured. Photographs should be taken for later reference.

Locations where most problems occur include:

(a) Penetrations;
(b) Door frames;
(c) Overlap between wall shielding and windows;
(d) Corners;
(e) Overlap between wall shielding sheets.

This method, while the best, requires good cooperation and timing between the builder and the person performing the inspection. All shielding must have been installed, yet not covered by other non-shielding building materials.

24.7.13.2. Transmission measurements

If a visual survey cannot be performed until construction is complete, then radiation transmission methods must be used. It is difficult to check every point on all barriers by transmission methods. The tester should select critical locations to test in the first instance, and add more as necessary.

Transmission methods can be used to:

(a) Detect any shielding faults (qualitative) using a radioactive isotope, or X ray equipment, as the source;
(b) Measure radiation transmission (quantitative) using a radioactive isotope, or X ray equipment, as the source.

The detection of shielding faults can be achieved with a Geiger counter or scintillation detector using the audible signal to indicate the level of radiation. The most appropriate radiation source is a radioisotope with an energy similar to the mean energy of a diagnostic beam at high kV. Americium-241 (60 keV) can be used for this purpose, but this isotope is not always available and transport can raise issues. Other higher energy isotopes have also been used, including ^{99m}Tc (140 keV) and ^{137}Cs (662 keV). When a radioactive source is used, the tester must be aware of safety issues and select an activity which is high enough to allow transmission detection, without being at a level that is hazardous. Remote controlled sources are preferable.

Use of X ray equipment as the source can be difficult. For radiographic units of any type, the exposure times are so short as to make a thorough survey almost impossible unless many exposures are made. A distinction also has to be made between surveying for primary and secondary radiation barriers. If the room contains a fluoroscopy unit only, then the unit itself, with tissue-equivalent scatter material in the beam, can make a useful source. In both cases, a reasonably high kV and mAs/mA should be used to increase the chance of detecting any faults in the shielding. The use of radiographic film can also be beneficial if the shielding material is thought to be non-uniform (as might be the case with concrete block construction).

Quantitative transmission methods require the measurement of the incident and transmitted radiation intensities (with correction for the inverse square law where appropriate) to allow calculation of barrier attenuation. For monoenergetic radiation such as that from ^{241}Am, a good estimate of lead or lead equivalence may then be made using published transmission data or in-house calibrations. Technetium-99m can also be used to determine lead thickness. However, if used to determine lead equivalence, the user should be aware of the pitfalls of using a nuclide with an energy of 140 keV, as the K absorption edge of lead is at 88 keV. In addition, and for the same underlying reason, because the photon energy range over which barium has higher absorption than lead is only between 37 and about 80 keV, a ^{99m}Tc source will not quantify the X ray shielding provided by barium plaster. Transmission through walls can be measured with X ray equipment, usually at 100 kV. While taking potentially more time than the radioactive source method, analysis can be easier if the composition of the wall is not known. Measurements can be made using a mobile radiographic unit or, if available, a ceiling mounted X ray tube.

Comprehensive information can be found in Refs [24.16, 24.17, 24.21].

24.7.13.3. Rectification of shielding faults

Any faults detected in shielding must be rectified. The most easily fixed problems are gaps. Figure 24.7 gives examples of how they can occur and how they can be rectified. Further information can be found in Ref. [24.21].

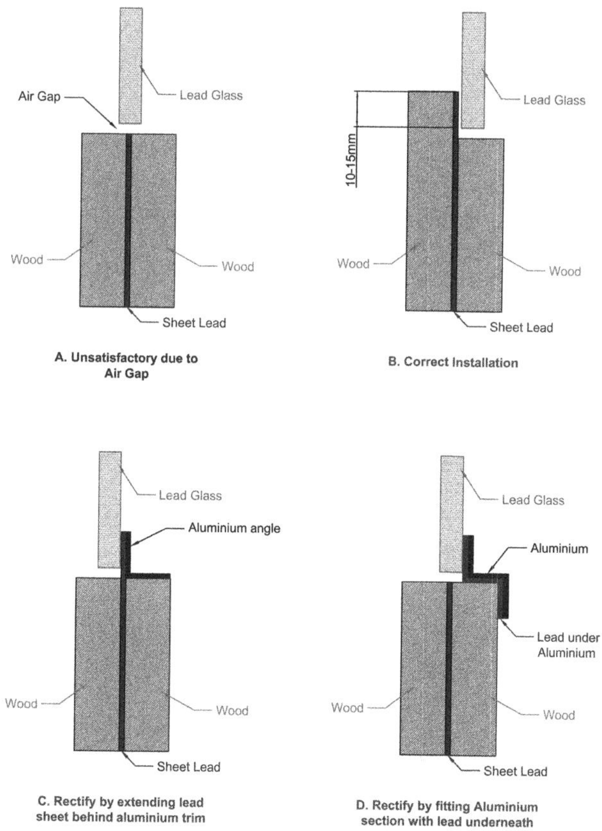

FIG. 24.7. Examples of shielding faults and rectification. Reproduced from Ref. [24.21] with permission.

REFERENCES

[24.1] INTERNATIONAL COMMISSION ON RADIOLOGICAL PROTECTION, The 2007 Recommendations of the International Commission on Radiological Protection, ICRP Publication 103, Elsevier (2008).

[24.2] UNITED NATIONS, Sources and Effects of Ionizing Radiation, Report 2000, Vol. 1: Sources, Scientific Committee on the Effects of Atomic Radiation (UNSCEAR), UN, New York (2000).

[24.3] UNITED NATIONS, Sources and Effects of Ionizing Radiation, Report 2008, Scientific Committee on the Effects of Atomic Radiation (UNSCEAR), UN, New York (2010).

[24.4] INTERNATIONAL ATOMIC ENERGY AGENCY, Radiation Protection and Safety of Radiation Sources: International Basic Safety Standards, Interim Edition, IAEA Safety Standards Series No. GSR Part 3, IAEA, Vienna (2011).

[24.5] INTERNATIONAL ATOMIC ENERGY AGENCY, Applying Radiation Safety Standards in Diagnostic Radiology and Interventional Procedures Using X Rays, Safety Reports Series No. 39, IAEA, Vienna (2006).

[24.6] INTERNATIONAL COMMISSION ON RADIOLOGICAL PROTECTION, Radiological Protection and Safety in Medicine, ICRP Publication 73, Pergamon Press, Oxford and New York (1996).

[24.7] INTERNATIONAL COMMISSION ON RADIOLOGICAL PROTECTION, Radiological Protection in Medicine, ICRP Publication 105, Elsevier (2008).

[24.8] INTERNATIONAL ATOMIC ENERGY AGENCY, Radiological Protection for Medical Exposure to Ionizing Radiation, IAEA Safety Standards Series No. RS-G-1.5, IAEA, Vienna (2002).

[24.9] INTERNATIONAL COMMISSION ON RADIOLOGICAL PROTECTION, 1990 Recommendations of the International Commission on Radiological Protection, ICRP Publication 60, Pergamon Press, Oxford and New York (1991).

[24.10] INTERNATIONAL COMMISSION ON RADIOLOGICAL PROTECTION, Pregnancy and Medical Radiation, ICRP Publication 84, Elsevier (2000).

[24.11] INTERNATIONAL COMMISSION ON RADIOLOGICAL PROTECTION, Radiation and Your Patient: A Guide for Medical Practitioners, ICRP Supporting Guidance 2, Elsevier (2001).

[24.12] INTERNATIONAL ATOMIC ENERGY AGENCY, Occupational Radiation Protection, IAEA Safety Standards Series No. RS-G-1.1, IAEA, Vienna (1999).

[24.13] INTERNATIONAL ATOMIC ENERGY AGENCY, Assessment of Occupational Exposure Due to External Sources of Radiation, IAEA Safety Standards Series No. RS-G-1.3, IAEA, Vienna (1999).

[24.14] McVEY, S., SANDISON, A., SUTTON, D.G., An assessment of lead eyewear in interventional radiology, J. Radiol. Prot. **33** (3) (2013) 647–659.

[24.15] INTERNATIONAL COMMISSION ON RADIOLOGICAL PROTECTION, ICRP Statement on Tissue Reactions and Early and Late Stage Effects of Radiation in Normal Tissues and Organs: Threshold Doses for Tissue Reactions in a Radiation Protection Context, ICRP Publication 118, Elsevier (2012).

[24.16] NATIONAL COUNCIL ON RADIATION PROTECTION AND MEASUREMENTS, Structural Shielding Design for Medical X-Ray Imaging Facilities, NCRP Report 147, NCRP, Bethesda, MD (2004).

[24.17] SUTTON, D.G., MARTIN, C.J., WILLIAMS, J.R., PEET, D.J., Radiation Shielding for Diagnostic Radiology, 2nd edn, British Institute of Radiology, London (2012).

[24.18] MARTIN, C.J., et al., Derivation of factors for estimating the scatter of diagnostic X-rays from walls and ceiling slabs, J. Radiol. Prot. **32** (4) (2012) 373–396.

[24.19] WORRALL, M., McVEY, S., SUTTON, D.G., Proposals for revised factors in intra-oral radiography shielding calculations, J. Radiol. Prot. **32** (3) (2012) 243–249.

[24.20] ARCHER, B.R., THORNBY, J.I., BUSHONG, S.C., Diagnostic X-ray shielding design based on an empirical model of photon attenuation, Health Phys. **44** (5) (1983) 507–517.

[24.21] OFFICE OF ENVIRONMENT AND HERITAGE, Radiation Guideline 7: Radiation Shielding Design, Assessment and Verification Requirements, State of NSW and Department of Evironment, Climate Change and Water, Sydney (2009), http://www.epa.nsw.gov.au/resources/radiation/09763ShieldingGuideline.pdf (accessed on 30 May 2014).

Appendix

ANATOMICAL NOMENCLATURE

A radiographic examination is conventionally described in terms of the projection, the patient's position and the direction and centring of the X ray beam, according to the terminology described below.

A.1. Anatomical position and body planes

The patient's posterior aspect is that observed when viewing the patient from the back, the anterior aspect is that observed from the front and the lateral aspect is that observed from the side.

There are three body planes, each at right angles to the other two (Fig. A.1). A patient is split into right and left halves along the median sagittal plane. The coronal plane is at right angles to the median sagittal plane and divides the body into posterior and anterior parts.

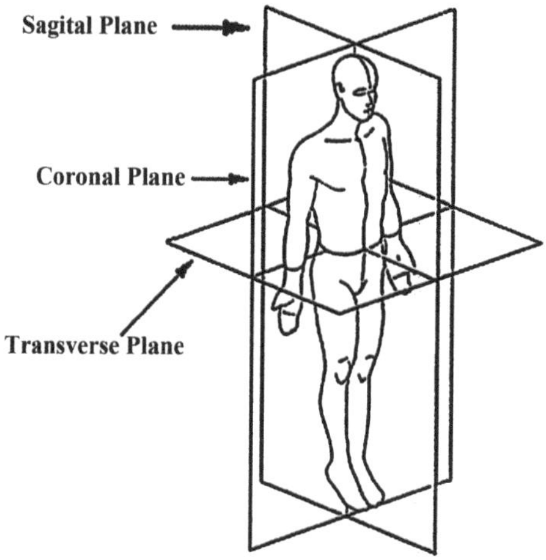

FIG. A.1. Anatomical planes.

A.2. Nomenclature for radiographic projections

Radiographic projections are referred to using a standard terminology. They are described by the direction of the central ray in relation to the anatomical body planes detailed in Fig. A.1.

For example, chest radiography is usually performed with the patient in the erect position (either standing or sitting), with the anterior aspect against the detector. The central ray enters the back of the body and emerges through the front (posterior–anterior image). If the central ray passes from one side of the body to the other, it is called a lateral projection. For the chest right lateral projection, the central ray enters the patient on the left hand side and exits on the right. Similarly, in a left lateral projection, the X ray enters the patient's right hand side and exits on the left.

Screening mammography typically involves taking views of the breast from above (cranial–caudal view) and from an oblique or angled view (mediolateral oblique) as shown in Figs A.2(a) and (b). For this examination, supplemental views may also be taken, tailored to a specific problem. These include views from each side (lateromedial: from the side towards the centre of the chest, and mediolateral: from the centre of the chest out), as shown in Fig. A.2(c).

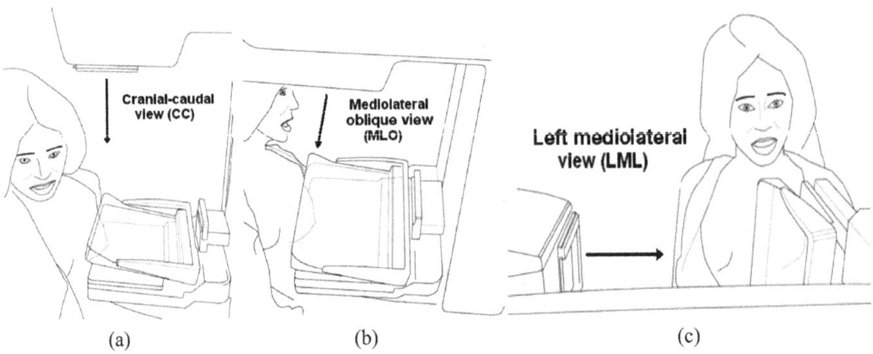

FIG. A.2. Mammography views: (a) cranial–caudal view, (b) angled or mediolateral oblique view and (c) left mediolateral view.

For descriptions of more views for specific applications, the reader is referred to the bibliographic reference.

BIBLIOGRAPHY

STEWART WHITLEY, A., SLOANE, C., HOADLEY, G., MOORE, A.D., ALSOP, C.W., Clarke's Positioning in Radiography, 12th edn, Hodder Arnold, London (2005).

ABBREVIATIONS

aBMD	areal bone mineral density
AC	alternating current
ACR	American College of Radiology
AEC	automatic exposure control
AFC	alternative forced choice
AFROC	alternative free response receiver operating characteristic
AP	anterior–posterior
a-Se	amorphous selenium
AUC	area under the receiver operating characteristic curve
BEIR	Biological Effects of Ionizing Radiation
BIR	British Institute of Radiology
BSS	Basic Safety Standards[a]
CAD	computer aided detection
CBCT	cone beam computed tomography
CCD	charge coupled device
CD	compact disc; cumulative dose
CDF	contrast degradation factor
CDRH	Center for Devices and Radiological Health
CHESS	chemically specific saturation
CIF	contrast improvement factor
CNR	contrast to noise ratio
CPE	charged particle equilibrium
CR	computed radiography
CSF	cerebrospinal fluid
CT	computed tomography
CTDI	computed tomography dose index

[a] International Atomic Energy Agency

CW	continuous wave
1-D	one dimensional
2-D	two dimensional
3-D	three dimensional
DC	direct current
DDREF	dose and dose rate effectiveness factor
del	detector element
DICOM	Digital Imaging and Communications in Medicine
DICOS	Digital Imaging and Communications in Security
DICONDE	Digital Imaging and Communications in Non-Destructive Evaluation
DLP	dose–length product
DNA	deoxyribonucleic acid
DQE	detective quantum efficiency
DR	digital radiography
DRL	diagnostic reference level
DSA	digital subtraction angiography
DSB	double strand break
DXA	dual energy X ray absorptiometry
EAR	excess absolute risk
ECG	electrocardiogram
EHP	electron–hole pair
EM	expectation minimization
EPI	echo planar imaging
ERR	excess relative risk
ESD	entrance surface dose
ESF	edge spread function
EU	European Union
FID	focus to image distance; free induction decay

FLASH	fast low angle shot
FN	false negative
FNF	false negative fraction
FOV	field of view
FP	false positive
FPF	false positive fraction
FROC	free response receiver operating characteristic
FSD	focus to skin distance
FT	Fourier transform
FWHM	full width at half maximum
GSDF	Greyscale Standard Display Function (DICOM)
GSF	National Research Centre for Environment and Health (Germany)
H&D	Hurter and Driffield (curve)
HL7	Health Level 7
HPA	Health Protection Agency
HU	heat unit or Hounsfield unit
HVL	half value layer
IAK	incident air kerma
IAKR	incident air kerma rate
ICNIRP	International Commission on Non-Ionizing Radiation Protection
ICRP	International Commission on Radiological Protection
ICRU	International Commission on Radiation Units and Measurements
IEC	International Electrotechnical Commission
IHE	Integrating the Healthcare Enterprise
IOD	information object definition
IRP	interventional reference point
ISL	inverse square law

ISO	International Standards Organization
JND	just noticeable distance; just noticeable luminal distance
JPEG	Joint Photographic Experts Group
KAP	kerma–area product
kerma	kinetic energy released in matter
KLP	kerma–length product
LROC	location receiver operating characteristic
LSF	line spread function
LSI	linearity and shift invariance
MDCT	multidetector row computed tomography
MGD	mean glandular dose
MI	mechanical index
MOSFET	metal oxide semiconductor field effect transistor
MPEG	Moving Picture Experts Group
MR	magnetic resonance
MRI	magnetic resonance imaging
MSH	message header segment (HL7)
MTF	modulation transfer function
NEQ	noise equivalent quanta
NHEJ	non-homologous end joining
NMR	nuclear magnetic resonance
NPS	noise power spectrum
NPV	negative predictive value
NRPB	National Radiological Protection Board
NTSC	National Television System Committee
OD	optical density
OID	object to image distance
OPG	orthopantomograph
OSL	optically stimulated luminescence

PA	posterior–anterior (projection)
PACS	picture archiving and communications systems
PC	proportion of correct answers
PDI	Portable Data for Imaging (IHE)
PET	positron emission tomography
PHS	pulse height spectrum
PL	photoluminescence
PMMA	polymethylmethacrylate
p-p	parallel plate
ppm	parts per million
PPV	positive predictive value
PRF	pulse repetition frequency
PSD	peak skin dose
PSDL	primary standards dosimetry laboratory
PSF	point spread function
PSL	peak side lobe level
PZT	ferroelectric ceramic lead zirconate titanate
QA	quality assurance
QC	quality control
QDE	quantum detection efficiency
QMS	quality management system
RF	radiofrequency
ROC	receiver operating characteristic
ROI	region of interest
RPO	radiation protection officer
RPP	radiation protection programme
RR	relative risk
SDNR	signal difference to noise ratio
SF	scatter fraction

SKE	signal known exactly
SNR	signal to noise ratio
SOP	service–object pair
SPR	scan projection radiograph; scatter to primary ratio
SSDL	secondary standards dosimetry laboratory
STEAM	stimulated echo acquisition mode
STUK	Radiation and Nuclear Safety Authority (Finland)
TE	echo time
TGC	time gain compensation
TI	inversion time; thermal index
TL	thermoluminescent
TLD	thermoluminescent dosimeter/dosimetry
TN	true negative
TP	true positive
TPF	true positive fraction
TR	repetition time
UNSCEAR	United Nations Scientific Committee on the Effects of Atomic Radiation
USB	universal serial bus
UV	ultraviolet
XRII	X ray image intensifier

SYMBOLS

Roman symbols

a — area

A — ampere (SI unit of current)

A — age; aperture function; atomic mass number;

A_Q — quantum detection efficiency

A_r — atomic weight

A_S — Swank factor

B — backscatter factor; barrier transmission; broad beam transmission factor; field strength; image brightness

$\boldsymbol{B}$ — magnetic field

c — dose conversion coefficient; speed of electromagnetic waves — speed of light; speed of sound

C — coulomb (SI unit of charge)

C — contrast; contour

$<C>$ — average cost

$C_{a,100}$ — CT kerma index

C_W — weighted CT kerma index

d — distance; width

d' — detectability

d_{FID} — focus to image receptor distance

D — absorbed dose; density; diffusion coefficient of water

D_{max} — maximum optical density

D_v — directional derivative of a function in the direction $\mathbf{v}$

e — charge on an electron ($= 1.602 \times 10^{-19}$ C)

E — effective dose; energy; energy fluence; retinal illumination

E_K — K atomic absorption edge

E_s — binding energy of an electron shell

f — frequency; function; input to a system

f_D	Doppler frequency
f_N	Nyquist frequency
F	coherent form factor; field area
g	average fraction of energy transferred to charged particles that is lost to photons when the charged particles are slowed down in the same medium as they were released; output of a system; gain; gradient
g_I	speed function
G	Gaussian filter; gradient amplitude
h	Planck's constant
$\hbar$	reduced Planck's constant: $(h/2\pi)$
H	transfer function/system response function
$H^*(d)$	ambient dose equivalent
$H'(d,\Omega)$	directional dose equivalent
$H_p(d)$	personal dose equivalent
I	electron current; intensity; mean excitation energy
$\mathbf{I}$	nuclear spin
J	Joule (SI unit of energy)
k	Boltzmann constant; proportionality constant from Coulomb's law; signal to noise ratio
k	wave number
K	kerma
K_i	incident air kerma
$\dot{K}_i$	incident air kerma rate
K_{scat}	scatter kerma
K_{sec}	secondary kerma
kVp	peak voltage
l	azithumal quantum number
L	aperture length; luminance
L_Δ	linear energy transfer

m	magnetic quantum number; magnification; mass; noise
m_0	rest mass
M	atomic mass; demagnification; moment
M_{opt}	optical modulation transfer function
n	bit depth; number; principal quantum number; refractive index
$\mathbf{n}$	noise
n_i	initial principal quantum number
n_f	final principal quantum number
N	number of neutrons in an atom; number
$\vec{N}$	unit normal vector to contour C
N_A	Avogadro's number
N_a	number of interaction centres (atoms) per unit volume
N_{am}	number of atoms per unit mass
N_K	dosimeter calibration coefficient
N_0	number of X ray quanta
p	projection; luminous flux to photon conversion factor
$\boldsymbol{p}$	angular momentum
P	power; power rating; pressure; target dose; wave amplitude
P_{It}	current–exposure time product
P_{KA}	air kerma–area product
P_{KL}	air kerma–length product
Q	charge; heat capacity; quality; quantity
r	grid ratio
r_0	'classical radius of the electron'
R	radiant energy; reflection coefficient; relaxation rate; voltage ripple
R_S	ratio value for soft tissue
s	signal; spin quantum number

S	incoherent scattering function; scatter factor; sensitivity; signal; stopping power; survival
S_{ion}	ionizational mass stopping power
$\bar{\bar{S}}_{g}^{w}$	ratio of average stopping powers
Sv	Sievert (unit of equivalent dose and unit of effective dose)
t	time; exposure time; thickness
t_{e}	exposure time
T	tesla (SI unit of magnetic flux density)
T	kinetic energy; modulation transfer function; occupancy; temperature; thickness; threshold; time; transmission; transmission coefficient
$<T>$	expectation value of energy converted to secondary electrons
$T_{1/2}$	half-time
T_{1}	time constant for spin–lattice relaxation
u	spatial frequency
U	tube voltage; unsharpness; field; use factor
U_{A}	kinetic energy of electrons bombarding the anode
$\|v\|$	reflector speed
v	velocity
V	volume
w	weighting coefficient/factor
w_{i}	normalized weight fraction of element i
W	workload
$\overline{W}_{\text{air}}$	mean energy spent in air to form an ion pair in dry air
x	momentum transfer quantity; thickness
X	exposure
X_{F}	focal spot size
Z	acoustic impedance; atomic number; nuclear charge

Greek symbols

α	(subscript) radiation emitted for transition between neighbouring cells; frequency dependent amplitude attenuation coefficient
$A(\xi)$	aperture function
β	(subscript) radiation emitted for a transition between non-neighbouring cells
γ	film gamma; gyromagnetic ratio
Γ	greyscale characteristic
δ	density correction factor
Δ	energy
ε_{tr}	energy transferred
$\bar{\varepsilon}_T$	energy imparted to a tissue
η	orthogonal dimension; quantum detection efficiency
θ	angle; azimuth angle; projection angle
θ_D	Doppler angle
θ_E	Ernst angle
Θ	angle
λ	decision variable; wavelength
κ	compressibility; curvature of the contour C
μ	linear attenuation coefficient ($N_a\sigma$); unified atomic mass unit
$\boldsymbol{\mu}$	nuclear magnetic moment
μ_0	rest mass of electron
ν	frequency; photon frequency
ξ	areal density; lateral dimension within aperture plane
ρ	density
ρ_o	undisturbed mass density of a medium
σ	cross-sectional area; standard deviation
σ_{Th}	cross-section for Thomson scattering
σ^2	variance

τ	cross-section for a photon to interact
Φ	number of photons; fluence
Φ_0	neural noise
Φ_{ph}	photon noise
Ψ	energy fluence
ω	fluorescent yield; frequency; radian frequency; resonance frequency
$\lvert\omega\rvert$	ramp filter
ω_o	Larmor frequency
Ω	solid angle

CONTRIBUTORS TO DRAFTING AND REVIEW

Alm Carlsson, G.	Linköpings University, Sweden
Båth, M.	Sahlgrenska University Hospital, Sweden
Bradley, D.	University of Surrey, United Kingdom
Castellano, I.	The Royal Marsden Hospital, United Kingdom
Christofides, S.	New Nicosia General Hospital, Cyprus
Clunie, D.	CoreLab Partners, United States of America
Collins, L.T.	Westmead Hospital, Australia
Cunningham, J.R.	Camrose, Canada
Dance, D.R.	Royal Surrey County Hospital, United Kingdom
Dawson, P.	International Society of Radiology, United Kingdom
Delis, H.	International Atomic Energy Agency
Faulkner, K.	North East Strategic Health Authority, United Kingdom
Geleijns, J.	Leiden University Medical Centre, Netherlands
Gray, J.E.	Chicago, United States of America
Hendry, J.H.	Arlington, United Kingdom
Hiles, P.A	Glan Clwyd Hospital, United Kingdom
Hourdakis, J.C.	Greek Atomic Energy Commission, Greece
Järvinen, H.	Radiation and Nuclear Safety Authority, Finland
Kesner, A.	International Atomic Energy Agency
Kremkau, F.W.	Wake Forest University, United States of America
Kyle Jones, K.A.	University of Texas MD Anderson Cancer Center, United States of America
Lacefield, J.C.	University of Western Ontario, Canada

Le Heron, J.	International Atomic Energy Agency
Leach, M.O.	The Institute of Cancer Research and The Royal Marsden Hospital, United Kingdom
Maidment, A.D.A.	University of Pennsylvania, United States of America
McLean, I.D.	International Atomic Energy Agency
Neitzel, U.	Philips Healthcare, Germany
Ng, K.-H.	University of Malaya, Malaysia
Nowotny, R.	Medical University of Vienna, Austria
Plewes, D.	University of Toronto, Canada
Poletti, J.L.	UNITEC Institute of Technology, New Zealand
Reiser, I.	University of Chicago, United States of America
Rowlands, J.A.	Thunder Bay Regional Research Institute, Canada
Sandborg, M.	Linköping University, Sweden
Shepherd, J.A.	University of California at San Francisco, United States of America
Soong, H.K.	Hospital of The University of Pennsylvania, United States of America
Sutton, D.	Ninewells Hospital, United Kingdom
Wondergem, J.	International Atomic Energy Agency
Yaffe, M.J.	University of Toronto, Canada
Yoshimura, E.M.	Universidade de São Paulo, Brazil
Yushkevich, P.A.	University of Pennsylvania, United States of America

IAEA
International Atomic Energy Agency

No. 23

ORDERING LOCALLY

In the following countries, IAEA priced publications may be purchased from the sources listed below or from major local booksellers.

Orders for unpriced publications should be made directly to the IAEA. The contact details are given at the end of this list.

AUSTRALIA

DA Information Services
648 Whitehorse Road, Mitcham, VIC 3132, AUSTRALIA
Telephone: +61 3 9210 7777 • Fax: +61 3 9210 7788
Email: books@dadirect.com.au • Web site: http://www.dadirect.com.au

BELGIUM

Jean de Lannoy
Avenue du Roi 202, 1190 Brussels, BELGIUM
Telephone: +32 2 5384 308 • Fax: +32 2 5380 841
Email: jean.de.lannoy@euronet.be • Web site: http://www.jean-de-lannoy.be

CANADA

Renouf Publishing Co. Ltd.
5369 Canotek Road, Ottawa, ON K1J 9J3, CANADA
Telephone: +1 613 745 2665 • Fax: +1 643 745 7660
Email: order@renoufbooks.com • Web site: http://www.renoufbooks.com

Bernan Associates
4501 Forbes Blvd., Suite 200, Lanham, MD 20706-4391, USA
Telephone: +1 800 865 3457 • Fax: +1 800 865 3450
Email: orders@bernan.com • Web site: http://www.bernan.com

CZECH REPUBLIC

Suweco CZ, spol. S.r.o.
Klecakova 347, 180 21 Prague 9, CZECH REPUBLIC
Telephone: +420 242 459 202 • Fax: +420 242 459 203
Email: nakup@suweco.cz • Web site: http://www.suweco.cz

FINLAND

Akateeminen Kirjakauppa
PO Box 128 (Keskuskatu 1), 00101 Helsinki, FINLAND
Telephone: +358 9 121 41 • Fax: +358 9 121 4450
Email: akatilaus@akateeminen.com • Web site: http://www.akateeminen.com

FRANCE

Form-Edit
5 rue Janssen, PO Box 25, 75921 Paris CEDEX, FRANCE
Telephone: +33 1 42 01 49 49 • Fax: +33 1 42 01 90 90
Email: fabien.boucard@formedit.fr • Web site: http://www.formedit.fr

Lavoisier SAS
14 rue de Provigny, 94236 Cachan CEDEX, FRANCE
Telephone: +33 1 47 40 67 00 • Fax: +33 1 47 40 67 02
Email: livres@lavoisier.fr • Web site: http://www.lavoisier.fr

L'Appel du livre
99 rue de Charonne, 75011 Paris, FRANCE
Telephone: +33 1 43 07 50 80 • Fax: +33 1 43 07 50 80
Email: livres@appeldulivre.fr • Web site: http://www.appeldulivre.fr

GERMANY

Goethe Buchhandlung Teubig GmbH
Schweitzer Fachinformationen
Willstätterstrasse 15, 40549 Düsseldorf, GERMANY
Telephone: +49 (0) 211 49 8740 • Fax: +49 (0) 211 49 87428
Email: s.dehaan@schweitzer-online.de • Web site: http://www.goethebuch.de

HUNGARY

Librotade Ltd., Book Import
PF 126, 1656 Budapest, HUNGARY
Telephone: +36 1 257 7777 • Fax: +36 1 257 7472
Email: books@librotade.hu • Web site: http://www.librotade.hu

INDIA
Allied Publishers
1st Floor, Dubash House, 15, J.N. Heredi Marg, Ballard Estate, Mumbai 400001, INDIA
Telephone: +91 22 2261 7926/27 • Fax: +91 22 2261 7928
Email: alliedpl@vsnl.com • Web site: http://www.alliedpublishers.com

Bookwell
3/79 Nirankari, Delhi 110009, INDIA
Telephone: +91 11 2760 1283/4536
Email: bkwell@nde.vsnl.net.in • Web site: http://www.bookwellindia.com

ITALY
Libreria Scientifica "AEIOU"
Via Vincenzo Maria Coronelli 6, 20146 Milan, ITALY
Telephone: +39 02 48 95 45 52 • Fax: +39 02 48 95 45 48
Email: info@libreriaaeiou.eu • Web site: http://www.libreriaaeiou.eu

JAPAN
Maruzen Co., Ltd.
1-9-18 Kaigan, Minato-ku, Tokyo 105-0022, JAPAN
Telephone: +81 3 6367 6047 • Fax: +81 3 6367 6160
Email: journal@maruzen.co.jp • Web site: http://maruzen.co.jp

NETHERLANDS
Martinus Nijhoff International
Koraalrood 50, Postbus 1853, 2700 CZ Zoetermeer, NETHERLANDS
Telephone: +31 793 684 400 • Fax: +31 793 615 698
Email: info@nijhoff.nl • Web site: http://www.nijhoff.nl

Swets Information Services Ltd.
PO Box 26, 2300 AA Leiden
Dellaertweg 9b, 2316 WZ Leiden, NETHERLANDS
Telephone: +31 88 4679 387 • Fax: +31 88 4679 388
Email: tbeysens@nl.swets.com • Web site: http://www.swets.com

SLOVENIA
Cankarjeva Zalozba dd
Kopitarjeva 2, 1515 Ljubljana, SLOVENIA
Telephone: +386 1 432 31 44 • Fax: +386 1 230 14 35
Email: import.books@cankarjeva-z.si • Web site: http://www.mladinska.com/cankarjeva_zalozba

SPAIN
Diaz de Santos, S.A.
Librerias Bookshop • Departamento de pedidos
Calle Albasanz 2, esquina Hermanos Garcia Noblejas 21, 28037 Madrid, SPAIN
Telephone: +34 917 43 48 90 • Fax: +34 917 43 4023
Email: compras@diazdesantos.es • Web site: http://www.diazdesantos.es

UNITED KINGDOM
The Stationery Office Ltd. (TSO)
PO Box 29, Norwich, Norfolk, NR3 1PD, UNITED KINGDOM
Telephone: +44 870 600 5552
Email (orders): books.orders@tso.co.uk • (enquiries): book.enquiries@tso.co.uk • Web site: http://www.tso.co.uk

UNITED STATES OF AMERICA
Bernan Associates
4501 Forbes Blvd., Suite 200, Lanham, MD 20706-4391, USA
Telephone: +1 800 865 3457 • Fax: +1 800 865 3450
Email: orders@bernan.com • Web site: http://www.bernan.com

Renouf Publishing Co. Ltd.
812 Proctor Avenue, Ogdensburg, NY 13669, USA
Telephone: +1 888 551 7470 • Fax: +1 888 551 7471
Email: orders@renoufbooks.com • Web site: http://www.renoufbooks.com

United Nations
300 East 42nd Street, IN-919J, New York, NY 1001, USA
Telephone: +1 212 963 8302 • Fax: 1 212 963 3489
Email: publications@un.org • Web site: http://www.unp.un.org

Orders for both priced and unpriced publications may be addressed directly to:

IAEA Publishing Section, Marketing and Sales Unit, International Atomic Energy Agency
Vienna International Centre, PO Box 100, 1400 Vienna, Austria
Telephone: +43 1 2600 22529 or 22488 • Fax: +43 1 2600 29302
Email: sales.publications@iaea.org • Web site: http://www.iaea.org/books

Printed and bound by CPI Group (UK) Ltd, Croydon, CR0 4YY

13/09/2026

14961052-0001